Quark-Gluon Plasma, Heavy Ion Collisions and Hadrons

World Scientific Lecture Notes in Physics

ISSN: 1793-1436

---

*Published titles**

Vol. 69: Deparametrization and Path Integral Quantization of Cosmological Models
C Simeone

Vol. 70: Noise Sustained Patterns: Fluctuations and Nonlinearities
Markus Loecher

Vol. 71: The QCD Vacuum, Hadrons and Superdense Matter (2nd ed.)
Edward V Shuryak

Vol. 72: Massive Neutrinos in Physics and Astrophysics (3rd ed.)
R Mohapatra and P B Pal

Vol. 73: The Elementary Process of Bremsstrahlung
W Nakel and E Haug

Vol. 74: Lattice Gauge Theories: An Introduction (3rd ed.)
H J Rothe

Vol. 75: Field Theory: A Path Integral Approach (2nd ed.)
A Das

Vol. 76: Effective Field Approach to Phase Transitions and Some Applications to Ferroelectrics (2nd ed.)
J A Gonzalo

Vol. 77: Principles of Phase Structures in Particle Physics
H Meyer-Ortmanns and T Reisz

Vol. 78: Foundations of Quantum Chromodynamics: An Introduction to Perturbation Methods in Gauge Theories (3rd ed.)
T Muta

Vol. 79: Geometry and Phase Transitions in Colloids and Polymers
W Kung

Vol. 80: Introduction to Supersymmetry (2nd ed.)
H J W Müller-Kirsten and A Wiedemann

Vol. 81: Classical and Quantum Dynamics of Constrained Hamiltonian Systems
H J Rothe and K D Rothe

Vol. 82: Lattice Gauge Theories: An Introduction (4th ed.)
H J Rothe

Vol. 83: Field Theory: A Path Integral Approach (3rd ed.)
Ashok Das

Vol. 84: Foundations of Quantum Field Theory
Klaus Dieter Rothe

Vol. 85: Quark-Gluon Plasma, Heavy Ion Collisions and Hadrons
Edward Shuryak

*For the complete list of published titles, please visit
http://www.worldscientific.com/series/wslnp

World Scientific Lecture Notes in Physics – Vol. 85

Quark-Gluon Plasma, Heavy Ion Collisions and Hadrons

Edward Shuryak
State University of New York, Stony Brook, USA

World Scientific

NEW JERSEY • LONDON • SINGAPORE • BEIJING • SHANGHAI • HONG KONG • TAIPEI • CHENNAI • TOKYO

Published by

World Scientific Publishing Co. Pte. Ltd.
5 Toh Tuck Link, Singapore 596224
USA office: 27 Warren Street, Suite 401-402, Hackensack, NJ 07601
UK office: 57 Shelton Street, Covent Garden, London WC2H 9HE

British Library Cataloguing-in-Publication Data
A catalogue record for this book is available from the British Library.

World Scientific Lecture Notes in Physics — Vol. 85
QUARK-GLUON PLASMA, HEAVY ION COLLISIONS AND HADRONS

ISBN 978-981-12-8234-8 (hardcover)
ISBN 978-981-12-8235-5 (ebook for institutions)
ISBN 978-981-12-8236-2 (ebook for individuals)

For any available supplementary material, please visit
https://www.worldscientific.com/worldscibooks/10.1142/13570#t=suppl

Desk Editor: Rhaimie Wahap

Typeset by Stallion Press
Email: enquiries@stallionpress.com

PREFACE

This book grew out of lecture notes from courses at Stony Brook and multiple conference talks over the last 40 years. Its three big subjects are a Quest for Quark-Gluon Plasma (QGP), understanding the structure of the QCD Vacuum, as well as its reflection in Spectroscopy of Hadrons.

This is the third book on these subjects, written with approximately 20 year intervals. Needless to say, all were completely rewritten as the subjects changed a lot. Suggestions become real experiments, and their results become well established facts. Yes, the QGP was discovered, with rather unusual properties. Yes, the fields in QCD vacuum are "topologically nontrivial", with vortices, magnetic monopoles and (4-dimensional solitons) called "instantons". Even hadronic spectroscopy, the most traditional field of those, is getting a true revival during the last decade, as large number of new hadrons were discovered, including novel tetra- and pentaquarks.

More generally, this book is intended to put the reader from the first encounter with the subject to the front lines of research, as quickly as possible (but not quicker than that, as Einstein once remarked). The logical and pedagogical considerations takes priority over historic ones.

Quantum Chromodynamics (QCD), the theory of interacting quarks and gluons, was discovered 50 years ago. As a theoretical construction QCD is even more logical and much more self-contained that its older relative, Quantum Electrodynamics (QED), which provided amazingly accurate description of all atomic phenomena.

Original optimism was however rather naive, since true Lagrangian written at the blackboard does not by itself explain all intricacies of that theory. While for "hard processes" one could still use familiar perturbative tools, the Feynman diagrams, in order to understand the non-perturbative phenomena one had a lot of theory/experiment development to do.

The QCD vacuum is a very complicated matter, made of strongly interacting quark and gluon fields. Furthermore, QCD has various phases as well. Those are revealed at extreme conditions, like heavy ion collisions studied at high energy colliders, RHIC at Brookhaven National Laboratory and LHC at CERN. Yet, in fact, the Quark-Gluon Plasma phase at high temperatures is the "normal" phase of QCD, unlike the vacuum we live in.

Other phases, such as Color Superconducting ones, are expected at high density inside neutron stars and revealed in their mergers. The book also had connections to Cosmology: flow harmonics in "fireballs" produced in accelerator labs are very similar to those we see on the sky, in Cosmic Microwave Background radiation. Interesting studies are also discussed of possible consequences of QCD and electroweak phase transitions, much earlier in the Big Bang.

Like atomic and nuclear physics before, the QCD-related hadronic physics is no longer at the edge of the highest available energies. Nevertheless, understanding of the non-perturbative QCD (and strongly coupled fields theories in general) continues to provide significant intellectual challenges, to theorists, phenomenologists and experimentalists. This rather large book is full of excited methods and results, but also of new puzzles to be solved by the next generation.

Edward Shuryak
Stony Brook University, USA

CONTENTS

Part 1

QUARK-GLUON PLASMA

Chapter 1

INTRODUCTION

1.1 What this Book is About

The central story of this book is the *quest for quark-gluon plasma* (QGP), summarizing several decades of theoretical and experimental efforts to prove the existence of this state of matter at high[1] *energy density* and understand its properties. It is based on quantum chromodynamics (QCD), a fundamental theory of strong interactions. The QGP is often referred to as its *exotic* phase, yet it is, in fact, the only "normal" phase of QCD. The low-temperature phase and the QCD vacuum (in which we live) are, on the other hand, very complicated and possess a number of "condensates," related to "color confinement" and "chiral symmetry breaking" phenomena. Theoretical and experimental studies of the phase transition to QGP, therefore, shed light on these important phenomena.

In Part 1 of the book, we discuss the properties of QGP, while nonperturbative phenomena in the QCD vacuum and the theory of its lowest excitations — the hadrons — are delegated to Part 3.

This book is the third in a series of my books, all done by World Scientific. The publications of the first (Shuryak, 1988a), the second (Shuryak, 2004), and the third, which is scheduled to appear in 2023, are separated by about two decades. While the first was kind of a "manifesto," proclaiming the birth of a new field of physics research, the second reported QGP discovery in laboratory conditions, and this book reports on the field reaching its maturity. As the reader will soon see, active work in many directions is going on, and, as is always the case, more theory and experiments produce plenty of new questions and puzzles, waiting to be explained.

[1] In relative terms, the nucleon mass divided by its volume. I do not mention the square of the speed of light, as in this book, the units are $c = \hbar = 1$.

The main part of the book is Part 2, in which we focus on heavy ion collisions. The theoretical developments along this line that began in the 1970s were complemented by extensive experimental programs involving high-energy heavy ion collisions. In the 1990s, these experiments started with fixed target kinematics, conducted both at the SPS accelerator at CERN (Geneva) and the AGS at Brookhaven National Laboratory (Upton, New York). Two colliders, the Relativistic Heavy Ion Collider (RHIC) at Brookhaven and the Large Hadronic Collider (LHC) at CERN, were then constructed and have been operating since the years 2000 and 2010, respectively. The (center-of-mass) energy per nucleon, $\sqrt{s_{NN}}$, of these experiments cover about four orders of magnitude, from a few GeV to about 10^4 GeV. Recently, the STAR experiment at the RHIC reinstated a fixed target program, to study the high baryon density domain at few-GeV-per-nucleon collisions.

These experiments resulted in an unprecedented amount of information: billions of events with thousands of secondary particles were recorded, with the type and momentum of each of them determined. The total amount of information about the "Little Bangs" (as such events are often called) we now have is comparable to that collected about the Big Bang that formed our Universe. Both start as relativistic explosions of very hot/dense matter, undergoing a series of stages separated by "freezeouts" of particular reactions. (We will emphasize existing parallels between the two as we proceed.)

By now, the existence of QGP as a novel phase of matter is well established. Its properties turned out to be rather unusual. While it agrees with theoretical expectations based on perturbative QCD (pQCD) in many qualitative aspects (e.g. its *equation of state*, EOS), it is in the *strong coupling regime* (denoted as strong QGP or sQGP) in the domain in which experiments are performed. In simplified terms, quarks and gluons in QGP are not "good quasiparticles," as they turned out to have unusually small mean free paths; therefore, it is perhaps the most ideal fluid ever observed. Understanding the consequences of this, as well as its roots, has demanded the development of quite novel theoretical tools.

This book describes how this theoretical and experimental program unfolded, from general ideas to specific details. The target audience of the book can therefore be both members of the community involved in this research as well as a wider circle of people interested in it. We specifically targeted young theorists and experimentalists, postdocs and graduate students interested in the field.

1.2 Quark-Gluon Plasma

The ordering of the material in this book thus follows its central logical flow and is different from what we normally do, for example, in teaching atomic or nuclear physics. The hydrogen atom is the simplest object on which, naturally, quantum mechanics courses focus. Similarly, a deuteron is the simplest nucleus, so all textbooks on nuclear physics include a discussion of it. Metals, insulators or nuclear matter are more complicated and are taught much later, after the tools of many-body theory are developed. Even more remote, in "advanced" teaching, are phases of matter with condensates, e.g. superconductors.

This order should actually be *inverted* in the case of QCD. The simplest perturbative theory based on Feynman diagrams and other tools of QFTs is applicable not in a vacuum but at high temperatures, where interactions become weak. QGP is the "normal phase," and its quasiparticles have quantum numbers of quarks and gluons, quanta of the very fields in the Lagrangian. We start with this theory in Part 1 of the book.

In fact it is the low-T phase and the *QCD vacuum* which are quite "exotic." In it, certain symmetries of the Lagrangian are broken, either explicitly (by quantum anomalies) or spontaneously. It possesses certain "condensates" in the same sense as superconductors do. Its excitations (hadrons, e.g. protons and neutrons) are complicated: the simplest models describing them operate in terms of "constituent quarks," while quarks and gluons appear only at a much deeper level, as "partons." We discuss all of them in Part 3 of the book.

1.3 Heavy Ion Collisions

When teaching about this subject, one inevitably meets the following question: Why study them in the first place? Isn't it more logical to study *pp* collisions first and then gradually go from smaller to larger systems?

This reasoning (widespread before year 2000, with some adhering to it even now) does not appreciate the enormous *power of entropy*. Yes, nuclei are made of nucleons, and nucleons are made of partons. So, one might think of their collisions as a very complicated process involving thousands of partons (connected, as we will discuss later, by thousands of QCD strings). To describe it, one needs to know zillions of parameters: where exactly each "parton" was located at the moment of collision, and how it then moved.

Yet, it is not at all so. In a short duration — a fraction of fm/c — the matter in question reaches *local equilibrium*. Entropy "eats up" all the details, and from that time on, the evolution of matter can be approximately described by a few functions of coordinates, such as the local temperature $T(x)$ or matter velocity field. Subsequent explosions are quite accurately described by (various versions of) *relativistic hydrodynamics*, followed through until (chemical and kinetic) *freezeouts*, at which the final secondaries are formed. There is compelling evidence that this happens in a purely *statistical* manner. In fact, macroscopically large systems are actually *much simpler* to describe than smaller ones!

In the same spirit, going from simpler theory to more complex, our presentation follows the collision in *inverse* order. We first discuss the conditions at freezeouts, at which thermal description is quite accurate, then proceed to hydro evolution, and only later turn to a discussion of the initial stage. In fact, not all details are erased, some survive and are detected, for example, in the form of small-amplitude sound waves.

Discussing experiments, we also start with heavy ion (AA) collisions, turning to "smaller systems," pA and pp, later. We discuss at what multiplicity they do display *collective phenomena* — radial, elliptic and higher-order harmonics of the flows. One would like to quantify the location of the regime change experimentally and theoretically.

Where exactly is the boundary between the micro and macro theories? The textbook answer is that one should compare the micro or "mean free path" scale l to the size of the system. If the l/L ratio is small

$$\text{(macro scale)}\ L \gg l\ \text{(micro scale)} \tag{1.1}$$

one can use the macroscopic toolbox.

The small phenomenological viscosity of sQGP, if interpreted as the mean free path, is found to be a few times *smaller* than the inter-particle distance! It makes no sense. A "mean free path" concept is indeed defined in kinetic theory, and it is valid only for weakly coupled systems. QGP is not one of them, as it is strongly coupled.

Finally, about the so called *initial stage* of heavy/light ion collisions. It is believed that "large systems" start with a state called GLASMA, a highly excited classical state of glue. Small systems are described instead *via* a set of multiple QCD strings connecting quarks (the color charges).

Transition to QGP in a strongly coupled regime happens via quark pair production in strong fields in the former case and via "string breaking" in the latter. The initial stage is far from equilibrium, and one needs certain theoretical tools to describe it. Fortunately, we have one of them, known as the *holographic approach*, which can be used both for "out of equilibrium" and "strongly coupled" settings. We will see how this approach can describe the equilibration and the onset of hydrodynamical flows.

Now, the power of entropy has been exaggerated slightly. Not all information about the initial state is "eaten up" by entropy and forgotten. Some remnants of "primordial fluctuations" survive and can be observable at the end. It is a result of hydrodynamical sound modes, which, due to the small QGP viscosity, are long lived and visible via correlations between secondaries in the detectors. These phenomena in the Little Bangs are quite similar in essence to what is seen across the sky in the microwave background radiation, also providing evidence on the magnitude of the primordial fluctuations in the Big Bang.

The state of the matter at the chemical freezeout stage is (i.e. in its local rest frame) quite well described by just two parameters, the temperature and baryon chemical potential, $T_{\rm ch}, \mu^B_{\rm ch}$, being constant at the freezeout surface. In the first approximation, all we need to know about hadrons are their masses and statistical weights. In the next approximation, the theory of the "resonance gas" also includes certain integrals over scattering phases. All particle yields, from pions to light nuclei, are well described by statistical mechanics over about eight orders of magnitude!

The so-called *kinetic freezeout* pertains to the properties of matter at the moment of the last collisions of secondaries. Its experimental understanding is helped by the so-called *femtoscopy*, which provides direct information about the space-time shape of the system at this stage. The growing number of studies of event-by-event fluctuations show that they are also consistent with thermal theory, in which those are related to the so-called *susceptibilities*.

Then, we turn to *relativistic hydrodynamics* and its role in describing the explosive stage of collisions. We start with the several analytic solutions available and then proceed to the actual description of experimental data, which starts with the so-called *ideal hydrodynamics* using only the local properties of matter. The only ingredient at this level is the EOS of matter (which by now is known and understood from lattice simulations). Predictions based on it *prior to the RHIC experiments* were shown to be accurate, describing both the particle spectra (radial flow) and their quadrupole component, known as the *elliptic flow*. This agreement led to "press releases" announcing "QGP discovery," etc., soon after the start of the RHIC era.

In nonrelativistic hydrodynamics, the nonlocal terms containing *first-order gradients* include few kinetic coefficients, such as shear and bulk viscosities. In the relativistic case, there are some conceptual issues, and in general, one introduces more kinetic coefficients. One of the issues here is causality during explosion. Some "practical formulations" developed based on *second-order* codes can handle even stronger gradients in event-by-event matter distributions. Whether going to second order is indeed necessary is under debate.

The phenomenology of collisions, due to unprecedented levels of statistics, has now included higher azimuthal harmonics of flow, up to $n \sim 10$. It has also been extended to higher $p_\perp$ of secondaries, where effects have maximal magnitudes. Theoretically, these harmonics are basically sounds and, as such, are well described by linearized hydrodynamics. The propagation of the *sound waves* on top of the exploding fireball is, of course, a complex but solvable problem. The propagation of these sounds, from initial perturbations to freezeout, is in fact similar to what is observed at the grand scale in the background microwave radiation in the Universe. These studies gave us a quantitative value for the viscosity.

After collective flows were understood for collisions of heavy $AuAu, PbPb$ nuclei, one naturally asked what happens in collisions of smaller systems, at higher impact parameters, lighter nuclei and even high-multiplicity *pp* collisions. Indeed, in a certain domain, such "small systems," also display collective flows, which is quite consistent with flow systematics. Moreover, in a way, those are even more "explosive" than "large" systems.

Another tool to be discussed is the so-called "penetrating probes," such as photons and dileptons, which can carry direct information about them, and experimentally, the studies of those are rather challenging. Another set of probes is provided by heavy quarks, which are produced promptly in partonic processes and then travel through QGP before forming the observed hadrons.

1.4 Nonperturbative QCD and Hadrons

Only after discussing all aspects of heavy ion collisions in the second part of the book, in Part 3, do we provide a more fundamental discussion of the QCD. The physics of its ground state — the QCD vacuum — requires an understanding of *gauge topology*. Monopoles and instantons have a role in such nonperturbative phenomena as *confinement and chiral symmetry breaking*. Since it is a vast subject, it was reviewed in my other book (Shuryak, 2021), which needs to be consulted by the reader for technical details. Here, we only explain the main ideas "on a physics level," skipping technical details. We are focusing here not on derivations but on consequences of various phenomena, in what can be broadly called "hadronic physics."

In Chapter 25, we discuss the elements of hadronic spectroscopy, including quark models and forces between quarks. The simplest cases are not protons and other baryons, but mesons, especially heavy-quark "quarkonia." Here, the QCD flux tubes (or QCD strings) are naturally introduced. Their role in high-energy hadronic collisions is discussed later, in Chapter 26, devoted to Reggeons and Pomerons.

A large part of hadronic physics, related to deep inelastic scattering and similar processes, leads to a description in terms of *partons*, point-like quarks and gluons. An important step toward bridging spectroscopy with partonic observables is now being taken through a novel quantum mechanical description of hadronic spectroscopy *on the light front*, which we discuss in Chapter 28.

The so-called chiral symmetry, which is spontaneously broken in a vacuum, is restored in QGP. It means that left- and right-polarized light quarks in it do (approximately) conserve their "handiness." So, if by a fluctuation their numbers do not match, the QGP in such events becomes an example of *chiral matter* with unbalanced chirality, in which new interesting physical phenomena, such as chiral magnetic and chiral vortical effects, can be detected.

Theorists worked hard to try to understand how these fascinating phenomena could lead to experimental observations. First of all, lattice gauge theory simulations have reproduced the hadronic spectrum and many important thermodynamic properties of QGP. These large-scale efforts have resulted in collaborations between theorists running large-scale operations at supercomputer centers. Multiple effective models of confinement/deconfinement and chiral transitions were proposed and studied based on solitonic objects of different dimensions: 2D flux tubes, 3D magnetic monopoles and 4D instantons.

Unfortunately, extrapolation from imaginary (Euclidean) time to real time proved difficult in practice, so the kinetic properties are hard to extract from the lattice measurements. Therefore, theorists have invented other tools, specifically designed to study strongly coupled gauge theories.

One set of these theories constitutes the so-called *holographic* models, using the *gauge–string* (or gauge–gravity) duality. The strongly coupled 4D gauge theories are studied via correspondence to weakly coupled effective theories in more dimensions, collectively known as AdS/QCD models. Those can also provide a certain derivation (or, rather, economic parameterization) of the hadronic spectrum and thermodynamics. It includes deconfinement and chiral symmetry breaking phase transitions. Unlike lattices, holographic models can also be applied in dynamical (time-dependent) settings. So, it is possible to derive the hydrodynamical equations with certain kinetic coefficients, account for deviations from equilibrium, and even describe settings which are completely out of equilibrium. A particular approach to hadronic physics is called *holographic QCD*. The original AdS/CFT correspondence developed for conformal theories is complemented by certain terms in bulk Lagrangians, which make it *confining*. We discuss that in Chapter 29.

Another set of models is based on the *electric–magnetic duality*, which treats sQGP as a dual plasma, containing both "electric" quasiparticles — quarks and gluons — and *magnetic monopoles*. The dynamical studies of the *dual plasmas* do indeed suggest large transport cross-sections of charge-monopole scattering, and even dynamical trapping of charges in certain cells. There is also evidence that charge-monopole scattering significantly contributes to jet quenching as well.

Our discussion of hadrons includes the ongoing program aiming to *bridge the gap* between hadronic spectroscopy and partonic observables. Its core part involves the redevelopment of quark models of hadrons at the light front.

Chapter 2

INTRODUCTION TO QUANTUM CHROMODYNAMICS

2.1 Gauge Theories and QCD

This section introduces QCD at a rather pedestrian level and can thus perhaps be skipped by readers familiar with it.

Quantum chromodynamics (QCD, for short) is the "younger brother" of *quantum electrodynamics* (QED), with which it shares the main underlying principle, namely *local gauge invariance.* Briefly, it means that the local values of the phase of the wave function of charged objects in QED as well as the orientation of color coordinates in QCD are not physical and can be selected at will. The word "local" is central here: One can use any color coordinates, arbitrary oriented at any space-time points. Substantial discussions of the meaning of local gauge invariance can be found in QFT textbooks and will not be taken up here in any detail. Nonetheless, this principle is so central that it needs some general comments.

The first comment is that the theoretical origins of local gauge invariance are related to the local symmetry of Einstein's general relativity (GR), allowing for *arbitrary coordinate transformations.* In local (differential) form, it means that one can use arbitrary sets of Lorentz comoving reference frames at different space-time points. Gauge symmetries are huge. Why are these symmetries in all three components of the Standard Model?

It has been emphasized by Weinberg[1] that the representation of the so-called "small Lorentz group" (leaving the state of massless particles invariant) requires only *two* polarizations (in three plus one dimensions).

[1]In perhaps his last Harvard lecture done via Zoom in the summer of 2021, he spoke about it, although his principal research on the issues was done many years ago in the mid-1960s. This shows how much he valued these ideas.

Indeed, this is true for photons, gluons and gravitons, as they are on-shell particles or waves. On the other hand, Lorentz invariance requires that photon and gluon fields are described by vector potentials A_μ with four components, and gravity by a symmetric metric tensor $g_{\mu\nu}$ with ten components. These facts imply that *not all* components of those fields are actually physical, since *some* of them can be chosen arbitrarily for the convenience of a particular problem. This is commonly called "fixing the gauge" in QED and QCD or "selecting a coordinate system" in GR.

Physical quantities should be "gauge invariant" and cannot be changed by gauge transformations. Examples include the squares of the field strength $G^2_{\mu\nu}$ in QCD and the scalar curvature in GR. Unsurprisingly, those precisely serve as the fundamental Lagrangians of both theories.

Another (standard but important) comment is that the local gauge symmetry is not a "symmetry" in the usual sense. It is quite different from, for example, translational, rotational or flavor symmetries. All physical observables are *singlets* under gauge rotations; therefore, they are not really transformed. Thus, there cannot be any relations between them to be deduced from this symmetry. Rather, it tells us which degrees of freedom of the fields are *redundant* variables.

In QED, we are used to expressing the electromagnetic field strength in terms of the vector potential:

$$F_{\mu\nu} = \partial_\mu A_\nu - \partial_\nu A_\mu \tag{2.1}$$

The left-hand side of the above equation is physical, but the potential can be "gauged transformed" by adding a gradient of some function

$$A_\mu \to A_\mu + (i/e)\partial_\mu \phi \tag{2.2}$$

without changing the field strengths. Soon after the discovery of quantum mechanics, in 1927, it was noted (by Fock and London) that *the phase of the wave function can be chosen arbitrarily at all space-time points* by the transformation

$$\psi \to \psi e^{i\phi(x)} \tag{2.3}$$

Respectively, the momentum operator $p_\mu = i\partial/\partial_\mu$ should be modified as well since the derivative of $\phi(x)$ is nonzero. This gradient can be *compensated* by a gauge transformation of the electromagnetic field, with a change in its redundant part above. Thus, the momentum is defined in terms of the *covariant* (or "long") derivatives

$$i\partial_\mu \to iD_\mu = i\partial_\mu + gA_\mu \tag{2.4}$$

to get gauge-invariant quantum mechanics. The principle of local gauge symmetry was extended by Yang and Mills (1954). Let the matter field $\psi_i(x)$ have some internal quantum number[2] $i = 1,\ldots,N$. Let the axis in this "internal" space be chosen arbitrarily at any space-time point. The phase rotation (2.3) is generalized to the unitary matrix rotation

$$\psi_i(x) \to \Omega_{ij}(x)\psi_j(x) \tag{2.5}$$

with a local symmetry, in which the matrix $\Omega(x)$ is now different at different points. By analogy to electromagnetism, in order to maintain invariance, Yang and Mills introduced vector *compensating fields*, which are matrix-valued $A_\mu = A_\mu^a T^a$, with T being generators of the color algebra.

In QCD, quarks are in the fundamental (spinor) representation of $SU(3)$, and pertinent form of the generator is written as $T = t^a/2$, where t are 3×3 Gell-Mann $SU(3)$ matrices. The 1/2 is historical, following the relation between spin and Pauli matrices, indicating the fundamental (spin 1/2) representation of the rotational group $O(3) \approx SU(2)$. Useful relations of color algebra with the fundamental generators of the Gell-Mann $SU(3)$ algebra are

$$t^A_{ab}t^A_{bc} \equiv C_F\delta_{ac}, \quad C_F = \frac{N_c^2-1}{2N_c}, \quad t^A_{ab}t^B_{ab} \equiv T_R\delta_{AB}, \quad T_R = 1/2 \tag{2.6}$$

where, as usual, repeated indices are summed over, $A = 1\ldots N_c^2 - 1$ counts the generators and $a = 1,\ldots,N_c$ counts the colors. The generators of the adjoint color representation are the structure constants f_{ABC} of the group, which are normalized so that

$$f_{ACD}f_{BCD} \equiv C_A\delta_{AB}, \quad C_A = N_c \tag{2.7}$$

We would like it to complement the ordinary derivatives in a "covariant" combination:

$$iD_\mu = i\partial_\mu + gA_\mu^a T^a \tag{2.8}$$

The expected property is that this combination can be transformed with only a rotation:

$$D_\mu\psi \to \Omega(D_\mu\psi) \tag{2.9}$$

[2]As a particular example of "internal quantum number," Yang and Mills have considered the isotopic spin, so their gauge particle was supposed to be the ρ meson. It is the so-called weak isospin with $N = 2$ in electroweak theory and color with $N = 3$ in QCD.

This requirement is met by the following non-Abelian gauge transformation of the potential

$$A_\mu \to \Omega A_\mu \Omega^{-1} - (2i/g)(\partial_\mu \Omega)\Omega^{-1} \tag{2.10}$$

This transformation of A_μ leads to the fact that the field strength $G_{\mu\nu}$ is simply rotated by Ω, and the QCD Lagrangian, which contains its square, is gauge-invariant.

QCD has another close relative, originally known as the Weinberg model (Weinberg, 1971), which describes the electroweak sector of the Standard Model. In this case, the symmetry group is $SU(2)_W$ of "weak isospin," which rotates, for example, a doublet of u and d quarks. Unlike QCD, this theory also has a fundamental scalar, the famous Higgs boson, a doublet of $SU(2)_W$, which has a vacuum expectation value (VEV) v, breaking this symmetry spontaneously. As a result, the gauge bosons W, Z acquire large masses, and the effective coupling of this theory, $\alpha_W(v) = g_W^2(v)/4\pi$, is so small that one need not worry about nonperturbative electroweak phenomena in the laboratory. We only discuss it briefly in Chapter 13, comparing the electroweak and QCD phase transitions in the early Universe.

Among the other relatives of QCD we will be discussing are two theories with extended supersymmetries (SUSYs). One of them, with two ($\mathcal{N}$=2) supersymmetries, also has a scalar field, which may have a nonzero VEV v, as in the electroweak theory. Depending on its value, this theory is either weakly or strongly coupled. We briefly discuss it in Section 24.2 in connection with the electric–magnetic duality.

Another one is the $\mathcal{N} = 4$ supersymmetric gauge theory, which has four fermions and six scalars. It is a celebrated conformal theory with a vanishing beta function, which is a *non-running* coupling constant. Due to this huge symmetry, it was the first theory using which the so-called gauge–string (or gauge–gravity) duality has been discovered. We briefly discuss it in Chapter 21, devoted to the holographic models of QCD.

2.2 The "Asymptotic Freedom" and Perturbative QCD

In the case of QED, the Lagrangian contains a single term, $eA_\mu\bar{\psi}\gamma_\mu\psi$, describing the interaction of the electron field ψ with the gauge potential. It generates Feynman rules for the perturbative diagrams, according to which the vertices include $e\gamma_\mu$. (The Dirac matrices act on the spinor indices of the electrons.)

In QCD, there is a similar vertex that includes additionally the color matrices $t^a/2$, acting on color indices of the quark propagators or external lines. The index $a = 1, \ldots, 8$ counts the color states of exchanged gluons.

Unfortunately, the quantization of the gauge part of the QCD path integral is not so simple: We will not describe it here and refer the reader to textbooks on quantum field theories (QFTs). Quantization can be either done non-covariantly[3] or covariantly, using the so-called Faddeev–Popov "ghosts." Using the latter method, 't Hooft and Veltman have proven the renormalizability of the Yang–Mills theory, i.e. all divergences can be absorbed by a "running" invariant charge g and quark mass renormalization.

The actual calculation[4] of the QCD charge renormalization led Politzer (1973) and Gross and Wilczek (1973) to the beta function of QCD-like theories:

$$\beta(\alpha_s(\mu)) \equiv \frac{d\alpha_s}{d\log(\mu^2)} = -(b_0\alpha_s^2 + b_1\alpha_s^3 + b_2\alpha_s^4 + \cdots) \tag{2.11}$$

$$b_0 = \frac{11C_A - 4n_f T_R}{12\pi}, \quad b_1 = \frac{17C_A^2 - n_f T_R(10C_A + 6C_F)}{24\pi^2}$$

(The subsequent coefficients b_2, b_3 had also been calculated, but they depend on specific details, such as the renormalization scheme, and we will not need those in this book.) The QCD quantization and derivation of the beta function are rather technical issues, well covered by QFT textbooks, and they will not be discussed here.[5]

The negative value of the beta function was a rather unexpected discovery since, in QED and scalar theories, its sign is positive. Consequently,

---

[3]That is, with gauge fixing involving a particular reference frame. We discuss the finite-T version of QCD in Chapter 3, in which the rest frame of the heat bath is the natural choice.

[4]Precursors to this discovery, which, unfortunately, were not followed, were work by Vanyashin and Terentev (1965) on constant gauge field backgrounds and by Khriplovich (1969) in a non-covariant Coulomb gauge. The dates of both early works — 1965 and 1968, respectively — are very important; they occurred *before* the famous SLAC deep inelastic experiments of 1969 and the parton model.

[5]Perhaps only one comment: At large enough numbers of quark flavors n_f, both b_0 and b_1 change signs. For $N_c = 3$, the former does so at $n_f = 33/2$, and theories with larger numbers of flavors do not have asymptotic freedom. Furthermore, there is a region in which $b_0 > 0, b_1 < 0$, a sign that the beta function may have a zero. If it does, the zero implies the presence of an infrared fixed point or that those theories become conformally invariant at large distances. The exact critical value of n_f at which this regime occurs remains unknown, despite the dedicated efforts of lattice practitioners.

the strong charge $\alpha_s(Q) = g^2(Q)/4\pi$ *decreases* when the corresponding momentum scale Q grows.

This theoretical discovery is connected to what is known as the Bjorken scaling, which has been confirmed by the SLAC deep inelastic experiment, and Feynman's parton model, opening the way to the perturbative form of QCD (pQCD) for "hard" processes. We further discuss it in the chapter related to jets (in Section 19.3).

An important general idea is *factorization*, which is based on the separation of scales. The "hard scale" $Q \gg 1\,\text{GeV}$ (e.g. an invariant mass of produced di-jets) is tens or hundreds of GeVs. Physics at such scales is accurately described with perturbative diagrams using quarks and gluons. Over the years, there has been significant progress, allowing us to calculate, say, the square of n-gluon matrix elements for various n and with an increasing number of loop corrections.

The physics of the incoming colliding hadrons and outgoing jets corresponding to "soft scales" is described phenomenologically, by the so-called *parton distribution functions* (PDFs) and the *fragmentation functions*, respectively. Those cannot be calculated perturbatively and need phenomenological inputs.

The theory of hard processes, based on pQCD, had matured over the years. The first stage involved resumming logarithms, appearing in the renormalization of individual operators or groups of operators mixed in the RG transformation. For example, an operator can be just a single field (e.g. $q(x)$ of a quark), quark bilinears ($\bar{q}\Gamma\mathbf{q}$) or currents. The structure functions of deep inelastic scattering can be related via an operator product expansion (OPE) series, to the tensor operators:

$$T^{(n)}_{\mu,\nu_1,\ldots,\nu_n} \equiv (\bar{q}\gamma_\mu D_{\nu_1}\ldots D_{\nu_n} q) \tag{2.12}$$

averaged over the target (the nucleon or some nucleus). Now, if the outgoing quark lines (and gluon ones, which originated from the long derivatives) have the virtuality scale μ, the radiative corrections coming from loop diagrams with extra gluons produce powers of $\log(\mu)$. Their summation leads to *powers* of $\log(\mu)$, known as the *anomalous dimensions*[6] of the operators, which have been calculated. Thus, their matrix element over

[6]For conformal theories (such as $\mathcal{N}=4$ SYM or QCD with many flavors and conformal IR fixed points), the spectroscopy of the anomalous dimensions (instead of masses) are the main object of calculations.

the nucleon, measured in DIS, predicts a dependence on the momentum transfer:

$$\langle N|\hat{O}^{(i)}|N\rangle|_Q \sim \left[\log\left(\frac{Q^2}{\Lambda^2}\right)\right]^{\gamma_i} \tag{2.13}$$

These matrix elements are related to moments of structure function obtained from experiments: Thus, one was able to test if the pQCD values of the anomalous dimensions were indeed correct. That was how the first tests of pQCD were performed in the 1970s. Nowadays, one needs to know the anomalous dimensions (and mixing matrix if there is a group of operators) if the results of lattice calculations with different lattice spacings need to be compared.

The alternative representation of the pQCD radiative corrections and the RG dependence on the scale took the form of Dokshitzer–Gribov–Lipatov–Altarelli–Parisi (DGLAP) evolution equations, written directly for the PDFs themselves. We will not present those here and only comment that one should use them to "evolve" PDFs from the scale μ, at which the phenomenological input has been made, to the scale Q, at which the hard process in question takes place.

In Fig. 2.1, we show the so-called "phase diagram" (which is, in fact, a map) of pQCD. In the vertical (virtuality scale) direction, the evolution is described by the DGLAP theory. In the horizontal direction — the fraction x — one needs a different resummation of perturbative $\log(1/x)$ terms, which is known as the Balitsky–Fadin–Kuraev–Lipatov (BFKL) theory. We briefly discuss its properties later in the chapter; for now, it is enough to know that multiple splittings of partons into two lead to their density at small x growing at a certain power of $1/x$ toward small x. When the gluon density becomes high enough, one needs to proceed to a nonlinear regime known as color glass condensate (CGC), which is approximated by *classical* gauge fields; for review, see, for example, Iancu *et al.* (2002). The boundary line is also known as the *saturation scale* $Q_s^2(x)$.

The dependence of PDFs is calculated perturbatively; however, the *intial conditions* at some μ^2 remain phenomenological, extracted either from experiment or (now with much less precision) from the lattice. The Hamiltonians and the corresponding wave functions on the light front are in the process of being calculated in certain approximations. We discuss the current attempts to do so in Chapter 28.

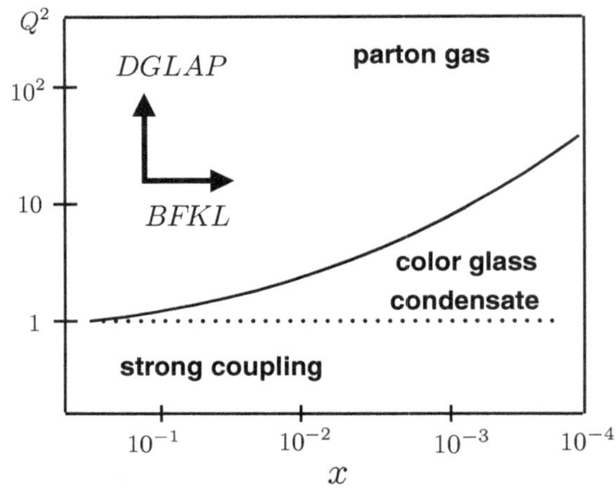

Fig. 2.1. The "phase diagram" of pQCD on the plane x (fraction of the total momentum) and virtuality Q^2 (GeV^2). At small Q^2, the coupling is strong, but going up in the plot, it becomes weaker. The evolution in this direction (indicated by an arrow with the "DGLAP" label) leads us to small-size partons, which form a dilute gas. Moving toward smaller x (indicated by the arrow with the "BFKL" label) leads to the radiation of more and more gluons. Eventually, in this way, we meet the boundary of CGC, marked by the solid line and indicating the saturation scale $Q_s(x)$.

2.3 The Nonperturbative QCD

In contrast to the term "perturbative QCD," defined via Feynman diagrams, by its "nonperturbative" versions, researchers belonging to different sub-communities mean quite different things. Let us mention some of them in an order of increasing complexity:

(i) One meaning of "nonperturbative" calculations is those in which a certain series of diagrams are resummed. For example, one can resum the leading logarithms, $\log(1/x)$, in ladder diagrams, leading to the BFKL Pomeron.
(ii) Another meaning of "nonperturbative" effects is those associated with QCD solitons, such as instantons and monopoles, which are treated in weak coupling via semiclassical approximation and lead to effects $\sim\exp[\text{const}/g^2(Q)]$. Since such functions have zero Taylor expansion in g^2, they are invisible to the pQCD to any order.
(iii) Last but not least is the lattice formulation of QCD. While at the lattice scale a (typically, $\sim$0.1 fm $= 1/(2\,\text{GeV})$) the coupling is assumed to be weak, numerical simulations provide the dynamics at larger

scales, at which the coupling can be arbitrary strong. That is why lattice QCD provides physical answers, such as the values of hadronic masses and other hadronic observables.

In this book, we will not discuss perturbative resummations (except in hot quark-gluon plasma). Our attention will be on topological solitons and their role in the generation of the main properties of the QCD vacuum: *color confinement* and *chiral symmetry breaking*. The latter provides "constituent quark masses," and the former provides the main forces between such quarks: Taken together, they are the basis of most of the models describing hadronic spectra and other properties.

2.4 Open Issues in QCD and Heavy Ion Physics

Despite the fact that QCD is about 50 years old, there are many open issues to be discussed in this book. Here, we start with some general questions, which are common to the whole field of strong interaction physics, and provide comments on them.

I. **Can one locate the "soft-to-hard" boundary scale**? Is it *universal*, i.e. the same for *all* observables under consideration, or does the transition from weak to strong coupling regimes depend on the particular observable?

II. Can one locate the **"micro-to-macro" boundary**, at which the ratio of the mean free path to system size passes a certain threshold beyond which a regime change occurs from large (ballistic) to small (hydrodynamic or "fireball") regimes? In particular, which experimental observables display this transition most clearly?

As soon as the applicability of the macroscopic regime is established, one may describe matter in terms of just a few thermal parameters — temperature and chemical potentials. In terms of those, one can relate the setting to the location of singularities, the **phase transition** lines or points. Heavy ion physics is particularly interested in the QCD phase transition from hadronic matter to the QGP.

At low baryon chemical potential, this transition is *a crossover*, whereas at larger values, a transition to the first order is expected. One of the ongoing RHIC programs is the Beam Energy Scan, which is searching for an answer to the following question:

III. Can we experimentally locate **the QCD critical point** by following higher-order fluctuations/correlations? Does it exist at all in certain regions of parameters?

Brief comments on all these "large questions" are as follows:

(Ia) Generally, perturbative QCD should be applicable when the so-called "dimensional transmutation" parameter Λ_{QCD}, with the dimension of a momentum or mass[7] is much smaller than the "hard" scale Q:

$$Q \gg \Lambda_{QCD} \sim .2\,\text{GeV} \sim \text{fm}^{-1} \tag{2.14}$$

However, in practice, for particular quantities, the meaning of "much larger" varies considerably.

The simplest quantity is a static quark–antiquark potential $V(r)$. It transitions from the perturbative $\sim -1/r$ (with coefficient displaying a pQCD running) to a stringy linear regime $\sim r$. The transition is well defined due to accurate lattice studies, and it occurs at $r \approx 0.5\,fm$. We discuss this potential, as well as spin-dependent ones, in Part 3 of the book.

More generally, there exists a large variety of correlation functions with different quantum numbers, which were studied, both phenomenologically and on the lattice. In each of them, one may locate the transition between pQCD and nonperturbative regimes. It turned out that in various channels, very different results were obtained. This is because topological phenomena are coupled differently to different operators/states. In particular, all spin-zero channels have large instanton-induced nonperturbative effects, e.g. for the $\langle G^2(x)G^2(0)\rangle$ correlator, the nonperturbative scale is as large as $Q^2 \sim 10\,\text{GeV}^2$.

The simplest experimentally testable pQCD predictions (diagrams) are hard exclusive processes, e.g. the pion or the nucleon form factors. Up to the highest $Q^2 \sim 5\,\text{GeV}$ measured so far, neither of them had reached quantitative agreement with the pQCD predictions. Because of their importance, the experimental studies should be continued at higher momenta, until such agreement might be observed.

pQCD does work well in hard, *inclusive* processes, such as di-jet production. (Now, at the LHC, people are working with jets of energy $\sim$100 GeV, two orders of magnitude above Λ_{QCD}.) The question is where this nice description stops. In the heavy ion field, there exists a hotly debated "mini-jet" issue. While the identified jets have rather large momenta, $p_\perp > 20\,\text{GeV}$ or so, it is generally assumed that the parton description is valid down to much smaller momenta. How much smaller? Following DGLAP evolution toward small Q^2 all the way to $\sim 1\,\text{GeV}^2$, one

[7]The precise definition of Λ_{QCD} depends on the exact renormalization scheme: the most commonly used is the so-called $\bar{MS}$ one. Let me recall again that, in this book, $c = \hbar = 1$.

eventually reaches a negative gluon density. This and other arguments tell us that at a scale of 1 GeV, the pQCD cannot be used. The lower boundary of pQCD at the scale of $Q_{\min}$ should be defined by the "higher twist effects," which are not yet studied to the extent necessary to provide a quantitative answer.

Chapter summary and connections: QCD is the fundamental theory of strong interactions. Perturbative QCD is applicable for "hard processes" at which, due to asymptotic freedom, the coupling is weak. By now, perturbative scattering amplitudes extend well beyond what we discussed, to multi-gluon diagrams and up to several loops. Processes producing very heavy objects, such as W, Z, H bosons or jets with momenta $O(100\,\mathrm{GeV})$, are well described by it. Its applicability perhaps starts at scales of a few GeV, but the details of that are still debated. Related to this in the book are the descriptions of jet quenching in QGP in Section 19.3 and the discussion of hadronic form factors in Chapter 28.

Chapter 3

PERTURBATIVE THEORY OF COLD/HOT QGP

As QGP is the "normal phase" of QCD, one may expect the standard order of presentation, starting with the Lagrangian and Feynman diagrams. One of the first technical questions appearing along the way is the issue of *gauge fixing*.[1] Unfortunately, in perturbation theory, gauge fixing is unavoidable, which brings about a large number of technical questions. Standard QFT courses spend a lot of attention on those, teaching about the details of quantization in covariant gauges, Faddeev–Popov ghosts and Gribov copies. These are all needed for certain goals, but fortunately not for ours so far. As we see shortly, the main perturbative results for finite-temperature QCD can be (and were, in fact, first) obtained in a Coulomb gauge, which is simpler and more directly related to thermal physics.

We begin with Section 3.2 by defining three scales of QGP: the thermal, electric and magnetic screening scales. At a high enough T, those are separated by powers of the coupling, which are small due to asymptotic freedom $g^2(T) \ll 1$. The first diagrams are evaluated in Sections 3.3 and 3.4.

A certain resummation of the diagrams, similar to what was done for electron gas and nuclear matter, is discussed in Section 3.5. Finally, in Section 3.6, we discuss the infrared divergences and show that, even at very high T, there remains a nonperturbative *magnetic* sector of the theory for which a perturbative approach cannot be used. It can be studied through dedicated lattice studies of a 3D magnetic effective theory.

[1]The full nonperturbative Wilson's formulation of QCD on the lattice uses *averaging* over all gauges rather than gauge fixing. Although universally considered correct, it is also not without issues; for example, it is related to counting the so-called Gribov gauge copies.

3.1 Simple Estimates of the Quark-Gluon Plasma Phase Boundary

Historically, the question of superdense/hot matter has been discussed starting with the *cold* ($T = 0$) case. It has been known since the 1930s that there should be neutron stars at the end of stellar evolution for a certain mass range, and it was realized that the density at their centers should reach several times the nuclear matter density. Obviously, the question of its properties and EOS has been raised.

The standard parameterization of the energy density of nuclear matter is

$$\epsilon(n_B) = m_N n_B + \frac{3}{2} n_B T + \frac{K n_0}{18}\left(1 - \frac{n_B}{n_0}\right)^2 \tag{3.1}$$

with a minimum at $n_0 \approx 0.16\,\mathrm{fm}^{-3}$ and compressibility $K \sim 200$ MeV.

Using hadronic language, people first thought about the presence of excited states of the nucleon and (strange) hyperons as Fermi energy (or μ_B) grows. To my knowledge, the first paper considering cold quark matter and "quark stars" was by Itoh (1970). The early discussions of those have been closely related to bag models of hadronic structure, which suggested the EOS of the form (for N_f massless quarks)

$$p = N_f \frac{p_F^4}{4\pi^2} - B, \quad \epsilon = 3N_f \frac{p_F^4}{4\pi^2} + B \tag{3.2}$$

where the first terms correspond to the ideal Fermi gas with the Fermi momentum p_F (the same as Fermi energy and μ in quark normalization) and the last, known as *the bag term*, was included to separate the physical and "perturbative" vacua.

For a couple decades, the issue has been related to the hypothetical *stable strange quark matter* (SQM), corresponding to $p = 0$. For a while, people used the value for the bag constant of $B \sim 50\,\mathrm{MeV/fm}^3$ from the MIT bag model, observed that such a model would lead to deeply bound dihyperon $H = (uds)^2$ and other "strangelets," and declared that our nucleons may not be the ground state but somehow strangely survived in the Big Bang, waiting for "strangelets" from supernovae or neutron star collisions/mergers to convert us all to SQM.

However, this was over by the late 1970s, as documented by Shuryak (1980). The so-called scale anomaly relates the nonperturbative energy density to the so-called gluon condensate, known from QCD sum rules;

therefore, the nonperturbative vacuum energy density must be one order of magnitude larger:

$$\epsilon_{\text{vac}} = -B = -\frac{b}{128\pi^2}\langle (gG^a_{\mu\nu})^2\rangle \approx -500\,\text{MeV/fm}^3 \tag{3.3}$$

This value of the "bag constant" B put the MIT bag model and the SQM issue into the dustbin of history.[2]

The same ideas were first applied to superhot matter at high temperatures. Representing QGP as an ideal gas, one gets

$$\begin{aligned} p &= T^4\frac{\pi^2}{45}\left[(N_c^2-1)+\frac{7}{4}N_cN_f\right]-B,\\ \epsilon &= T^4\frac{\pi^2}{15}\left[(N_c^2-1)+\frac{7}{4}N_cN_f\right]+B \end{aligned} \tag{3.4}$$

Note that standard thermodynamics relates these two equations. If there are only T and zero chemical potentials, as assumed, the free energy is simply related to pressure $pV = -F$; therefore,

$$\epsilon + p = Ts, \quad s = \frac{dp}{dT} \tag{3.5}$$

Note further that in both these formulae, the bag term cancels out: This is how it should be because vacuum energy shift cannot contribute to matter entropy! Indeed, the vacuum energy density requires a specific definition to be given, but the vacuum entropy must be zero.

Before delving further into quantitative theory, let us present a sketch of the QCD phase diagram in Fig. 3.1. A vertical blue line indicates the cooling of matter during the cosmological expansion of the Universe. In it, there is a very small baryon–antibaryon asymmetry, and the corresponding nonzero μ_B is not shown to scale. Our main interest resides in heavy ion collisions, or "Little Bangs" as we will often call them. Depending on the collision energy, they lead to paths covering QCD phase diagrams (to be discussed later). The existence of a *QCD critical point*, as shown by a red dot, is expected, and its location has been searched for but not found yet. Finally, a large-density (and thus μ_B) region at near-zero T should be present in the cores of neutron stars, while a rather noticeable T should be created during neutron star mergers.

[2]Also, the idea that hadrons are "perturbative" or "empty" inside leads to the wrong dependence of hadronic masses on the number of colors.

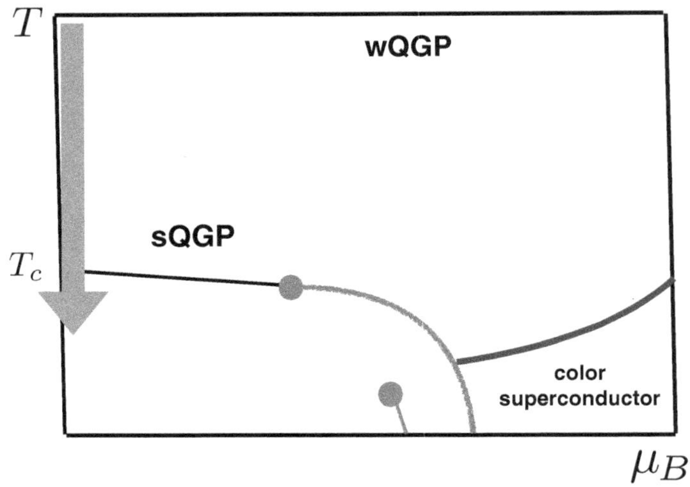

Fig. 3.1. Schematic QCD phase diagram in coordinates of baryon chemical potential μ_B and temperature T. Weakly and strongly coupled versions of QGP are indicated by wQGP and sQGP, respectively. The matter below the phase transition line is hadronic, to be described as hadron resonance gas (HRG). The thick red part of the line indicates the expected first-order transition, ending at the QCD critical point (the red dot). Another short line separates the vacuum from nuclear matter, with its red dot corresponding to the "multifragmentation" phenomenon. A high density region is expected to be a color superconductor. Finally, the vertical blue arrow corresponds to matter evolution in cosmology after the Big Bang.

Now, we proceed to a "first attempt" estimate of the location of the phase transition. Let us use a crude expression for the pressure of the HRG pressure (to be detailed in the following) at zero baryon density, $\mu_B = 0$:

$$p \approx (20\,\mathrm{GeV}^{-2})T^6 \tag{3.6}$$

Transition is expected when pressures (= free energies) of both (QGP and HRG) phases are equal. The intersection of two different curves obviously leads to a jump in the derivatives — the first-order transition. How pressure and energy density behave is shown in Fig. 3.2.

A final note: It is because ϵ_{vac} is so large — the vacuum is very dense — that we had to use the high-energy colliders RHIC and LHC to create the QGP.

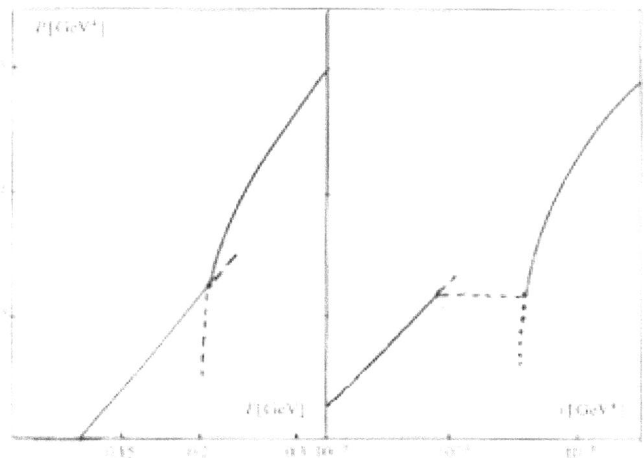

Fig. 3.2. Pressure and energy density (GeV^4) versus temperature T (GeV) of QCD matter using the bag model for QGP and power fit for HRG. While the pressures of both phases at the critical temperature are the same, the energy density (the right-hand plot) has a jump (as it happens in many transitions, e.g. the familiar liquid–gas boiling point).

3.2 Overview and Scales of QGP

Perturbative calculations for hot (high T) QGP, including gluons, were initiated by Shuryak (1978b) and Kapusta (1979). They included the first perturbative diagrams and their resummations. The first reviews were carried out by Shuryak (1980) and Gross *et al.* (1981), establishing this new field of high-T QCD.

Ignoring the small masses of u, d, s quarks and the effects associated with heavy c, b, t quarks completely, one finds a situation in which T is the only scale in the problem. Thus, the (free) energy density and pressure should simply be

$$\epsilon \sim p \sim T^4 \tag{3.7}$$

In the *zeroth* approximation, in which quarks and gluons are not interacting at all, the QGP is just an ideal gas. So, the coefficients in the expressions above simply follow from the usual Bose–Einstein distributions for gluons and the Fermi–Dirac distributions for quarks and antiquarks.

In perturbative QCD, the dimensionless coupling g usually appears in the combination $g^2/4\pi = \alpha_s$. So, naively, one may expect F/T^4 to be some

Taylor series in powers of α_s. Yet, it is only true for the first-order $O(\alpha_s)$ corrections to thermodynamical observables. Since it is straightforward to calculate these lowest-order diagrams, I will skip them and move on to the next issue: diagram resummations. We will soon see that the results are more complicated than Taylor series with nontrivial powers of α_s.

The first theoretical issue involving QGP was the question of whether this kind of plasma has *screening* or *antiscreening* of the charge. It is related to the running of the coupling. Cold quark matter has the same diagrams as QED plasmas, so the *screening* regime in it was obvious. The gluonic part — appearing at finite T only — was new. In the mid-1970s, when I approached this problem, the following facts were known:

- In QED plasma, the charge is screened by matter particles.
- In QED vacuum, the charge is screened by vacuum polarization.
- In QCD vacuum, the charge at small distances is *antiscreened*; this is, of course, the celebrated asymptotic freedom.
  The remaining question was then:
- *What happens in a QCD heat bath at a high enough T at large distances?* The answer came from a calculation (Shuryak, 1978b) of the polarization tensor $\Pi_{\mu\nu}(T,q)$:
- In QCD plasma, the charge is *screened* at large distances.

That is why the quark-gluon plasma was called a *plasma*. A schematic diagram illustrating the behavior of the effective electric charge is shown in Fig. 3.3. So, a combination of screening and antiscreening keeps it small at all distances, provided T is high enough.

What this means is that apart from the obvious scale provided by the temperature, or, more accurately, by the Matsubara frequency:

- the "hard" scale $M_T \sim \pi T$
  there is also the so-called
- electric or Debye scale $M_E \sim g(T)T$
  Furthermore, it was found in the same work (Shuryak, 1978b) that static magnetic fields are not screened and power infrared divergences remain in the magnetic sector. Polyakov suggested that there exists a "magnetic mass" with:
- a magnetic mass scale of $M_M \sim g^2(T)T$

Linde (1980) suggested that the magnetic screening mechanism could be due to magnetic monopoles. Years later, Kajantie and collaborators

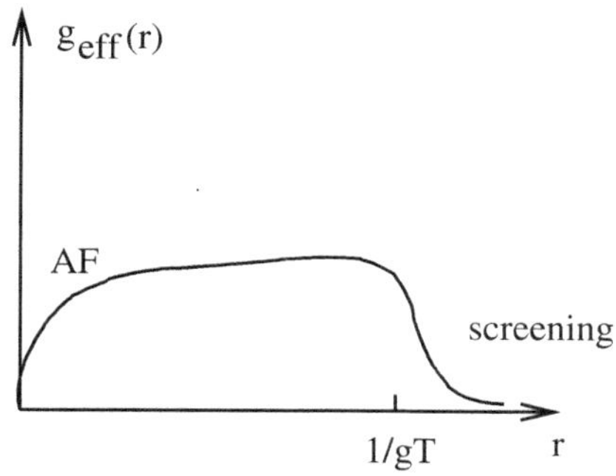

Fig. 3.3. Schematic diagram illustrating the behavior of the effective electric interactions in QGP as charge versus distance. "AF" on the left means "asymptotic freedom," a logarithmic decrease in effective charge at small distances.

included the zero Matsubara frequency sector of the vector potential $\vec{A}$ and the temporal component A_0, which plays the role of an adjoint Higgs field. So, it is a three-dimensional (3D) version of the Georgy–Glashow model, with the dimensional coupling $g_3 = g^2T$. Two other couplings are the coefficients of mass term $\mathrm{Tr}(A_0)^2$ and the quartic coupling $(Tr(A_0)^2)^2$. The A_0 mass term from the four-dimensional (4D) calculation (outlined in the following section) is positive. Therefore, this magnetic theory is *not* in the broken phase, $\langle A_0 \rangle = 0$; therefore, there are no magnetic monopoles in it. This theory is, however, a *confining* one with a mass gap: Studies on it have been made using 3D lattice simulations; see Hietanen *et al.* (2009) for details.

Outside of the magnetic sector, direct pQCD evaluation of multi-loop diagrams is possible; see, for example, Arnold and Zhai (1995), which we follow. Those have demonstrated that the convergence of the pQCD series is not reached in the temperature region of practical interest, $T \sim 1\,\mathrm{GeV}$. Different resummations have been done by Andersen *et al.* (2000), Blaizot *et al.* (1999), and those claimed to be applicable to lower T.

So far, we have mentioned pQCD applications to thermodynamical quantities, free energy, etc. Of course, they are not the only quantities which one would like to know. Kinetic quantities, such as viscosity, are crucial to understanding equilibration and matter evolution in heavy ion collisions: We will turn to their discussion later. Proceeding to the main logic of the book, let me present the main scales (normalized to T) in the table, for weak and strong coupling regimes.

Table 3.1. A comparison of the basic scales in high-T gauge theories in (a) the weak coupling with (b) the strong coupling limit of the $N = 4$ SUSY gauge theory. (All constants are ignored except in the last entry).

Quantity	Small g, finite N	Large g^2N
Inter-particle $n^{-1/3}$	~ 1	$\sim 1/N^{2/3}$
Debye screening radius R_D	$\sim 1/g$	~ 1
Magnetic screening radius R_M	$\sim 1/g^2$	~ 1
Sound attenuation time Γ_s	$\sim 1/g^4 \log(1/g)$	$1/3\pi$

3.3 Elements of Perturbative Theory of QGP

The continuous version of the Yang–Mills action in the perturbative normalization

$$S = \int d^4x \frac{1}{4}(G^a_{\mu\nu})^2, \quad G_{\mu,\nu} = \partial_\mu A_\nu - \partial_\nu A_\mu + ig[A_\mu A_\nu] \tag{3.8}$$

may look similar to the gauge part of action of electrodynamics. Yet, there is a crucial difference here because the field strength contains the commutator. So, it is not just Gaussian path integrals due to nonlinear terms of the order $\sim gA^3, g^2A^4$. We should now expand $\exp(-S)$ in powers of the nonlinear terms, or powers of the coupling g. The quadratic part of the action, as usual, defines the Gaussian operator and propagators. Since it has two derivatives, or two momenta in momentum representation, one finds $G \sim (\text{numerator})/k^2$ as usual. The numerator depends on the choice of gauge. These and other details of Feynman rules for gauge theory I do not want go into for now. The cubic and quartic vertices can be read off the Lagrangian directly.

In order to master this formalism, it is convenient to consider some simple examples. The single-body observable $O_{mn}(\omega, k)$ (such as energy) of two transversely polarized gluons[3] can be written as follows:

$$\epsilon = T\sum_n \int \frac{d^3k}{(2\pi)^3} O_{mn}(\omega, k) D_{mm}(\omega_n, k) \tag{3.9}$$

where $D = (\delta_{mn} - k_m k_n/k^2)/(\omega_n^2 + k^2)$ is the usual zeroth-order gluonic propagator. Evaluating the sum first, one gets the famous "bosonic

[3] In the Coulomb $\vec{\partial}\vec{A} = 0$ gauge, we use the fact that there is no longitudinally polarized one.

summation" formula:

$$\sum_n \frac{1}{x^2+n^2} = \frac{2\pi}{x}\left(\frac{1}{2} + \frac{1}{\exp(2\pi x)-1}\right) \tag{3.10}$$

which leads to the following result:

$$\epsilon(T) = 2(N_c^2-1)\int \frac{d^3kk}{(2\pi)^3}\left(\frac{1}{2} + \frac{1}{\exp(k/T)-1}\right) = 2(N_c^2-1)\frac{\pi^2 T^4}{30} \tag{3.11}$$

The first term — the divergent energy of zero-point oscillations of all oscillators in the vacuum — should be omitted as usual, while the second term — due to thermal gluons — must be retained. Note that the gluon field periodicity on the Euclidean time circle leads to integer n in the denominator, which directly leads to the Plank formula for blackbody thermal radiation.[4]

It is also worth repeating the evaluation for the zeroth-order single-body observables for fermions, again by using the propagator $S = (\gamma_\mu p_\mu + m)/(p^2+m^2)$. Recall the following generic *fermionic* identity:

$$\sum_n \frac{1}{x^2+(2n+1)^2} = \frac{\pi}{2}\left(\frac{1}{2} - \frac{1}{\exp(\pi x)+1}\right) = \frac{\pi}{2x}\tanh\left(\frac{\pi x}{2}\right) \tag{3.12}$$

The right-hand side in our context has some rather transparent physical meaning that we already discussed. Note also how the extra factor of two — from the two poles on the left-hand side — appears, accounting for the equal number of quarks and antiquarks in the heat bath.

In addition, note the following curious detail: Unlike in the bosonic case, with its standard combination, familiar from the harmonic oscillator $\langle n_B(\text{quanta}) + 1/2\rangle$, the fermionic one is $\langle 1/2 - n_F(\text{quanta})\rangle$. Furthermore, at very high temperatures $T \to \infty$, the exponent $\exp(\omega_k/T) \to 1$; therefore, $\langle 1/2 - n(\text{quanta})\rangle \to 0$. It will have an important effect on the high-T behavior of the fermionic propagator.

An explicit calculation of all closed two-loop diagrams includes all qq, qg, gg scattering processes of order g^2. In hot quark-gluon plasma, it leads to the following result: which is positive and thus corresponds to attraction. (F is the free energy, related to the partition function $\log(Z) = \exp(-F/T)$).

[4]From time to time in the literature, there appear claims that periodicity can be substituted by periodicity *modulo gauge transformation*. If this were true, the Plank formula would be modified by some unphysical phase factors, which is not the case.

3.4 The Polarization Operator

The heat bath defines a preferred Lorentz frame in which matter is at rest. So, Lorentz symmetry is broken anyway, and thus, the zeroth component of vector potential A_0 is not the same as others. The calculations are easier to perform in the "physical" non-covariant gauges than in the covariant ones, which are used in the QFT textbooks. It was performed first by Shuryak (1978b) using the Coulomb gauge. In this case, not only are there no Faddeev–Popov ghosts, but one can also show that there is no vertex renormalization of the static charge. As a result, all one has to do in this gauge is calculate the polarization tensor $\Pi_{\mu\nu}(\omega, k)$, in close parallel to what is done in the electron gas.

As usual, the polarization tensor has two Lorentz structures corresponding to the electric (Coulomb) and magnetic (transverse) fields. The former is described by $\Pi_{00}(\omega, k)$ and the latter by the spatial part of the polarization tensor

$$\Pi_{mn}(\omega, k) = \left(\delta_{mn} - \frac{k_m k_n}{k^2}\right) \Pi_\perp(\omega, k) + \frac{\omega^2 k_m k_n}{k^2} \Pi_{00}(\omega, k) \quad (3.13)$$

These two functions enter the corresponding "dressed" (resummed) Green functions as follows:

$$D_{00} = \frac{1}{\vec{k}^2 + \Pi_{00}(\omega, k)} \qquad D_{mn} = -\frac{\delta_{mn} - k_m k_n / k^2}{\omega^2 - \vec{k}^2 - \Pi_\perp(\omega, k)} \quad (3.14)$$

The logarithmic contributions from the three diagrams shown in Fig. 3.4 are sufficient to obtain the invariant charge renormalization, as in this gauge, there is no renormalization of the Coulomb vertex function. Including the usual contribution from quarks (analogous to that in QED) plus these three gluonic diagrams, one arrives at the following respective contributions to

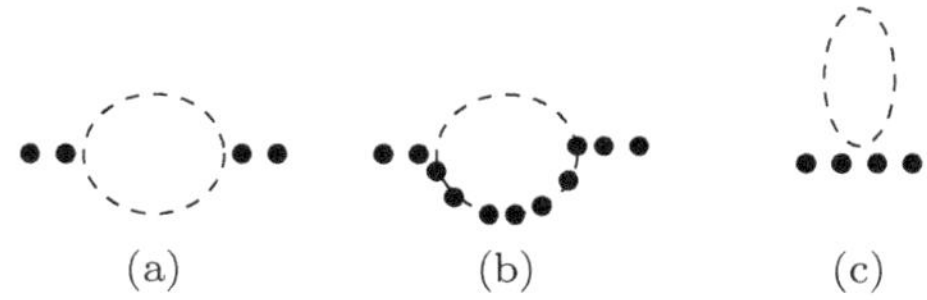

Fig. 3.4. Three diagrams in the Coulomb gauge for the polarization operator. The dashed lines indicate the transverse (magnetic) gluon propagators, and the dotted lines denote the $\langle A_0 A_0 \rangle$ propagator.

the gluon polarization operator:

$$\Pi_{00}(\vec{q}) = \frac{g^2\vec{q}^2}{\pi^2}\log(\vec{q}^2/\Lambda^2)\left[+\frac{1}{2} - 6 + 0 + \frac{N_f}{3}\right] \tag{3.15}$$

(The fourth term is a contribution from the quark loop, added for completeness.) Note that the only negative contribution comes from diagram (b). Indeed, it is only possible for this diagram because it has no imaginary part — the Coulomb line shown by the dotted line is instantaneous and has no pole in energy, so the sign of the log is not restricted by the unitarity. An explicit calculation shows that it has a log, $\log(\vec{q}^2)$, which contains not the usual square of the 4D momentum but only its spatial components.

Thus, resumming the Coulomb field of a charge via a sequence of polarization operators, one finds

$$\begin{aligned}&(1/\vec{q}^2) + (1/\vec{q}^2)\Pi_{00}(1/\vec{q}^2) + \cdots \\ &\quad = (1/\vec{q}^2)\left(1\Big/\left(1 + \frac{g^2}{\pi^2}\log(\vec{q}^2/\Lambda^2)[(N_f/3) + (1/2) - 6]\right)\right)\end{aligned} \tag{3.16}$$

which agrees with the asymptotic freedom formula by Gross, Politzer and Wilczek. This derivation is, in fact, much simpler than theirs, done in a covariant gauge and is due to Khriplovich (1969), performed as early as 1968.[5]

I mentioned all that for completeness. However, in this course, we never discussed the UV completion of the theory, leaving these issues to QFT courses. Our focus should always be on the behavior of the dressed propagators in the opposite limit, soft momenta $Q \to 0$, or the IR.

Let me begin with the electric polarization tensor in the "soft" limit $\omega, k \ll \mu$ for cold quark matter, in which case it comes from the fermion loop:

$$\Pi_{00} = -\frac{g^2}{2\pi^2}\left(\sum_f \mu_f^2\right)\left[1 - \frac{\omega}{2k}\ln\left|\frac{\omega+k}{\omega-k}\right| + \frac{i\pi\omega}{2k}\theta(\omega)\theta(k-\omega)\right] \tag{3.17}$$

[5]Khriplovich was a good friend and, for a few years, even a roommate of mine. So, I had a full five years before the work of Gross, Politzer and Wilczek to connect this sign to the parton model and the weakness of strong interaction at small distances, known since the famous SLAC experiment of 1969. I can only blame myself for understanding next to nothing about the renormalization group from the QFT course at the time. Later, David Gross told me that it was Wilson's lectures about indices which were crucial to their thinking at the time.

where the sum runs over all quark flavors if their chemical potentials are all different.

Let us try to understand what this expression means, starting with its real part. Note that the function within the square brackets is a function of ω/k, so there are different "long-wave" limits depending on this ratio. One is *the static limit* $\omega = 0, k \to 0, \omega/k = 0$ in which

$$D_{00}(\omega, k) \to -\frac{1}{k^2 + \kappa_D^2}, \quad \kappa_D^2 = \frac{g^2}{2\pi^2}\left(\sum_f \mu_f^2\right) \tag{3.18}$$

The Fourier transform of this Coulomb-field propagator is the screened potential between two charges $V(r) \sim \exp(-\kappa_D r)/r$, so the inverse $R_d = 1/\kappa_D$ is the so-called *Debye screening radius.*

The opposite limit $k = 0, \omega \to 0, k/\omega = 0$ is called *the plasmon limit.* Using the expansion at small k,

$$1 - \frac{\omega}{2k}\ln\left(\frac{\omega + k}{\omega - k}\right) \approx -\frac{k^2}{3\omega^2} + \cdots$$

one finds

$$D_{00}(\omega, k) \to \frac{1}{k^2(1 - \omega_D^2/\omega^2)}, \quad \omega_D^2 = \frac{g^2}{6\pi^2}\left(\sum_f \mu_f^2\right) = \frac{\kappa_D^2}{3} \tag{3.19}$$

where ω_D is the frequency of the longitudinal long-wavelength plasma oscillations. We do not present here an analogous expression for the magnetic polarization tensor and only note that in this case, the most interesting physical limit is the so-called *gluon effective mass* defined by the third "on shell" limit $k = \omega \to 0, k/\omega = 1$

$$m_g^2 = \Pi_\perp|_{\text{onshell}} = \frac{g^2}{4\pi^2}\left(\sum_f \mu_f^2\right) = \frac{\kappa_D^2}{2} \tag{3.20}$$

The last important point is that the static magnetic field in quark Fermi gas is not screened because, in the static limit $\omega = 0$, $k \to 0$, $\omega/k = 0$, the magnetic polarization $\Pi_\perp \to 0$ vanishes.

The imaginary part of the tensor originates from the log in (3.17), when its argument is negative. Its physics is the so-called *Landau damping* of waves in plasma, which may occur if their ω/k matches the longitudinal velocity of some particles inside the Fermi sphere. If this happens, the particle may "surf the wave," continuously draining its energy.

Now, it is time to present some results of my calculation for the *hot* QGP. Rather surprisingly, they lead to the polarization tensor being the same function of $\omega, \vec{k}$ as the cold quark matter: The difference was only the coefficient! The Debye screening parameter obtains the following contributions from the three diagrams (a–c) of Fig. 3.4 plus that of the fermionic loop

$$\kappa_D^2 = g^2T^2\left(\frac{1}{2} + 0 + \frac{1}{2} + \frac{N_f}{6}\right) \tag{3.21}$$

Note that the unusual diagram in Fig. 3.4(b), with a transverse-Coulomb loop responsible for the unexpected sign and the asymptotic freedom, does not contribute at all!

The relations between this parameter, the plasma oscillation frequency, and the effective on-shell gluon mass are the same as above at $T = 0$. Very importantly, perturbatively, there is *no static magnetic screening* again, as in QED.

3.5 Ring Diagram Resummation

In higher orders, at high T, there appear soft (IR) divergences of the perturbation theory, which are absent in a vacuum. The simplest example of that are the "ring" diagrams exemplified in Fig. 3.5. Their electric contribution for Matsubara frequency zero is

$$\delta\Omega|_n = \frac{1}{n}\int \frac{d^3k}{(2\pi)^3}\left(\frac{\Pi_{00}}{k^2}\right)^n \tag{3.22}$$

and for $n > 2$, those are clearly divergent at small k, at which $\Pi_{00} = \kappa_D^2$.

However, following the treatment of cold QED plasma of electrons by Gell-Mann and Bruckner in the 1950s, one may first sum all *ring diagrams*

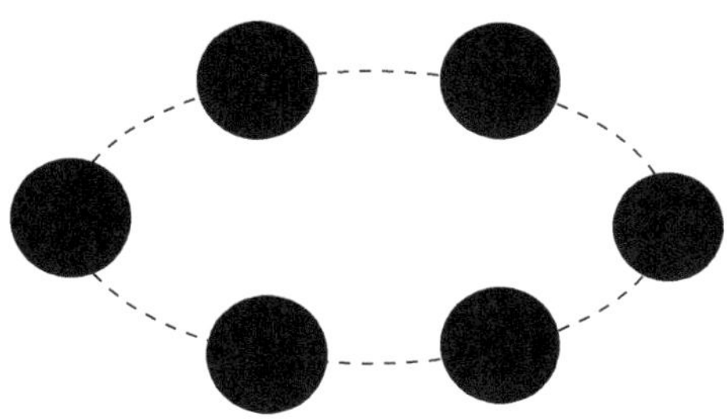

Fig. 3.5. Example of a ring diagram with $n = 6$ gluon polarization operators in one single-gauge field loop.

over n from 2 to infinity (Shuryak, 1978b; Kapusta, 1979). This gives the following result:

$$\delta\Omega^{\rm ring} = \frac{1}{2}\int\frac{d^3k}{(2\pi)^3}[\ln(1+\Pi_{00}/k^2) - \Pi_{00}/k^2] \tag{3.23}$$

which is infrared-finite. The resulting correlation energy, or the "plasmon term," at high T and μ is equal to

$$\delta\Omega^{\rm plasmon} = -\frac{N_c^2-1}{12\pi}g^3T^4\left[\frac{N_c}{3}+\frac{N_f}{6}+\sum_f\frac{\mu_f^2}{6\pi^2T^2}\right]^{3/2} \tag{3.24}$$

This nice example of resummation shows that, sometimes, the divergence of diagrams is the only way Nature tells us that the answer is not necessarily an analytic function of g^2 but g^3. Of course, it is still small if g is small.

The physical nature of this correlation energy is the interaction of a charge with its Debye cloud, and its magnitude per particle is $\sim g^3T$, as one can get from dividing this contribution by particle density. The same magnitude has a correlation energy for a static quark in a plasma. Following Debye, one can derive it in a much simpler manner by writing the potential as

$$4\pi V(r) = -\frac{4}{3}\frac{ge^{-\kappa_D r}}{r} \approx \frac{4}{3}\frac{g}{r} - \frac{4}{3}g\kappa_D + \cdots \tag{3.25}$$

and noting that the second constant potential times g of a point static charge is the correlation energy

$$E_{\rm corr} = -\frac{4}{3}\alpha_s\kappa_D \tag{3.26}$$

3.6 IR Divergences in Magnetic Sector

Unfortunately, the resummation of IR divergent diagrams to the finite answer just demonstrated is not the general case. The bosonic nature of the gluon field together with its Abelian (or "charged") properties leads to IR problems being absent, for example in the QED plasma. Its presence can, in fact, be seen already on dimensional grounds. Each loop at the nonzero temperatures contains the factor T and a sum over discrete Matsubara frequencies. Consider only the sector with zero frequencies (which corresponds to the classical limit, the vanishing Plank constant, or 3D gauge theory). The factor T enters any loop and, in fact, makes the coupling dimensional, and its power grows with the number of loops.

Something should compensate for it in the denominator, which can only be the particle momenta. The growing powers of momenta in the denominator must eventually produce infrared divergences. In fact, the four-loop or eighth-order diagrams already have such power divergences. Therefore,

$$\delta\Omega^{(8)} \sim g^6 T^4 \left(\frac{g^2 T}{q_{\min}}\right) \tag{3.27}$$

where a cutoff is set explicitly.

Polyakov (1979) suggested that this affliction can only be cured by the development of some magnetic screening length for the gluomagnetic field of the order of $M_M \sim g^2 T$. If so, using it as an IR cutoff in the formula above, we would conclude that *all diagrams* above the eighth order may contain a contribution of the same magnitude, namely $\delta\Omega \sim g^6 T^4$. So, terms of such order *cannot be perturbatively calculated.*

One can compare the situation in high-T QCD with that of a star, e.g. the Sun. If we would like to evaluate its total thermal energy, normal plasma physics would be enough. It is, however, true that the Sun has a very complicated magnetic field structure, with spots of opposite polarities and magnetic fluxes protruding from those into space. We do not know how to calculate those. Nonetheless, however interesting those can be, their total energy is negligible compared to the total thermal energy of the Sun.

One can separate the soft fields of a *magnetic scale* $g^2 T$ from all harder scales. This has been done by Braaten (1995). In principle, an effective theory could be constructed to deal with this particular problem by combining analytical techniques (to determine the coefficients in the effective Lagrangian) and numerical ones (from the nonperturbative solution of this 3D effective theory). The resulting effective theory is a 3D theory of static fields, which has the Lagrangian

$$L_{\text{eff}} = \frac{1}{4}(F_{ij}^a)^2 + \frac{1}{2}(D_i A_0^a)^2 + \frac{1}{2} m_D^2 (A_0^a)^2 + \lambda (A_0^a)^4 + \delta L \tag{3.28}$$

with $D_i = \partial_i - ig\sqrt{T} A_i$. This strategy has then been applied to the calculation of the free energy of the quark-gluon plasma at a high temperature (Kajantie *et al.*, 2001). This technique of dimensional reduction places special weight on the static sector (it singles out the contributions of the zero Matsubara frequency), and a major effort is devoted to the calculation of the coefficients of the effective Lagrangian (which contain the dominant contribution to the thermodynamic functions).

3.7 At Which T Do the Perturbative Series Converge?

Later, we discuss QGP produced experimentally, and the region of temperatures we can hope to reach is never larger than 3–4 times the critical temperature $T_c \approx 150\,\text{MeV}$. The corresponding momenta are about $3T$, at which scale, $\alpha_s \sim 1/3$ or $g \sim 2$. So, one may doubt the convergence of the perturbative series.

Due to the efforts of many people, which took many years, the calculable terms are now known. Following Arnold and Zhai (1995), we present the results for the free energy $F(T)$ (the same as Ω at zero density or μ):

$$\begin{aligned}
F = d_{\rm A}T^4\frac{\pi^2}{9}\Bigg\{&-\frac{1}{5}\left(1+\frac{7d_{\rm F}}{4d_{\rm A}}\right)+\left(\frac{g}{4\pi}\right)^2\left(C_{\rm A}+\frac{5}{2}S_F\right)\\
&-\frac{16}{\sqrt{3}}\left(\frac{g}{4\pi}\right)^3(C_{\rm A}+S_FS_F)^{\frac{3}{2}}-48\left(\frac{g}{4\pi}\right)^4C_{\rm A}(C_{\rm A}+S_F)\\
&\times\ln\left(\frac{g}{2\pi}\sqrt{\frac{C_{\rm A}+S_F}{3}}\right)\\
&+\left(\frac{g}{4\pi}\right)^4C_{\rm A}^2\left[\frac{22}{3}log\left(\frac{\bar{\mu}}{4\pi T}\right)+\frac{38}{3}\frac{\zeta'(-3)}{\zeta(-3)}-\frac{148}{3}\frac{\zeta'(-1)}{\zeta(-1)}-4\gamma_E+\frac{64}{5}\right]\\
&+\left(\frac{g}{4\pi}\right)^4C_{\rm A}S_FS_F\left[\frac{47}{3}log\left(\frac{\bar{\mu}}{4\pi T}\right)+\frac{1}{3}\frac{\zeta'(-3)}{\zeta(-3)}-\frac{74}{3}\frac{\zeta'(-1)}{\zeta(-1)}-8\gamma_E\right.\\
&\left.+\frac{1759}{60}+\frac{37}{5}\ln 2\right]+\left(\frac{g}{4\pi}\right)^4\\
&\times S_F^2\left[-\frac{20}{3}log\left(\frac{\bar{\mu}}{4\pi T}\right)+\frac{8}{3}\frac{\zeta'(-3)}{\zeta(-3)}-\frac{16}{3}\frac{\zeta'(-1)}{\zeta(-1)}-4\gamma_E-\frac{1}{3}+\frac{88}{5}\ln 2\right]\\
&+\left(\frac{g}{4\pi}\right)^4S_{2F}\left[-\frac{105}{4}+24\ln 2N_f\right]+O(g^5)\Bigg\}
\end{aligned}\tag{3.29}$$

where for theory with N_c colors and N_F flavors $d_A = N_c^2-1$, $d_D = N_f$, $C_A = N_c$, $S_F = N_f/2$, $S_{2F} = N_F(N_c^2-1)/4N_F$. Evaluated numerically for QCD with N_f quark flavors, it simplifies to the numerical form

$$\begin{aligned}
F = -\frac{8\pi^2T^4}{45}\Bigg\{&1+\frac{21}{32}N_f-0.09499\,g^2\left(1+\frac{5}{12}N_f\right)\\
&+0.12094\,g^3\left(1+\frac{1}{6}N_f\right)^{3/2}
\end{aligned}$$

$$+ g^4 \left[0.08662 \left(1 + \frac{1}{6} N_f\right) \ln \left(g\sqrt{1 + \frac{1}{6} N_f} \right) - 0.01323 \left(1 + \frac{5}{12} N_f\right) \right.$$
$$\left. \times \left(1 - \frac{2}{33} N_f\right) \ln \frac{\bar{\mu}}{T} + 0.01733 - 0.00763\, N_f - 0.00088\, N_f^2 \right] + O(g^5) \Big\} \tag{3.30}$$

For QED with N_f massless charged fermions with charges $q_i e$, the free energy is

$$F = -\frac{\pi^2 T^4}{45} \left\{ 1 + \frac{7}{4} N_f - 0.07916\, e^2 \sum q_i^2 + 0.02328\, e^3 \left(\sum q_i^2\right)^{3/2} \right.$$
$$\left. + e^4 \left[\left(-0.00352 + 0.00134 \ln \frac{\bar{\mu}}{T} \right) \left(\sum q_i^2 \right)^2 + 0.00193 \sum q_i^4 \right] + O(e^5) \right\} \tag{3.31}$$

With such results at hand, one can investigate whether the perturbative expansion of the QCD free energy is well behaved for physically realized values of couplings. Figure 3.6 shows the result for a six-flavor QCD when $\alpha_s(T) = 0.1$ (which corresponds to scales of order a few hundred GeV). The free energy is plotted versus the choice of renormalization scale $\bar{\mu}$. We have

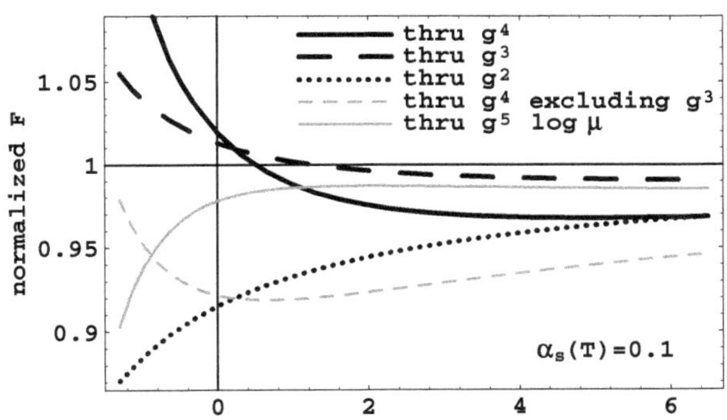

Fig. 3.6. The dependence of the free energy F on the choice of renormalization scale $\bar{\mu}$ for six-flavor QCD with $\alpha_s(T) = 0.1$. The free energy is normalized in units of the ideal gas result $-(\frac{1}{45} d_{\rm A} + \frac{7}{180} d_{\rm F})\pi^2 T^4$. The thick solid, dashed and dotted lines are the results for F, including terms through g^4, g^3 and g^2, respectively. The light solid curve is the g^4 result plus the $g^5 \ln(\bar{\mu}/T)$ term required by renormalization group invariance. The light, dashed curve is the g^4 result minus the g^3 term.

taken the expression for running charge up to two loops

$$\frac{1}{g^2(\bar{\mu})} \approx \frac{1}{g^2(T)} - \beta_0 \ln\frac{\bar{\mu}}{T} + \frac{\beta_1}{\beta_0}\ln\left(1 - \beta_0 g^2(T)\ln\frac{\bar{\mu}}{T}\right), \tag{3.32}$$

where

$$\beta_0 = \frac{1}{(4\pi)^2}\left(-\frac{22}{3}C_{\rm A} + \frac{8}{3}S_F\right), \quad \beta_1 = \frac{1}{(4\pi)^4}\left(-\frac{68}{3}C_{\rm A}^2 + \frac{40}{3}C_{\rm A}S_F + 8S_{2F}\right) \tag{3.33}$$

If the expansion is well-behaved, the result for F should become more independent of $\bar{\mu}$, as higher-order corrections are included. Unfortunately, it obviously does not.

3.8 HTL Resummations and the Quasiparticle Gas

The ring diagram resummation above is an example in which the hard momenta $p \sim T$ inside the small loops (polarization tensor) are different from the softer scale $p \sim gT$ in the large loop. The program initiated and carried out by Braaten and Pisarski (1990) was to identify *all* diagrams containing hard thermal loops (HTL) and resum them, obtaining some effective Lagrangian to be consistently used at this softer scale.

Instead of going into this rather technical discussion, we note that it can be alternatively explained by classical kinetics. The fluctuations at the *soft* scale $k \sim gT \ll T$ can be described using classical fields because the associated occupation numbers $N_k \sim T/E_k \sim 1/g \gg 1$ are large. As emphasized by Blaizot and Iancu (1993), these soft excitations can be described in terms of average fields, which obey the classical equations of motion. In QED, these equations are the Maxwell equations:

$$\partial_\mu F^{\mu\nu} = j^\nu_{\rm ind} + j^\nu_{\rm ext} \tag{3.34}$$

with a source term composed of an external perturbation, $j^\nu_{\rm ext}$, and an extra contribution, $j^\nu_{\rm ind}$, referred to as the *induced current*, obtainable using linear response theory:

$$j^\mu_{\rm ind}(x) = \int dy \Pi_{\mu\nu}(x-y)A^\nu(y). \tag{3.35}$$

which generalizes the usual dielectric and diamagnetic constants. In leading order in weak coupling, this polarization tensor is given by the one-loop approximation we discussed above: $\Pi \sim g^2T^2 f(\omega/p)$.

In a non-Abelian theory, a linear response is not sufficient. Constraints due to gauge symmetry force us to take into account specific nonlinear effects. The relevant generalization of the Yang–Mills equation reads (Blaizot and Iancu, 1993)

$$D_\nu F^{\nu\mu} = \Pi^{ab}_{\mu\nu} A^\nu_b + \frac{1}{2}\Gamma^{abc}_{\mu\nu\rho} A^\nu_b A^\rho_c + \cdots \tag{3.36}$$

where the induced current on the right-hand side is nonlinear. When expanded in powers of A^μ_a, it generates an infinite series of HTMs, not only in self-energy but also in vertices with any number of soft gluonic legs.

Physically, the need for resummation arises from the existence of collective excitations in the system, whose properties are not well captured by perturbation theory. One may hope that a "gas of quasiparticles" would do a better job of reproducing the results. The main spirit of these works is that quasiparticles have effective masses that are functions of g and T, which one can evaluate perturbatively first and only then include their motions (e.g. Peshier *et al.*, 1996) and interactions (Andersen *et al.*, 2000).

As an example of such an approach, I present a calculation by Blaizot *et al.* (1999), which will be compared to the lattice data. For some technical reasons, these authors prefer to focus on the entropy $s(T)$ rather than the pressure $p(T)$: At least, the former is defined without an arbitrary constant (the bag term).

The entropy density of their *quasiparticle gas* is defined as the entropy of the ideal gas of quasiparticles:

$$s = \int \frac{d^3k}{(2\pi)^3} \left[(1+N_k)\log(1+N_k) - N_k \log N_k\right] \tag{3.37}$$

where $N_k = \frac{1}{e^{E_k/T}-1}$ and $E_k = \sqrt{M^2(T)+k^2}$. Note that at this stage, the (T-dependent) effective quasiparticle mass is not yet defined.

The pressure p and the energy density ϵ are then given by the corresponding ideal gas expressions, corrected by a function $B(T)$ adjusted so as to satisfy thermodynamical identities. That is, one defines the pressure and energy density as

$$p = -T\int \frac{d^3k}{(2\pi)^3} \log\left(1 - e^{-E_k/T}\right) - B(T) \tag{3.38}$$

$$\epsilon = \int \frac{d^3k}{(2\pi)^3} N_k E_k + B(T) \tag{3.39}$$

Note that such a parameterization obviously satisfies the thermal identity $\epsilon + p = Ts$. The function $B(T)$ is then determined by requiring that $s = \frac{dp}{dT}$.

The effective mass used is the HTL one. The quasiparticle gas can be viewed as an approximation to a more general approach known as the *hard thermal loop perturbation theory* (Andersen *et al.*, 2000), in which one adds the HTL mass term into a tree-level Lagrangian and then subtracts it from the "perturbation" part.

Blaizot and Iancu noted that for the special case of entropy, the calculations can be pushed one order further to make better "Φ-derivable" self-consistent approximations. The stationarity property of free energy entails important simplifications in the calculation of entropy. Indeed, we have

$$S = -\frac{\mathrm{d}F}{\mathrm{d}T} = -\left.\frac{\partial F}{\partial T}\right|_D \tag{3.40}$$

where the last derivative is at a fixed propagator. Due to that expression, the temperature dependence of the propagator can be ignored here. Explicitly, one gets

$$\begin{aligned} S = &-\int \frac{d^4k}{(2\pi)^4}\frac{\partial N(\omega)}{\partial T}\mathrm{Im}\log D^{-1}(\omega,k) \\ &+\int \frac{d^4k}{(2\pi)^4}\frac{\partial N(\omega)}{\partial T}\mathrm{Im}\Pi(\omega,k)\mathrm{Re}D(\omega,k) + S' \end{aligned} \tag{3.41}$$

where $N(\omega) = 1/(\mathrm{e}^{\beta\omega} - 1)$, and

$$S' \equiv -\left.\frac{\partial(T\Phi)}{\partial T}\right|_D + \int \frac{d^4k}{(2\pi)^4}\frac{\partial N(\omega)}{\partial T}\mathrm{Re}\Pi\,\mathrm{Im}D$$

At two loops in the skeleton expansion, the entropy then takes the simple form

$$S = -\int \frac{\mathrm{d}^4p}{(2\pi)^4}\frac{\partial N}{\partial T}\left\{\mathrm{Im}\ln D^{-1} - \mathrm{Im}\Pi\,\mathrm{Re}D\right\} \tag{3.42}$$

The simplifications discussed above have led to important cancelations, leaving for the entropy an expression which is effectively a one-loop expression, thus emphasizing the direct relation between the entropy and the quasiparticle spectrum. Residual interactions start contributing on the order of three loops. This expression is also manifestly ultraviolet-finite.

Now, in the regime where the loop momenta are soft, we can use as an approximation for Π that corresponds to a HTL. As an illustration of the quality of the results which were obtained in this direction, let me show the entropy of pure SU(3) gauge theory (normalized to the ideal gas limit). The agreement with the parameterized lattice results at large $T > 2.5\ T_c$ is quite good. Note also that in going from one level of approximation to

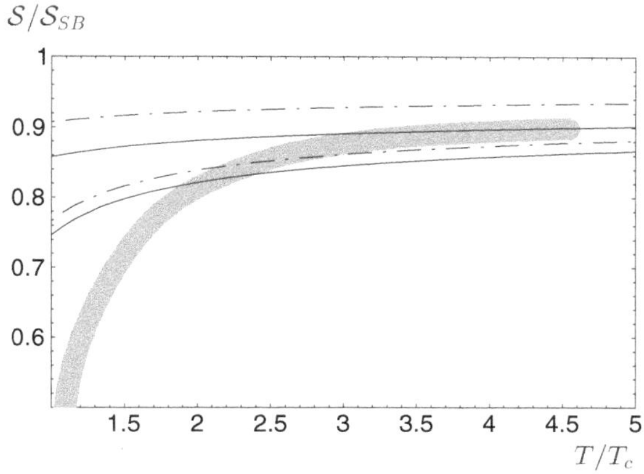

Fig. 3.7. The entropy of pure SU(3) gauge theory normalized to the ideal gas entropy, S_{SB}. full lines: S_{HTL} from Blaizot *et al.*; dashed-dotted lines: S_{NLA}. For each approximation, the two lines correspond to the choices $\bar{\mu} = \pi T$ and $\bar{\mu} = 4\pi T$ of the running coupling constant $\alpha_s(\bar{\mu})$ in the $\overline{\mathrm{MS}}$ renormalization scheme. The dark-gray band represents the lattice results.

the next (i.e. from S_{HTL} to S_{NLA}), the changes are moderate, in contrast to what happens in ordinary perturbation theory.

On the other hand, a radically different approach — the strong coupling limit of $N = 4$ SUSY theory (Gubser *et al.*, 1998) — gives the value $s/s_0 = 3/4$, which is also not very far off. So, based on the QGP thermodynamics alone, it is difficult to conclude which of the two limits — weak or strong coupling — is more appropriate at $T \sim 2$–$3\ T_c$. The ultimate choice, to be discussed in the following, would then be based on kinetic quantities.

Chapter summary and connections: Perturbative QCD at finite temperatures/densities has been well developed. It describes a "normal" phase, the QGP, with electric screening and quark/gluon quasiparticles. This theory is supposed to apply at a very high T. The question of whether this is the case for QGP produced in real-world heavy ion collisions will be discussed in other chapters devoted to QGP kinetics. In particular, we will discuss another type of quasiparticle — magnetic monopoles — in Chapter 24 and strong coupling holographic methods in Chapter 29.

Chapter 4

HADRONIC MATTER

4.1 Hadron Resonance Gas

One very well-studied form of matter made of hadrons—nucleons, to be precise — is *cold nuclear matter*, whose properties closely resemble those of matter inside heavy nuclei.[1] It is a relatively dilute *Fermi liquid*, with rather complicated binary forces approximated by meson exchanges. Nuclear binding also depends on (small but still essential) three-body forces. This book is not the place to discuss it in any detail.

Highly compressed nuclear matter, however, does belong in this discussion. It exists in the cores of neutron stars. A thermally excited, compressed version of this type of matter appears during supernova explosions, especially during neutron star mergers, which we discuss in Chapter 16.

Approaching hot hadronic matter with zero baryon density (and zero chemical potential, $\mu_B = 0$), one finds symmetry with respect to quarks and antiquarks.[2] At a relatively low temperature T, one would think of matter made of the lightest hadrons, the pions π^+, π^0, π^-, with a mass of only about $m_\pi \approx 138\,\text{MeV}$.[3]

Near-massless pions are close relatives of other Nambu–Goldstone modes, such as sound modes. According to the Goldstone theorem, they are not allowed to interact at the low momentum limit $k \to 0$. For the same reason, in a *cool* (low $T < f_\pi$) pion gas, they must consist of very large

[1]To consider an infinite volume limit, one has to switch off electromagnetism.

[2]It does have a certain density, say, of entropy or energy, s or ϵ, subject to hydrodynamical flows, which we study later.

[3]Why pions are so light has been explained by Nambu and Jona-Lasinio (1961b) and is related to the spontaneous breaking of $SU(N_f)_a$ chiral symmetry, which we discuss in Chapter 23.

mean free paths. (Explicit low-energy pion scattering amplitudes follow from chiral Lagrangians, and corresponding effects for thermal pion gas have been calculated, as given in the appendix to Shuryak (1980). We will not discuss those here since they are too small to be important in practice.)

Therefore, it would be very hard to create such a pion gas in an experiment. Indeed, thousands of pions produced in heavy ion collisions do not create cool "pion matter" but kinetically freeze out at $T \sim 100\,\mathrm{MeV}$, after which they simply freestream toward the detectors. Since they are cooling set of Bosons, one may consider their Bose–Einstein condensation; however, so far, appropriate conditions have not yet been created, and it remains a hypothetical possibility.

However, interactions between pions grow rapidly with their relative momenta and $s = (p_1 + p_2)^2$. In some channels, they become so strong that there appear $\pi\pi$ resonances: The lowest are σ in the $I = 0, J = 0$ channel and ρ in the $I = 1, J = 1$ channel.[4]

In scattering amplitudes of pions on nucleons, many more resonances were found.

These excited states of the nucleons, called N^*, Δ depending on spin-isospin, were experimentally discovered as early as the 1950s. By the end of the 1960s, many more hadronic resonances had been discovered. The question was what would be the properties of *hadronic matter* consisting of a mixture of all those resonances.

A famous observation by Hagedorn (1965) was that, *if* the density of resonances were to grow exponentially, i.e. $\rho(m) \sim \exp(m/T_H)$, convoluting it with the Boltzmann factor

$$Z \sim \int dm \rho(m) e^{-m/T} \tag{4.1}$$

one finds that the partition function converges only if $T < T_H$ (called the Hagedorn temperature). Hagedorn's original conclusion was that there should exist a *maximal possible temperature* in nature, which, from the fit to the resonance data, he found to be $T_H \approx 160\,\mathrm{MeV}$. Needless to say, this statement attracted much debate in the popular press.[5]

But why should the density of hadrons grow exponentially? Hagedorn's motivation came from a philosophy, popular at the time, called *the dual*

[4]There are no $I = 2$ resonances because in this channel, $\pi\pi$ interactions are repulsive.

[5]Partly because, exactly at that time, the background radiation was discovered, and the Big Bang cosmology became a hot subject. Of course, in reality, there is no direct connection between them at all.

bootstrap, which considered hadronic resonances to be made of other resonances in a certain way. Its physical basis was rather murky, corresponding to then prevailing views that duality and unitarity would somehow be sufficient to build a strong interaction theory, bypassing QFT, Lagrangians and all that. Eventually, the motivation changed after Veneziano and collaborators worked out the famous expression for scattering amplitudes, and the exponentially rising density of states was related to *QCD strings*.

During the decade following that famous paper, Hagedorn and collaborators have tried to make practical applications of the "statistical bootstrap model" to various hadronic reactions. Yet, their main conclusion remained the same: No matter how much energy is put into the system, a huge number of available degrees of freedom would "eat it up," so that temperature would never exceed T_H. If true, it would imply that no new phases of matter can be created and that high-energy collisions of hadrons and/or heavy ions would be a pointless exercise.

I got involved[6] and soon argued (Shuryak, 1972a) that one can also fit the mass spectrum using a power law: $\rho(m) \sim m^{\text{power}}$. If so, a very simple EoS for the resonance gas results. My fit gave $\epsilon \sim p \sim T^6$, with the speed of sound $c_s^2 = 1/5$ independent of temperature. It is this simple EOS which we used in our initial discussion of the QCD phase transition in Fig. 3.2. Such EOS is very convenient for hydrodynamical applications and was widely used to model explosions in heavy ion collisions at relatively low energies when the matter is the resonance gas.

After the discovery of QCD, it became clear that Hagedorn's scenario could not be applied at high temperatures, at which there should be a QGP phase made of quarks and gluons. Its mass density is not exponential. Eventually, Hagedorn (1985) conceded that QGP should exist and that he got to such an idea "by trial and error." Indeed, his famous T_H was decreased from the maximal temperature possible in nature to just the maximal temperature of the *hadronic phase*.

Further developments of the theory and lattice gauge simulations have shown that, in QCD with light quarks, the critical temperature is indeed

[6]Under the following circumstances: G. Budker, who was a director of the institute which now carries his name, called me, an aspirant (graduate student), in the early 1970s and asked me to work out some parameterization of the antiproton spectra in proton collisions with heavy nuclei. Budker was thinking about the optimization of the antiproton yield, and he needed antiprotons for a proton-antiproton collider project, a hot issue at the time. While such a collider was not created at Budker's institute, it was built a decade later at CERN.

close to the original $T_H \approx 160\,\text{MeV}$, and the degrees of freedom of the vibrating QCD strings do not really play any substantial role.[7]

4.2 Theory of the Hadron Resonance Gas

As the temperature grows, mesons heavier than pions (e.g. vector mesons, ρ, ω) should be thermally excited. Naively, one may think that happens when the temperature reaches the mass of the lowest state, e.g. $T \sim m_\rho \sim 800\,\text{MeV}$ in QCD and when $T \sim m_{\text{scalar glueball}} \sim 1.6\,\text{GeV}$ in gluodynamics without quarks. Not at all! It happens much earlier because the next excitations have large statistical weights.[8]

The theoretical approach widely used for $T = 100$–$160\,\text{MeV}$ is that of *hadron resonance gas* (HRG). The main idea was suggested very early by Belenkij and Landau (1955). They have shown, using the lowest-order virial expansion, that resonances[9] observed in scattering phases, in fact, contribute to thermodynamical parameters exactly as much as stable particles with the same mass.

Binary interactions in statistical mechanics are described by the Beth–Uhlenbeck formula, which gives a correction to the partition function in terms of scattering phases $\delta_l(p)$:

$$\Delta Z = \frac{1}{\pi}\int dp \sum_l (2l+1)\frac{d\delta_l(p)}{dp}\exp\left(-\frac{p^2}{mT}\right) \tag{4.2}$$

where the sum is over the integer angular momenta $l = 0\ldots$. Note that, here, we have mT rather than $2mT$ in the exponent because p is the *relative* momentum of the two particles, assumed to have the same mass m. The scattering phase $\delta_l(p)$ (defined in quantum mechanics) modifies the boundary condition of the wave function, and by doing this, it may create extra states. Suppose particles are in a sphere of radius R, then the quantization of momenta is performed via the equation

$$\sin(k_n R + \delta_l(k_n)) = 0, \quad k_n R + \delta_l(k_n) = n\pi \tag{4.3}$$

---

[7]In the phase transitions of pure gauge theories and glueballs, the Hagedorn phenomenon plays some role, but the Hagedorn temperature is much higher: $T_H \approx 270\,\text{MeV}$. String theorists are well aware of the Hagedorn phenomenon, and if their beloved *string theory* is indeed the Theory of Everything, Hagedorn's ultimate temperature should make a comeback, being now many orders of magnitude larger, of the order of the Planck mass.

[8]One more example of the kind: The Sun is made of plasma even up to its photosphere, which has $T \approx 6000^\circ \sim 0.5$ eV$\ll \delta E \sim 10$ eV the gap between lowest levels.

[9]Those should be narrow enough: $\Gamma \ll T$.

For large R, one can think of n as continuous, and the number of extra states is obtained by differentiating this equation over n.

If the interaction is such that it creates a resonance, the scattering amplitude as a function of collision energy behaves as

$$f_l \sim \frac{1}{E - m_{\rm res} + i\Gamma/2}$$

where $m_{\rm res}, \Gamma$ are the resonance mass and width. ($1/\Gamma$ is the resonance lifetime.) In the contribution of this resonance part of the amplitude to the scattering phase

$$f(\theta) = \sum_{l=0}(2l+1)f_l P_l(\cos(\theta)), \quad f_l = \frac{1}{2ik}(e^{2i\delta_l} - 1) \tag{4.4}$$

one observes that the phase makes a circle, as the collision energy passes through a resonance. If its width Γ is small, one can approximate

$$\delta_l(E) \approx \pi\theta(E - m_{\rm res})$$

After plugging it into the formula above, one finds that it literally adds one more state to the statistical sum.

Another comment about the Beth–Uhlenbeck formula from Shuryak (1991) refers to the cases when there are both *smooth non-resonance* and narrow resonance parts of the scattering phase. The former defines the so-called *optical* or collective potential, and the latter sums up the new states. The question is whether such a procedure does or does not include certain double counting.

Note that the pertinent sum can be identically rewritten as follows:

$$\sum_l (2l+1)\frac{d\delta_l(p)}{dp} = \frac{d}{dp}[pRef(p,\theta=0)] + \frac{i}{4\pi}\int d\Omega\left(f\frac{\partial f^*}{\partial p} - f^*\frac{\partial f}{\partial p}\right) \tag{4.5}$$

The first term related to the real part of the forward scattering amplitude is a "collective potential." The second term (which is, of course, also real) is the one which generates an extra contribution from narrow binary resonances. So, there is no "double counting," and it is safe to include both potentials and narrow resonances together.

Rather detailed calculations based on the Beth–Uhlenbeck formula with realistic scattering phases have been initiated by Prakash *et al.* (1993), who also extended those to kinetic quantities of hadronic matter. In practical applications, one issue is whether one knows all the relevant resonances since their source — the Particle Data book — is constantly growing. Whether a particular resonance appears there or not and with what number of stars (meaning the degree of reliability) depends, of course, on many experimental issues, including even funding and workforce availability.

Instead of discussing those, let me note that the use of realistic scattering phases reveals that, in fact, only a few resonances are *narrow*, with $\Gamma < T$. If the resonance is not narrow, the Beth–Uhlenbeck formula shows that its contribution is reduced. Furthermore, smooth "potential" parts are not negligible. The so-called "proton anomaly" has been recently cleared by the account of the non-resonant (and mostly repulsive) contribution from the scattering phases: We will discuss it in Chapter 8 on chemical freeze-out.

So far, we have only discussed binary collisions. It is possible to generalize the theory of non-ideal gases to a general case, including S-matrix elements for $n \leftrightarrow m$ particle collisions. The practical problem, however, remains: Only data for binary collisions are experimentally available. So, in practice, people are often simply pretending that multi-hadron reactions do not exist! It somehow works for thermodynamical quantities, but seems to be completely failing for the kinetics of hadronic gas.

In summary, the HRG model approximately reproduces the QCD thermodynamics for $T < T_c$ but fails to show any precursors for the phase transition. (Contrary to Hagedorn's claims.) We will also see in the following that it fails to describe the kinetics of hadronic matter, in particular the sharp chemical freeze-out observed in heavy ion experiments.

4.3 Do Hadronic Masses Depend on the Temperature?

Our discussion of finite-T QCD thermodynamics will begin in Chapter 6 on event-by-event fluctuations. We will cover more basic issues related to gauge theory set on the lattice, etc., in Chapter 5.

Yet, now, we jump over those systematic presentations, focusing on a question asked in the title of this section. With the traditional usage of the HRG model being not at $T = 0$ but rather at temperatures $T = 100$–$150\,\text{MeV}$, should we derive and include the *temperature dependence* of all hadronic masses?

To answer this question, one should recall that the Beth–Uhlenbeck formula includes "potentials," the smooth part of scattering phases, as well as "resonances" due to pole singularities at some distance from the real axes. The notion of "T-dependent masses" corresponds to certain resummations, including the potentials and "correcting" the original positions of the resonant poles. While such procedures may be theoretically complex, one can define the quasiparticles and their T-dependent masses by measuring correlation functions on the lattice.

Indeed, in the literature, there were multiple works suggesting downward shifts in the masses. Generic arguments stem from the basic notion that at the critical temperature $T_c \approx 155\,\text{MeV}$, both the *quark condensate* and the *string tension* need to vanish, and those are two basic parameters affecting hadronic masses. In experiments, huge efforts have been made to detect such mass shifts for vector mesons ρ, ω, ϕ via invariant mass distribution in the dilepton $(e^+e^-, \mu^+\mu^-)$ channels, to which we turn in Chapter 18.

Not going into the large theoretical literature on "shifting masses," let me instead jump to lattice results shown in Figs. 4.1 and 4.2 on the issue (from the FASTSUM collaboration). Note that at $T \to T_c$, the states with opposite parity must have identical masses, as indeed was expected from chiral symmetry restoration at $T > T_c$.

What was *not* expected is the manner in which it happens: The lowest mesons and baryons apparently show no T dependence (within the errors), while the opposite-parity states are indeed shifting.

Why are the masses of the lowest hadrons so T-independent? From a strict standpoint, we do not know. While the "constituent quark masses" $\sim \langle \bar{q}q \rangle$ must "melt" at $T \to T_c$ as they violate chiral symmetry, one has

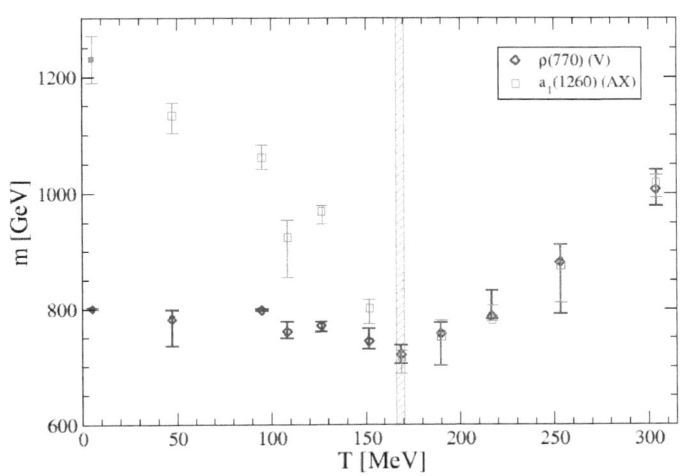

Fig. 4.1. Temperature dependence of the vector and axial meson masses.

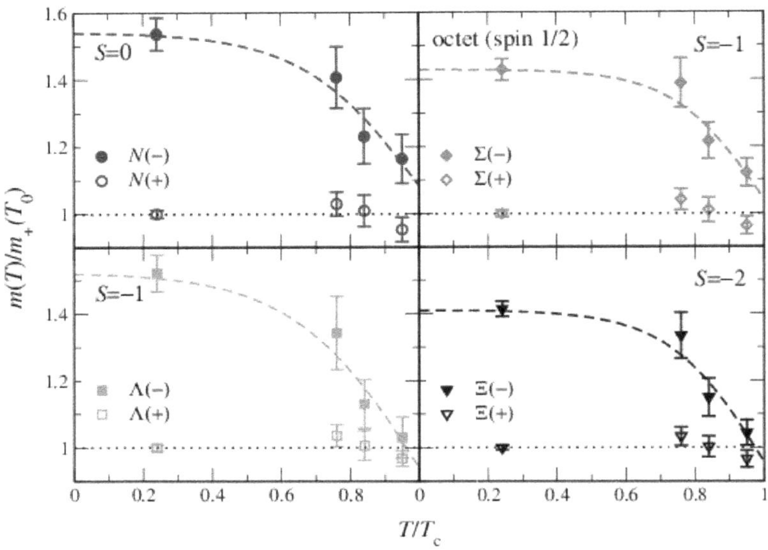

Fig. 4.2. Temperature dependence of the masses of octet baryons. Plus and minus here indicate the spatial parity P.

additional "thermal energy" $\sim \langle \bar{q}\gamma_0 q\rangle$, which grows with T and apparently compensates for this melting. In fact, the upper plot is continuous at $T > T_c$, showing that the latter is linear in T.

The states of opposite parity need to have odd angular momentum, $L = 1$, in nonrelativistic classification; therefore, they are of larger size than $L = 0$ states and thus should depend more on the string tension. So far, I am not aware of the inclusion of their shifted masses in thermodynamic modeling.

Summary and connections: HRG is a popular model describing hadronic matter at temperatures below but not too far from $T < T_c$. Yet, it has open issues: The overwhelming fraction of resonances is not narrow enough, and the smooth "potential" part of scattering phases is not generally negligible. Furthermore, lattice studies find that hadronic masses have T dependence, which depends on their quantum numbers. A more general issue is including multi-hadron scatterings. We return to those in Chapter 8 in the discussion of hadron kinetics near a chemical equilibration surface.

Chapter 5

QFTS AT FINITE TEMPERATURES

5.1 QFT in Euclidean Time and Lattice Gauge Theories

The standard formulation of quantum mechanics and quantum field theories (QFTs) is given by the Feynman path integral over the space-time "paths" (or rather "histories") of the fields involved; for an original pedagogical introduction, see Feynman and Hibbs (1965). A standard introduction to perturbation theory and Feynman diagrams can be found in QFT textbooks, which are too numerous to mention.

A significant simplification of path integrals is obtained by analytic continuation from the Minkowskian time t to the Euclidean time $\tau = it$, as the oscillating phases in the weight are changed to non-oscillating ones. What we will focus on is the formulation of QFT at a finite temperature T, which takes both quantum and thermal fluctuations of the fields into account. This is achieved by defining the Euclidean time on the circle $\tau \in [0, \hbar/T]$ with enforced periodicity. An original discussion of statistical mechanics in such terms is available in Feynman (1972). A pedagogical discussion of its applications to the many-body theory is given, for example, in Shuryak (2019b).

All these elements are used in Wilson's definition of lattice gauge theory, in which three spatial coordinates and the fourth (Euclidean time) $\tau = x^4$ are discretized into a four-dimensional lattice of points with some universal step a. The quark fields are defined at the lattice points, while the gauge fields are defined at the *links* between them. Gauge transformations are arbitrary gauge rotations at all lattice points. Gauge-invariant objects are formulated via *closed loops* built out of the links, such as elementary squares or "plaquettes." Gauge-invariant lattice actions are constructed out of those. Another important object we will discuss is the *Polyakov loop*, which loops around the thermal circle.

There is a huge amount of literature in which all the technical details of lattice gauge theory are described in detail, so I skip it. Instead, let us just remember here the main parameters and scales of the problem. The lattice input is the value of the coupling $g(a)$ at its ultraviolet (UV)-end scale a, the lattice step.[1] It is assumed that in UV, the theory is weakly coupled, so the standard gauge action is adequate.

The temporal size of the lattice is defined by the temperature

$$N_\tau a = \frac{1}{T}$$

where N_τ is the number of steps in time direction. In most studies, it ranges from approximately half of fm to a few fm. The spatial size of the lattice $L = Na$ provides the infrared (IR) cutoff. Typical numbers are $\sim$3 fm and $N \sim 30$.

The physics phenomena under investigation must be located *between* the two:

$$a \ll (\text{physical} \quad \text{scale}) \ll L \tag{5.1}$$

From the numbers I mentioned, one can see that "much less" in reality means a factor of $\sqrt{N} \sim 5$ at best. In this relatively narrow interval of scales, the effective coupling constant g must "run," as prescribed by the RG flow, from weak in the UV regime to strong in the IR regime.[2] Indeed, such a transition is observed, providing for a reasonably accurate representation of nonperturbative physics, e.g. confining potentials and hadronic spectra.

The numerical simulation lattice gauge grew into a large enterprise, occupying a good portion of the largest supercomputers. It took decades to develop adequate algorithms, correct quark fields respecting chiral symmetry and perform actual calculations. One major achievement is that it currently reproduces most of the hadronic phenomenology, such as hadronic masses, reasonably well. Its discussion, however, will take us too far away from our main interests. Therefore, I would only briefly discuss the finite T results; for a review, see, for example, Soltz *et al.* (2015).

---

[1]For orientation, typically, $a \sim 0.1\,\text{fm} \sim 1/(2\,\text{GeV})$, which is considered to be much smaller than hadronic sizes.

[2]Compare it to Cooper pairing, which also has logarithmic running of the effective coupling. It takes typically several decades of such running, from atomic scale to the sizes of the Cooper pairs, to make the interaction strong enough for pair binding. Fortunately, for the gauge theories, the rate of the RG flow is enhanced by the rather large coefficients involved.

The main finding in the finite-T lattice is the discovery of the phase transition from the so-called hadronic phase to the QGP phase at higher T. For pure $SU(3)$ gauge theory (no quarks), it is the first-order *deconfining* transition at $T_c \approx 270$ MeV. In QCD with three massless quarks, it is also the first-order transition; however, in QCD with physical quark masses, it happens to be a *crossover chiral/deconfinement* transition at $T_c \approx 155 \pm 5\,\mathrm{MeV}$.

Figure 5.1(a) shows how the order of this transition depends on quark masses. The physical point is outside of all the regions of the singular phase transitions. Thus, it is a smooth crossover. This is also seen from the plot in Fig. 5.1(b), which shows the dependence of the chiral-order parameter called the *quark condensate* on T. Although it does switch from a large value to near zero, there are no jumps, and the transition follows a smooth curve. The critical point is defined by an *inflection point* (a zero of the second derivative), or the maximum of the corresponding susceptibility.

Figure 5.2(c) shows a compendium of lattice calculations of the thermodynamical quantities of energy density, pressure and entropy. The first two are also shown as the "interaction measure" $\epsilon - p$. The "inflection point" of the $p(T)$ curve, or the maximum of the specific heat, which is the second derivative of p, is usually used as a definition of T_c. The gray curves have been obtained by the Wuppertal–Budapest collaboration with the 2stout action (Borsányi *et al.*, 2015a), whereas the colored ones have been obtained by the HotQCD collaboration with the HISQ action (Bazavov, 2014). These lattice results are used as equations of state (EOSs) in heavy ion physics applications.

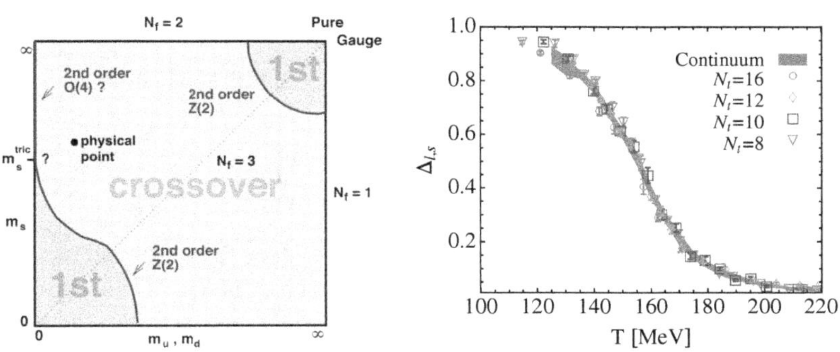

Fig. 5.1. (left) Map of the phase transitions as a function of light and strange quark masses. (right) The temperature dependence of the quark condensate in high-T QCD.

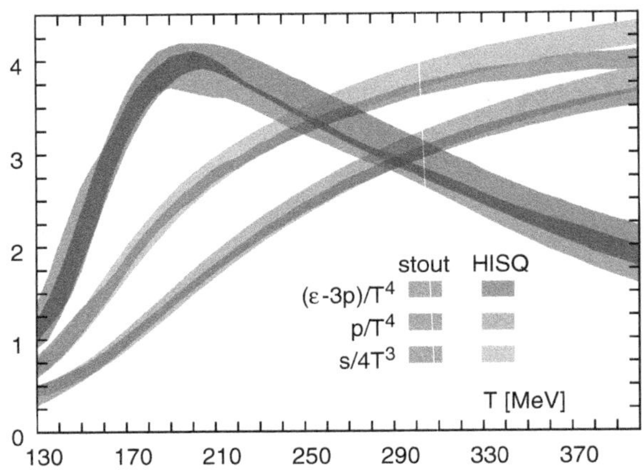

Fig. 5.2. The continuum-extrapolated thermodynamical observables versus the temperature from lattice gauge theory simulations.

5.2 Higher-Order Susceptibilities Versus Event-by-Event Fluctuations

Thermodynamical fluctuations in general are described by the expansion of thermodynamical quantities to the second order of fluctuations. Einstein famously inverted Boltzmann's relation for entropy into the form $P \sim \exp(S)$ and used the expansion of entropy near its maximum to higher orders. The coefficients are, in general, various susceptibilities.

In heavy ion physics, this tactic was suggested by Shuryak (1998). The event-by-event fluctuations in heavy ion collisions were also related to susceptibilities. The first example chosen was fluctuations of the temperature

$$P \sim \exp\left[-\frac{C_v}{2}\left(\frac{\delta T}{T}\right)^2\right] \tag{5.2}$$

whose fluctuations are related to specific heat, C_v. (Similar to entropy, it is an extensive quantity, so the fluctuations scale as $1\sqrt{N}$, as expected.) In that paper, it was suggested to use fluctuations as a "thermometer" via their comparison to calibrated $C_v(T)$ for a hadron resonance gas.

However, the observation of temperature fluctuations turned out to be difficult to access in practice, but another idea (suggested in the same paper

Shuryak, 1998) was related to event-by-event *particle number fluctuations*:

$$\langle \Delta N^2 \rangle = T \frac{\partial N}{\partial \mu} \tag{5.3}$$

Note that the particle number is by itself a derivative of $\log(Z)$ over μ, so the right-hand side is the *second* derivative over μ, or the susceptibility over chemical potential.

While nontrivial fluctuations can happen even in ideal quantum gases,[3] these types of fluctuations turned out to be quite sensitive to physical interactions between particles.

Generalized susceptibilities for all three conserved charges (baryon number, electric charge and strangeness) can be defined with any number of the derivatives

$$\chi_{lmn}^{BQS} = \frac{\partial^{l+m+n} P/T^4}{\partial(\mu_B/T)^l (\mu_Q/T)^m (\mu_S/T)^n} \tag{5.4}$$

Many of them have been calculated on the lattice by now, starting with the pioneering work by Karsch and Redlich (2011). This activity has flourished in the past decade due to consistent calculations of those in lattice simulations; see, for example, Goswami *et al.* (2022). Their comparison to observed fluctuations in the heavy ion collisions became a large enterprise; see, for example Karthein *et al.* (2021) and references therein. Here, we give just one example of such a comparison, Fig. 5.3 from Bellwied *et al.* (2022). While the errors remain relatively large, is there clearly agreement between lattice-based susceptibilities and experimental cumulants. (Note, however, that the rightmost C_4/C_2 (kurtosis) point deviates from the lattice-based Taylor series by about two sigmas, hinting at a different phenomenon. We will return to this issue in Chapter 6.)

Several derivatives of the pressure (minus free energy) over μ at $\mu = 0$ can also be used via Taylor series over the chemical potential of the baryon number, electric charge and strangeness[4]:

$$\frac{1}{VT^3} \log[Z(T, V, \mu_i)] = \frac{p}{T^4} = \sum_{i,j,k} \frac{\chi_{ijk}^{BQS}}{i!j!k!} \hat{\mu}_B^i \hat{\mu}_Q^j \hat{\mu}_S^k \tag{5.5}$$

where hat denotes μ normalized to T.

[3]For example, Bose (Fermi) statistics of mesons (baryons) already cause deviations from the Poisson distribution.

[4]These three conserved quantities can alternatively be related to the conservation of u, d, s quarks.

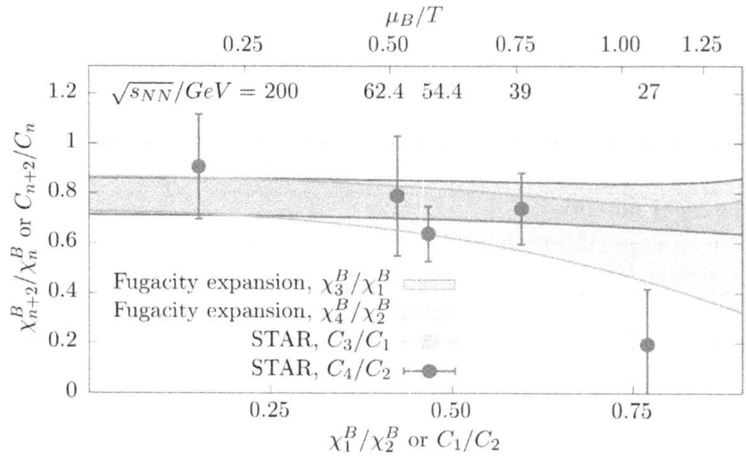

Fig. 5.3. Shaded green and yellow bands correspond to the ratios of lattice susceptibilities as a function of baryonic chemical potential μ/T (see upper scale for corresponding collision energies). The points show pertinent ratios of baryon number cumulants, obtained from experimental distributions by the STAR collaboration.

Currently, the susceptibilities are calculated on the lattice to an order of eight. Using the Taylor series (or some possible improvements using the Pade approximations), it was possible to extrapolate to the portion of the finite μ part of the phase diagram. The region covered reaches an approximate value of $\mu/T \sim 1$, as one can see from Fig. 6.4. We can see from the lattice results thus reported that, in this part of the phase diagram, the QGP/hadron boundary remains a crossover: No obvious change is suggested by the series.[5]

The obvious question to ask is, *what happens at higher values of* μ? In strict terms, we do not know yet, despite the fact that higher values of μ are achieved at *lower* collision energies, corresponding to fixed-target experiments performed as early as the 1980s.

Many theoretical models predict that, at some critical value μ_c, the crossover line can change into the first-order transition. If so, at the critical point, it would be the *second-order* transition, presumably of the simplest universality class with a single critical parameter and thus with the same indices as the Ising model.

[5]The argument here is a bit circular in the sense that the Taylor series perhaps indeed gives us a *non-singular* part of thermodynamics. It would be interesting to use the same logic for known second-order transitions and see if they can or cannot be found this way.

5.3 Hadronic Screening Masses at High T

One important consequence of the perturbative calculation of the polarization operator (Shuryak, 1978b) is that electric and magnetic fields behave quite differently: the electric fields are screened, while the magnetic fields remain *unscreened*.[6] Therefore, interaction between *charges* (quarks propagating along the time direction) interacting electrically are different from *currents* (quarks propagating along space directions). A pair of quark and antiquark propagating along time has mass spectra of ordinary mesons. A similar pair propagating in the space direction have what is called "screening masses." Although they are usually labeled with the same mesonic symbols, they have very different meanings. We would like to clarify the confusion between these two in this section.

Let me first recall that, in a Matsubara finite-T setting, the difference arises solely from the fact that Euclidean time is defined on a finite circle, while space axes have infinite extents. The existence of a zero Matsubara frequency for bosons implies that the gluonic sector in the limit $T \to \infty$ smoothly goes into the 3D gauge theory as the circle shrinks to nothing.

The situation is different for fermions since they are antiperiodic on the circle. Thus, their Matsubara frequencies are $\omega_n = 2\pi T(n+1/2)$, which are always nonzero for all n. For a quark–antiquark pair propagating in *spatial* directions, the nonzero Matsubara frequency plays the role of a mass. Note that at high T, this mass becomes large.

A confusion started with the work of Detar and Kogut (1987), who decided to calculate on the lattice the rho-meson screening mass at certain temperatures, and instead of the propagation of a quark and an antiquark, they found meson-like behavior, with their quarks being together as a bound state. Moreover (by some coincidence of numbers), in the vector channel, the rho meson mass was close to its PDF value in the vacuum $T = 0$. While well aware that their "screening masses" are not really masses, DeTar and Kogut still ended the paper by remarking that "[...]their appearance in the screening spectrum deals a serious blow to the naive deconfinement picture[...]."

Surprisingly, the same wrong idea has resurfaced again and again. In Fig. 5.4 from Glozman (2021), we show the correlation functions of various mesonic correlators separated by a *spatial* distance of z. As one

[6] We will return to this point in our discussion of color superconductivity at very high T.

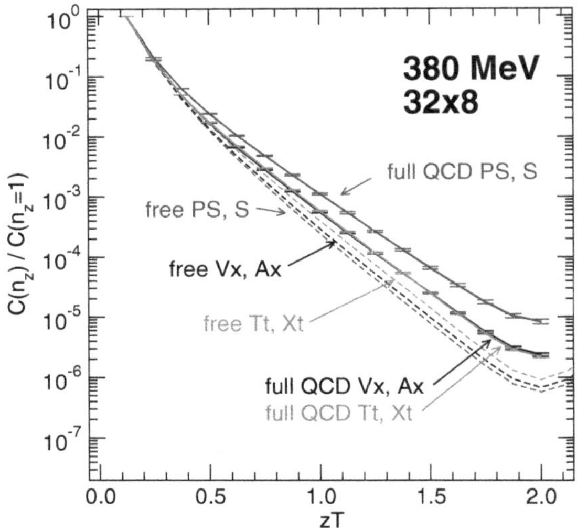

Fig. 5.4. Correlation functions of $\bar{q}q$ operators in the z direction for scalar (S), pseudoscalar (PS), vector (V) and axial (A), versus zT, where the temperature $T = 380\,\text{MeV}$. The dashed lines show the same correlators for free quarks (in an "empty" lattice).

can see, for all quantum numbers, the functions become exponential at large distances, indicating a certain value of "screening masses." These data were interpreted as an argument for "confinement" at the corresponding temperature of $T = 380\,\text{MeV}$, more than twice as large as the known deconfinement transition of $T_c \approx 155\,\text{MeV}$. Furthermore, at $T = 960$ MeV (not shown here), the differences between free and correlated motion were no longer observed in their data. This led Glozman (2021) to speculate that there is *no binding above* $T \sim 1\,\text{GeV}$, thus leading to the discovery of a deconfined "true QGP" regime.

This conclusion is wrong on several counts. First of all, the existence of a bound state does not require confinement (i.e. linear potential) since any strong enough attraction would be able to do so. Second, as pointed out in our paper (Koch *et al.*, 1992), the phenomenon obviously should persist until $T \to \infty$. The lowest Matsubara frequency would make quarks heavier and heavier, which would make them easier and easier to bind.

At $T_{\text{upper}} \sim 1\,\text{GeV}$, the effective quark mass (actually the Matsubara energy) is as large as $\pi T \sim 3\,\text{GeV}$. So, the effective quarkonium-like mesons have a mass of about 6 GeV between charmonia and bottonia in real QCD (but, of course, in its 2+1-dimensional world, with the spatial string tension instead of the usual one).

Recently, screening masses were calculated for a wider range of temperatures/couplings by Bazavov *et al.* (2019) and then by Laudicina *et al.* (2022). Their results are reproduced in the upper and lower plots of Fig. 5.1. Note that the latter results are for quite a wide temperature range, from $T = 1 - 160\,\text{GeV}$. The corresponding coupling range is $\hat{g}^2 = 2.7$–1.0, so, naively, one may doubt the validity of the perturbative analysis. Yet, as we show in the following section, it actually works quite well.

It is convenient to write the 2D Schrodinger equation using as units the Debye mass/length, for which I use the original perturbative form (Shuryak, 1978b)

$$M_E^2 = \left(1 + \frac{N_f}{6}\right) g^2 T^2 \tag{5.6}$$

with $N_f = 3$. (The inclusion of the fourth charm quark is debatable at such a T, and it is not very important.) So, the wave function depends on $x = r \cdot M_E$, with r being the radial distance, and the equation takes the form

$$\begin{aligned} -\psi'' - \frac{1}{x}\psi' + \rho\big(\log(x/2) + \gamma_E - K_0(x)\big)\psi \\ = \hat{E}_0\psi \end{aligned} \tag{5.7}$$

Note that the coefficient in front of the potential

$$\rho = \frac{p_0 g_E^2 C_F}{2\pi m_E^2} = \frac{3(N_c^2 - 1)}{2N_c(2N_c + N_f)} \tag{5.8}$$

is *universal*, independent of both T and the coupling $\hat{g}^2$. For the case we need, $N_c = N_f = 3$, $\rho = 4/9$. Here, $C_F = 4/3$ is the standard color factor for quark–antiquark interactions.

The “quarkonium” mass is close to twice the Matsubara frequency πT; however, in this equation, it is without two because, as usual, it is the reduced mass for relative motion. The universal eigenvalues of this operator are related to the quarkonium mass via

$$M_{\text{full}} = 2\pi T + g^2 T \frac{C_F}{2\pi}\left(\frac{1}{2} + \hat{E}_0\right) \tag{5.9}$$

so the actual deviation from the asymptotic value is linear in coupling $g^2(T)$, for which the usual two-loop expression is used

$$\frac{1}{g^2(T)} = \frac{9}{8\pi^2}\log\left(\frac{2\pi T}{\Lambda}\right) + \frac{4}{9\pi^2}\log\left(2\log\left(\frac{2\pi T}{\Lambda}\right)\right) \tag{5.10}$$

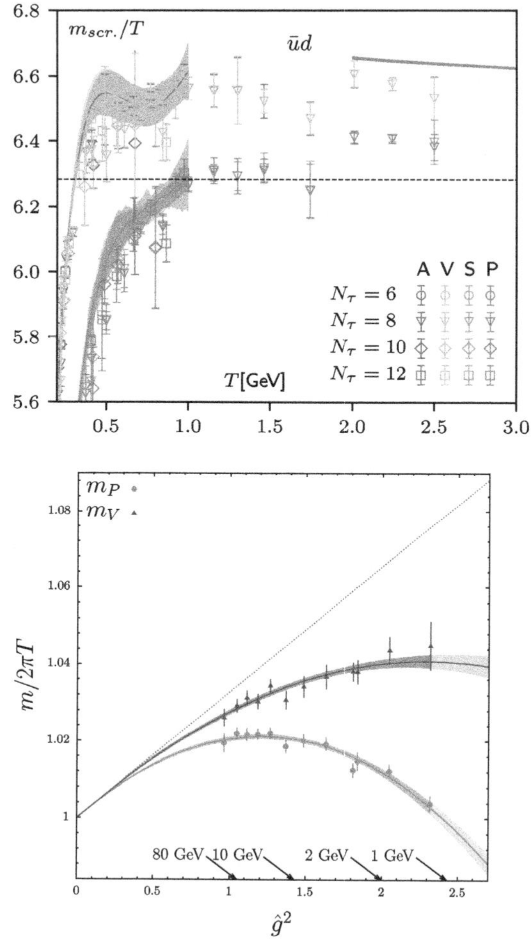

Fig. 5.5. In the upper plot, mesonic screening masses are given normalized to the temperature $m_{\rm scr}/T$ versus T (GeV). In the lower plot, the nomalization is $m_{\rm scr}/2\pi T$, plotted versus the effective coupling $\hat{g}$. Blue points and curve are for vector/axial channel, and lower red ones are for scalar/pseudoscalar channels.

The value $\Lambda \approx 0.5$ GeV leads to coupling values, as used by Laudicina *et al.* (2022).

Equation (5.7) does not have an analytic solution but can easily be solved numerically (we do not know why that was not done by Laine and Vepsalainen, 2004). It was done in my paper (Shuryak, 2022a), which we follow from this point until the end of this section.

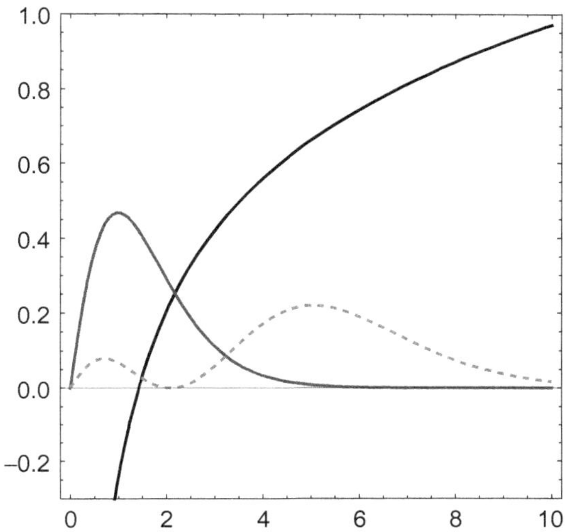

Fig. 5.6. The effective potential is shown by the black solid curve, shown together with the probability distribution for the ground state $2\pi r|\psi_0(\hat{r})|^2$ (blue solid) and for the first excited state $2\pi r|\psi_1(\hat{r})|^2$ (red dashed).

In Fig. 5.6, we show the 2D Coulomb potential, together with the probabilities to find systems in the ground and first excited states. It shows that those spatial distributions are quite different. While for the ground state, the maximum is close to the zero of the potential, this is not true for other states. As a result, the energies of these states are quite different:

$$\hat{E}_{0,1,2} = 0.208618, 0.836333, 1.07963, \ldots \tag{5.11}$$

with that of the ground state much smaller than one because of partial cancellation of large and small $\hat{r}$ regions, where the potential is of the opposite sign. According to (5.9),

$$\frac{M_{\text{full}}}{2\pi T} = 1 + g^2 \frac{C_F}{4\pi^2}(0.5 + 0.208618) = 1 + g^2 0.0239327 \tag{5.12}$$

which does not agree with the numerical coefficient suggested by Laine and Vepsalainen (2004), $1 + g^2 0.032739961$ (based on some estimates of the wave function near the origin). Therefore, we find that the dashed line in the lower part of Fig. 5.5(lower) should have a different, significantly smaller, slope. As we will see soon, this change actually brings it much closer to the lattice data.

Using the obtained wave functions, one can also perturbatively evaluate the effects of a higher order in coupling in the Schrodinger equation

$$\hat{V}_{\text{higher}} = C_4 g^2 \hat{r} + C_6 g^4 \hat{r}^2 + \cdots \tag{5.13}$$

(Note that we use as the index for coefficients not the power of coupling in this equation but the power of g in the full mass.)

In the subsequent orders, there should appear spin-dependent forces, which would split vector and scalar quarkonia. One may also introduce (in the simplest local form), the spin-dependent potential, producing a shift in the mass of scalar/pseudoscalar states:

$$\hat{V}_{SS} = C_{\text{spin}} g^2 \left(\frac{1 - \vec{\sigma}_1 \vec{\sigma}_2}{2} \right) \delta^2(\hat{r}) \tag{5.14}$$

Table 5.1 gives the corresponding values of mean $\hat{r}, \hat{r}^2, \delta^2(\hat{r})$ for the first two states. Using those and the data from Laudicina *et al.* (2022), one can fit the coefficients of the higher-order terms. For the spin term, we follow a fit from Laudicina *et al.* (2022):

$$\frac{m_V - m_S}{2\pi T} = 0.00704 \times g^4 \tag{5.15}$$

from which and Table 5.1, the C_{spin} value follows.

A revised version of the plot, compared to the same lattice data (Laudicina *et al.*, 2022), is shown in Fig. 5.7 for the ground and first excited states in both the V and S channels. The predictions for excited states are obtained perturbatively based on the mean values given in Table 5.1. The extracted coefficients of the higher-order terms in the effective action used in this plot are

$$C_4 = 0.095 \quad \text{and} \quad C_6 = -0.027 \tag{5.16}$$

Note that in our parameterization, the string tension (the coefficient C_4 of $\hat{r}$) is positive, unlike the one in the parameterization used by Laudicina *et al.* (2022).

Table 5.1. The average values of the reduced $\hat{r}$, $\hat{r}^2$ and 2D delta function for the first two levels of 2+1-dimensional quarkonia.

	$\langle n\vert\hat{r}\vert n\rangle$	$\langle n\vert\hat{r}^2\vert n\rangle$	$\langle n\vert\delta^2(\hat{r})\vert n\rangle$
$n = 0$	1.58652	3.49549	0.147316
$n = 1$	5.32307	33.0913	0.0320479

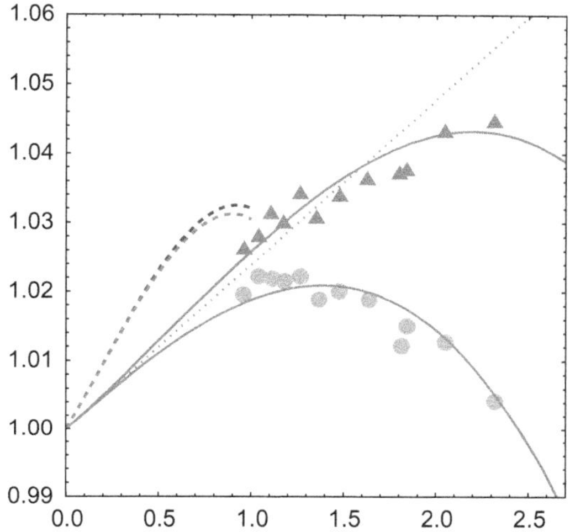

Fig. 5.7. Normalized mesonic screening masses $m_{\text{scr}}/2\pi T$ versus the square of effective coupling $\hat{g}^2$. The dotted straight line is the corrected $O(\hat{g}^2)$ asymptotic term. Blue lines denote vector (upper) and scalar (lower) ground states, with parameters given in the text. Blue triangles and red discs show vector and scalar masses, respectively: They are the same lattice data as in Fig. 5.1. Finally, the blue- and red-dashed lines on the left side of the plot show our predictions for the first excited vector and scalar states, respectively.

Summary and connections: In this chapter, we introduced the finite temperature formalism based on Feynman path integrals in Euclidean space-time. Its practical realization on a 4D lattice led to very successful numerical simulations, providing the thermodynamical quantities of hot matter. We will return to the discussion of higher susceptibilities in Chapter 6 in the context of event-by-event fluctuations and the search for a QCD critical point. At the end, we have shown that at very high T, the theory effectively reduces to 2+1 dimensions: We have calculated mesonic screening masses in this way and shown that the perturbative theory reproduces mesonic screening masses well.

Chapter 6

EVENT-BY-EVENT FLUCTUATIONS AND SEARCHES FOR THE QCD CRITICAL POINT

6.1 Theory of Second-Order Phase Transitions

This section describes historic developments in the theory of critical phenomena. Unlike the first-order transitions, which were basically described by the van der Waals model of liquid–gas transitions more than a century ago, the second-order transitions are characterized by quite nontrivial behavior in the IR limit due to the correlation length tending to infinity.

6.1.1 *Landau model, broken symmetry and critical indices*

One reason second-order phase transitions are so important is the notion of *a change in symmetry*, inducing analytic singularity at the phase transition points. The central notion of the *order parameter* has been introduced in the *Landau model*. Since we only discuss transitions belonging to the simplest universality class of the 3D Ising model, they are represented by a single-valued scalar field $\phi(x)$.

Landau suggested that an effective action (free energy) near the critical point can be represented as

$$\Omega = \int d^d x \left(\frac{1}{2} \left(\frac{\partial \phi}{\partial x_m} \right)^2 + \frac{m^2}{2} \phi^2 + \lambda_4 \phi^4 \right) \tag{6.1}$$

Note that Landau included only even powers of the field: He did so to preserve the presumed Z_2 parity symmetry $\phi \to -\phi$.

There are supposed to be two distinct phases with different symmetries: In one of them (*unbroken*), the mean value of the field is zero, while in the *broken* phase, it is *nonzero* $\langle \phi \rangle \neq 0$. The Z_2 field-reflection symmetry in it is broken.

The main assumption by Landau was that the parameters m^2 and λ_4 are smooth functions of temperature and other external parameters. He then argued that, at the critical point, $m^2(T, \mu \ldots) = 1/\xi^2$ changes sign, while the quartic coupling λ_4 is assumed to *remain finite and positive.* If so, thermal distribution $\exp(-\Omega)$ over ϕ changes from a single to double-maxima shape. When the parameter $m^2 > 0$, the minimum of the potential is at $\phi = 0$; however, if at some temperature it changes sign and becomes negative, $m^2 < 0$, the potential possesses *two* minima. One can obtain them from $dV/d\phi|_{\phi_*} = 0$ or

$$\phi_* = \pm\sqrt{\frac{-m^2}{4\lambda}} \tag{6.2}$$

A macroscopically large system randomly selects one of them. This is called *spontaneous symmetry breaking.* If a system is very large and its space dimension is not too small, the transition between them is strongly suppressed, so once the choice is made, it stays in the whole system.[1]

Expanding the potential near the new minimum, one finds that a new effective mass is still positive, as it needs to be

$$m_{\text{eff}}^2 = \frac{d^2V}{d\phi^2}|_{\phi_*} = -2m^2 > 0 \tag{6.3}$$

For convenience, let us introduce a dimensionless temperature indicating closeness to the critical point:

$$t = \frac{T - T_c}{T_c} \tag{6.4}$$

We already derived that in the asymmetric phase

$$\phi_* \sim t^{1/2} \tag{6.5}$$

This power of $1/2$ is the first Landau prediction for the *critical index* called β. The primary name α among the indices is, for historical reasons, reserved for that of the second derivative of the free energy

$$\frac{\partial^2 F}{\partial T^2} \sim t^{-\alpha} \tag{6.6}$$

It is easy to check that Landau theory predicts a *finite jump* in this quantity. The predictions of Landau's theory which we have worked out so far are

$$\alpha_L = 0, \quad \beta_L = 1/2$$

[1] Of course, one can also find domain structure depending on initial conditions and the kinetics of the transition.

Two more important critical indices are related to the limits of the Green function. At $t \neq 0$, it is expected to fall off exponentially with some correlation length related to the index ν

$$G(r) \sim e^{-\frac{r}{\xi(t)}}, \quad \xi(t) \sim |t|^{-\nu} \tag{6.7}$$

In the Landau theory, it is given by the inverse effective mass, and thus, as we evaluated it already, Landau's prediction is $\nu_L = 1/2$.

At $t = 0$, there is no exponent, the correlation length is infinite, $\xi \to \infty$, and the correlator has to be power-like. This power is the definition of the (last index we discuss) η, namely

$$G \sim \frac{1}{r^{d-2+\eta}} \tag{6.8}$$

where d is the space dimension. In the $d = 3$ case under consideration, Landau's theory leads to a Coulomb-like field correlator $G \sim 1/r$, so the predicted value is $\eta_L = 0$.

Landau's theory is based on an analyticity assumption which seems so natural that it was generally accepted for decades as almost obvious. Yet, over time, some clouds start appearing at the horizon, showing that the problem of the second-order phase transitions in $d = 2$ and 3 dimensions is more complex.

In 1944, Onsager had solved the $d = 2$ Ising model. His values for these indices *disagreed* with Landau's prediction (see the values in Table 6.1).

In experiments during the 1950s and 1960s, and eventually in computer simulations performed for the $d = 3$ Ising model and other similar systems, bringing, as usual, a mixture of good and bad news. The good news was an emerging *universality* conclusion: Indices for all systems with the same number of order parameters (boiling liquids at their critical points, magnetics, restructuring crystals, etc.) converged to the same

Table 6.1. The values of major critical indices predicted by Landau's theory, Onsager's solution to the two-dimensional Ising model and empirical values for the three-dimensional Ising model from numerical simulations.

	Landau	2D Ising	3D pheno
α	0 (jump)	0 (log)	0.08
β	1/2	1/8	0.33
ν	1/2	1	0.629
η	0	1/4	0.04

universal values.[2] The bad news was a persistent disagreement with Landau's theory. The main assumption of the analyticity of the effective action coefficients does not seem to be correct.

6.1.2 Wilson's theory based on ϵ expansion

A huge amount of theoretical and experimental work has been done on this problem, especially in the 1960s. The feasibility of a self-consistent solution with nontrivial powers of t was gradually realized, and some important tools, such as the operator product expansion and representations of the conformal group, were developed. Unfortunately, we do not have sufficient space for it here,[3] and I will just jump to the end of the process in the 1970s.

Second-order phase transitions occupy a special place in theoretical physics because of the correlation length $\xi \to \infty$. This naturally leads to applications of the *renormalization group* (RG), which follow interaction in matter to its infrared (IR) limit.

Kenneth Wilson famously connected this problem with RG in $d = 4$ theories by his expansion in $\epsilon \equiv 4 - d$. While in his original papers, Wilson discussed the space dimension $d = 3.99$ or so, it later turned out that the expressions he got for critical indices remain in fact reasonable even for $\epsilon = 1, d = 3$. Even at $\epsilon = 2, d = 2$, these formulae were qualitatively correct! In his theory, there exists a *fixed point of RG* at which both effective m^2 and λ_4 are nonzero but numerically small, allowing for perturbative treatment.

The idea of an RG appeared in the QFT context as a tool to do resummation of certain diagrams. Its continuous version, due to Murray Gell-Mann and Frances Low, dating back to the 1950s, we will discuss later on in other applications. For simplicity of presentation, we now use its discrete version in a form worked out by Kadanoff and then famously used by Wilson.

Let us introduce here the lattice representation of the field: Space is discretized with points arranged in a simple cubic lattice with step a. Any scalar field configuration is represented by a set of discrete variables $\phi(x_i)$, and the field values on the d-dimensional grid.[4] The RG "doubling

---

[2] We are referring here to one-order parameter examples. Higher symmetry breaking leads to higher-order parameters and thus other universality classes.

[3] I need to apologize for that in particular to my personal friends, A. Patashinsky, V. Pokrovsky, A. Polyakov and A. Migdal, who made significant contributions to this theory.

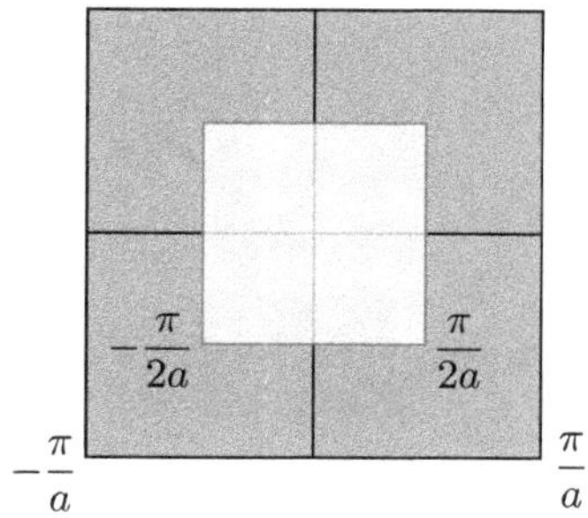

Fig. 6.1. The "shell" in momentum space, which should be integrated out in one step of the renormalization group procedure.

operation" is defined as a transition to twice more coarse lattice, with the step size doubled, $a \to 2a$. A sequence of such steps will eventually lead to very large lattices, appropriate to studies of large correlation length scale $\xi(t)$, at which the physics of the phase transition near the critical point is concentrated. To reach this goal, one may need many — $\log_2(\xi(t)/a) \gg 1$ — such doubling steps, during which the original atomic-scale specific action would *flow into some universal effective action.*

In momentum space, the doubling procedure appears as follows. The original lattice has momenta located inside the square domain shown in Fig. 6.1, and the doubling of the lattice step a means that new momenta are located only inside the white square at the center, with half size in each direction. The gray area around it would represent momenta no longer present there, so all effects associated with such momenta need to be accounted for and "integrated out." Their effect is not neglected but preserved via appropriate *renormalization* of the effective action parameters.

Wilson suggested a rather crude but instructive way of performing this doubling transformation. Let us include only two diagrams, to be referred to as the "bubble" and the "fish," shown in Fig. 6.2. Those are needed because they renormalize two parameters of the effective Lagrangian, the mass and the coupling. Here are the expressions for them according to

[4]In the problem under consideration, time development is very slow and classical. This formally translates into keeping only the zeroth Matsubara frequencies in all sums or ignoring the time variable completely. That is why time is not discretized or even mentioned here.

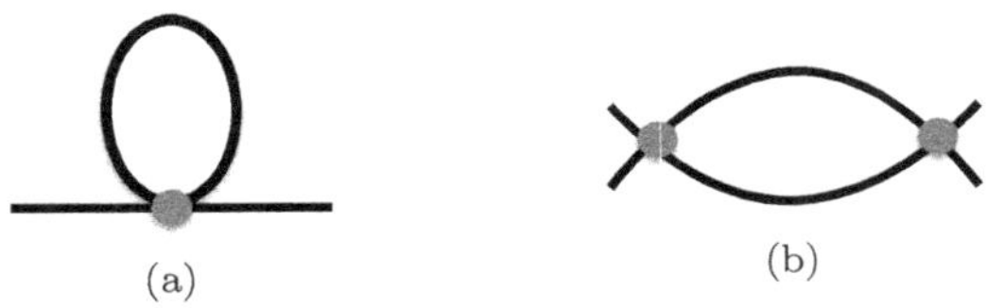

Fig. 6.2. Diagrams for the mass squared renormalization (a), the "bubble," and coupling renormalization (b), the "fish."

Wilson's approximation:

$$\text{“bubble”} = \lambda \int_{1/2<|p|<1} \frac{d^D p}{(2\pi)^D} \frac{1}{p^2+m^2} \approx \frac{\lambda}{1+m^2} C \tag{6.9}$$

$$\text{“fish”} = \lambda^2 \int_{1/2<|p|<1} \frac{d^D p}{(2\pi)^D} \frac{1}{p^2+m^2} \frac{1}{(\vec{p}-\vec{q})^2+m^2} \approx \frac{\lambda^2}{(1+m^2)^2} C \tag{6.10}$$

where the constant

$$C = \int_{1/2<|p|<1} \frac{d^D p}{(2\pi)^D}$$

is the volume of the integrated (shaded) region. Note that, in his units, the size of the momenta is normalized to $\pi/a = 1$. At the end of this chapter, we will check the accuracy of his approximation, but for now, we will just accept it and see what it leads to.

The RG transformation, for one doubling step

$$(m_n^2, \lambda_n) \to (m_{n+1}^2, \lambda_{n+1})$$

is complemented by stretching of the units, so that the new lattice has the same size. "The doubling relations" take the following form:

$$(m)_{n+1}^2 = 4\left[m_n^2 + 3\frac{\lambda_n C}{1+m_n^2}\right], \quad \lambda_{n+1} = 2^{4-D}\left[\lambda_n - 9\frac{\lambda_n^2 C}{(1+m_n^2)^2}\right]$$

The factors in front of brackets are related to the dimension of the quantity: the square of mass has a dimension of 2, thus 2^2; the coupling is dimensionless if $D = 4$, although, in general, its dimension is $4 - D$; therefore, 2^{4-D}.

The central idea is that the sequence of such transformations leads to a *fixed point* m_*, λ_*. Its values correspond to a point at the m, λ plane, which is no longer changing, from n to $n+1$. At this fixed point, the values of the parameters on the l.h.s. and the r.h.s. of the RG relation above must coincide! This idea leads to the equations

$$(m)_*^2 = 4\left[m_*^2 + 3\frac{\lambda_* C}{1+m_*^2}\right], \quad \lambda_* = 2^{4-D}\left[\lambda_* - 9\frac{\lambda_*^2 C}{(1+m_*^2)^2}\right]$$

From the second equation, one gets $\lambda_* = (2^{4-D}-1)(1+m_*^2)/9C$. Its form rather naturally brings us to Wilson's next brilliant idea: Let us consider the dimension of space to be a variable parameter, and consider

$$4 - D \equiv \epsilon \tag{6.11}$$

to be small. This leads to the fixed-point coupling

$$\lambda_* = \epsilon\frac{1+m_*^2}{9C}\log(2) \tag{6.12}$$

being $O(\epsilon)$, which is thus also small!

How small? Well, in the title of their early paper (Wilson and Fisher, 1972), the authors mentioned the space with $d = 3.99$ dimensions, or $\epsilon = 0.01$. And yet, as we will see in the following, one can eventually use it at $\epsilon = 1$, and some people did it even for $\epsilon = 2$, with surprising degrees of success. The reason is the case is seen from the first equation:

$$\lambda_* C = \epsilon \times \log(2)/9, \quad m_*^2 = -\epsilon\frac{4}{9}\log(2) \tag{6.13}$$

A small value of the coupling justifies the perturbative procedure. Furthermore, for $d = 3$, or $\epsilon = 1$, one finds the l.h.s. to be just $\lambda_* C = 0.077 \ll 1$, still a really small value! Even for $d = 2$, or $\epsilon = 2$, it is not bad at all! Near the final fixed point, the RG transformation can be linearized. There are two eigenvalues of the resulting matrix: one less than 1, which in large power leads to an approach to the fixed point, and one larger than 1, creating a departure from it. Playing with the RG transformation in the form given above or in its simplified linear form, one finds that it is difficult but not impossible to tune the initial values of the system's action, so that it will eventually approach the fixed point.

For a detailed pedagogical discussion of further calculations, the reader may consult the original papers or a pedagogical review (Wilson and Kogut, 1974). Let me illustrate the results using just one example of the index ν of the mass, which is the critical correlation length. Here are the subsequent terms of the epsilon expansion:

$$\nu = (1/2)(1 + 0.167\epsilon + 0.04\epsilon^2 - 0.016\epsilon^3 + \cdots) \tag{6.14}$$

Compare it to the value in the table above, for $\epsilon = 1$ and even $\epsilon = 2$ (Onsager's d = 2 Ising), and you will be surprised at how good the results are. Similar expressions for other indices were also derived.

It is indeed simply a miracle: The use of only two key diagrams, evaluated rather crudely, plus a very clever idea of implementing the RG transformation, was shown to go a very long way! The lesson is that as $t \to 0$ and with the infinite scale of the correlation length, the mass parameter does *not* tend to zero, as Landau assumed, but to a certain value. It was exactly such a value that is able to delicately balance a charge renormalization. Wilson was awarded the 1982 Nobel Prize for these works, an inspiration to many theorists ever since.

6.1.3 *Modified Landau model in three dimensions*

The previous material was based on the scaling/dimension relations for $\epsilon \approx 0$ or $d \approx 4$ dimensions. In the action, one finds the "kinetic" term

$$d^d x(\partial_\mu \phi)^2,$$

and for $d = 4$, one immediately sees that the field dimension is $[\phi] = 1/L$. As a result, $[\phi]^4 = 1/L^4$; therefore, the coupling λ_4 is a dimensionless constant.

However, since time plays no role in second-order transitions, the effective action is a three-dimensional volume integral, $d = 3$. Then, the derivative term tells us that the field dimension must be different, $[\phi] = 1/\sqrt{L}$. If so, the dimension of $[d^3 x \phi^4] = L$; therefore, quartic coupling λ_4 has a dimension of $1/L$. It thus must vanish in the $\xi \to \infty$ limit. Now, it is the *sixth* power term $[d^3 x \phi^6] = L^0$, which has a dimensionless constant λ_6! It is tempting to modify the Landau model, ignoring the λ_4 term and instead keeping λ_6 finite.

It turns out that such a modification gives indices *closer* to the observed ones, although of course not exactly. The minimum of the effective potential shifted by $J \neq 0$ is given by the solution of

$$\frac{d\Gamma}{d\phi} = J. \tag{6.15}$$

For $m \to 0$ and the Landau theory, when the only nonlinear term is ϕ^4, one finds $\delta = 3$, which is not close to the true value. The closest integer to the empirical value of 4.78 is 5. Note that it indeed corresponds to neglecting the $\eta \to 0$ in the general expression above, which we assumed anyway for propagators.) If so, it implies that $\Gamma \sim \phi^6$.

Multiple numerical studies of the Ising model suggested various forms of the effective action $\Gamma(\phi)$, describing the distribution of fluctuations near or at the critical point. Those were fits to simulations at many lattices and J values. In particular, good quality fits were reached by Tsypin (1994), after the pre-exponent factor $\sqrt{d^2V/d\phi^2}$ was included. We follow this paper, in which the suggested fit of effective action takes the form

$$\Gamma(\phi) = \int d^3x \left[\frac{(\partial_\mu \phi)^2}{2} + \frac{m^2\phi^2}{2} + m g_4 \phi^4 + g_6 \phi^6 \right]. \tag{6.16}$$

with *universal dimensionless coupling values*

$$g_4 = 0.97, \quad g_6 = 2.05 \tag{6.17}$$

6.1.4 *RG flow in the form of Wetterich functional equation*

There were discussions in the literature whether the effective potential $\Omega(\phi)$ near CP can be truncated to some finite powers of ϕ. In particular, whether one can just keep only one extra term of order 6 above the Landau model. The "magnetization" $\langle\phi\rangle(J)$ index δ is

$$\langle\phi\rangle(J) \sim J^{1/\delta}, \quad \delta = \frac{d+2-\eta}{d-2+\eta} \approx 4.78. \tag{6.18}$$

while for $\eta = 0$, it would be five, not a big change. Setting the propagator index η to zero as a numerical approximation has been discussed in the literature, such as by Tetradis and Wetterich (1994).

The next step of RG development was based on RG equations in the form worked out by Wetterich and his group (for a review, see, for example, Berges *et al.*, 2002; Dupuis *et al.* 2021). We will not be able to go into it here in any detail, but let me just provide a few theoretical and empirical hints that predated this marvelous development. Subsequent studies have shown that no truncations of the series for potential expansion can be justified, and in deep infrared $t \to 0$, it contains infinitely many terms.

Let us introduce a few technical formulae before jumping to Wetterich RG. The effective action $\Gamma(\phi)$ is defined by integral over fluctuations χ around some *background* field ϕ:

$$\exp[-\Gamma(\phi)] = \int D\chi \exp\left[-S(\phi+\chi) + \int d^dx \frac{\delta\Gamma}{\delta\phi}\chi \right] \tag{6.19}$$

This equation can be used for the generation of perturbative diagrams. In the zeroth "classical" order, $\Gamma(\phi) = S(\phi)$. In the one-loop order, there is a linear term (which vanishes) plus the second-order term

$$S[\phi + \chi] = S[\phi] + \frac{1}{2}\int d^d x d^d y \chi(x) \frac{\delta^2 S}{\delta\phi(x)\delta\phi(y)} \chi(y) \tag{6.20}$$

In this order, the integral over fluctuation is Gaussian, and one gets a determinant of the operator of the second derivative. For a constant background $\phi(x) = \phi_0$, the effective action adds to the initial action the term

$$\Gamma_{(1)} = \frac{1}{2}\int \frac{d^d q}{(2\pi)^d} \log\left[q^2 + m^2 + (3/2)\lambda_4\phi_0^2\right] \tag{6.21}$$

Introducing standard simplifying notations for the field of this approach and renormalized couplings

$$\rho = \phi^2/2, \quad m_R^2 = \frac{\partial U}{\partial \rho}(0), \quad \lambda_R = \frac{\partial^2 U}{\partial \rho^2}(0)$$

one can calculate those by explicitly calculating the one-loop integral above. Since at $d = 3$ it diverges in UV, one needs to explicitly cut-off Λ as the upper limit of momentum integration. The results are

$$m_R^2 = m^2 + \frac{3\lambda\Lambda}{4\pi^2}, \quad \lambda_R = \lambda - \frac{9\lambda^2}{16\pi m} \tag{6.22}$$

Wilson's approach was to look for a fixed point, which means solutions of these equations in which both in the l.h.s. and r.h.s. there appears the same coupling of λ_* and mass m_*. And, of course, one can include higher-order effects in the expansion of integral over fluctuations.

The Wetterich method is based on the introduction of the *scale-dependent* potential $U_k(\phi)$, which has a scale of k. All momentum integrals are done for momenta larger than k (and of course smaller than Λ). Here is the simplest form of the RG equation:

$$\partial_k U_k(\rho) = -\frac{k^2}{4\pi^2} \log\left[1 + U_k'/k^2 + 2\rho U_k''/k^2\right] \tag{6.23}$$

where prime means derivative over the argument of U_k, $\rho = \phi^2/2$. The initial condition is the classical action $S = m^2\rho + \lambda\rho^2/2$, and then, it can be solved toward IR, $k \to 0$.

This equation is obtained by adding one additional term to effective action:

$$\Delta S_k(\chi) = \frac{1}{2}\int \frac{d^d q}{(2\pi)^d} R_k(q)\chi(q)\chi(-q) \tag{6.24}$$

A convenient form of an IR cutoff inserted in all propagators is

$$R_k(q) = \frac{q^2}{\exp(q^2/k^2) - 1} \tag{6.25}$$

For modes with momenta $q^2 \ll k^2$, the exponent is expanded and the formula gets inverted, so their "effective mass" becomes k^2, and such modes are effectively cut off. In the opposite limit, $q^2 \gg k^2$, this term is exponentially small and irrelevant.

The solution to the equation is complicated and done numerically. The equations are nonlinear, resembling, in this respect, those of hydrodynamics. Lately, using the so-called Galerkin solution method, it has been found that the generated solutions may possess discontinuities resembling hydrodynamical shocks; see, for example, Grossi and Wink (2019).

Some RG solutions were summarized by Berges *et al.* (2002). In their treatment of the Ising model in Section 4.4, they have concluded that "in terms of the rescaled field [...] the potential is almost a polynomial ϕ^6-potential." More specifically, from their polynomial fit to $\partial\Omega/\partial\phi$, we have recalculated their potential in the form

$$\Gamma = 0.5042m^2\phi^2 + 0.798175m\phi^4 + 1.61793\phi^6 + \frac{0.06495}{m}\phi^8 + \cdots \tag{6.26}$$

containing a term with ϕ^8, singular at the critical point where $m = 1/\xi \to 0$. The dots, of course, hide even more singular terms. Note that the m-independent combination of couplings from it is 2.5, while for the Tsypin fit, $g_6/g_4^2 = 2.17$.

We return to this result in the following while discussing a search for a critical point in heavy ion collisions. We then argue that since we are not expected to be too close to the singular point, further terms of the effective potential would not be needed.

6.2 Event-by-Event Fluctuations and Thermodynamical Susceptibilities

We know that at the chemical freeze-out stage, the matter is well equilibrated, so its description can be done using the standard tools of thermodynamics. So far, we have discussed only the average quantities, such as the yields of different types of secondaries.

Yet, thermodynamics should be able to do much more than just describe the average quantities: As we already showed at the beginning of this chapter, it can provide much more detailed information about the distribution of various final states. Indeed, the probability of the deviations from the average — known as *event-by-event fluctuations* — can also be calculated thermodynamically, provided one knows how to calculate the appropriate function governing the description of this final state. We already argued that for a microcanonical description, this probability is related to the entropy $S(E, N)$ by

$$P \sim e^{-S(E,N)} \tag{6.27}$$

Note that it is basically the famous Boltzmann formula, inverted by Einstein in 1910 into this useful form. Of course, one should use in it the free energy $F(T, V)$ for a canonical ensemble or the grandcanonical potential $\Omega(T, \mu, V)$ for the corresponding case.

A further step[5] is to expand the thermodynamical potentials in powers of the deviations of the variables from their mean values. Because the mean always corresponds to an extremum of the thermal potentials, this expansion does not contain the first-order term and starts in the second order. This means that small thermal fluctuations are always Gaussian in shape. The width of the Gaussian is thus dictated by the *second* derivative of the potential. The third- and fourth-order terms, producing the *skewness* and the *kurtosis* of non-Gaussian distributions, are given by the higher-order *susceptibilities*.

Our theoretical tools, especially the lattice gauge theory at finite temperatures, which we discuss in Section 5.1, can calculate these susceptibilities. By comparing those with event-by-event fluctuations in heavy ion collisions, one can further test the assumed thermal equilibration at the freeze-out stage. A description of the corresponding activity is the main content of this section. However, before we proceed to this comparison, let us briefly comment on their magnitudes and then describe the history of this subfield.

Because the global potential scales as the number of particles $O(N)$, the relative width of small Gaussian fluctuations of some quantity x scales as

$$\frac{\delta x}{\langle x \rangle} \sim \frac{1}{\sqrt{N}} \tag{6.28}$$

[5]Although it was also used by Einstein, the idea actually goes back to Abraham de Moivre in 1718. Later, Gauss developed a theory of small fluctuations in full glory. Thus, we do call those "Gaussian."

Furthermore, particle number eliminate ranges from $N \sim 10^4$ in central heavy ion collisions to about $N \sim 10^2$ in peripheral ones, or "small systems," to be discussed in detail in the following. Therefore, the typical scale of fluctuations is between one and ten percent.

In our field, the first papers on this subject (Stodolsky, 1995; Shuryak, 1998) initially focused on event-by-event fluctuations of the freeze-out temperature. The standard expansion of the free energy to the second order in δT has the form

$$P \sim \exp\left[-\frac{C_V(T)}{2}\left(\frac{\delta T}{T}\right)^2\right] \tag{6.29}$$

where $C_V = \frac{\partial^2 F(T,V)}{\partial T^2}$ is the *specific heat* at a fixed volume. It was argued that the measure of the temperature is the mean transverse momenta $\langle p_\perp \rangle_{\text{event}}$ *in the event.* It was further argued, using the resonance gas with constant speed of sound as an example $pV = -F \sim T^{1+\frac{1}{c_s^2}}$, that $C_V(T) \sim T^5$ at $\mu = 0, c_s^2 = 1/5$, or even $\sim T^7$ at μ, corresponding to SPS/AGS collisions, that temperature fluctuations "make a thermometer" capable of measuring the freeze-out temperature independently of the mean quantities, maybe even more accurately.

Fluctuations of $\langle p_\perp \rangle_{\text{event}}$ have been experimentally studied by the NA49 experiment at SPS, triggered by the paper by Gazdzicki and Mrowczynski (1992). One significant result of that study was that it ruled out a simplistic model, in which *PbPb* collisions were represented as a simple statistical superposition of many independent *pp* collisions. The data were consistent with the thermodynamics of the resonance gas model.

From the 1990s until now, comparisons of lattice susceptibilities with event-by-event fluctuations have become a real industry. Because the chemical freeze-out line is close to the transition line (the QGP/hadronic phase boundary), one should be interested in the (quasi) singular behavior of the free energy. Note that if light quarks are truly massless, the chiral symmetry would be exact and the transition line would be a true second-order transition line, with a singular part of the free energy expected to have a universal scaling form:

$$\frac{F_{\text{sing}}(T,\mu,m_q)}{T^4} \sim h^{1+\frac{1}{\delta}} f\left(\frac{t}{h^{\frac{1}{\beta\delta}}}\right) \tag{6.30}$$

where $t = |\delta T|/T_c + \kappa\mu/T_c, h \sim m_q/T_c$, and $\delta T, \mu, m_q$ are all considered to be small, κ is a finite parameter. The powers contain the critical indices, with their values depending on the *universality class* of the phase transition. This, in turn, depends on the number of critical parameters, which for massless QCD with two light quarks is the model with the $O(4)$ symmetry: the four order parameters are massless $\sigma, \pi^+, \pi^0, \pi^-$ mesons. The values of the relevant indices (Engels *et al.*, 2003) are

$$\beta = 0.380, \delta = 4.824 \tag{6.31}$$

The interest has now shifted from temperature fluctuations to susceptibilities obtained by derivatives over the baryon chemical potential[6]: μ

$$c_n \equiv \frac{\partial^n (f/T^4)}{\partial(\mu/T)^n} \tag{6.32}$$

The singular part of it is thus expected to be

$$(c_n)_{\text{sing}} \sim h^{1+\frac{1}{\delta}-\frac{n}{2\beta\delta}} \tag{6.33}$$

and using the values for the $O(4)$ model, one finds that the first susceptibility for which the power is negative, i.e. divergence as $h \to 0$, occurs for $n = 6$. This implies that large correlations are expected to start with six baryons in the experiment!

Lattice gauge theory is a numerical approach, and it is very hard to define the thermodynamical potentials so accurately that their sixth derivative can be reliably measured. What is achieved in practice for now for basic thermal potentials will be shown in Fig. 5.2(c) as part of the section on lattice gauge theory. Going ahead of that discussion, we show here in Fig. 6.3 from a brief review by Ratti (2017), the corresponding susceptibilities with two, four and six derivatives over the chemical potential. The calculation corresponds to physical quark masses.

While the pressure itself (the free energy) looks rather smooth and perfectly consistent with the resonance gas model (the line), its second

[6]For the three light quark flavors u, d, s, one can define three independent chemical potentials, which are usually grouped under the baryonic one we here simply call μ, the one associated with electric charge and strangeness. Therefore, there are susceptibilities numbered by three integers: n_B, n_Q, n_s, according to the corresponding number of derivatives. In the text, however, we keep only one chemical potential for brevity.

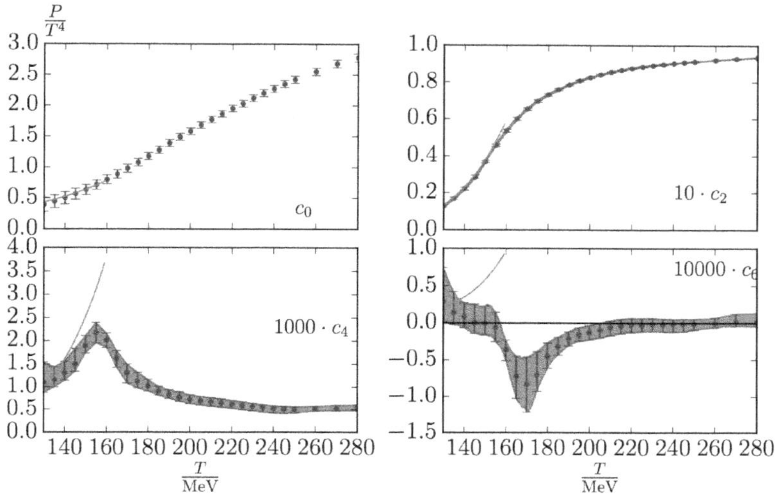

Fig. 6.3. Pressure versus temperature and its second, fourth and sixth derivatives over the baryon chemical potential at $\mu = 0$. The red line shows predictions from the resonance gas model.

derivative already shows a departure from it and a constant value at large T, corresponding to QGP. The fourth derivative already possesses a peak, after which there is a very rapid departure from the resonance gas. The sixth derivative — although with larger errors — shows an oscillation, signaling a singularity (which would fully appear only when the quark masses were going to zero). These results exemplify the idea of "better thermometers" provided by higher-order lattice susceptibilities.

Those were compared to experimental (STAR) data on the proton number fluctuations at different collision energies. According to Ratti (2017), the resulting freeze-out points on the phase diagram are as shown in Fig. 6.4. Although it is based on "preliminary" lattice data, we would like to point out certain interesting features of this comparison. While in general the method works, in the sense that the obtained freeze-out points are overall consistent with those obtained from the particle yields, one should still note that all of them are several MeV *lower* than those from the particle yields. It is perhaps too early to discuss why this could be the case.

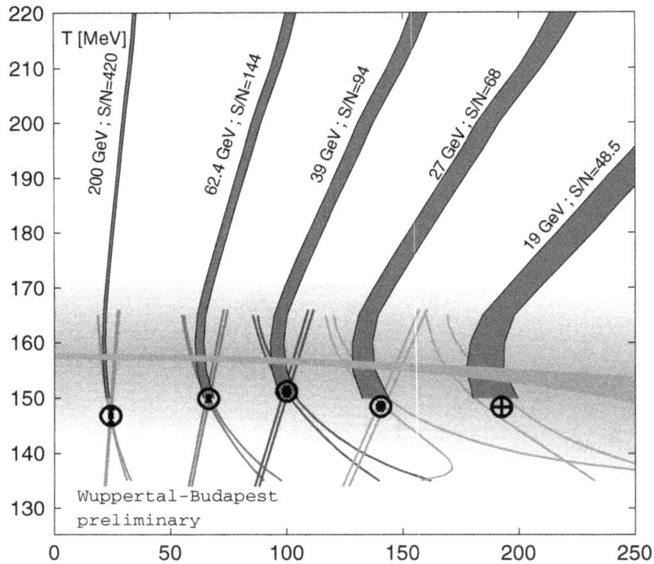

Fig. 6.4. The μ–T phase diagrams, both in MeV. The wide gray lines are adiabatic lines with the values of the ratio of the entropy-to-baryon number. The points are from a comparison of the lattice susceptibilities to the STAR data on baryon number (protons) fluctuations. The thin red line and pink band around it are the phase boundary and its error, respectively.

6.3 The Search for Enhanced Fluctuations at the QCD Critical Point

The phenomenon of *critical opalescence* was observed in various settings as early as the 19th century, usually in binary fluid mixtures. As the temperature crosses the critical line, transparent fluid appears milk-like (or, put on an overhead projector, black). Its origin is based on an increase in the correlation length ξ, from the usual microscale $\sim$nm to a scale comparable to optical wavelength $\sim\mu$m, enhancing the scattering of light. Its qualitative explanation was due to Smoluchowski (1908), with quantitative theory provided by the already mentioned paper by Einstein (1910).

A particular strategy to look for a hypothetical critical point of QCD at sufficiently large μ_B has been proposed by us (Stephanov *et al.*, 1998, 1999). The suggested strategy was to perform the *energy scan* at the RHIC, *downward* from its full energy, recording carefully the fluctuations. It was later accepted as a Beam Energy Scan (BES) program for the RHIC, and at the time of this writing, its first stage, BESI, was completed, while BESII is still on its way.

Technically, correlators of the various quantities, such as the baryon number, are given by diagrams, which can be classified according to the number of propagators of the critical modes and a certain number of vertices. At zero momentum, each propagator is $1/m^2 \sim \xi^2$. For example, the correlation of four baryonic currents, i.e. the susceptibility with four derivatives over μ called above c_4, in the lowest order of quartic coupling is shown in Fig. 6.5(a) from Stephanov (2009). The crosses in circles with solid circles stand for the baryonic density operators and baryon propagators, respectively. The wavy lines are propagators of the critical field: At zero momentum, each is $\sim 1/m_\sigma^2 \sim \xi^2$, so, in total, the four wavy lines give ξ^8. The quartic coupling of critical modes, shown by the black dot at the center, is not a constant in the infrared limit. From RG renormalization, one can see that it by itself vanishes as the first power of ξ. So, in total, this diagram is $\sim\xi^7$. Another similar diagram (not shown) with two triple vertices has five propagators, giving ξ^{10} power, but by including a proper scaling of the triple coupling, it returns to the same $\sim\xi^7$ behavior as the first diagram. Stephanov further argued that the contribution of these diagrams is such that corrections to the four-baryon correlator should change sign when the critical point is passed.

To illustrate the results of the first stage of the beam energy scan, let us look at the STAR data in Fig. 6.5(b). The kurtosis — four-proton integrated correlator — has an interesting energy dependence: a minimum at $\sqrt{s}$ =20–30 GeV and perhaps a maximum at a lower energy. Antiprotons have a similar shape but much smaller amplitudes. Smaller systems produced in peripheral collisions (right plot) do not show non-monotonous behavior: This can be attributed to the smaller lifetime of the fireball, which is insufficient for development of critical fluctuations.

The structure qualitatively agrees with the theoretical predictions of a sign-change behavior of the kurtosis. However, these data are preliminary and the conclusion of such importance clearly requires further scrutiny (see also the following discussion).

Before we go into the details of the calculations, let us present qualitatively the main idea of this paper. Suppose the CP indeed exists and is located in the part of the phase diagram near the freeze-out line of the BES program. Furthermore, while scanning this line, for some (yet unknown) beam energy, the freeze-out conditions happen to reach the

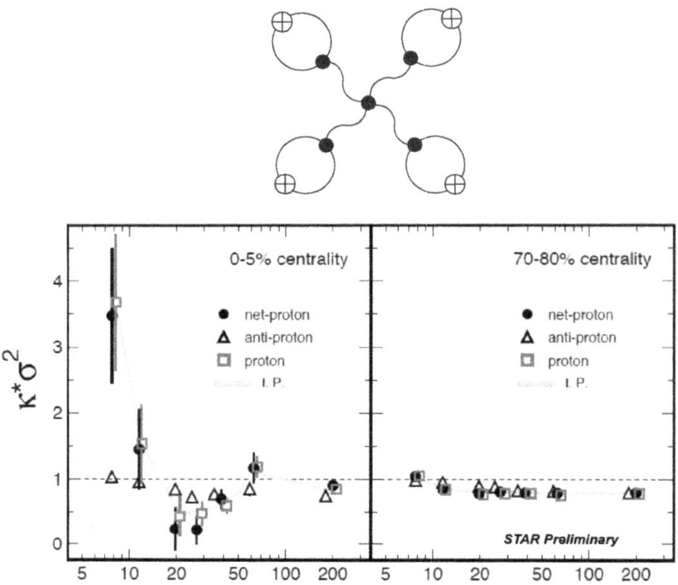

(a) (b)

Fig. 6.5. (a) Diagram producing an enhanced contribution to the four-particle correlator, the wavy lines are propagators of the critical field. (b) The kurtosis — four-particle correlator — in units of the width of the distribution, as a function of the collision energy $\sqrt{s}$, GeV from the RHIC low-energy scan and the STAR collaboration.

maximal value of the correlation length ξ, as compared to all other collision energies. What are the observables sensitive to ξ? And, more specifically, on what scale of ξ would they show observable signals?

One possibility actively discussed is related to hydrodynamic (sound-like) fluctuations of the density, with a wavelength comparable to ξ. One is expecting their enhancement due to critical opalescence near CP.

In heavy ion collisions — due to an unprecedentedly small viscosity — we can indeed observe several harmonics of sound. They are numbered by the harmonic number n in azimuthal angle ϕ. The maximal harmonic number observed is currently at $n_{\max} = 9$ (ALICE) at LHC energies, while at BES energies, it is about $n_{\max} = 6$ (STAR).[7] These harmonics can be viewed as sound propagation *around* the fireball surface, inducing correlations in ϕ of secondaries emitted from this surface. The maximal

[7]We will discuss the dependence of harmonic amplitude on n in the chapter on sounds.

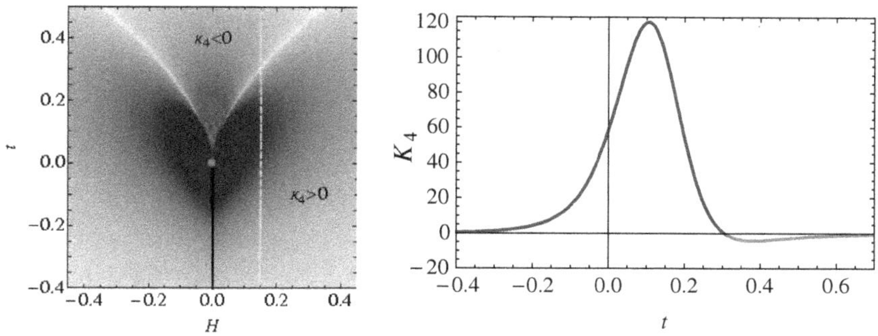

Fig. 6.6. The sign and magnitude of kurtosis on the Ising critical plot. The right figure traces the magnitude along the blue vertical line, corresponding to the constant magnetic field H.

harmonic number $n_{\max} = 6$ corresponds to the minimal sound wavelength:

$$\lambda_{\min} = \frac{2\pi R}{n_{\max}} \sim 6\,\text{fm} \tag{6.34}$$

where R is the fireball radius. Unfortunately, the critical correlation length ξ on such a large scale is unlikely to be reached in reality. Therefore, more sophisticated correlations would be needed, perhaps combining azimuthal and rapidity correlations, aiming at the yet unobserved tails of sounds.

A discussion of the *sign* of kurtosis was reported by Stephanov (2011a) using a certain Ising critical plot (see Fig. 6.6) based on the "modified Landau model" with ϕ^6 coupling we discussed above. If the map of the freeze-out line on the QCD phase diagram corresponds to the constant magnetic field H line (blue vertical line in the left plot), then the predicted shape of kurtosis dependence is as shown in the right plot.

6.4 Baryon Clustering at Freeze-out and Light Nuclei Production

Another observable, sensitive to smaller scale

$$\xi_{\max} \sim 1.5 - 2\,\text{fm}$$

was proposed by Shuryak and Torres-Rincon (2019, 2020a). This scale is the natural size of few-nucleon correlations, called *preclusters*. Their existence is due to ordinary nuclear forces, and their experimental manifestations are:

(i) higher moments (e.g. kurtosis, the fourth moment) of the proton multiplicity distribution;

(ii) yields of light nuclei — d, t, ^3He, ^4He — due to additional feed-down from precluster decays.

As we show in the following, the interplay of attractive binary and repulsive many-body forces is expected to show strong non-monotonous behavior in precluster formation probability during the BES. The idea is illustrated in Fig. 6.7. The left one, far from CP, has the usual short range ($\sim 1/m_\sigma \sim 0.4$ fm) of nuclear forces. Near the CP, where correlation length is of the order of cluster size (Fig. 6.7 (right)), the correlations of nucleons in the cluster become stronger. As we will show, evaluating the magnitude and even the sign of the effect is rather nontrivial.

Stephanov *et al.* (1998) and Stephanov (2011b) related the moments of the critical field fluctuations to those of the nucleons, under the crucial assumption that *nucleons are uncorrelated* by any other effects. If the locations of the nucleons can be integrated independently, each external line of these diagrams becomes simply a propagator integrated over space, namely

$$\int d^3r \frac{\exp(-r/\xi)}{4\pi r} = \xi^2$$

Unfortunately, this simplifying assumption is incorrect in reality. Conventional nuclear forces do create significant correlations between them. Rather nontrivially, they survive even at the freeze-out stage of heavy ion collisions, with a temperature $T \sim 100$ MeV, much larger than conventional bindings of light nuclei. As shown by Shuryak and Torres-Rincon (2019, 2020a),

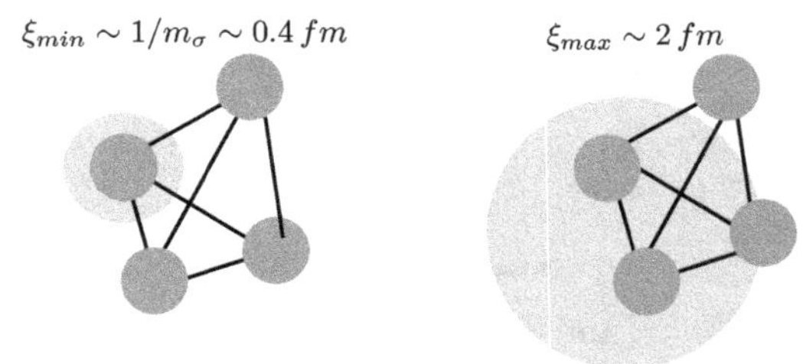

Fig. 6.7. Preclusters of four nucleons, shown by blue circles. Six lines connecting them indicate binary potentials. The gray area indicates the range of forces between them for standard nuclear forces (left) and near the critical point (right). In the latter case, the interactions are not only binary, but many-body ones also appear.

there exists a phenomenon of nucleon *preclustering* starting from four-nucleon systems. One needs six (or more) pair potentials for correlations to remain appreciable at these high temperatures.

Clustering of the baryons can also be described differently, using the modification of the inter-baryon potential. In our discussion of chemical freeze-out in Section 8.5, it is shown that yields of light nuclei — d, t, He^3, He^4 — sit on a Boltzmann exponent with one common T_{ch} with other species surprisingly accurately. This plot[8] covers about nine orders of magnitude, demonstrating that equilibration is really well established.

Let me remind the reader that those data were from ALICE at the LHC, corresponding to practically zero baryon chemical potential $\mu_B \approx 0$. At increasing μ_B (i.e. decreasing collision energies), the corresponding factor $\exp\left((\mu_B - M)/T_{\text{ch}}\right)$ becomes less and less punishing, and eventually, the production of light nuclei becomes quite common. Here are a few puzzling questions that appear while describing their production.

Production of weakly bound states from the freeze-out stage is sometimes called "snowflakes jumping out of a hot oven." Indeed, the ambient temperature of $T_{\text{ch}} \sim 100\,\text{MeV}$ is much larger than binding energy B, typically only a few MeV. Yet, statistical mechanics insists that in equilibrium, *all* states should be occupied according to statistical weight.

A related problem is *the size issue*. Small binding energies B imply large spatial sizes. For example, for a deuteron, $B_d = 2.2\,\text{MeV}$. A champion seems to be a bound state $pn\Lambda = He^3_\Lambda$, for which B is so small that it is not yet well known.[9] Such objects should be extremely fragile, and naively, with hundreds of pions flying out from the fireball, they seem to be inevitably destroyed. Any cascade code indeed predicted that to happen. Yet, the yields are not small and do agree with the statistical model.

Another question is whether one has to take into account *all nuclear states*? Deuterons, tritiums and He^3 have just a single bound state, but (as we will show in the following section), starting from He^4, nuclides possess multiple excited states as well as low-lying resonances. It would be logical to include those in a statistical sum, as we do for all hadronic resonances. This implies that accounting for "feed-down" from their decay to lighter systems needs to be done.

[8]Of course, these yields were divided by the statistical weight of each state $(2J+1)(2I+1)$.

[9]Two experiments yield quite different values.

A novel concept of *preclusters* was introduced by Shuryak and Torres-Rincon (2019, 2020a). They are different from "nuclear clusters" (the bound states of few nucleons in vacuum, such as $d, t, \ldots$) being just statistical correlations in multidimensional space of few-nucleon coordinates. Their size is not given by binding energy but by potential range and is therefore 1.5–2 fm. One can think of them as small-sized wave packets with indefinite energy. With time, they spread to multiple bound states, but this time is typically so long that all other secondaries are already far away.

"Preclustering" phenomena were studied by a number of theoretical tools:

(i) classical molecular dynamics (Shuryak and Torres-Rincon, 2019),
(ii) semiclassical "flucton" method at finite temperatures (Shuryak and Torres-Rincon, 2020a);
(iii) quantum mechanics in hyperspherical coordinates (Shuryak and Torres-Rincon, 2020a);
(iv) the (first-principle) path-integral Monte Carlo (PIMC) (DeMartini and Shuryak, 2021d).

Before we go into these papers, note that clustering is a rather complex but very generic phenomenon. *Out-of-equilibrium* examples from classical mechanics include, for example, "globular clusters" of stars inside galaxies. Many other examples occur near phase transition. Consider, for example, the production of "artificial snow" from freezing water, widely used at ski slopes. The preclustering we discuss is different, taking place at (very accurate) thermal equilibrium.

Omitting the applications of the first two methods in the list above, let us present some results from the last two. The so-called method of K-harmonics was developed in the 1960s; its main idea is that for the ground state, the multi-dimensional Schrödinger equation can be approximated by ignoring all angular variables and using only the radial coordinate. In certain cases (including ^4He), a rather good approximation can be obtained.

Recall that the first standard step in *few-body physics* is the exclusion of center-of-mass motion. It is usually done using Jacobi coordinates, which for the $A = 4$ case are

$$\vec{\xi}_1 = \frac{\vec{x}_1 - \vec{x}_2}{\sqrt{2}} \quad \vec{\xi}_2 = \frac{\vec{x}_1 + \vec{x}_2 - 2\vec{x}_3}{\sqrt{6}} \tag{6.35}$$

$$\vec{\xi}_3 = \frac{\vec{x}_1 + \vec{x}_2 + \vec{x}_3 - 3\vec{x}_4}{2\sqrt{3}}. \tag{6.36}$$

The radial coordinate, or *hyperdistance*, is defined by

$$\rho^2 = \sum_{m=1}^{3} (\vec{\xi}_m)^2 = \frac{1}{4}\left[\sum_{i\neq j} (\vec{x}_i - \vec{x}_j)^2\right]. \tag{6.37}$$

For four nucleons, there are six distances on the r.h.s., so the mean distance between nucleons is $R_{NN} \approx \sqrt{4/6}\rho$.

The radial part of the Laplacian in these Jacobi coordinates is $\psi''(\rho) + 8\psi'(\rho)/\rho$, and using the substitution

$$\chi(\rho) = \psi(\rho)\rho^4 \tag{6.38}$$

one arrives at a conventional-looking "radial" Schrödinger equation:

$$\frac{d^2\chi}{d\rho^2} - \frac{12}{\rho^2}\chi - \frac{2m_N}{\hbar^2}[W(\rho) + V_C(\rho) - E]\chi = 0 \tag{6.39}$$

where W is the projection of the potential to hyperdistance

$$W(\rho) = \frac{315}{4}\int_0^1 V_{NN}(\sqrt{2}\rho x)(1 - x^2)^2 x^2 dx \tag{6.40}$$

where $V_{NN}(r)$ is the two-body nuclear potential.

Note that the $1/\rho^2$ term looks like a centrifugal potential, but it is not: In fact, we assume that the lowest states have no dependence on angles, or $\vec{L} = 0$. Using it, Shuryak and Torres-Rincon (2020a) reproduced the ground-state binding energy $B(^4He) = 28.3$ MeV, as it was done in the original papers. Furthermore, to our surprise, we found that literature from the 1970s missed the second bound state, with few MeV bindings. In order to see if it really exists, we looked into nuclear databases which list the known states of He^4 (listed in Table 6.2). There is one 0^+ excited state in its first line. The first column is excitation from the ground state, the actual energy is $E_{0+} \approx -8\,\text{MeV}$.

In the spirit of the statistical thermal model, one should assume that all $N_{\text{states}} = \sum_i (2J_i + 1) = 49$ states in this energy strip should be populated about *equally* with all other hadronic and nuclear states in the heat bath. After freeze-out, the feed-down from these states to observed spectra should be included; however, it has never been done before in statistical models!

Performing quantum decomposition of preclusters into these states and using their decays (indicated in the table as $p + t, n +^3 He$, $d + d$ exclusive channels), one finds that decays of a single *ppnn* precluster should produce, on average, 0.30 (p + tritium), 0.22 (n + ^3He) and 0.96 deuterons

Table 6.2. Low-lying resonances of the ^4He system, from BNL properties of nuclides. J^P are the total angular momentum and parity, and Γ is the decay width. The last column is the decay channel branching ratios, in percent. p, n, d correspond to the emission of proton, neutron or deuterons, respectively.

E (MeV)	J^P	Γ (MeV)	Decay modes (%)
20.21	0^+	0.50	$p = 100$
21.01	0^-	0.84	$n = 24$, $p = 76$
21.84	2^-	2.01	$n = 37$, $p = 63$
23.33	2^-	5.01	$n = 47$, $p = 53$
23.64	1^-	6.20	$n = 45$, $p = 55$
24.25	1^-	6.10	$n = 47$, $p = 50$, $d = 3$
25.28	0^-	7.97	$n = 48$, $p = 52$
25.95	1^-	12.66	$n = 48$, $p = 52$
27.42	2^+	8.69	$n = 3$, $p = 3$, $d = 94$
28.31	1^+	9.89	$n = 47$, $p = 48$, $d = 5$
28.37	1^-	3.92	$n = 2$, $p = 2$, $d = 96$
28.39	2^-	8.75	$n = 0.2$, $p = 0.2$, $d = 99.6$
28.64	0^-	4.89	$d = 100$
28.67	2^+	3.78	$d = 100$
29.89	2^+	9.72	$n = 0.4$, $p = 0.4$, $d = 99.2$

(0.48 *dd* pairs). Such decays can be experimentally observed but have not yet been studied.

Path-integral Monte Carlo is a numerical method applied to many-body (and few-body) physics at finite temperatures. It is a *first-principles* method based directly on internuclear potentials. DeMartini and Shuryak (2021d) considered the potentials to be their simplified version, known as the Serot–Walecka potential (Serot and Walecka, 1986):

$$V_{SW}(r) = -\alpha_\sigma \frac{e^{-m_\sigma r}}{r} + \alpha_\omega \frac{e^{-m_\omega r}}{r}, \tag{6.41}$$

with $\alpha_\sigma = 6.04$, $m_\sigma = 500$ MeV, $\alpha_\omega = 15.17$, and $m_\omega = 782$ MeV. While this potential describes the properties of infinite nuclear matter well, it does not possess a bound state in the $4N$ system. Therefore, the repulsive term was slightly modified to $\alpha_\omega = 11.02$, which reproduces the desired binding energy. In Fig. 6.8 from DeMartini and Shuryak (2021d), one sees the resulting normalized distribution in hyperdistance ρ (red squares), projected to the ground state (blue disks) and the "precluster" part (green triangles). As one can see, the precluster sizes peak around $\rho \sim 1.5\,\text{fm}$.

Apart from affecting the production of light nuclei, such preclustering affects the statistical distribution of nucleons, which is another important

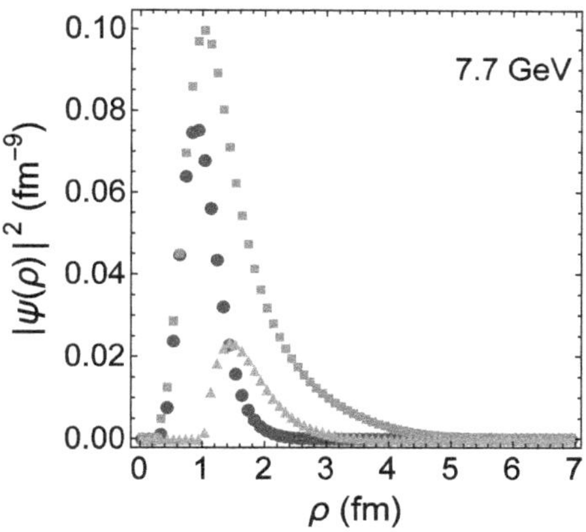

Fig. 6.8. (Color online) Normalized probability distribution in hyperspace ρ of the ground state (blue •), kinetic freeze-out distribution (red ■) and the precluster (green ▲) for the $\sqrt{s} = 7.7\,\text{GeV}$ beam energies considered.

experimental observable. The potential part of the partition function (of a single species system) of N particles can be rewritten in the form

$$Z_{\text{pot}} = 1 + \frac{1}{V^N} \int d^3x_1 \cdots \int d^3x_N \left[e^{\left(-\sum_{i>j} V(\vec{x}_i - \vec{x}_j)/T\right)} - 1 \right] \tag{6.42}$$

by adding and subtracting 1. Since we focus on clusters of distinguishable four particles, the coordinates of all others can be integrated out, as can the coordinates of their centers of mass. What is left is

$$Z_{\text{pot}} = 1 + \left(\frac{N}{4}\right)^4 \left(\frac{V_{\text{cor}}}{V^3}\right) \tag{6.43}$$

where the so-called nine-dimensional correlation volume is

$$V_{\text{cor}}^{(9)} = \frac{32}{105}\pi^4 \int d\rho \rho^8 (P(\rho) - 1). \tag{6.44}$$

Here, $P(\rho)$ is the probability distribution in the nine-dimensional hyperdistance ρ normalized to that of a non-interacting ideal gas, with the prefactor in front being the solid angle in nine dimensions. We neglect repulsion and integrate over the region in which the integrand is positive. The addition to the free energy is then $\Delta(-T\log(Z)) = -Tn^3 V_{\text{cor}}^{(9)} \frac{N}{4^4}$, same as to the grand partition sum. Differentiating it with respect to μ, present in each N, one finds the addition to particle number $\Delta N/N = n^3 V_{\text{cor}}^{(9)}/4^3$.

The magnitude of this effective volume depends on the temperature and density of the matter, and it was calculated in our PIMC simulations

(DeMartini and Shuryak, 2021d). For example, at kinetic freeze-out conditions of $\sqrt{s} = 7.7$ GeV, we found

$$V_{\rm cor}^{(9)}(7.7) \approx 4.3 \cdot 10^4 \,\text{fm}^9 \tag{6.45}$$

To put it in proper perspective, one can define the "density of the cluster" as

$$n_{\rm cl} \equiv \frac{4}{\left(V_{\rm cor}^{(9)}\right)^{1/3}} \tag{6.46}$$

which for $\sqrt{s} = 7.7$ GeV is $n_{\rm cl} \approx 0.114/\text{fm}^3$. This value is about three times the density of ambient matter $n_B(7.7) \approx 0.037/\text{fm}^3$.

Let us now return to the partition function and point out that, by general statistical mechanics rules, it describes not only the density of nucleons but also their distribution. All one needs to do is calculate higher derivatives over μ_b. "Kurtosis" of the distribution is given by the *fourth* derivative of $-T\log(Z)$ over μ, as given above. Since correction due to four-nucleon clusters is proportional to $\exp(4\mu_b/T)$, one finds that four derivatives lead to a large factor of 4^4. Therefore, preclusters contributing to the mean nucleon density at a subpercent level contribute to kurtosis at the $O(1)$ level. The prediction for the precluster contribution to kurtosis by DeMartini and Shuryak (2021d) is shown in Fig. 6.9.

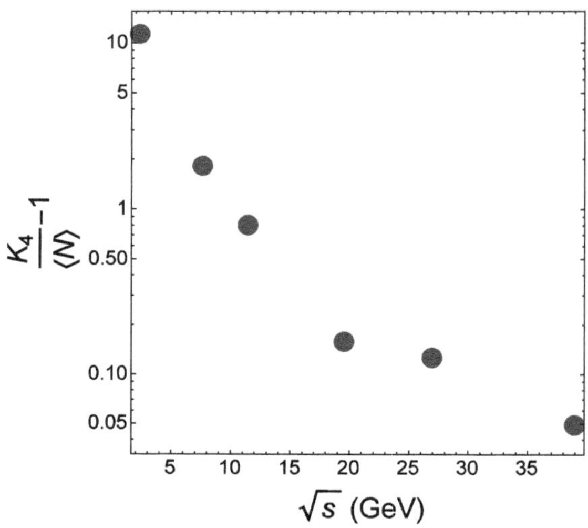

Fig. 6.9. The normalized fourth cumulant versus the collision energy, based on effective volumes $V_{\rm cor}^9$ derived in PIMC simulations.

6.5 Many-Body Forces Near the Critical Points

Let me start with a *paradox* (pointed out in Shuryak, 2006b): Consider the effect of *binary* forces induced by long-range critical mode *at* the critical point. If the critical order parameter is a scalar, its coupling to nucleons should generate *attractive* Yukawa-like potential,

$$V(r) \sim \frac{\exp(-r/\xi)}{r}$$

At the critical point $\xi \to \infty$, and this potential becomes infinite-ranged. A little thought shows that it should lead to catastrophic consequences. Indeed, if all $N(N-1)/2 \sim 10^4$ pairs of nucleons in the fireball were attracted to each other, by a Newton-like potential, the fireball would *implode*, like stars in a gravitational collapse.

A milder version of the same phenomenon has been extensively discussed. Various models of chiral symmetry restoration predict that at $T \to T_c$, the sigma meson mass is expected to decrease, as it is related to quark condensate. If so, the range of effective nuclear forces $1/m_\sigma$ should grow. Now, calculations (similar to those described in the previous section) show that it should lead to a drastic increase in nucleon clustering. For example, the binding of He^4 ($ppnn$) changes from its vacuum value (28 MeV) to the order of GeV if m_σ is reduced by a factor of two. If that were indeed true, nearly all nucleons would be clustered into He^4 and lighter nuclei.[10] We do not see it happening at any of the collision energies of the BES program.

It is thus clear that such paradoxical conclusions are in fact wrong. But why? It is because *multibody forces* were ignored.

Before we come to many-body forces, we need to calculate the effective potential *near* the critical line. Let us introduce the nonzero "current" term $J(x)\phi(x)$. Such a shift modifies the effective potential. In particular, the maximal value of the correlation length $\xi = 1/m$ becomes limited. Also, the $\phi \to -\phi$ symmetry is broken and odd powers of ϕ appear. As we now detail, it turned out to be very important for the estimated many-body forces. As an example, for $\tilde{J} = 1/100$, we perform this procedure for various values of t. In particular, the real roots of this equation are

$$\tilde{\phi}_0(t = 0.01) \approx 0.224, \quad \tilde{\phi}_0(t = 0.41) \approx 0.031$$

[10]This is indeed happening in the Big Bang nucleosynthesis era, but at much lower temperatures $T \sim 0.1\,\text{MeV}$.

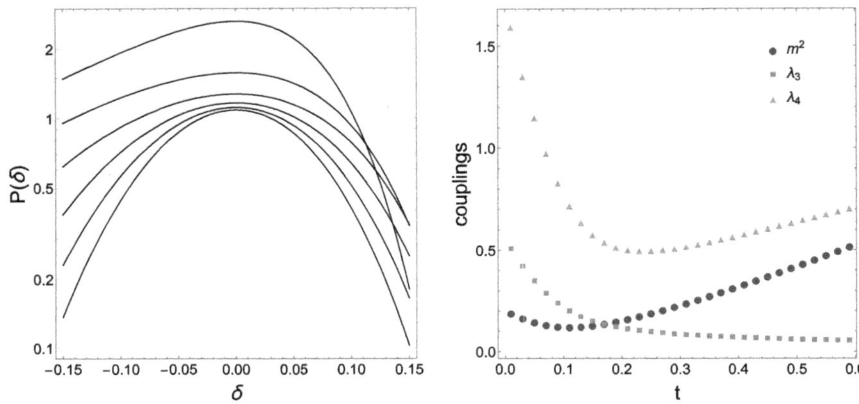

Fig. 6.10. (Left) Probability distributions of the deformed action $P(\delta)$ distorted by $\tilde{J} = 0.01$ in the four-nucleon cluster volume. The six curves, from top to bottom, correspond to values of $t = 0.01, 0.09, 0.17, 0.25, 0.33, 0.41$, respectively. The 3D volume and mass prefactors are explained in the text. (Right) The dependence of $m^2 = 1/\xi^2$ and the triple and quartic couplings for the effective action deformed by $J = 1/100$ on scaled temperature $t = (T - T_c)/T_c$.

We then rewrite the fluctuation field in the form

$$\tilde{\phi} = \tilde{\phi}_0 + \delta \tag{6.47}$$

The dependence of $m^2 = 1/\xi^2$ and the triple and quartic couplings from the deformed effective action on t show significant growth of these couplings near CP ($t \to 0$). Note also that at small t, the inverse correlation length m does not tend to zero, although it remains small.

The calculation of these diagrams for four-nucleon clusters has been made by DeMartini and Shuryak (2021c), from whom we borrowed Fig. 6.12. One can see that many-body repulsive interactions overcompensate the binary attraction, killing the clustering very close to the critical point.

Let us now proceed to the status of experimental observables. The summary of kurtosis data for the net proton distribution is shown in Fig. 6.13. The upper plot is an earlier summary from Adamczewski-Musch *et al.* (2020); the lower one is from recent a STAR publication (Adam *et al.*, 2021). More strict event selection has basically modified one point, the central bin of 7.7 GeV run.

When interpreting summary plots, one should keep in mind that the detectors involved in these and the following plots — STAR at BNL

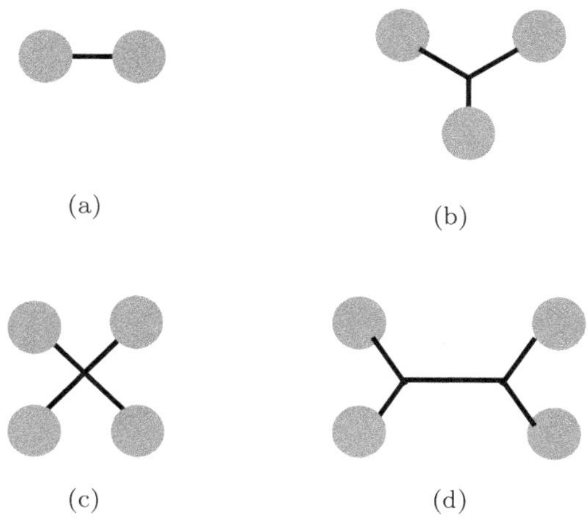

Fig. 6.11. Diagrams representing the many-body interactions of the four-nucleon cluster. Blue circles are nucleons, and black lines are propagators of the ϕ fields.

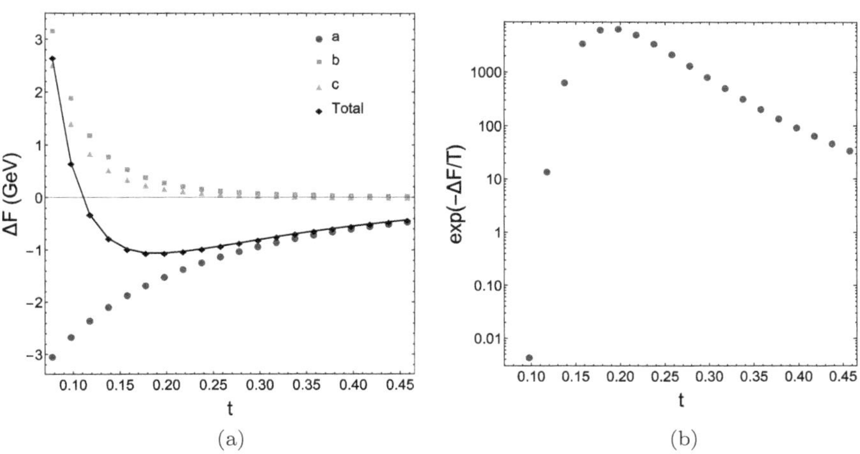

Fig. 6.12. (Left) Contributions to the change in free energy ΔF of a four-nucleon cluster of size $\rho = 2$ fm due to individual diagrams (a, b, c) and their total combined contribution, with coupling $g_c^2/4\pi = 10$ versus $t = T/T_c - 1$. Diagram (a) has had the unmodified binary interaction subtracted out, as described in the text. (Right) Boltzmann factor of the change in potential $\exp(-\Delta F/T)$ with $T = 120$ MeV versus t.

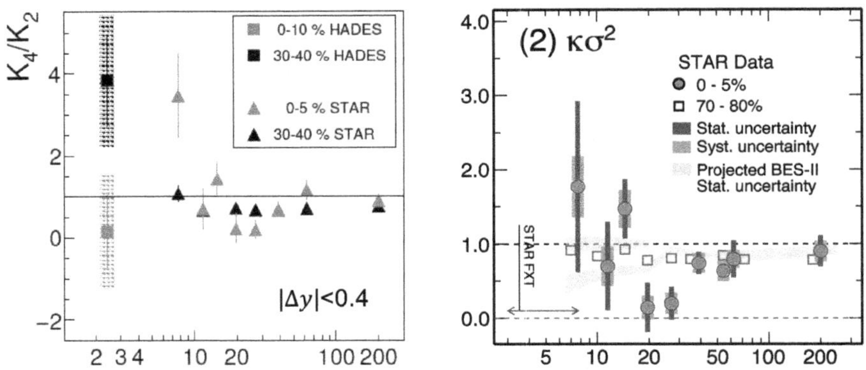

Fig. 6.13. (Upper plot) The kurtosis ratio K_4/K_2 from STAR and Hades experiments versus $\sqrt{s}$: Red points show the most central bins, and black points show the mid-central one, 30–40%. (Lower plot) More recent STAR results.

RHIC, HADES at GSI, NA49 at CERN SPS and ALICE at LHC — have completely different kinematic settings and acceptances and use different extrapolation procedures. Therefore, the comparison of their points needs to be done with care. As one can see, the errors are still large. The right plot indicates the projected accuracy of the BES-II program (green shaded area near 1).

On the other hand, *inside each group*, the centrality bins are supposed to be processed in exactly the same way. If one trusts the centrality dependence of each set, one finds a striking *reversal* between the STAR centrality dependence at 7.7 GeV and that reported by HADES at 2.4 GeV. Also, both plots show clear depletion near 20 GeV in the central bins relative to the peripheral ones. The behavior is changed in recent STAR BES data, shown in the lowest plot (from STA, 2022b), with specific centralities and smaller errors. Now, there is a clear *maximum* near 20–30 GeV, with a statistical significance of about 4.6 sigma.

Another observable sensitive to pre-clustering is additional feed-down into the production of light nuclei. The most sensitive to four-nucleon clusters are t and ^3He. The compilation of the data for the ratio of yields N_tN_p/N_d^2 from Shuryak and Torres-Rincon (2020b) is shown in Fig. 6.14.

This ratio was selected because, in it, the main driver of light nuclei yields — the factors of fugacity $\exp(\mu/T)$ — cancel out. The upper-left plot, from Zhao *et al.* (2020), is a theoretical result from a state-of-the-art cascade/coalescence code, which reproduces many features of heavy ion

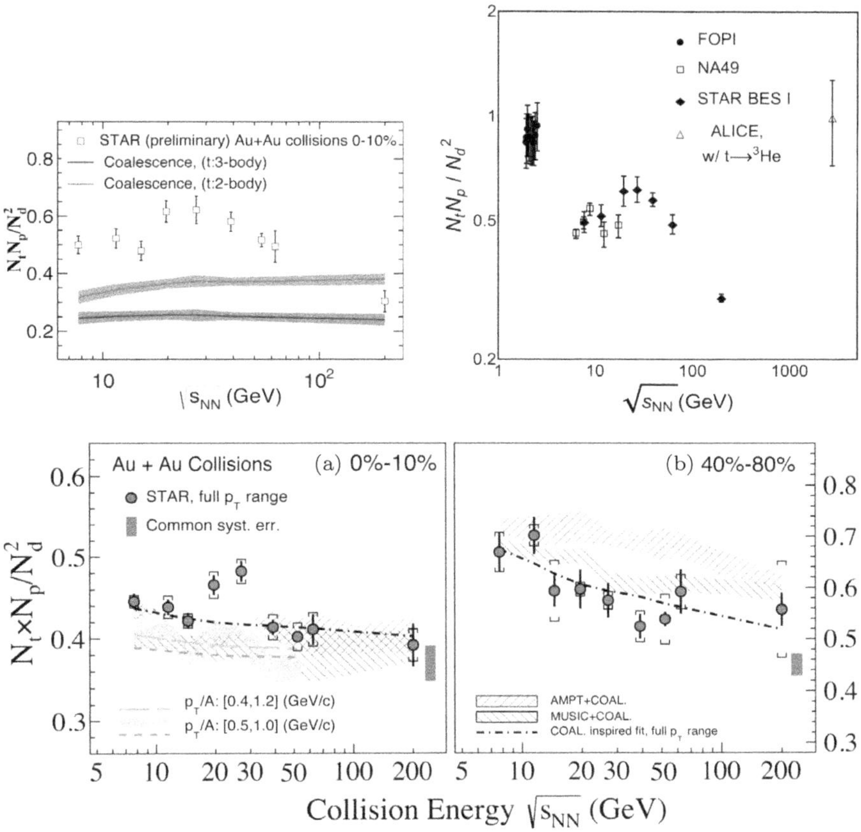

Fig. 6.14. Energy dependence of the $N_t N_p/N_d^2$ ratio. Predictions from a cascade code and preliminary STAR data (upper plot); Compilation from several experiments is indicated (upper-right plot), latest BES data for the central (a) and peripheral (b) sets (lower plot).

collisions. As one can see, it indeed predicts basically no dependence of this ratio on the collision energy.

However, its predictions are well below STAR data points, except the rightmost point, which is consistent with the simple thermal value for these nuclei. We recall that t, p, d have 2, 2, 3 spin states, respectively, and also that the Maxwellian distribution integral for a particle of mass M is $\sim M^{3/2}$. Combining these factors, one expects

$$\frac{N_t * N_p}{N_d^2} = \frac{2*2}{3*3} * \frac{3^{3/2}}{2^3} = \frac{1}{2\sqrt{3}} \approx 0.288 \tag{6.48}$$

Furthermore, we know that the direct thermal production of each of these species is also subject to *feed-down* from resonance decays. The best studied of those are feed-down corrections for the proton yield: It should increase the ratio by about a factor of two at high energies, and less so at low energies (since chemical temperature decreases there). The *feed-down* contribution to tritium is what we discussed above: The decays of 4-N resonances are supposed to increase it by another factor of two, especially at small energies. The deuteron is expected to have little feed-down; however, its yield is complicated by issues related to its large size.

Let us now return to the experimental data in the last figures. The upper-right plot is a compilation of data on this ratio from several experiments. The extrapolations and other experimental procedures are different; therefore, the systematic errors at those points are difficult to estimate, so the pattern is unclear.

The lower plots are from the latest STAR BES data: All systematic corrections must be common to all points, so it should be better suited for our purposes.

The first feature seen is that the ratio *grows* to the left, toward the lower collision energies. This is how it should be if it is due to an *increase in tritium production*[11] via feed-down from four-nucleon resonances. Indeed, their thermal production is not $\sim\exp[3(\mu_b - m_N)/T]$ but $\sim\exp[4(\mu_b - m_N)/T]$, and the feed-down correction thus has an additional fugacity, increasing at small collision energies.

The second feature is two outstanding points for central collisions (lower-left plot) at a collision energy of 20–30 GeV: Their total statistical significance is rather large, about 4.6 σ. Note further that their energy locations correspond to the *softest point* of hydrodynamics and the maximal emission time we have seen in Fig. 14.6. The location and width of the peak crudely match.

Summary and connections: Critical phenomena near second-order phase transitions are much more interesting than those near the first-order ones. We reviewed the history and current status of its theory and applied it to a search for the QCD critical point. Higher-order susceptibilities are related to event-by-event fluctuations in heavy ion collisions. The original suggestion to use this effect was to use the *Beam Energy Scan* and look for

[11]The ratio, of course, can also grow due to a *decrease* in deuterium yield in the denominator: A verification of what is happening is still lacking.

any non-monotonous pattern in fluctuation observables. The data are still not accurate enough to say with confidence whether or not it is found.

The second observable sensitive to the existence of the critical mode is the production of *light nuclei*. There seems to be a clear enhancement of the $N_t * N_p/N_d^2$ ratio for central collisions at $\sqrt{s} \sim 20$–$30\,\text{GeV}$, perhaps correlated with a smaller dip in kurtosis. In this region, $\mu_b \sim 300$ MeV, perhaps too small for a critical point. Yet, it fits well well what we know about the *softest point* of the EOS.

Part 2

HEAVY ION COLLISIONS

Chapter 7

HEAVY ION COLLISIONS

7.1 Heavy Ion Terminology

In an ideal course, the field is presented as a logical tree and all terminology introduced gradually, with a simultaneous discussion of the corresponding physical ideas. However, when describing such complex phenomena as heavy ion collisions, it is difficult to avoid the language used in the field, "prematurely." In order to resolve this chicken-or-egg problem, we begin with a brief introduction of the terminology.

"Ion" in physics generally refers to atoms with some of their electrons missing. While at various stages of the acceleration process the degree of ionization varies, electrons are unimportant for the nuclear collisions to be discussed. Therefore, one should think of collisions performed with nuclei fully stripped of all electrons. In short, in the following, *ions* are simply nuclei.

By **"heavy ions,"** we mostly mean gold Au^{197} (the only stable isotope of natural gold, a favorite at BNL accelerators) or lead Pb^{208} (the double magic nucleus used at CERN). Some experiments with the even heavier uranium U^{235} have been done, not so much because of its size but rather to observe the consequences of its strong deformation.

Light ions with which collisions have been studied are d, He^3 (RHIC) and C, O (SPS). The SHINE collaboration is studying what they call the *two-dimensional phase diagram*, with one of the coordinates being the system's atomic number A. They also investigated (central) collisions of intermediate nuclei and found that many features of $BeBe$ ($A = 9$) are very close to pp, while the Ar beam ($A = 40$) is already close to $PbPb$ ($A = 208$). This is often called "the onset of a fireball."

"Isobars" are nuclei with the same number of nucleons A but different charges Z. In order to identify the effects of electric charge (such as the chiral magnetic effect), RHIC operated with two beams: $^{96}_{44}Ru$ and $^{96}_{40}Zr$. We will discuss what we learned from a comparison of these two isobars.

Collision energy needs to be defined uniquely to facilitate comparison between "fixed target" and "collider" settings. Whatever ions are colliding, we use the relativistic invariant

$$\sqrt{s_{NN}} = \sqrt{(p_1 + p_2)^2} \tag{7.1}$$

the total energy of *one pair of nucleons* at the center of mass.

Collision centrality, in theory, is defined via an *impact parameter* b, a displacement from a line going through the centers, defined far from the collision region. It is a classical concept: the angular momentum $L = p * b$. In quantum mechanics, it is quantized to the integer $l = L/\hbar$ (in units of the Plank constant). However, collisions at the very high energies we discuss have such a large angular momentum that the uncertainty in b is negligibly small.

In experiments, one defines *centrality classes*, usually given by a percentage of the total cross-section. One divides events into histogram bins with a predetermined fraction of events. For example, we will see many plots with captions including something like "centrality 20–30%": This means that the total sum $\sum_n P_n$ is taken to be 100%, the events are split into say 10 bins, numerated 0–10%, 10–20%, 20–30%, etc., and only events from the particular one are used on the plot under consideration. The most central bins have the largest multiplicity and are always recorded, while peripheral ones at the impact parameter $b \approx 2R$ (denoted as, for example, 80–100%) are often not even recorded. While the observables, such as mean multiplicity, decreases monotonically with centrality b, this is not true for individual events. Another definition of centrality may use either distribution over *transverse energy*

$$E_\perp = \sum_{\text{particles}} \sqrt{p_\perp^2 + m^2}$$

or distribution over the observed number of *spectators*. Conflicting possible definitions of the centrality classes may seem confusing. Indeed, multiple dedicated studies via models (e.g. the Glauber model) of the initial state were needed to relate distributions over different observables. In practice, mapping between various centrality measures is well established, and any one of them can be safely used.

The number of "participant" nucleons N_{part} plus the number of "spectators" N_{spect} give the total nucleon number $2A$. Neutron spectators propagating along the beam direction are typically recorded by special small-angle calorimeters in one or both directions from the collision point. Two-dimensional distributions over signals from such calorimeters are cut into bins of special shapes, also in a way that each bin retains a fixed percentage of the total. Small corrections for nucleons suffering only small-angle elastic and diffractive scatterings not counted as "participants" are also made.

The **overlap region** is a region in the transverse plane in which the participant nucleons are located at the moment of the collision. Note that due to strong relativistic contraction, high-energy nuclei can be viewed as 2D objects, with the longitudinal size reduced by the Lorentz factor (by 2–3 orders of magnitude at RHIC/LHC). Therefore, the "collision moment" is quite well defined and is the same for all nucleons. Crudely, one may think of the overlap region classically as an almond-shaped intersection of two circles. Note that its shape changes from a circle for central collisions to a highly deformed one for peripheral collisions.

Azimuthal harmonics are Fourier coefficients of the expansion in azimuthal angle ϕ_s of the distribution of *secondaries* in *transverse* momentum $\vec{p}_\perp$:

$$\frac{dN}{dydp_\perp^2 d\phi_p} = \frac{dN}{dydp_\perp^2}\left[1 + 2\sum_m v_m(p_\perp)\cos\left(m\phi_s - m\phi_{\text{event}}\right)\right] \tag{7.2}$$

The most important harmonics are the elliptic ($m = 2$) and triangular ($m = 3$), although there is meaningful dataset up to $m = 10$ harmonics.

In theory, ϕ_{event} is the direction of the impact parameter vector $\vec{b}$. The direction of $\vec{b}$ and the beam define the *collision plane*, traditionally denoted by x, with the orthogonal direction denoted by y and the beam direction by z.

In experiments, one has to find it on an event-by-event basis, and it may differ according to the harmonic number m, which reflects the fluctuations in the initial state. The angle ϕ_{event} is defined using all (or several) other secondaries. More precise experimental definitions involve two or more particle correlators at azimuthal angles. In practice, the event plane detection originates either from separate "near beam" calorimeters, recording "spectator" nucleons, or from correlation with other particles.

The azimuthal harmonics describe collective flows in the system, introduced as a response to the *spatial asymmetry* parameters ϵ_m, the Fourier components of the matter distribution at the collision moment. Note that v_m relates to momentum distribution at the end of the process and ϵ_m to spatial distribution at the initial time: Establishing a connection between them is the main issue, which will be discussed in detail.

Rapidity is a relativistic generalization of velocity. Unlike velocity, which does not exceed the speed of light and is defined between -1 and 1, rapidity is unlimited. Longitudinal rapidity y is defined as

$$y = (1/2)\log\left[\frac{E+p_l}{E-p_l}\right] \tag{7.3}$$

with the energy-to-mass ratio $E/M = \cosh(y)$ known as the gamma factor, and $p^l/M = \sinh(y)$, where p_l is the longitudinal momentum. Rapidity is related to the longitudinal velocity by $y = \tanh^{-1}(v_l)$.

In a few instances, we also use *transverse* rapidity, defined similarly as $\kappa = \tanh^{-1}(v_\perp)$.

Rapidity related to momenta should not be confused with the so-called *space-time rapidity*, a similar combination of coordinates

$$\eta = (1/2)\log\left[\frac{t+z}{t-z}\right]$$

we use in relativistic hydrodynamics.[1] With the (longitudinal) proper time $\tau = \sqrt{t^2 - z^2}$, space-time rapidity η form a convenient $1+1$-dimensional set of coordinates, appropriate for the description of ultrarelativistic collisions.

The rapidity transforms additively under the longitudinal Lorentz boosts

$$y \to y + Y, \quad \eta \to \eta + Y,$$

This is so because it is the (hyperbolic) angle of the Lorentz transformation

$$L = \begin{pmatrix} \cosh(Y) & \sinh(Y) \\ \sinh(Y) & \cosh(Y) \end{pmatrix} \tag{7.4}$$

acting on (p^4, p^3) or (t, z).

The simplest longitudinal flow pattern is "Lorentz invariant $1+1$-dimensional Hubble flow," for which $y = \eta(t, z)$.

Collectivity, in a non-technical language, means that most secondaries (if not even *all* of them) participate in a certain dynamical phenomenon.

[1] Unfortunately, shear viscosity is also designated by η; hopefully, those two cannot be confused.

Collectivity of flow implies that the multi-particle distributions approximately factorize into the product of similar single-particle ones, e.g.

$$\frac{dN}{d\vec{p}_\perp^1 d\vec{p}_\perp^2 \dots d\vec{p}_\perp^N} \sim \prod_{i=1,\dots,N} \frac{dN}{d\vec{p}_\perp^i} \tag{7.5}$$

so that any particle has the same momentum distribution as any other.

In more technical terms, flow harmonics were originally derived from the *two-particle correlations in relative angle*, to which they enter as a mean squared value:

$$v_n^2\{2\} = \langle e^{\mathrm{i}n(\phi_1-\phi_2)}\rangle = \langle |v_n|^2\rangle \tag{7.6}$$

and v_n thus defined are the r.m.s. of this quantity. Alternatively, it can be derived from certain multi-hadron correlation functions, such as those for four and six particles which are most commonly used, defined via

$$v_n^4\{4\} = 2\langle |v_n|^2\rangle^2 - \langle |v_n|^4\rangle \tag{7.7}$$

$$v_n^6\{6\} = \frac{1}{4}(\langle |v_n|^6\rangle - 9\langle |v_n|^2\rangle\langle |v_n|^4\rangle + 12\langle |v_n|^2\rangle^3) \tag{7.8}$$

By "collectivity," one means an approximation according to which *all* of such measurements produce nearly the same values:

$$v_n \approx (v_n\{2\})^{1/2} \approx (v_n\{4\})^{1/4} \approx (v_n\{6\})^{1/6}$$

In contrast to that, the "non-flow" effects, such as the production of hadronic resonances like $\rho \to \pi\pi$, basically affect mostly the binary correlator $v_n\{2\}$ but not the others. This fact is used to "remove the non-collective effects."

When we mention **soft and hard secondaries**, we refer to their *transverse* momenta, not the total. They have different dynamical origins. The "soft" ones come from the thermal heat bath conditions at "freeze-out," boosted by collective flows, while the "hard" ones come from hard partonic reactions and subsequent jet fragmentation. The boundary is not well established and depends on the reaction. For orientation, by "soft," we will mean $p_\perp < 4\,\mathrm{GeV}$, while "hard" are perhaps those with $p_\perp > 10\,\mathrm{GeV}$. (Yes, there is "a nobody's land" in between.)

Fireball, in common language, is something bright and shining. Yet, in the terminology presented, it is actually something *dark* — the region of space-time from which no secondaries originate. One can relate it to the interior of the Sun or even black holes. They may be very hot, but all we see outside is the radiation from *their surface*.

Yet, some particle species interact so weakly that they easily get out of the *bulk* of the fireball. For example, solar neutrinos reach us from the Sun's center, while photons cannot. The described fireball also has *penetrating probes*: photons and dileptons. Their emission yield is given by an integral over space-time four-volume rather than the 3D surface.

Chemical and kinetic freeze-outs refer to stages of the explosion at which the rates of the *inelastic* and *elastic* collisions become *smaller than the rate of expansion*, so that certain reactions are effectively stopped. These ideas are also applied to many other explosions. For example, in the Big Bang, the background radiation photons stopped their scattering when the Universe got much cooler than the atomic scale of $T \ll 1\,\text{eV} \approx 10^4\text{K}$, and plasma recombined into the neutral atoms we see today. Since that time, photons have mostly propagated freely, only changing their wavelength due to the expansion of the Universe. Today, they are at temperature $T = 2.7\,\text{K}$, the red-shifted ones were emitted billions of years ago at the freeze-out stage.[2]

At the *chemical freeze-out*, particle composition (somewhat resembling the chemical composition of matter) is finalized. The *kinetic freeze-out* is the surface where the final scatterings happen. It is similar to the photosphere of the Sun. These three-dimensional surfaces are usually approximated by *isotherms* corresponding to certain values of temperature. The final particle spectrum is usually defined as (the so-called Cooper–Frye) integral of the local thermal distribution over these surfaces.

Femtoscopy, also known as *HBT interferometry*, is a method of fireball size detection using particle correlations. It originally came from radio astronomy: HBT is an abbreviation for Hanbury-Brown and Twiss, who developed it for measurements of stars' radii.[3]

This method allows us to measure the fireball's dimension and even "see" its explosion as a function of time.

The influence of Bose symmetrization of the wave function of the observed mesons in particle physics was first emphasized by Goldhaber *et al.* (1960) and applied to proton–antiproton annihilation. Its use for the determination of the size/duration of the particle production processes

[2]This temperature is measured with about seven digits now; it is perhaps the most accurately measured temperature ever.

[3]A recent achievement of the consortium of astronomers using 1 mm waves at different locations on Earth has resulted in a famous picture of a giant black hole at the center of our nearby galaxy.

had been proposed back in the 1970s (Kopylov and Podgoretsky, 1973; Shuryak, 1973a). With the advent of heavy ion collisions, this "femtoscopy" technique has grown into a large industry. Early applications for RHIC heavy ion collisions were in certain tension with the hydrodynamical models, although this issue was later resolved; see, for example, Pratt (2009).

The **initial state** is defined as a state in which (relativistically contracted) disks have just *passed* each other. One of the models describing it is called the *GLASMA* model, which describes matter as a highly excited near-classical chaotic gluon field. It is followed by the **equilibration** period, in which the occupation numbers of gluons and quarks become $O(1)$ and matter motion approaches certain solutions to hydrodynamics. Although its duration is the subject of debate, the order of magnitude is about 1 fm/c.

Hydrodynamical explosions mark the next period: They last for a longer time, up to $\tau \sim$ 10–20 fm/c and gets terminated by the two subsequent freeze-outs mentioned above, *chemical* and *kinetic*.

7.2 Geometry of the Collisions

Let us begin the discussion of heavy ion collisions with simple geometrical considerations. For simplicity, one can first view colliding nuclei as spheres with a homogeneous distribution of matter inside. In the rest frame of the nucleus, the nuclear matter density is approximately constant over the volume, but for a moving frame, one has to project to the 2D plane transverse to the beam. An ultrarelativistic nucleus can thus be viewed as a disk with a variable density tending to zero at the edges.

Generic collisions have a nonzero *impact parameter* (the shortest distance between two centers) $\vec{b}$, which changes from zero (central collisions) to twice the nuclear size $2R_A$ (highly peripheral collisions).

A sketch of non-central collisions is shown in Fig. 7.1, which explains how one divides nucleons into "spectators," which fly by without collisions, and "participants," which are indeed colliding. Note that all participants are located in the almond-shaped *overlap region*. It is from this shape which the explosion of superdense matter originates.

Only for *central* collisions $\vec{b} = 0$, the system is azimuthally symmetric, whereas for non-central ones, the initial almond shape generates a certain anisotropy of the resulting flow of matter. They are characterized by expansion in harmonics in azimuthal angle ϕ, most notably the "elliptic" flow with the distribution $\sim \cos(2\phi)$. The left–right symmetry of the picture tells us that only even harmonics would be present.

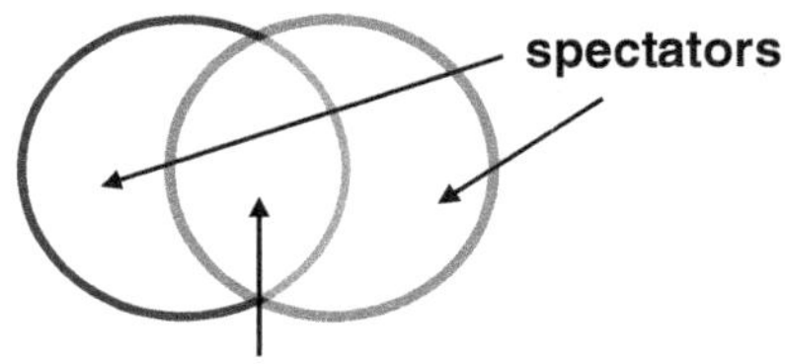

Fig. 7.1. A sketch of the non-central collision of two identical spherical nuclei in the transverse plane. Participants are located in the overlapping almond-shaped region.

One can change this simple picture to a significant extent by using *asymmetric collisions* of two different nuclei, with $A_1 \neq A_2$. Naturally, those have different radii. In this case, there is no left–right symmetry, and odd harmonics in the azimuthal angle ϕ are also generated.

An extreme case is the pA collisions between one proton and heavy nucleus. Both RHIC and LHC have studied those. Depending on the impact parameter, the number of "participants" varies, with the maximal number at $b = 0$ being about $N_{\text{part}} = 16$. In this case, the excited matter has a transverse extent of only one nucleon, and we call it a "small system." The interesting question is whether such systems are still "macroscopic," displaying collective flows and other features similar to those in "large systems."

In fact, one can also play with the initial geometry, changing it greatly. At the RHIC, dAu collisions have been studied. The deuteron d consists of two nucleons that are relatively far from each other. So, one can have "two-center" explosions, with the initial shape enhancing the transverse quadrupole moment and elliptic flow. It still creates a common fireball, but a very deformed one in the transverse plane. He^3Au has also been done with the idea of enhancing the "triangularity" of fireballs.

Even more interesting/complex geometries appear if one collides strongly deformed nuclei (Shuryak, 2000; Heinz and Kuhlman, 2005). Possible collision geometries are illustrated in Fig. 7.2. The first one, called head–head, is a central collision in which, in the transverse plane, one finds a *minimal* cross-section πR^2_{min}. Those can be selected because they produce the highest energy density (about 30% higher than in $PbPb$ collisions). The last is a *maximally deformed* collision, in which the longest axes are parallel to each other, and the transverse plane area is $\pi R_{\text{min}} R_{\text{max}}$.

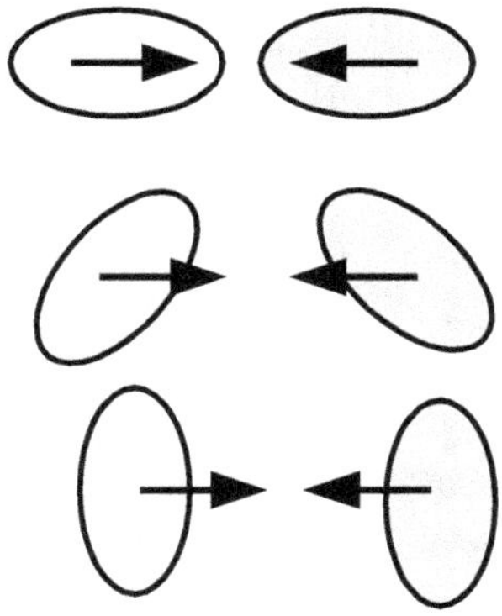

Fig. 7.2. A sketch of possible geometries for the collision of two deformed nuclei.

The first and last sketches are central collisions $\vec{b} = 0$, which can be selected by applying a veto condition to spectators.[4] For comparison, the second picture, corresponding to some generic collisions, has them. Using different cuts in the three observables of the events — $N_{\text{spectators}}, N_{\text{scondaris}}, v_2$ — one can single out particular geometries.

This was indeed done with an isotope of uranium ^{235}U about 30% deformed); see, for example, Huang (2013).

The relation between the initial sizes and shapes of the fireball and the observed distribution over particle momenta will be the subject of subsequent chapters. Note that detectors directly measure momenta and their directions. The angles to the beam direction $\theta_{\mathbf{p}}$ are related to longitudinal *rapidities* y. The azimuth angles are called $\phi_{\mathbf{p}}$, and we will be mainly discussing distributions in those.

Generally, for causality reasons, the constituents of colliding systems — *partons* (quarks or gluons) — have an isotropic initial moment distribution. Indeed, they do not yet "know" about an almond-shaped fireball region. It is only due to subsequent interactions, eventually leading to *hydrodynamical collective flow*, that any anisotropy in ϕ is *converted* into the observed anisotropy.

[4]Note that it is the spectators who are responsible for a strong magnetic field at the moment of the collision. By vetoing those, one removes any effects associated with the magnetic field.

7.3 Initial State and Its Fluctuations

In reality, nuclei are far from homogeneous spheres. Let us briefly recall some properties of *nuclear matter*. It is, in fact, rather dilute since the relative scale rather than the absolute one is important. Multiplying the nuclear matter density $n \approx 0.16\,\text{fm}^{-3}$ by the volume of the nucleon $V_N = 4\pi r_N^3/3 \approx 2.76\,\text{fm}^3$ for $r_N = 0.87\,\text{fm}$, one finds only 0.44 of the volume occupied, and the rest is empty. Also, the probability of a nucleon interacting is close to one only inside the "black disk,"[5] which, even at LHC energies, is not so large. Most of its transverse area is "gray," with a probability of interaction between zero and one.

As a result, using the so-called *Glauber model* of multiple NN collisions, one finds initial states which look like the two examples shown in Fig. 7.3 (from Bozek *et al.*, 2011). The points indicate the random positions of *wounded nucleons* at the collision moment. Initial "matter distribution" is a certain smoothed function reflecting these positions. The standard procedure now is that, for each simulated event, one defines two sets of parameters ϵ_n, Φ_n, the amplitudes and phases of the initial state

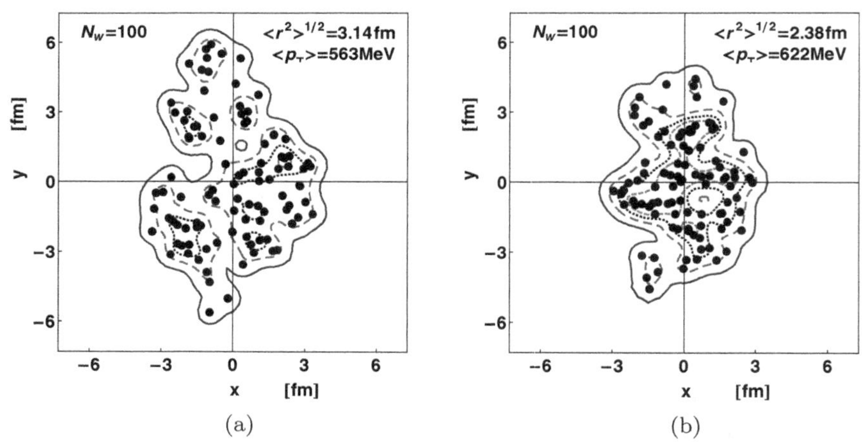

Fig. 7.3. Two typical wounded nucleon events, for $Au + Au$ collisions with $N_w = 100$ wounded nucleons. Lines indicate contours of constant entropy density after smearing the wounded nucleons with Gaussian distributions.

[5] To be discussed in the chapter on Pomerons.

deformation, via

$$\epsilon_n e^{\mathrm{i}n\Phi_n} = -\frac{\int drr^{n+1}d\phi e^{\mathrm{i}n\phi}e(r,\phi)}{\int drr^{n+1}d\phi e(r,\phi)} \tag{7.9}$$

where $e(r, \phi)$ is the energy density distribution. Note that the r.h.s. can also be understood as a sum over the wounded nucleons, as $e(r, \phi) \sim n_w(r, \phi)$. Note also that this definition includes an extra factor r^n, so that the higher harmonics are dominated by the fluctuations at the fireball's *surface*.

The average value over the ensemble of events is shown in Fig. 7.4(a) (from Heinz and Snellings, 2013) as a function of n, the harmonic number.

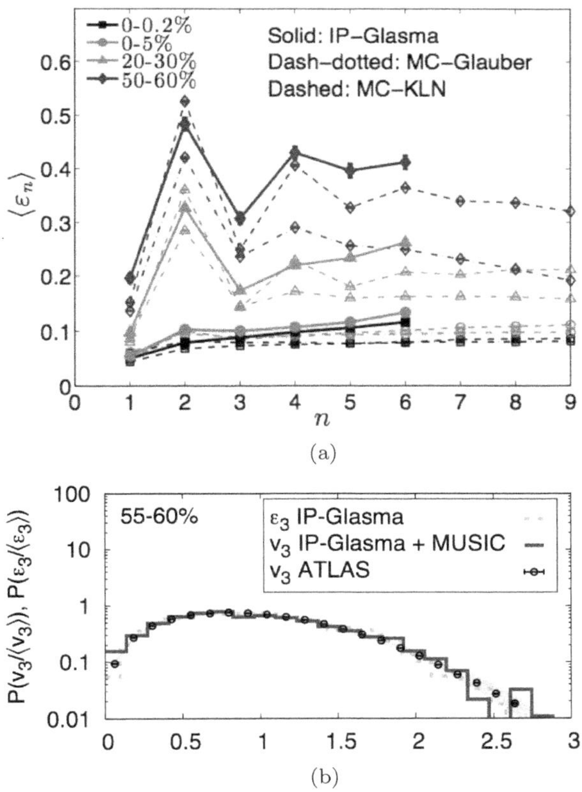

Fig. 7.4. (a) Average deformations $\langle \epsilon_n \rangle = \langle \cos(n\phi) \rangle$ for various centralities using the models indicated in the figure. (b) Data points correspond to the event-by-event distribution of v_3 measured by the ATLAS collaboration, compared to the distributions of initial eccentricities in the IP-Glasma model and the distributions of v_3 from fluid dynamic evolution with IP-Glasma initial conditions.

The peaks at $n = 2, 4$ for peripheral collisions are reflections of the geometry discussed above, and the overall flat distribution indicates that the fluctuations are caused by "local perturbations" with a small angular size. Recall that a flat harmonic spectrum is a consequence of the delta functions in coordinates.

The normalized distribution over one of the amplitudes, ϵ_3, and the corresponding observed distribution of the "triangular flow" v_3 are superimposed in Fig. 7.4(b) (from Schenke and Venugopalan, 2014). The flow harmonics are defined via

$$v_n(p_t)e^{\mathrm{i}n\Phi_n} = \langle e^{\mathrm{i}n\phi_p} \rangle_{p_t} \tag{7.10}$$

where the angle mean average over secondaries, which belong to a bin with a certain transverse momentum p_t.

Note that the shapes of those two distributions match well. So, v_n and ϵ_n are related on an *event-by-event basis*. The first lesson from this observation is that all fluctuations observed do come from the initial stage. The second lesson is that collective flows work "classically" (completely predictably), providing no noise of their own.

7.4 Why are High-Energy Heavy Ion Collisions Qualitatively Different from *pp* Ones?

The key idea driving this field is that *AA* collisions are *qualitatively different* from *pp* collisions. Since its birth, it has met enormous resistance in the high-energy community. Heavy ion collisions are "Little Bangs" in which (approximately) *equilibrated superdense matter* is produced, creating *collective explosions*. In contrast, *pp* and other hadronic collisions produce systems too small to be described in that way.

In fact, this debate started in the 1950s, with the "Little Bang" proposed for *pp* collisions first. It was rejected in the 1960s and has been replaced by a Reggeon exchange theory, eventually leading to the dynamics of QCD flux tubes. We will discuss it in the chapter on Pomerons later.

My personal story also began with the bold assumption that *pp* collisions *may* be explosive. In the 1970s, Zhirov (my first graduate student) and myself were looking for signs of collective effects in *pp* collisions using the first data from the first hadronic collider, ISR. We compared the transverse momentum spectra of various secondaries, π, K, p. These spectra turned out to be just thermal, with no sign of collective flows.[6]

This course is about *heavy ion physics and quark gluon plasma.* There were three different (but, of course, interrelated) aspects of its development. One was the *theoretical* path to QGP, coming from the 1970s after the discovery of QCD. It started first in its perturbative form and then in a nonperturbative theory. The development of QCD at finite temperature and/or density led to the realization that QGP is a *qualitatively new* phase of matter. As we will discuss later, QGP is a very peculiar plasma with rather unusual kinetic properties.

Now, work in this direction includes not only a certain number of theorists, specializing in QFTs and statistical physics but also a community performing large-scale computer simulations of *lattice gauge theories* and rather sophisticated models based on them. This activity has also expanded and now includes the collaborations of dozens of people.

The second (and now perhaps dominant) aspect of the quest for QGP is the *experimental* one. Let me here only mention that experimental activity is now dominated by five large collaborations: STAR and PHENIX (now under complete rebuilding of the detector) at the Relativistic Heavy Ion Collider (RHIC) in Brookhaven National Laboratory, and ALICE, CMS and ATLAS at the Large Hadron Collider (LHC) at CERN. The last two have been built by the high-energy physics community and are therefore designed for other purposes. Nonetheless, they work just fine with heavy ion collisions, adding complementary information on hard processes.

Each of these collaborations has hundreds of members, so the "Quark Matter" conference and other conferences on the subject have become huge in size and obviously dominated by hundreds of experimental talks. It is completely justifiable, as the list of discoveries — often puzzling or at least unexpected — continues to grow.

The third component of the heavy ion physics community — the glue which holds it together — are *phenomenologists*, who build models

[6] Heavy ions were never injected into ISR, only alpha-particles He^4 and, respectively, nothing significantly different from pp collisions has been detected. The ISR collider, which might have been an excellent prototype for RHIC, has been physically destroyed by a decision of the CERN management.

of the collisions. While, by now, those get very sophisticated, they are based on certain physical ideas, which we will discuss. In short, one can formulate their goal as converting experimental data on billions of detected events into the *properties of hadronic matter* under variable conditions.

A brief history of heavy ion physics can be divided, following a time-honored tradition of historians, into *three periods*, as follows.

7.4.1 *The "prehistory"*

The field dates back to the 1950s, when the very first observations of multi-particle production were seen in cosmic ray events. These first detectors were just stacks of emulsions exposed for a long time at some high mountains or even on the board of a commercial airplane, providing long exposures.

Three very famous physicists conjectured that a new form of matter can be produced, equilibrated and exploded in high-energy collisions. Fermi (1951) argued that a strongly interacting system of particles should rapidly equilibrate and predicted that multiplicity should grow with (CM) energy $\sqrt{s}$ as $N \sim s^{1/4}$. Pomeranchuk (1951b) immediately replied that if the interaction is strong enough for rapid equilibration, then no particles would be able to leave the system. He introduced what we now call the *freeze-out* concept. Landau (1953) elegantly connected the initial stage of Fermi with the final stage of Pomeranchuck via a description of *relativistic hydrodynamics* in between. Landau, in particular, pointed out that the expected quasi-adiabatic expansion leads to *entropy conservation*, thus preserving Fermi's prediction for entropy/multiplicity generation.

In a later review, Landau related these expectations of very strong coupling at high momenta scales, as expected from QED and other QFTs known at the time. This last argument was dramatically reversed two decades later, when the asymptotic freedom of non-Abelian gauge theories and QCD was discovered by Gross, Politzer and Wilczek in 1973. This discovery initiated studies of the high temperature, T, (or large chemical potential, μ, or both) limit of hadronic matter using perturbative diagrams and their resummations. It has been found that, unlike vacuum fluctuations of the gluon fields, which provide the asymptotic freedom or *anti-screening*, thermal fluctuations of both gluon and quark fields lead to *screening* in matter. Thus, it was called "quark-gluon plasma," or QGP for short. For early reviews on finite-T QCD, see Shuryak (1980) and Gross *et al.* (1981). Very hot QGP is now referred to as *weakly coupled* (wQGP).

Despite being the simplest from a theoretical perspective, it has not yet been attained even at the highest temperatures reached at the LHC collider.

While during the 1970s and 1980s the field and its conferences were mostly run by theorists, appropriate experimental observables were high on the agenda. Collective hydrodynamical *explosion in the transverse plane* and *penetrating probes* (photons and dileptons) were emphasized (Shuryak, 1980). Important QGP signals suggested in the 1980s were also the *jet quenching* by Bjorken (1982) and *sequential melting* of charmonium and bottonium states by Matsui and Satz (1986).

Experimental studies at Berkeley BEVALAC in the 1970s discovered thermal fireballs and some form of collective flows. Important experiments with heavy ion collisions in the 1980s used fixed targets at the CERN SPS and Brookhaven AGS accelerators. Yet, it is perhaps fair to say that only during this period did the experimental community start to be built. The large tasks awaiting it at the RHIC and LHC detectors were still just a dream then.

7.4.2 *The RHIC era*

By the end of the 1980s, the fateful decision had been taken by the Nuclear Science Advisory Committee (NSAC) to the DOE/NSF funding agencies, chaired by Professor Peter Paul (Stony Brook), to start building the Relativistic Heavy Ion Collider (RHIC) at Brookhaven National Laboratory. On the scale of nuclear experiments at the time, it was an unprecedented and monumental decision. One should also mention that, apart from its heavy ion program, it included scattering of *polarized* protons, with $\sqrt{s} = 500\,\text{GeV}$.

The start of RHIC operations in 2000 led to rapid discoveries. In a matter of days, it became apparent that particle spectra contained evidence of collective flows of magnitude compatible only with the QGP equation of state. The observed radial and especially *elliptic* flows confirmed nicely the predictions of hydrodynamics, which were by that time available for various EOSs. Most successful were the hydro codes supplemented by hadronic cascades (Teaney *et al.*, 2001a, 2001b; Hirano *et al.*, 2006), as they correctly took care of the final (near-freeze-out) stage of the collisions. Flows as a function of $p_\perp$, centrality, particle mass, rapidity and collision energy were found to be in good agreement with the explosion scenario. At the famous 2004 RBRC workshop in Brookhaven, with theory and experiment summaries collected in a special volume, *Nucl. Phys. A750*, the statements that QGP "is produced, and that it happens to be a *near-perfect liquid* which does flow hydrodynamically" were the main finding, which have since been repeated many times.

Another very important event that happened around 2005 was the recognition of a *strongly coupled regime* of QGP in these collisions, now called sQGP for short.[7] Hundreds of theoretical papers have been written, developing various approaches to gauge field dynamics at strong coupling.

A fortunate coincidence was that at the same time (late 1990s), the string theory community invented a wonderful theoretical tool, the AdS/CFT duality, connecting strongly coupled gauge theories to five-dimensional weakly coupled variants of supergravity. It shed entirely new light on the process of QGP equilibration, which is dual to the "collapse process" of (five-dimensional) black holes. The entropy produced in a Little Bang was, in particular, reinterpreted as the *information lost to outside observers*, as some part of a system is inside the so-called "trapped surfaces."

Another duality we will discuss in connection with sQGP is the *electric–magnetic duality*. This form of matter is in fact a "dual plasma," containing both *electrically charged* quasi-particles — quarks and gluons — and *magnetically charged* magnetic monopoles, such as certain solitons made of glue.

In addition, we briefly discuss other strongly coupled systems which have also been addressed by theorists and whose similarity to sQGP has been noticed. Those include strongly coupled classical QED plasmas at one end and quantum ultracold atomic gases in their "unitary" regime at the other. These studies focus on unusual kinetic properties, essentially the unusually small mean free paths which such systems display.

7.4.3 *The LHC era*

The last (and so far the shortest) era started in the year 2010, when the world's largest instrument of (high-energy/nuclear) physics, the *LHC at CERN*, started its operation, including a heavy ion program and the "quest for QGP."

---

[7]The ESQGP conference held at Stony Brook in the year 2008 was called "Extremely Strongly coupled QGP." In fact, there is no such thing, and it was a meeting related to my 60.

The collaboration, including nuclear physicists from many countries, developed a single detector, called ALICE, built specially for heavy ion conditions. It is capable of working with high multiplicities, exceeding 10,000 per event. Two other collaborations, CMS and ATLAS, although built mostly for high-energy physics goals, both include subgroups focused on heavy ions as well.

The first runs of the LHC were expected with obvious interest. The energy of this collider is about a factor of 20 higher than that of RHIC, which leads to about twice as much multiplicity. Hydrodynamics predicted collective flows to become even stronger. Indeed, at the LHC conditions, relatively "stiff" QGP plays a larger role than "soft" matter near and below the critical point T_c. On the other hand, entering the higher-T domain means increasing the momentum scale well over 1GeV; therefore, many colleagues expected the onset of asymptotic freedom and perturbative regime to moderate equilibration and explosions.

The first LHC data confirmed what has been learned at RHIC about enhanced collective effects. Due to their highly sensitive detectors and strong collaboration teams over the past decade, the LHC has added invaluable information to what we know about QGP properties. It is not restricted to heavy ion collisions. Perhaps, the most surprising discovery made at the LHC was that QGP explosions also happen in central pA and even high-multiplicity pp collisions. My review of the current status of the field (Shuryak, 2017) has recently appeared in the *Review of Modern Physics*. Many details can be found in it, a significant portion of which will be discussed in subsequent chapters.

Summary and connections: We introduced collision geometry and other terminology used in the description of heavy ion collisions. This chapter is only an overview of historical events. It will be followed by a discussion of the statistical models and *chemical* freeze-out in Chapter 8, in which the particle composition (fractions of different species and their absolute yields) is defined. It is followed by Chapter 9, which describes the collision end. We then describe some applications of relativistic hydrodynamics.

Chapter 8

CHEMICAL FREEZE-OUT AND THE COMPOSITION OF SECONDARIES

8.1 Introductory Remarks on Chaos, Entropy and "Equilibration" in Quantum Systems

Statistical mechanics textbooks begin with a *microcanonical distribution* based on the assumption that *all states in some selected energy interval* $E, E + \Delta E$ *occur with equal probability.* Accepting it, all one needs to do is calculate the *density of states* related to the thermodynamical entropy:

$$\rho(E) \equiv \frac{dN_{\text{states}}}{dE}(E) \equiv \exp(S(E)) \tag{8.1}$$

The *microcanonical formulation* of statistical mechanics leads to the introduction of the *temperature* defined by a derivative,

$$\frac{1}{T} \equiv \frac{\partial S}{\partial E} \tag{8.2}$$

and so on. It is important that ΔE is assumed to be relatively small and, at the same time, include a *very large* number of states:

$$\Delta E \ll E, \quad \Delta N_{\text{states}} = \rho(E)\Delta E \gg 1$$

If the number of particles in a system is macroscopic (say comparable to the Avogadro number $N \sim 10^{23}$, as is the case for traditional applications of statistical mechanics), it is indeed true that the number of states is huge and that any attempt to separate those and treat them individually would be completely hopeless.

The number of particles produced in hadronic collisions is typically not very large. In heavy ion collisions at semi-relativistic energies, it is $N \sim 10^3$. In what we will call in this book "small systems," multiplicities (per rapidity) are typically $N \sim 10^2$. These numbers are not so large.

Yet, what is important is not the number of particles but the *number of states* which can be estimated by

$$N_{\text{states}} \sim (\langle p \rangle R)^{3N} \tag{8.3}$$

where the typical momenta are $\langle p \rangle = 400$ MeV ≈ 2 fm^{-1}, with the fireball size at freeze-out R being several fm. Therefore, this number is $\sim 10^{3N}$. *Conclusion*: It is still so large that one may not worry about the applicability of statistical mechanics once conditions such as long-enough equilibration time are met.

Nonetheless, for pedagogical reasons, let me begin with a brief introduction of the foundations of statistical mechanics and the evolution in our understanding of its applicability limits. The starting point can be Boltzmann's kinetic theory of gases with his entropy growth H-theorem. Its explanation was given in terms of the *classical chaos* phenomenon. For example, if a system is initially confined to a small region of the phase space, dynamical instabilities exponentially increase its size in some directions $\sim \exp(\lambda t)$, so that with time, it can quickly reach and then much exceed the system's size. (Since, according to the Liouville theorem, the phase space volume must be conserved, in some other directions the sizes should exponentially decrease.) Therefore, the spectrum of λ, known as *Lyapunov exponents*, contains both positive and negative values. Chaotic behavior means that *any* initial distribution becomes so complicated that any predictability is lost. Applying a *coarse graining* (basically, a histogram) of the phase space distribution, one can substitute the actual complicated dynamical distribution for its smoother version. This substitution is what actually produces entropy. All this may not be well explained in standard statistical mechanics textbooks, but it is well documented in the literature on chaos.

What is typically not well explained is which systems are chaotic and which are not. This complicated dynamical issue has drawn the attention of the best mathematicians for centuries, but real experience/understanding has only been obtained since the 1950s, when computer simulations became possible. Fermi and collaborators famously used one to study whether or not a chain of coupled oscillators reaches thermal equilibrium.

The resulting picture revealed from simulations is that a generic dynamical system has a *split phase space*, with some parts showing deterministic quasi-periodic motion, typically surrounded by parts in which dynamics is chaotic. Even in systems which are examples of very stable deterministic motion, such as the solar system, there are certain chaotic regions. Classical chaos typically has very complicated boundaries with "fractals" (i.e. patterns repeated many times at smaller scales). It was also found that one does not need a high dimension of the phase space (=many particles). The "hydrogen atom" of chaos theory happens to be a periodically driven anharmonic oscillator, and the practical condition of the onset of chaos is the so-called Chirikov's criterion of *overlapping resonances.* The universal instability mechanism is the so-called Arnold diffusion; see, for example, Chirikov *et al.* (1980).

Since the 1980s, *quantum chaos* has been studied in various settings. On the one hand, quantum fluctuations make it impossible to follow the classical paths for too long, so it was clear that the beautiful classical theory with fractals and indefinite entropy growth must somehow be terminated at some scale. Indeed, the very existence of $\hbar$ sets a limit on the phase space cells which can be discussed. The number of states can be very large; however, in quantum theory, it is still limited. So, quantum mechanics must set certain boundaries for classical chaos; see discussion with examples in Shuryak (1976).

An early precursor of "quantum chaos" was Bohr's *compound nuclei model.* The original setting included a low-energy neutron captured by a heavy nucleus. Bohr reasoned that the excitation energy as the neutron falls into the potential well is $E > 8\,\text{MeV}$, while excitation resonances are separated by $\Delta(E) \sim 1\,\text{eV}$. So, a neutron excites several millions of states, which Bohr suggested should be described statistically as *thermal* states. The temperature was calculated by the Fermi gas model, producing a single-particle distribution:

$$\frac{dN}{d^3p} \sim \frac{1}{\exp\left((p^2/2M - \mu)/T\right) + 1} \tag{8.4}$$

The probability for a neutron to escape back is then predicted to be exponentially small, $\sim \exp(-E/T)$, with this T.

Let me not go into the phenomenological successes of the model but rather discuss the following central question: Suppose ΔE is taken to be increasingly smaller, so that *only one state* of the nucleus is selected.[1] Thus, absolutely no "ensemble of levels" over which to average is available. Can one still use the thermal description, for this one quantum state? It turned out the answer to it is "yes," because such nuclear states are very complicated in a "quantum chaos regime."

[1] This is indeed practically possible to do with cold or even "room temperature neutrons" with $\Delta E \sim 10^{-2}\,\text{eV}$.

What it actually means and where the boundary of quantum chaos is have been studied in many different model settings, sometimes even for realistic atoms or nuclei. Suppose we start with a "noble" atom, or a doubly-magic nucleus, so that the ground state contains just some closed shells. Its lowest excitations can be described in terms of particle–hole pairs, yet by increasing E, one soon gets into very complicated states.[2] The question is at what excitation energy — or number of states — one gets into the chaotic regime.

For atoms, one has a well-defined Hamiltonian, and one can find all its excited states just by brute-force diagonalization of the Hamiltonian, defined as a matrix on some convenient basis. One of the first studies of this kind has been done by Flambaum *et al.* (1994) for the (rare-earth) atom of cerium[3] Ce. The number of *valence* electrons is $four$ for this atom. The states analyzed all have the same global quantum numbers $J^P = 4^+$ (or 4^-), and the focus was on an excitation energy range in which there are *hundreds* of states of each kind. Yet, it was found that the exact eigenstates are basically a *random mixing* of all "simple" states (defined on a shell-model basis). The diagonal density matrix — probabilities to find quasiparticles of certain subenergy — was found to be in good agreement with the Fermi distribution (8.4) . The statistical properties of levels follow well the random matrix theory. The log of the number of mixed levels is the entropy, etc. In short, the rare-earth atom Ce at sufficient excitation energy $E > E_{\text{chaos}} \approx 2\,\text{eV}$ is already fully chaotic!

Similar behavior was found in many nuclear shell models: A sufficiently excited state does agree with Bohr's "compoud nuclei" statistical description. A nucleus can be simultaneously a perfectly "pure" quantum state, identified, for example, by a narrow resonance in neutron scattering, and, at the same time, internally described by a thermal single-body density matrix *traced over all particles but one.* This form of entropy, known as an *entanglement entropy*, was recently recognized and studied for quantum computers.

---

[2]The states with different global quantum numbers J^P do not mix, so one needs to in fact have many mixing states for each J^P separately.

[3]Discovered in 1803, two years after the discovery of the small planet Ceres after which it was named. Ceres itself was named after the Roman goddess of fertility and motherhood. The 58 electrons of the element are arranged in the ground state in the main configuration [noble gas Xe] $+ 4f^1 + 5d^1 + 6s^2$. The four outer valence electrons can easily be moved to other levels. The three single-particle orbits indicated have 14, 10 and 2 electron states, respectively, and thus, there are plenty of four-electron states even inside the Hilbert space based on them, without particle–hole excitations of the Xe core.

Summary: Originally, statistical mechanics was based on averaging over an ensemble of a huge number of quantum states, N_{states}, of multiparticle macroscopic systems. Yet, for many real systems, atoms and nuclei among them, a single atom or nucleus is, for sufficient excitation energy $E > E_{\text{chaos}}$, chaotic. A single multibody state can be treated statistically, with ascribed temperature and (entanglement) entropy. Those who are interested in learning more about classical and quantum chaos can, for example, refer the book by Reichl (1989). More recent references can be found in reviews such as those by Zelevinsky and Volya (2016).

8.2 Early Statistical Approaches to Multihadron Production

We now turn to hadronic physics. Fermi and others argued that even a few-hadron system may be described as a thermal one. In particular, a *single-particle* distribution has a thermal form of

$$\frac{dN}{d^3p} \sim \frac{1}{\exp(\epsilon_p/T) \pm 1}, \quad \epsilon_p = \sqrt{p^2 + m^2} \tag{8.5}$$

(as usual, "+" denotes Fermions and "−" denotes bosons). In order not to discuss complications related to the system's explosion, it is convenient to begin with the low-energy collisions in which, nevertheless, significant energy is released. The most convenient reaction of this type historically was proton–antiproton annihilation:

$$\bar{p}p \to \text{mesons}$$

The first applications of statistical mechanics have been to single-body or *inclusive* reactions, namely for $\bar{p}p \to \pi + \cdots$, where the dots stand for any number of other produced particles. The Bose distribution for mesons, in fact, turned out to be very accurate, see Fig. 8.1 from Hoang *et al.* (1971).

Note that here, chemical potential is absent — assumed to be zero $\mu = 0$ — as in blackbody radiation. So, the temperature is the only parameter which can be fitted to the observed single-pion spectrum. It turned out to be in the range of 130–140 MeV.

This is not the end of the story. Can a statistical description of proton–antiproton annihilation go much further to describe particular *exclusive* reactions? It was demonstrated in my work (Shuryak, 1972b) that one can indeed describe the *exclusive* reactions one by one. These reactions can either be characterized by the number of pions, N_+, N_0, N_- for π^+, π^0, π^-, respectively, or by some number of kaons plus pions, etc. *Each* reaction now

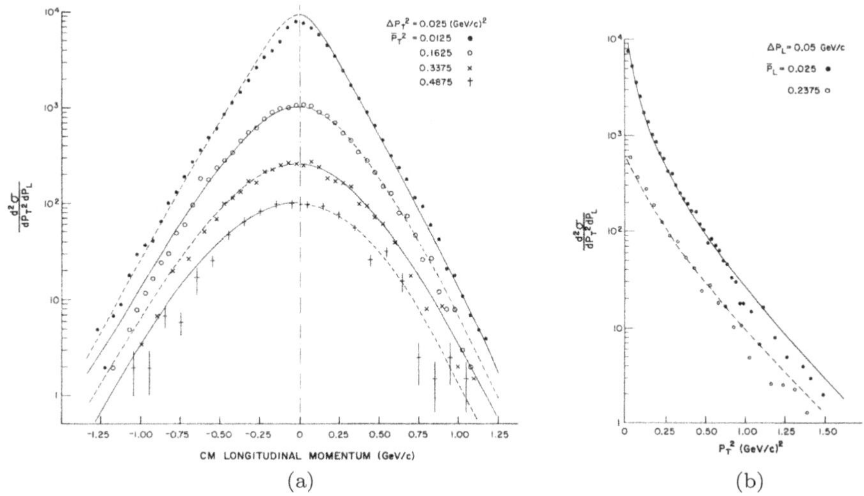

Fig. 8.1. Inclusive spectra of the reaction $\bar{p}p \to \pi + \cdots$ at $p_{\text{lab}} = 2.32\,\text{GeV/c}$. The curves are fitted by the thermal Bose distribution.

has its own T, μ pair of thermal parameters as well as its own "freeze-out volume" V.

We know the total particle number and the total energy available, so there are two equations:

$$E_{\text{tot}} = V \sum_i \int \frac{g_i d^3p}{(2\pi)^3} \frac{\epsilon_p}{\exp\left((\epsilon_p - \mu)/T\right) \pm 1} = VT^4 \sum_i g_i \Phi_\pm(M_i/T, \mu/T) \tag{8.6}$$

$$N_{\text{tot}} = V \sum_i \int \frac{g_i d^3p}{(2\pi)^3} \frac{1}{\exp\left((\epsilon_p - \mu)/T\right) \pm 1} = VT^3 \sum_i g_i \phi_\pm(M_i/T, \mu/T) \tag{8.7}$$

where $g_i = (2J_i + 1)(2I_i + 1)$ are the statistical factors, different for each particle species. Here, we have defined two functions, plotted in Fig. 8.2.

By dividing one equation by the other, one cancels the volume V and obtains an equation for T and μ. In addition, neglecting Bose's -1 in the denominator, which is not numerically important, one finds that μ drops out of the ratio, so one can determine T. Yet, there are not enough equations, so one does need to make some additional physical assumptions about the freeze-out conditions.

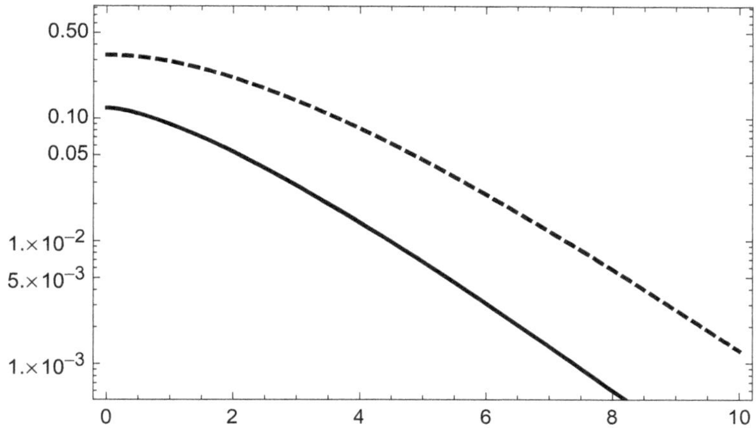

Fig. 8.2. Bosonic thermal functions $\Phi_-(M/T, 0)$ (dashed) and $\phi_-(M/T, 0)$ (solid) defined in the text for energy density and density versus M/T.

The central idea of Shuryak (1972b) was, following Pomeranchuck, that hadron formation happens at a *fixed universal density* of $n_0 = 1/V_0$

$$\frac{V}{N_{\text{tot}}} = V_0 \tag{8.8}$$

Now, with three conditions at hand, all three parameters can be determined for each exclusive reaction. The distributions over momenta can be calculated and compared to the experimental ones. Indeed, one then finds thermal spectra with variable temperatures for *exclusive* channels as well.

But this is not all: The final step is the following consideration that the number of states per particular (exclusive) final state i is proportional to the exponent of the corresponding *entropy*[4]:

$$P(a) \sim e^{S(V_a, T_a, \mu_a)} \tag{8.9}$$

where $a = N_+, N_0, N_-$ indicates a particular final channel.

Statistical mechanics tells us that entropy is the derivative of the grand canonical potential

$$S = -\left(\frac{\partial \Omega}{\partial T}\right)_{V,\mu},$$

[4]Note that the logarithm of this expression is what is written on Boltzmann's tomb. The first person who used it in this form for the description of a system's fluctuations was none other than Einstein.

and for a Bose gas of mesons,

$$\Omega = TV \int \frac{d^3p}{(2\pi)^3} \log\left[1 - \exp\left((\mu - \epsilon_p)/T\right)\right] \tag{8.10}$$

In the Boltzmann approximation, μ can be found directly, and the expression simplifies[5] to

$$P(a) \sim e^{E_{\rm tot}/T_a} \prod_i \frac{[V_a n_i(T_a)]^{N_i}}{N_i!} \tag{8.11}$$

where n_i is the density of the relativistic Boltzmann gas. Combining these formulae, one can calculate the probabilities of all exclusive channels. The results are shown in Fig. 8.3 from Shuryak (1972b). As one can see, for the number of pions $N_{\rm tot} > 5$, the statistical model does become accurate.

Standard thermodynamics use the Stirling formula for the factorial as $N! \approx (N/e)^N$. In this form, it is not really accurate for small N, e.g. for

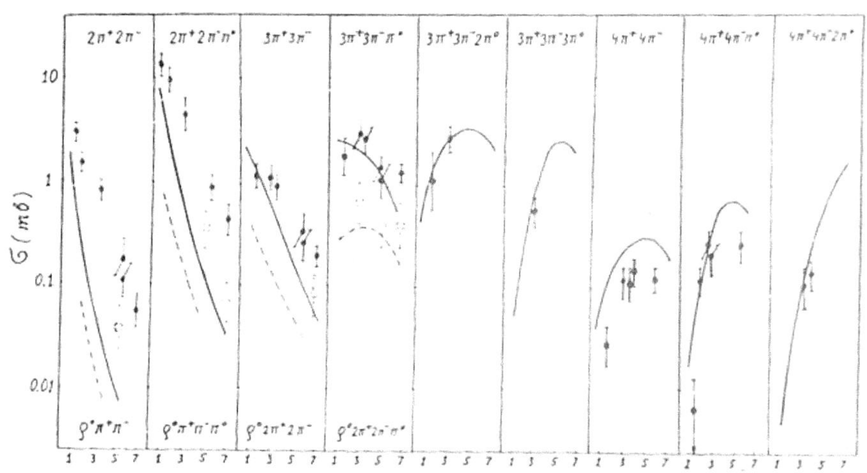

Fig. 8.3. Cross-section $\sigma_i(mb)$ of various exclusive channels of proton–antiproton annihilation into pions as a function of the lab momentum of antiprotons (GeV/c). The lines correspond to the statistical model, and the points are the experimental ones. Solid points and lines denote the reactions indicated at the top of the figures. The dashed lines and points denote subreactions, indicated at their bottom, in which two pions actually come from the decay of ρ meson.

[5]For clarity, it is not a Poisson distribution because T_a, V_a are different for different channels.

$N = 5$, the l.h.s. is 120 and the r.h.s. is 21.0. In subsequent work, I included the following Stirling term corresponding to the improved relation:

$$N! \approx \sqrt{2\pi N}\left(\frac{N}{e}\right)^N$$

This greatly improves the accuracy of statistical formulae (for $N = 5$, it gives 118.0, and even for $N = 2$, this expression is 1.919). It correspondingly improved the agreement between the data and the statistical model as well.

Finally, in this section, let me discuss the following two issues, which inevitably lead to questions.

As shown in Fig. 8.3, any of the observed exclusive channels — e.g. the simplest one with four particles $2\pi^2\pi^-$ — includes events with resonances — in this case, $\rho^0\pi^+\pi^-$, in which the invariant mass of one pair corresponds to m_ρ. Experimentalists count those as "subchannels," a part of the observed four-pion final state. How should a theorist, using statistical mechanics, treat it? Does it depend on whether ρ is considered to be a fundamentally different $\bar{q}q$ meson or just a $\pi\pi$ resonance? The answer is that in both formulations, the existence of resonances does correspond to *additional* states of the system. We will discuss it further later in this chapter.

Another confusion some people have with the statistical model is reflected in the following question: Does the thermal model simply mean that one calculates the *phase space* of the reaction considered while the matrix element is a constant? The answer is "no" at several levels. First of all, since we calculated the production probabilities of various channels, we obviously did include a certain statement about the matrix element $|M(\text{intial} \to \text{final})|^2$. Second, the relativistically covariant phase space is proportional to the product of

$$d^4p\delta(p^2 - M^2) = \frac{d^3p}{2\sqrt{p^2 + m^2}}$$

unlike the statistical integrals done above. And, last but not least, an assumption that the matrix element is a constant, independent of momenta, implies that there is only *one* state, so that the outgoing secondaries are *mutually coherent*. If this were true, their Bose symmetrization would be redundant, and femtoscopy (HBT interferometry) would not have a signal! In fact, the secondaries are *mutually incoherent*: We will discuss the evidence in the chapter on femtoscopy.

Summary: *Inclusive* production corresponds to $\mu = 0$ and some universal temperature. *Exclusive* channels have variable V, μ, T, although, to good accuracy, all of them have a fixed *freeze-out density* n_0.

8.3 Microcanonical Suppression of Flavor Production

In early heavy ion experiments, it has been noticed that the fraction of strange particles $K, \Lambda, \Sigma, \Omega$ in AA collisions was larger than in pp. It has even been proposed that "strangeness enhancement" be a "QGP signature."[6] However, there is *no* enhancement of strange particles in AA collisions, as compared to correct statistical expectations. Furthermore, even strangeness suppression in pp has been understood as a consequence of statistical mechanics applied appropriately.

Due to the conservation of various charges, not all final states are allowed. For example, in the discussion of $\bar{p}p$ annihilation above, we always kept the total electric charge of the final state at zero. Similarly, the total flavor charge (strangeness, charm, etc.) must be zero as well. Therefore, strange hadrons must be produced in pairs.

For a large system, for which $n_i(T)V \gg 1$, this restriction does not matter. However, if the opposite is the case, $n_i(T)V < 1$, the probability of such particle production is suppressed *quadratically*, $\sim(n_i(T)V)^2$, because any strange particle has to be produced with an anti-strange companion. The general expression for n pair production can be written as

$$P_n \sim \frac{(n(T,m)V)^{2n}}{n!n!} \tag{8.12}$$

Normalizing the sum using $\sum_{n=0}^{\infty} x^{2n}/(n!)^2 = I_0(2x)$, one can get a formula describing any size of the system, as done by Shuryak (1974a):

$$\langle N_i \rangle = \frac{xI_1(x)}{2I_0(x) - 1}, \quad x = 2n_iV \tag{8.13}$$

where I_n are Bessel functions and n_i includes statistical factors such as isospin factor 2 for kaons or nucleons. In Fig. 8.4 from Shuryak (1974a), it is shown how this expression describes the average number of kaons or antiprotons as a function of the average number of pions: A transition from quadratic to linear behavior is apparent in these plots.

This phenomenon, called *microcanonical suppression*, is naturally more severe if the particle in question has a larger flavor charge. For example, triple-strange Ω^- baryon requires the associated production of three anti-strange companions.

It has been further suggested that charge neutrality should not be enforced only globally, i.e. in the whole freeze-out volume V, but *locally*,

[6]Leading to a separate sequence of conferences, "Strange Quark Matter," which continue to be organized even today.

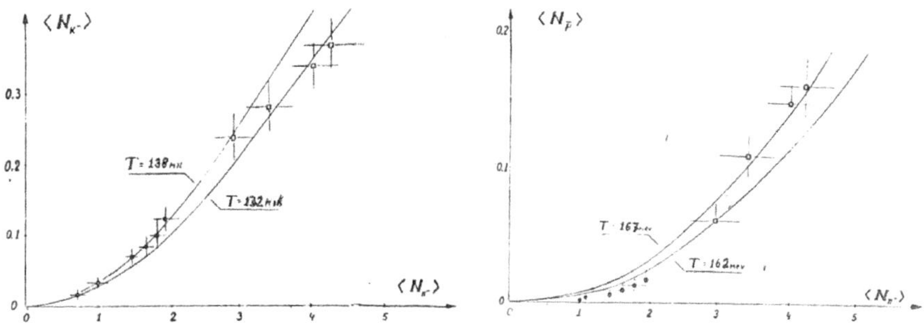

Fig. 8.4. The number of kaons (left) and antiprotons (right) as a function of the number of pions.

in the strangeness correlation volume $V_c < V$. The probability of creating a pair of strange (or charged, etc.) mesons is then defined as $P \sim VV_c[n(M/T)]^2$, not $V^2[n(M/T)]^2$.

Castorina *et al.* (2016) proposed the following evaluation of the "correlation volume" V_c. They argued that since distant parts of the freeze-out volume are causally disconnected, they cannot be correlated in their quantum numbers. A specific model assumes that there are two values of the proper times: the "initial equilibration" τ_0 when flavored quarks can be produced and the final freeze-out proper time τ_f. For the one-dimensional Bjorken expansion, they are related to the energy densities as

$$\frac{\tau_h}{\tau_0} = (\epsilon_0/\epsilon_h)^{3/4} \tag{8.14}$$

For the correlation distance, Satz *et al.* proposed

$$\frac{d}{\tau_0} = \left(\frac{\tau_h}{\tau_0}\right)^{1/2} \left(\frac{\tau_h}{\tau_0} - 1\right) \tag{8.15}$$

As shown in Fig. 8.5 from Castorina *et al.* (2016), this model provides a universal description of strangeness suppression for various collision systems.

Summary: Production of secondaries with a nonzero flavor quantum number — *strangeness* (or charm, which we will discuss a bit later) — is suppressed in smaller systems because of the necessity to produce their anti-flavor companion(s). This phenomenon is well understood in standard statistical mechanics. There is simply no need to devise models with "strangeness enhancement" in QGP.

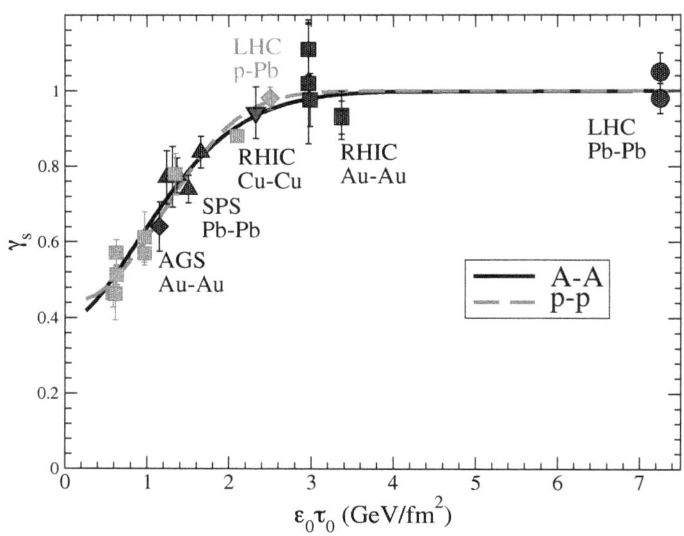

Fig. 8.5. Strangeness suppression as a function of initial transverse energy density for various collisions indicated in the figure.

8.4 Two Freeze-Outs

Looking at a fireball through a detector, one has the same problem as an astronomer has when looking at a star (e.g. the Sun): only the photosphere can be seen (with $T \sim 6000°$), but not the hotter interior. It took years to detect the elusive solar neutrinos and directly test the predictions of the "standard solar model."

Similarly, people studying the Little Bang have a large zoo of particle species available to them. Since they have quite different scattering cross-sections, one may think that, in principle, there would be multiple analogues of "photospheres" with different depths. Yet, taking away "penetrating probes," to be discussed separately, these cross-sections are divided into just two classes: inelastic and elastic. They define two freeze-out surfaces: *chemical* and *kinetic*.

The fireballs that we discuss are dynamical objects, evolving in four-dimensional space-time. As shown in Fig. 8.6, the freeze-outs correspond to certain three-dimensional surfaces surrounding the fireball. These surfaces typically have large (blue) time-like parts and smaller (white) space-like ones. Secondary particles we see are coming out of these surfaces. (Through the initial (red) surface, the matter is moving *into* the volume.)

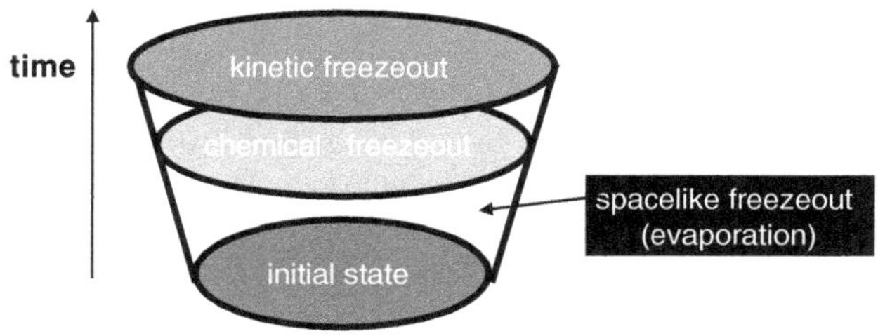

Fig. 8.6. A sketch of the freeze-out surfaces surrounding the fireball volume.

A general *freeze-out condition* can be formulated in several different ways. One is to say that the microscopic collision rate of a particle becomes comparable to the system expansion rate:

$$\nu_{\text{collisions}} = n\sigma v \sim \nu_{\text{expansion}} \tag{8.16}$$

where σ is the cross-section of the reaction considered, involving other particles with the density n and relative velocity v.

The usual approximation to the r.h.s. is typically defined in terms of the four-velocity divergence

$$\nu_{\text{expansion}} \equiv \partial_\mu u_\mu \tag{8.17}$$

They are both included because the freeze-out surface has, in general, both space-like and time-like parts. We will return to the exact definitions and examples of this surface later in this chapter.

What is extremely important to emphasize is that for low-energy hadronic reactions, *the rates for inelastic reactions are much smaller than for elastic ones.* (One of the reasons for that: Pions interact with derivatives due to their Goldstone nature, so it is difficult to rescatter or produce a soft pion. Thus, in a low-T pion gas, the elastic and inelastic reactions are proportional to different powers of T.)

As a result, we identify *two separate freeze-out temperatures*: (i) "*chemical*" (fixing the composition of secondaries) and (ii) "*thermal*" or kinetic, at which the last collisions happen and the particle momenta are finally fixed at their observed values. The two temperatures approximately[7] associated with them are denoted as T_{ch} and T_{th}, respectively.

[7]The freeze-out conditions are not coincidental with isotherms, as will be clear in the following.

8.5 Statistical Hadronization at the Chemical Freeze-Out

On the phase diagram, hadronic matter is in either QGP or *hadronic* phase. It is obviously assumed that freeze-outs are in the latter phase. (Although, as we will soon see, they can be located close to the transition line.)

The main idea is that the composition of secondary hadrons is determined by the rapid freeze-out of *inelastic* collisions. It is assumed that matter is in local equilibrium and can be approximated by the hadron resonance gas (HRG) model and an ideal gas of all hadronic species and resonances, which we discussed in Section 4.1. In this section, we follow a particular version of it, developed by Stachel *et al.* (2014), which is basically a fit of the LHC ALICE data and the resonance gas model shown in Fig. 8.7 from Andronic *et al.* (2018). The model assumes that all hadronic resonances are produced thermally with unmodified masses, as they are in the Particle Data Tables.

The conclusion from these plots is that the model is exceptionally successful. A simplistic statistical model with three parameters, T, μ_b, V, describes the production of very different species over nine orders of magnitude!

Note that secondaries from subsequent decays of all resonances *after* freeez-out — known as "feed-down" — are also included. The magnitude

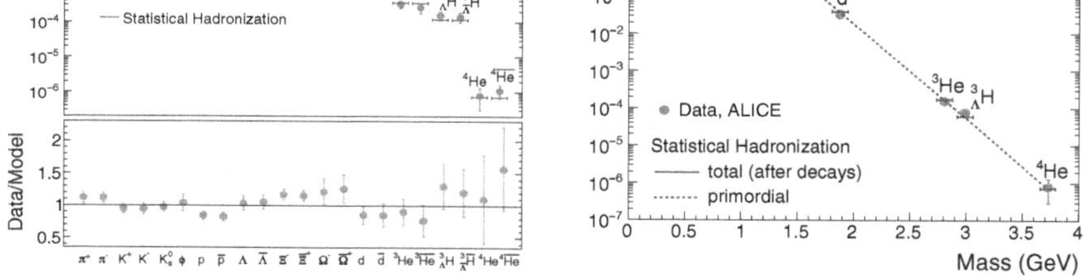

Fig. 8.7. (left) Hadron yields from ALICE at the LHC and fit with the statistical hadronization model for $T_c = 156.5 \pm 1.5\,\text{MeV}$, $\mu_B = 0.7 \pm 3.8\,\text{MeV}$, $V = 5280 \pm 410\,\text{fm}^3$. (right) The same data and theory, but with the statistical weight and antiparticle removed and plotted as a function of the particle mass. The horizontal blue lines are the decays of the resonances, and the dotted blue lines are primordial only.

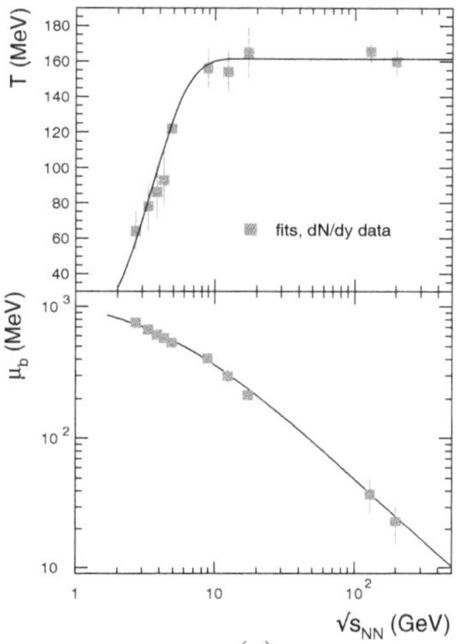

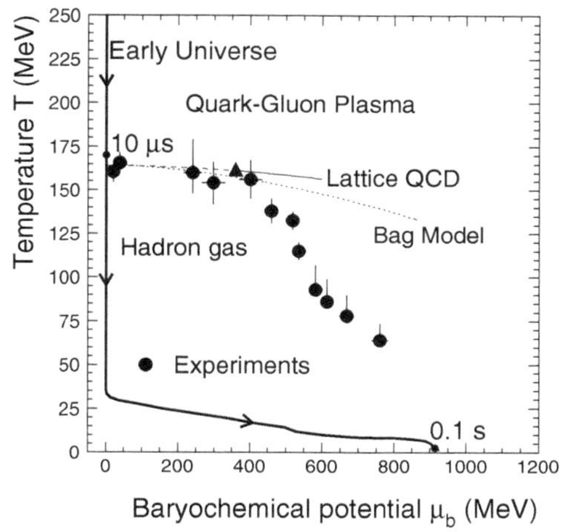

Fig. 8.8. (a) Energy dependence of T and μ and (b) their positions on the phase diagram. The points are extracted from a statistical model fit to particle ratios.

of the feed-down is seen from a comparison of the full predictions to the primordial only, shown by a blue dotted line on the right plot.

Note that even light nuclei, such as $d,^3_\Lambda H$, and their antiparticles are well described. One may wonder how d, with its $B = 2\,\text{MeV}$ binding energy, can be found in an environment with an ambient temperature of $T \sim 160\,\text{MeV} \gg B$. It is akin to finding a snowflake jumping out of a hot oven. Shouldn't d be instantly destroyed in it? The answer to this and similar questions is well known. Thermodynamics does not depend on the lifetime of d. As one deuteron is destroyed, perhaps at a certain large rate, in equilibrium, another one must be recreated by the inverse process with *the same* rate. The average population is conserved, and that is what the statistical mechanics gives us.

Note further that for light nuclei species, such as $d,^3 He,^4 He$, the points coincide with the blue dotted line. This means that there is no "feed-down" from large nuclei species into those. It is indeed so at high collision energies at the LHC, corresponding to a near-zero baryon density of $\mu_b \approx 0$. Yet, at lower collision energies, the suppression per baryon $\sim\exp[-(E-\mu_b)/T]$ is less strong, as μ_b grows, and such feed-down becomes important.

(In particular, *preclusters* made of four nucleons (Shuryak and Torres-Rincon, 2020a) lead to a population of nuclear bound states and resonances of ^4He, about 50 states(!), decaying into various channels with light nuclei in those. At $\sqrt{s} \sim$ few GeV, the feed-down contributes nearly half of the yield of tritium. We discuss nuclear clustering in the chapter devoted to searches for critical point.)

Let me comment on the definition of volume V. Its fitted value, indicated on the figure, is supposed to be the sum of the volumes of all hydro cells in their respective rest frames, taken over the freeze-out surface:

$$V = \int d\Sigma_\mu u^\mu$$

In the local approximation implied, the fact that different parts of this surface may move at large speeds relative to each other does not matter: Motion does not affect thermal fluctuations.

Even these successful plots show certain imperfections. One visible deviation between thermal theory and the data are seen for protons and antiprotons: between data points are below predictions. The origins of that are related to the fact that the *resonance gas model* used does not incorporate hadron–hadron interactions accurately. It does so for narrow resonances related to *attraction* between them, but channels in which there is *repulsive* scattering without any resonances are ignored. It was shown that accounting for full scattering phases in all channels does indeed correct this "proton anomaly," see more in Andronic *et al.* (2021a).

Another issue which needs to be discussed (not so much for its practical importance but for logical consistency) is the role of resonances *not yet* included in the statistical model. The standard version of the statistical hadronization model contains 555 established hadrons, as given by the PDG, with 223 mesons (123 non-strange, 32 strange, 40 with charm and 28 with beauty) and 270 baryons (104 non-strange, 96 strange, 56 with charm and 14 with beauty). It includes, in addition, 62 nuclei, hypernuclei and anti-(hyper)nuclei. Both lattice simulations and various quark models predict a much larger (if not infinite) number of states belonging to multiple Regge trajectories: By using them, one has some idea that masses grow as a function of radial quantum number and total angular momentum as

$$M_{nJ} \sim \sqrt{n+J}$$

but very heavy excitations are, of course, cut off by the Boltzmann factor $\exp(-M/T_{\rm ch})$, with $T_{\rm ch} \approx 155\,{\rm MeV}$. Under investigation are especially baryonic states with masses above 2 GeV, which, for whatever reason, are not listed as reliable by the Particle Data Group. In Andronic *et al.* (2021a),

N^* is extended to masses of 2.2 GeV and Δ^* to 2.5 GeV, increasing non-strange baryons to 284 states (from 123). Similar increases were included in other sectors. After their decay, one finds that the main effect is an *increase* in the number of nucleons, up to 50%. It thus moves the predictions of the statistical model *away* from the data. A full picture of what happens with extra states is shown in Fig. 8.9: Basically, all predictions move away from the data!

It is clear that one needs to approach the HRG model in a more consistent way, not just cutting off resonances arbitrarily. The answer comes from a better use of the Beth–Uhlenbeck expression in two ways: (i) include realistic widths of the resonances, instead of narrow resonance approximation; (ii) include (as complete as possible) S matrix, with phase shifts in channels with repulsive interactions, in which resonances are absent. This was done by Andronic *et al.* (2019) for the "case study" of πN channels with $N^*.\Delta^*$ resonances and the corresponding phase shifts. Here, we would not, of course, go into details of the phase shifts themselves but just show how their account resolves the "proton anomaly." In Fig. 8.10 from that paper, the results are compared with the original version of the HRG model (dashed lines): The conclusion is that baryon density and susceptibility are *reduced* due to repulsive channels, which, so to speak,

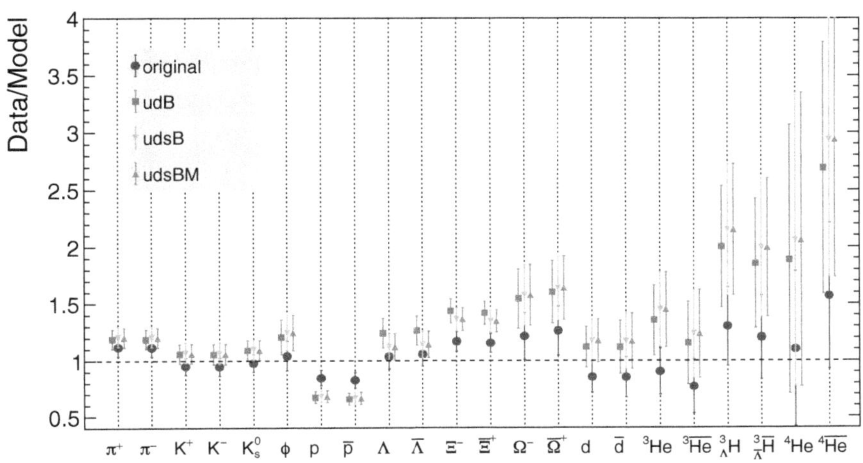

Fig. 8.9. Ratio of the ALICE experimental data over model fit for yields of all 22 fitted particle species for the original and the three modified hadron spectra.

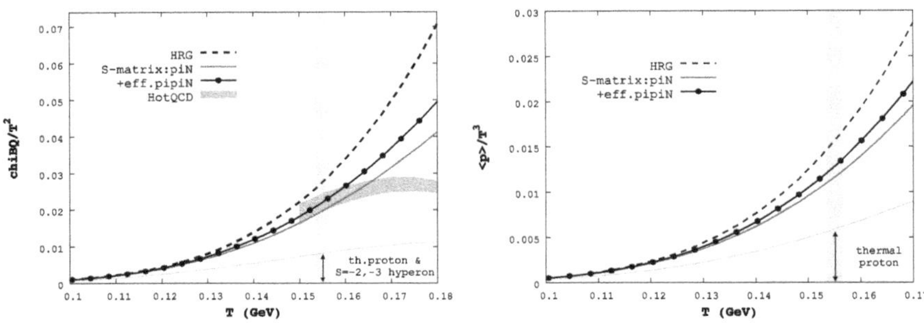

Fig. 8.10. The baryon-charge susceptibility χ_{BQ} (Z derivative over μ_b and μ_Q (left), and the proton yield (right) computed within the S-matrix formulation (with and without the effective N contribution) compared with the original predictions. The continuum extrapolated LQCD result is shown by the blue band. The vertical yellow band indicates the location of the crossover region, $T_C = 156.5 \pm 1.5\,\text{MeV}$.

make baryons effectively heavier. These results also agree well with *ab initio* lattice thermodynamics results (blue band).

Note one particular deviation between the statistical hadronization model and the data: K^*. The model predicts more than what is observed. This is expected, as it is a relatively short-lived resonance which decays when the density is still non-negligible, the products can be rescattered and their invariant mass moves out of the peak. Corrections to that can be made using any cascade code. This topic takes us to the following section.

Figure 8.8 from Braun-Munzinger and Wambach (2009) shows the value of the temperatures and chemical potentials extracted from similar fits at different collision energies. One can see that, at least at high collision energies, the chemical freeze-out conditions seem to become universal (independent of the collision energy). One can also observe that the line on the phase diagram seems to follow — at least at small μ — the line of the QCD phase transition obtained from the lattice. Why is this the case? We discuss it in the following.

8.6 Statistical Production of Charmed Hadrons

Note that in the plot of statistical hadron yields, one can see secondaries with strangeness but not with c or b quarks. The reason is that they are *not* produced thermally but from hard processes at the initial stage. If equilibrated, they have thermal distribution in momenta and among species, but their total number is normalized to production rates, included statistically via certain fugacities g_c (or chemical potentials).

Their thermal yields can be written as follows:

$$N_{c\bar{c}} = \frac{g_c}{2} V \sum_{i=\text{open charm}} n_i^{\text{th}} + \frac{g_c^2}{2} V \sum_{i=\text{hidden charm}} n_i^{\text{th}} + \frac{g_c^2}{2} V \times \sum_{i=\text{double open charm}} n_i^{\text{th}} + \cdots \tag{8.18}$$

where there is a term for quarkonia (states with "hidden" charm), the "double charm" (e.g. *ccq* baryons), etc. Of course, there can be triple-charm ones *ccc* as well, but they are too small numerically to be included here. Note the powers of charm fugacity g_c here: This relation is supposed to be an equation for g_c for a given l.h.s. Note also that the l.h.s. is an observable, and thus, g_c is *not* a free parameter. At the LHC energies, $g_c \sim 30$. Therefore, J/ψ and other charmonia production are very much enhanced, for example, by $g_c^2 \sim 10^3$, and for *ccc* baryon Λ_c, by $g_c^{\sim}3*10^4$. In 2022 LHCb collaboration, and then others, have found $cc\bar{c}\bar{c}$ all-charmed tetraquarks. One may wonder if, with such large enhancements, the statistical model would still work.

Following Andronic *et al.* (2021b), one can see that it actually does! In Fig. 8.11 we see a comparison for many single, hidden/double and even

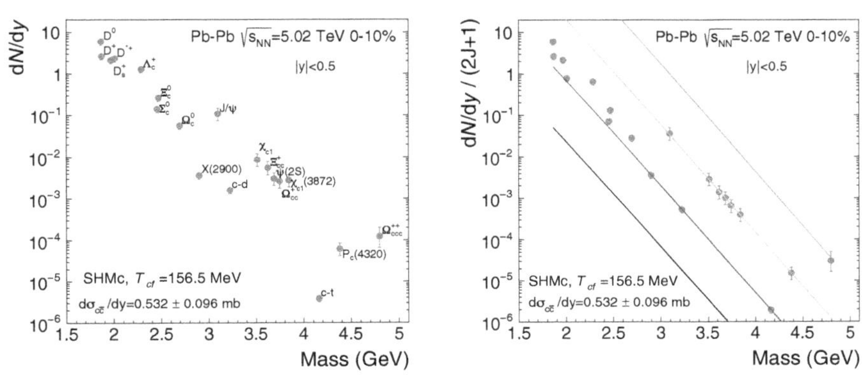

Fig. 8.11. Mass dependence of yields dN/dy for various hadron species for *Pb-Pb* collisions at mid-rapidity. The left panel is for absolute yields, and the right panel is for yields per degree of freedom $(2J + 1)$. In this plot also, the primordial (prior to decay) values are shown as lines, corresponding to hadrons with charm-quark or anti-quark content of 0, 1, 2 and 3.

triple-charmed hadrons. Four lines on the right plot correspond to powers of g_c from 0 to 3.

Looking at these plots, one would conclude that charmed states are chemically well equilibrated, as are u, d, s quark hadrons. Yet the number of charm quarks seems to be preserved.

8.7 Is Chemical Freeze-out Well Described by Hadronic Cascade Codes?

Thinking of multiple types of all kinds of secondary hadrons and their possible mutual rescattering processes, one can get the idea that each kind of secondary has its own freeze-outs and become frustrated. Among the available practical tools, people used cascade codes (RQMD, UrQMD or similar) switching to those from hydrodynamics.[8] Another argument in their favor goes as follows: Particle densities perhaps depend on space-time quite smoothly; therefore, the existence of a sharp (well-defined) decoupling surface is perhaps just a crude approximation. Running such codes, one indeed finds that the positions of the last scatterings do not form any well-defined surface.

The cascades predicted certain deviations from thermal scenarios, and naturally, there is a question of which of them one should trust. Initially, people relied on cascades, but then the detailed comparison with data revealed certain serious problems, and the trend was reversed.

Let me give just one example. In pre-RHIC 1990s, heavy ion collisions have been studied in a fixed-target mode at AGS and SPS accelerators. At such energies the fraction of baryons is large, comparable to that of mesons. With a large $\bar{N}N \to mesons$ annihilation cross section, $\sigma \sim 50\,\text{mb}$, one can expect most of the anti-baryons to be "eaten up" on the way out. Various cascade calculations confirmed that pessimistic conclusion. Yet, stubborn data showed a large yield of the antiprotons. Furthermore, the experimental ratio $\bar{p}/p = (5.5 \pm 1)\%$ turned out to be in good agreement with the thermal predictions of $\exp(-2\mu/T) \approx 5\%$.

The puzzle was resolved by Rapp and Shuryak (2001, 2002), who pointed out that, contrary to popular belief, inverse multi-pion collisions $n\pi \to p + \bar{p}$ with $n \sim 4 - 6$ do have rates sufficient to regenerate antiprotons. There are instructive reasons *why* it is the case. The rates of direct and inverse

[8] In fact, our own first description of the Little Bang (Teaney *et al.*, 2001b), which was very successful, used such a freeze-out.

reactions can be written, in Boltzmann approximation, as

$$R \sim \int d^3k_p d^3k_{\bar{p}} \Pi_{i=1}^n d^3k_i \delta^4(k_{\text{tot}})|M|^2 \left(z_p z_{\bar{p}} e^{-\frac{E_p+E_{\bar{p}}}{T}} - (z_i)^n e^{-\sum_i E_i/T} \right) \tag{8.19}$$

with fugacities $z = \exp(-\mu/T)$. The time reversal of strong interactions requires that the matrix elements be the same in both directions. In thermal equilibrium, energies in both exponents sum to the same total values, so that two sub-rates cancel each other. It is just an example of the so-called "detailed balance."[9]

All cascade codes turned out to be simply wrong for the same reason: They included the direct annihilation cross-section $p + \bar{p} \to n\pi$ but *not* their inverse ones. Their authors simply *believed* that inverse reactions were suppressed and never bothered to check that implicit assumption. Even after this apparent mistake was pointed out, I am not sure if multi-pion collisions have been included in those codes.

The main lesson one can get from this story is that there exists a huge number of multi-hadron reactions, e.g. $m\pi \leftrightarrow n\pi$, with experimentally unavailable rates if $m > 2$. Perhaps, those can be evaluated using some model. However, in practice, those are simply omitted from the existing codes. An inevitable conclusion: Any usage of those should be seriously suspect.

8.8 Why Does the Chemical Freeze-out Follow the Phase Boundary Line?

Inspecting the extracted statistical model parameters, shown in Fig. 8.8, as a function of center of mass energy, two characteristic features arise: (i) The baryon chemical potential drops monotonically with increasing energy, and (ii) the temperature initially increases but appears to level off at the center of mass energies per colliding nucleon pair of about 20 GeV at a temperature of around 160 MeV. The other plot shows their positions on the phase diagram, and it appears that the chemical freeze-out follows very closely the phase transition line.

A reason why this should be the case was suggested by Braun-Munzinger *et al.* (2004). It is based on the assumption that a chemical equilibrium is maintained by the *multiparticle reaction rates*, similar to those we discussed

[9] After chemical freeze-out, the back reaction becomes larger, as we discuss in the following.

in connection with the antiproton puzzle. (Note that this assumption is the opposite of what is assumed in cascade codes.)

Since those depend on a rather high power of density, they include very high powers of temperature. Therefore, they should all decouple very close to the critical line:

$$\frac{|T_{\rm ch} - T_c|}{T_c} \sim \frac{1}{\rm power} \ll 1. \tag{8.20}$$

How can one test where the particles come from? We do have femtoscopy data which tell us about the duration of the freeze-out process; however, those correspond not to the chemical freeze-out but to the kinetic one. Perhaps, the way to proceed is to study resonances. Indeed, there is a large variety of those with vastly different lifetimes, so in principle, the lifetime of the hadronic phase and the degree of sharpness of the chemical freeze-out surface can eventually be understood.

Summary and connections: The main conclusion of this chapter is that at the "chemical freeze-out" surface, the composition of secondary species corresponds to the chemical equilibrium between them. If the corresponding powers of fugacity are included, it also describes the yields of hadrons containing one or two charm quarks.

Another striking observation is that the parameters of chemical freeze-out $T_{\rm ch}, \mu_{\rm ch}$ quite accurately follow the critical line. We discussed what this observation tells us about the corresponding hadronic reaction rates.

Chapter 9

KINETIC FREEZE-OUT

9.1 Kinetic Freeze-Out

Returning to the subject of the second freeze-out, let me recall the setting. The separation of two freeze-outs is due to a large difference in magnitude of the elastic and inelastic (low-energy) hadronic reactions. Let me also recall that any freeze-out condition — expansion rate equal to the reaction rate — depends on the absolute system size L.

While we have not yet discussed this aspect of freeze-out, it is convenient to do so now. Let us start with the extreme cases. Very large systems, $L \to \infty$, would expand forever and never get frozen. So, in this limit, $T_{\rm kin} \to 0$. Very small systems, on the other hand, will freeze out as soon as hadrons can be formed, so for them,

$$T_{\rm ch} \approx T_{\rm kin} \approx T_c \approx 155\,{\rm MeV}$$

Now, heavy ion collisions naturally produce systems of all sizes because there are collisions of various centralities, from central (producing the largest fireball) to peripheral (also known as "grazing" collisions), producing small fireballs.

As it was already mentioned, the particle yields are represented as an integral of (approporiately boosted) thermal distributions over the freeze-out surface. A traditional way to discuss this issue is via its so-called "blast wave" form, in which the surface and collective velocities are parameterized in some simplified way.[1] For each centrality bin, one essentially fits two parameters: the matter temperature $T_{\rm kin}$ and the mean velocity of the transverse flow $\langle \beta_T \rangle$. A version of such a fit is compared with the ALICE LHC data in Fig. 9.1 as a function of these two parameters.

[1]The details vary from paper to paper, but their differences are not really important for us now to discuss such details.

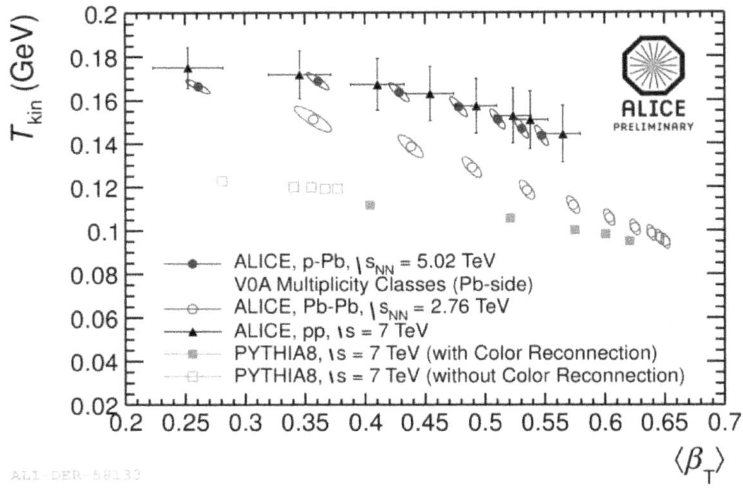

Fig. 9.1. The temperature of the kinetic freeze-out versus the mean velocity of the radial flow, fitted to the ALICE spectra of $\pi, K, p, \Lambda, \Xi, \Omega$.

We begin discussing these results with the heavy ion *PbPb* collisions, indicated by open red dots (ovals with rotated error bars). The most central *PbPb* collisions produce the largest systems, which generate the largest flow velocity. Therefore, these points are at the extreme right of the figure, reaching $\langle\beta_T\rangle \approx 0.67$. Note that while central collisions obviously start with the highest initial temperature $T_{\rm in}$ at the early stages, the displayed fit indicates that they produce the *coolest*[2] matter temperature $T_{\rm kin} < 0.10\,\text{GeV}$. Here is the lesson: The larger the system, the more effectively it transforms the initial thermal energy into flow and thus ends up being the coolest at freeze-out.

The opposite end of the red open points, at the left of the plot, corresponds to more peripheral collisions. The system is smaller, the flow velocity is much smaller $\langle\beta_T\rangle \approx 0.35$, and the temperature $T_{\rm kin} \approx 0.15\,\text{GeV}$ is close to the critical value. Since $T_{\rm kin}$ is basically the same as that chemical freeze-out, $T_{\rm ch}$, there is no space between the two freeze-outs: Such fireballs basically have no hadronic phase at all.

Note that, unlike $T_{\rm ch}$, the kinetic freeze-out $T_{\rm kin}$ depends on the centrality of *PbPb* collisions. The central-to-peripheral $T_{\rm kin}$ temperature

[2]Ignore the blue squares, open and closed, corresponding to an event generator, Pythia8. It clearly does not describe these data but is tuned to *pp* collisions.

difference, from 0.16 to 0.10 GeV, may not appear to be so drastic. However, one should remember that the physical properties of matter (such as the energy density) in this T region are proportional to a high power of the temperature, approximately $\epsilon \sim T^6$. So, we speak about a change in ϵ by more than an order of magnitude!

Although we will discuss "small systems" pA, pp later, while we discuss this plot, it is tempting to have a look at the pp, pA data also shown in Fig. 9.1. The corresponding data (blue closed points and black triangles, respectively) also extend to the right (i.e. produce a variable radial flow), although the temperature variation is much reduced. Both freeze-out temperatures, $T_{\rm ch}$ and $T_{\rm kin}$, are close to each other. What this implies is that these "small systems" also cannot support the hadronic phase. All the acceleration in them, producing radial flow, occurs in the QGP.

Summary: In central AA collisions, matter begins in the hottest state and then cools down deep into the *hadronic phase*, retaining kinetic — but not chemical — equilibrium for a long time. Let me mention a few relevant numbers: Fireballs created in central $PbPb$ LHC collisions have up to 10^4 particles, and their kinetic freeze-out time is ~15 fm/c. The highest multiplicities in pp, pA still correspond to fireballs with only hundreds of particles, and their freeze-out times are only about 2 fm. Yet, since those managed to get significant radial and directed flows, they are "macroscopic systems" as well!

9.2 Hadronic Rescattering Rates

The main processes in hadronic matter reached at RHIC/LHC are the low-energy elastic $\pi\pi, \pi K$ and πN scatterings. The estimates of $\pi\pi$ collision rates follow from the chiral Lagrangian; see, for example, early calculations by Shuryak (1980) and, with more details, by Goity and Leutwyler (1989). A consequence of the Weinberg lowest-order chiral Lagrangian is the pion scattering rate per time

$$\nu_{\pi\pi} = \frac{T^5}{12F_\pi^4} \tag{9.1}$$

where $F_\pi = 93\,\text{MeV}$ is the pion decay constant. Note the very strong T dependence (which can be obtained by dimensional considerations alone). In the resonance gas, its T dependence is, in fact, even stronger due to the large cross-sections produced by resonances, e.g. ρ, K^*, Δ.

One consequence: cooling the pion gas below a temperature of, say, $T = 0.1$ GeV would require systems much larger than any that are currently available — a pity since "cool pion liquid" can be Bose-condensed and have other interesting properties discussed in theoretical literature. Yet, since it is so far impossible to confront these ideas with any experiments, we will not discuss them.

The largest resonant cross-section is πN, reaching about 200 mb $\approx$ 20 fm^2 at the Δ peak. The corresponding radius of the interaction is $R = \sqrt{(\sigma/\pi)} \approx 2.6$ fm. In this volume, there is always more than one pion; therefore, one should question the ideology of cascades based on subsequent "scattering events." The Deltas therefore must collectivize and form "pi-sobars." Absolute scattering rates depend on the density of nucleons at the decoupling stage.

More quantitative information about the rates tuned in the AGS/SPS region can be found in Hung and Shuryak (1995a) and Prakash *et al.* (1993). It has been repeatedly performed for various hadronic cascades. The scattering rate for $\pi\pi$ can be obtained from the following expression, including resonances with the parameterization of the chemical potential term (the exponent):

$$\nu_{\pi\pi} = 40\,\text{MeV}\left(\frac{120\,\text{MeV}}{T}\right)^{3.1} \exp\left[190\,\text{MeV}\left(\frac{1}{T} - \frac{1}{T_c}\right)\right] \tag{9.2}$$

(which is not valid at T below 80–100 MeV, where it would predict the Bose condensation of pions).

9.3 The Shape of the Freeze-out Surface

A schematic shape called "conical cup" has already been shown in Fig. 12.9. It consists of the "top" part, taken to be at constant (longitudinal) proper time $\tau = \tau_f$, and the "outer edge," taken to be of conical shape, with radius linearly interpolating between the initial and final radii, R_i and R_f, respectively. The bottom part is where the initial conditions are to be defined; it is understood to be at a small but finite proper time τ_i, where one can begin the hydrodynamical description. We will not discuss the bottom part and only note that the time τ_i is typically a fraction of fm/c, negligible compared to $\tau_f \sim 17$ fm/c.

At the top part of the surface, the transverse rapidity dependence on r is linear, up to its maximal value:

$$\kappa(r) = \kappa_{\max}\frac{r}{R_f}, \quad (r < R_f) \tag{9.3}$$

At the outer edge, the transverse rapidity takes a nearly constant value on the whole wall, $\kappa = \kappa_{\text{wall}}$. (Justification to follow later.)

Keeping in mind the very strong dependence of rescattering rate on T, one may think that the freeze-out surfaces should be very close to *isotherms*, whatever the explosion in question. Those have been calculated using hydrodynamics, and although we will discuss them later, let us show the results now for kinetic freeze-out. Here, we follow Ramamurti and Shuryak (2018) based on the hydro code MUSIC (Ryu *et al.*, 2013). We use smooth (Glauber-based) initial conditions for LHC and take a small impact parameter, corresponding to a 0–5% centrality bin, the same as in the ALICE data. On the left of Fig. 9.2, we show the location of the isotherm $T = T_f = 100\,\text{MeV}$ in the r–τ plane. Two parts of the freeze-out surface, the "lid" and the "outer edge," are clearly visible. The "lid" is close to having a constant (and very large) proper time $\tau_{\text{f.o.}} \approx 27\,\text{fm/c}$. Note further that the "outer walls" have a very small vertical (i.e. no flow) part at $\tau < 3\,\text{fm/c}$ and then extend linearly, from the initial edge of the nuclei at $r \approx 7\,\text{fm}$ to about $r \approx 20\,\text{fm}$, the final fireball radius. The slope of this edge shows that it moves outward with the velocity of $v_{\text{edge}} \approx 1/3$.[3]

In the right-hand side plot of Fig. 9.2, we show the same isotherm but in a plot with transverse flow rapidity κ and radius r. It consists of a familiar Hubble-like linear rise on the "lid," complemented by a "loop" generated by the "outer edge" of the fireball.[4]

A rather striking fact is that the "loop" reaches a maximal transverse rapidity value $\kappa_{\max} \approx 1.5$, which is nearly flat (independent of r). The reason for that will be explained in Section 12.7 using the Riemann analytic solution between QGP and vacuum.

[3]The velocity of the freeze-out surface must not be confused with the collective flow velocity on this edge, which is much larger.

[4]These plots are to be contrasted with predictions of a Gubser flow, in which the outer side of the fireball extends further to large r and the "loop" is completely absent. This confirmed our conclusion in Section 10.7 that a Gubser flow cannot be used for the description of the outer edge of the fireball.

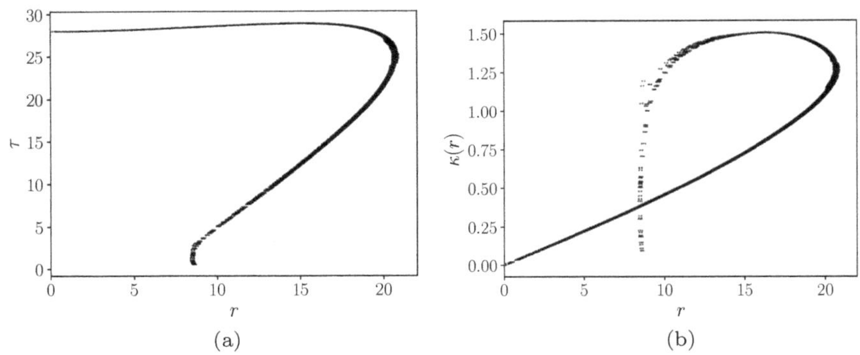

Fig. 9.2. The left plot (a) shows the freeze-out surface $T = 100\,\mathrm{MeV}$, in r, τ (fm) coordinates. The right plot (b) shows the transverse flow rapidity $\kappa(r)$ versus the radial distance r (fm). Individual dots represent hydro cells at the freeze-out surface, corresponding to numerical solution using MUSIC code for 0–5% centrality.

Although it is, of course, nice to observe consistency between numerical hydro and analytic knowledge, the real question is whether hydrodynamics is indeed applicable up to this edge in real heavy ion collisions. An affirmative answer, as advocated in our paper (Ramamurti and Shuryak, 2018), would mean that one can ascribe the high-$p_\perp$ spectra of secondaries to the radial flow up to, say, $p_\perp \approx 5\,\mathrm{GeV}$ for pions. (A more conservative mainstream view on it is that hydro effects only dominate until the maximum of flow harmonics, which is around $p_\perp \approx 2.5\,\mathrm{GeV}$.) It has a factor of 2 in $p_\perp$ but orders of magnitude for the spectra.

9.4 Between Chemical and Kinetic Freeze-outs: Resonance Timing

The most famous applications of radioactive nuclei is the dating of certain historical or archeological specimens. The same ideas are used for addressing timing issues in our field. To answer the question of whether two freeze-outs do happen successively, one uses mesonic/baryonic resonances with lifetimes of corresponding scales, which can act as clocks. With generic estimates, we relate two freeze-outs to inelastic and elastic cross-sections and find that the time between those two strongly depends on the size of the system, from several fm/c for central heavy ion collisions to basically zero for “small systems.”

Let us compare the expected duration of the intermediate hadronic phase to the lifetimes of the resonances. For definiteness, we focus on vector mesons, ρ, K^*, ϕ. The first of them is short lived $\tau_\rho \sim 1\,\text{fm}$, the last lived long as compared to the scale we need, while the semi-strange one, $K^* = s\bar{q}$, has a lifetime of $\tau_{K*} \approx 4\,\text{fm/c}$ and has exactly the time scale we need.

A KK pair produced by a ϕ decay appears too late to suffer significant rescattering on the fireball remnants. However, πK from K^* has significant chances of being rescattered and kicked out of the peak in invariant mass. So, certain additional suppression of the K^* yield should be observed. For simplicity, let us assume that all K^* before τ_{kin} are killed, while all after this time survive. Then, suppression depends exponentially on the difference between chemical and kinetic times: $\sim \exp(-(\tau_{\text{kin}} - \tau_{\text{chem}})/\tau_{K*})$.

Jumping over works, let us illustrate the effect using data from STAR Collaboration (2022d), which include BES-1 runs at several collision energies, all of which are AuAu collisions. Figure 9.3 from this paper compares the centrality dependence of the ratios $\phi/K, K^*/K$. On the left side of the picture are data from smaller systems, and on the right side are those from the largest ones, produced in central collisions. As one can see (inside the uncertainties), the ϕ/K ratio does not depend either on system

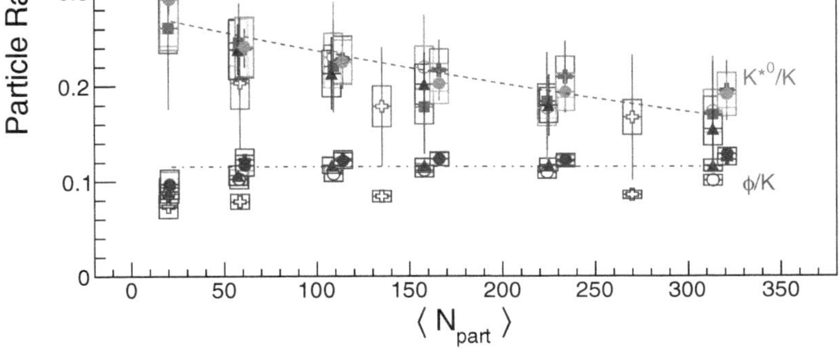

Fig. 9.3. Particle yield ratios for semi-strange K^* and $\bar{s}s$ ϕ resonance as a function of centrality of the collisions, defined via the number of the participant nuclei.

size or on collision energy. The value itself corresponds well to the chemical freeze-out temperature, as we discussed in the corresponding section. In contrast, the K^*/K ratio does show the expected suppression, which gets stronger for larger systems (the right-hand side of the plot). Although fits to these data are not very accurate, in any case, one may conclude that the time difference between two freeze-outs is limited as follows:

$$\tau_{\mathrm{kin}} - \tau_{\mathrm{chem}} > 3\,\mathrm{fm/c} \tag{9.4}$$

9.5 Between Chemical and Kinetic Freeze-outs: Mesonic Chemical Potentials

Before chemical freeze-out, in *chemical equilibrium*, chemical potentials (other than those associated with conserved quantities) should be zero.[5] The "thermal cost" of, say, adding a pion to the fireball is zero.

After chemical freeze-out, the composition-changing inelastic reactions are switched off, after which all numbers of secondaries of *each kind*, N_i, are preserved individually. We call it a transition from a "chemically equilibrated" to a "chemically frozen" regime.

As expansion of the system continues, the corresponding densities $n_i = N_i/V$ are all decreasing, and elastic rescatterings maintain thermal equilibrium. The thermal expressions for those densities

$$n_i = \int \frac{g_i d^3p}{(2\pi)^3} \frac{1}{\exp[(E_p - \mu_i)/T] \pm 1}$$

remain valid, provided the nonzero chemical potentials are tuned appropriately for *any particle kind.* All of them start at zero at T_{ch} and then grow in magnitude until T_{kin} is reached.

This was described in detail for pions by Bebie *et al.* (1992); see also Hung and Shuryak (1995a) and Teaney (2002). By using the thermodynamical relation in the following form

$$\frac{\epsilon + p}{nT} - \frac{s}{n} = \frac{\mu}{T} \tag{9.5}$$

one can further note that for slow (adiabatic) expansion, the s/n ratio is not changing, while in the first term on the left-hand side, the chemical potential *nearly* cancels (it does so provided that the Boltzmann approximation is used). So, neglecting it, one can read the T dependence of μ directly from the left-hand side.

[5] As they are in Planck's blackbody radiation.

(A notorious exceptional case worth mentioning is massless particles, for which the whole left-hand side is just a constant. Therefore, $\mu/T = \text{const}$, so if $\mu = 0$ at the beginning, it remains so for any T. This is what happens in the case of background radiation in the expanding Universe: Although the photons do not collide after the Big Bang, they still have the Planck spectrum, with $\mu = 0$.)

Accounting for the nonzero masses of mesons and Bose statistics, one finds from this relation the corresponding chemical potential $\mu_\pi(T)$. For example, if one assumes chemical freeze-out at T_c,

$$\mu_\pi(Tc = 160\ \text{MeV}) = 0$$

one finds that from a thermal freeze-out (which happens for $PbPb$ collisions at CERN SPS at $T = 110$–120 MeV), the pion chemical potential $\mu_\pi = 60$–80 MeV, see Fig. 9.4. Note that since the system undergoes cooling, the earlier time corresponds to the right-hand side of these figures and the final freeze-out to the left side.

In order to demonstrate that such an effect really occurs in experiments, we give two examples. One, from an early paper (Hung and Shuryak, 1995a), displays the difference between distributions at small momenta, sensitive to chemical potential, between large $PbPb$ and smaller SS collisions. We show in Fig. 9.5 that the ratio of p_t spectra for PbPb collisions (in which we expect thermal freeze-out at $T = 100$–120 MeV and thus the formation of significant pion chemical potential) to our reference point, central SS collisions. (For much smaller sulfur nuclei, the kinetic and chemical freeze-out should be much closer, and thus, the effect of chemical potential becomes negligible). The data sets both correspond to π^+ from the NA44 experiment are in the same kinematical settings, so detector efficiency should cancel out. The observed significant enhancement of this ratio at small p_t is real, indicating a pion chemical potential of about $\mu_\pi \approx 80\,\text{MeV}$.

This effect itself, as well as the previously discussed data confirming it, are from the 1990s. For about two decades, the issue was basically forgotten, but it has resurfaced at the LHC. In Fig. 9.6 from Begun *et al.* (2014), it is shown that the ALICE low-$p_\perp$ data also give a better fit with a nonzero — and a rather large — pion chemical potential.

It is interesting that the parameter of the fit in this last work gives $\mu_\pi \approx m_\pi$, so the authors even speculated whether conditions for Bose–Einstein condensation (BEC) were actually reached for pins at T_{kin}. If this indeed were true, we would see this directly from the value of the femtoscopy parameter λ, giving a "degree of coherence" of the pions.

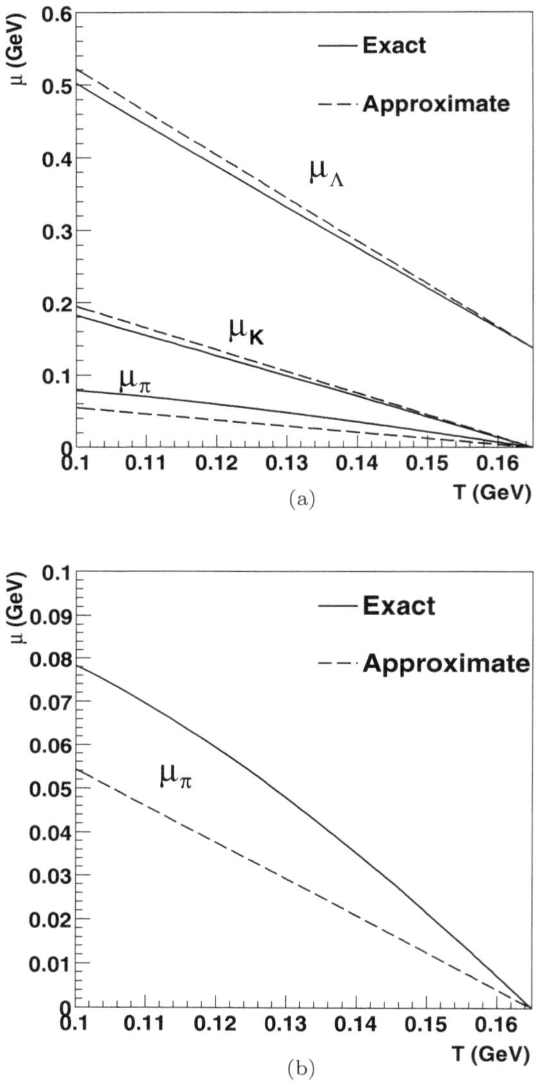

Fig. 9.4. Chemical potentials as a function of temperature at the SPS conditions ($s/n_B = 42$) for (a) π, K and Λ and (b) π only.

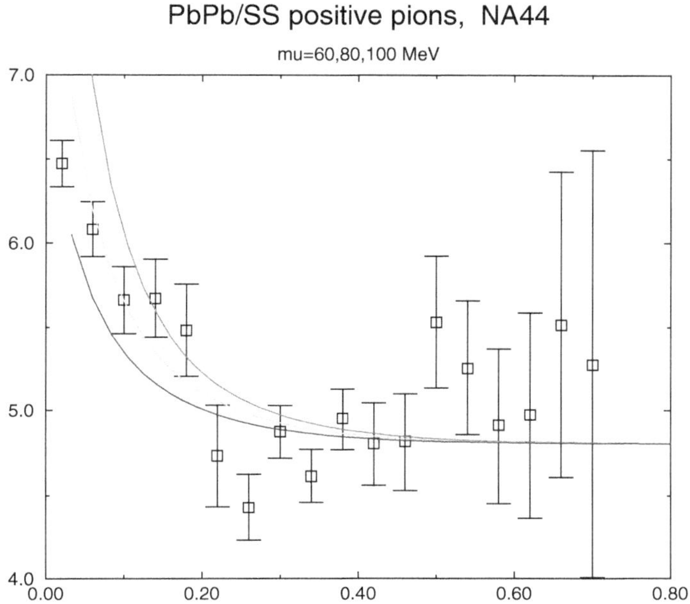

Fig. 9.5. The ratio of π^+ $p_\perp$ spectra for *PbPb* to *SS* collisions. Points denote the experimental data from the NA44 experiment, and the three curves correspond to pion chemical potentials, $\mu_\pi = 60$, 80 and 100 MeV (from bottom up).

Femtoscopy ALICE data, using two and three identical pion correlations, can indeed be fitted with a coherent source containing a fraction of pions as large as 20%.

So, do we actually witness the pion BEC formation at the LHC? In order to answer this question, it is useful to recall the BEC discovery in experiments with ultracold atoms.[6] As the atomic system cools down by evaporation, measurements of the momentum distribution (by switching off the trap) revealed the appearance of a new and much more narrow component. Unlike the usual thermal component, its width was independent of T and related to the inverse spatial size of the BEC cloud.

What is observed in heavy ion collisions is, in fact, similar: As the kinetic freeze-out temperature becomes small, $T_{\rm kin} < 100\,{\rm MeV}$, one finds a new

[6] For the first observation of the BEC of ultracold gases, Eric A. Cornell, Wolfgang Ketterle and Carl E. Wieman were awarded the Nobel Prize in Physics in 2001. One can find the history of this discovery and its details in their Nobel lectures.

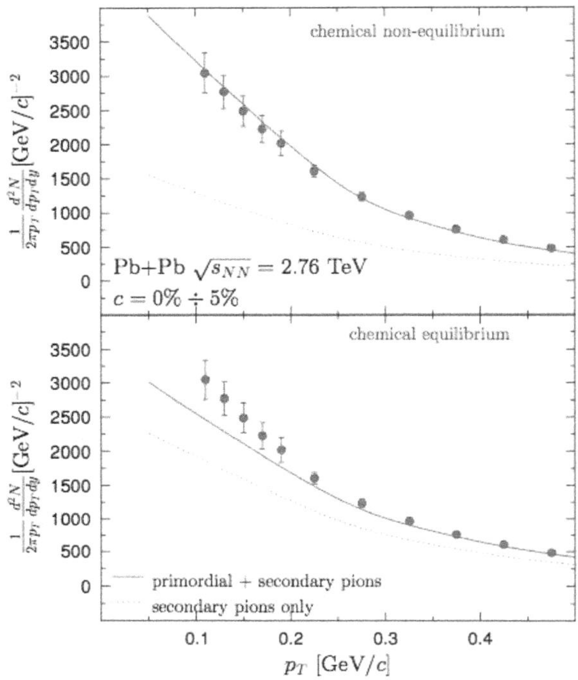

Fig. 9.6. Two plots show by points the same ALICE transverse momentum spectra of pions in the low-p_t region. They are compared with the model with (upper plot) and without (lower plot) pion chemical potential.

component (which we called enhancement) at very small p_t. The difference, however, is in the shape: We don't see an entirely new Gaussian as in the atomic experiments but only smooth deformation of the spectrum, induced by a nonzero $\mu \neq 0$.

However, if the BEC would indeed happen, the non-condensate pions should have the pion chemical potential[7] $\mu = m_\pi$, which does not agree with the fitted value. The in-condensate pions would have such small transverse momenta $p_\perp \sim 1/R$ that we would not be able to see them, either with ALICE or with any other LHC detector available.[8]

[7]Note that we use the relativistic definition of energy; therefore, the critical value of the chemical potential is not zero but a particle mass.

[8]This issue can perhaps be further clarified by a short dedicated run, in which the ALICE detector would use a reduced magnetic field to improve the low p_t acceptance.

Summary and connections: The kinetic freeze-out condition is that rescattering rates become small enough to reach the rate of the system's expansion. One way to observe it is via *femtoscopy*, to be discussed in Chapter 14. Another is via the hydrodynamical flows at their highest extents. The matter between the two freeze-outs is *not* in chemical equilibrium: It is "chemically frozen," and each species gets its own chemical potential. Estimates and observations show that the one for pions reach $\mu_\pi \sim O(100\,\text{MeV})$, but it is still believed to be smaller than the pion mass, $\mu_\pi < m_\pi$ and therefore insufficient for the BEC of the pions.

Chapter 10

RELATIVISTIC HYDRODYNAMICS AND HIGH-ENERGY COLLISIONS

10.1 History and Main Ideas

The theory of high-energy heavy ion collisions is based on three ideas, suggested by three great names in physics. The original paper (Landau, 1953) elegantly connected the initial conditions proposed by Fermi (1951) with the freeze-out stage proposed by Pomeranchuk (1951a). Entropy conservation connects the two stages together, preserving Fermi's prediction of particle multiplicity dependence on energy:

$$n(s) \sim s^{1/4} \tag{10.1}$$

In the paper, it was not mentioned but was subsequently explained by Landau that his motivation was related to the "Landau pole" in QED, suggesting that at small distances, all couplings become large. The main focus of Landau's paper was on the *angular* distribution of secondaries,[1] or (pseudo) rapidity distribution, and the main result was that it should have a near-Gaussian shape. Landau's high standing notwithstanding, the immediate reaction from the high-energy theory community was highly skeptical.

When I entered this field (around 1972), all those papers were 20 or more years old and considered by the leading theorists to be just wrong. Reggeons and Pomerons ruled the day, while any thermo/hydro approaches were completely abandoned. The discovery of QCD and its asymptotic freedom in 1973 further undermined Landau's motivation for matter at

[1]In the 1950s, multiple productions were studied with cosmic ray events detected by emulsion stacks exposed at high-altitude stations. Thus, no momentum measurements were possible.

strong coupling. Yet, as the first accelerator data on many-body production reactions appeared, it was noticed (Shuryak, 1974b)[2] that their rapidity distributions were in agreement with the Landau theory, provided the EOS of hot matter was appropriately modified to that of the "resonance gas."

The next set of data, from the first *pp* collider at CERN called Intersecting Storage Rings (ISR), provided also distributions of transverse momenta for individual secondaries. Those were used to investigate whether they indicated what we now call the "radial flow." It was found by Shuryak and Zhirov (1979) that π, K, p spectra were just thermal *without* any noticeable traces of the collective radial flow! It was rather unsettling.[3]

This setback did not discourage us then: One should perhaps move on to heavy ion collisions, we argued. By 1980, it started looking like an experimental program, which was put into a review (Shuryak, 1980). Fixed-target experiments at CERN and BNL in the 1980s provided encouraging motivation for the RHIC construction. The mainstream theory was the one successfully applied to nuclear collisions at lower energies, namely various cascade codes. After many species of meson and baryon resonances were included (RQMD, UrQMD, etc.), they provided a fair description of fixed-target data. The first applications of hydrodynamics to *radial flow* (Hung and Shuryak, 1995b, 1998) were also reasonably successful.

It was pointed out that for *some* observables, predictions by hadronic cascades and hydrodynamics were very different, or even *opposite*. One clear case was that of *elliptic flow*: The cascades suggested that it would be weakened by a factor of two or so at the RHIC, while hydro predicted its *increase* by a similar factor of two. In the year 2000, the RHIC era started, and hydro predictions turned out to be quite accurate. Moreover, the predicted linear rise of $v_2(p_\perp)$ went well beyond the predicted range of $p_\perp < 1$ GeV, all the way to $p_\perp \sim 3$ GeV, where the elliptic deformation of spectrum was $O(1)$, visible to the naked eye for individual events!

By 2004, all four RHIC experiments were convinced that QGP was seen and that it is not only flows but it does it exceptionally accurately! It turned out to be a "perfect fluid," with very small viscosity. The "parton

---

[2] My attempt to speak about it at the Gribov seminar was a disaster, as even mentioning any macroscopic approach was considered heretic.

[3] Although we came with an explanation: QGP pressure was compensated by "vacuum pressure." Today, we know that *pp* at ISR indeed produced QCD flux tubes, which are mechanically stabilized against the vacuum. Only many years later was it shown that a large number of flux tubes can also initiate the QGP fireball and explosion: we discuss it in Chapter 26.

cascades" (based on pQCD) failed to describe the elliptic flow,[4] thus the term strongly-coupled quark-gluon plasma (sQGP). Multiple applications of holography (see Chapter 21) followed, which theoretically described how strongly coupled matter flows into the hydro regimes.

Let me briefly remind the reader *why* one needs to study relativistic hydrodynamics. Unlike hadron–hadron collisions or e^+e^- annihilation into hadrons, heavy ion collisions produce *macroscopically large* systems. The absolute size does not matter: The size should be large compared to the correlation lengths (or mean free paths, in kinetic language) inside. If it is true, one can use a notion of "matter close to equilibrium" and therefore classify it by "phases" separated by "phase transitions."

Hydrodynamics is also a theory applicable to macroscopically large systems. In order to describe *collective flows*, one adds, on top of thermal fields such as $T(x)$, the flow field $u^\mu(x)$, and derives equations from which those can be obtained. Various versions of hydrodynamics depend on how many derivatives of these fields are included. Needless to say, in practice, one has to use one for which neglected higher derivatives do indeed produce a negligible effect. As we will see, the precise applicability conditions are not so simple, and the issue is actually the central one when hydrodynamics is used for what we call "small systems."

For clarity, let us mention that, while we often call heavy ion collisions the "Little Bangs," we do not mean such "small systems" at all. On the contrary, in such instances, we emphasize analogies between the Little Bangs and the Big Bang, which refers to the expansion of the Universe studied in cosmology. In both, one separates "smooth average behavior" and "small perturbations." The former is described by "full nonlinear relativistic hydrodynamics," while for the latter, its *linearized* version is sufficient.

10.2 "Ideal" Relativistic Hydrodynamics or Local Approximation

The conceptual basis of all versions of hydrodynamics are equations of *energy–momentum–particle number conservation*:

$$\partial_\mu T^{\mu\nu} = 0, \quad \partial_\mu j^\mu_B = 0 \tag{10.2}$$

[4]Some cascade practitioners gave up, some did not. Naively, all one needs to do is increase the cross-sections in a cascade code until all predictions of hydro are met. It did not work, especially at large $p_\perp$; see, for example, Molnar and Gyulassy (2002).

These equations are *always* true, while particular versions of hydrodynamics may or may not be applicable. This issue is somewhat confusing for beginners.

Hydrodynamics is an effective theory, in which $T^{\mu\nu}, j^{\mu}_{B}$ are *approximated*, typically by an expansion in space-time gradients presumed to be small.[5]

The *zeroth* order — known as *ideal* hydrodynamics — assumes that the stress tensor and relevant currents take the *local* form without any derivatives. In equilibrium, $T^{\mu\nu}$ and J^{μ}_{i} are related to the bulk properties of the fluid by the relations

$$\begin{aligned} T^{\mu\nu} &= (\epsilon + p)u^{\mu}u^{\nu} - pg^{\mu\nu} \\ J^{\mu}_{i} &= n_i u^{\mu} \end{aligned} \tag{10.3}$$

Here, ϵ is the energy density, p is the pressure, n_i is the number density of the corresponding current and $u^{\mu} = \gamma(1, \mathbf{v})$ is the proper velocity of the fluid. When matter is at rest (and this can sometimes be arranged by using the so-called *comoving* coordinates[6]) $u^0 = 1, u^m = 0$, the stress tensor is $\mathrm{diag}(\epsilon, p, p, p)$, and $j^0 = n$.

(In strong interactions, the conserved currents are isospin (J^{μ}_{I}), strangeness (J^{μ}_{S}) and baryon number (J^{μ}_{B}). For the hydrodynamic evolution, isospin symmetry is assumed, and the net strangeness is set to zero; therefore, only the baryon current J_B is considered in the following.)

Using such a local form, the set of conservation laws for the stress tensor ($T^{\mu\nu}$) and for the conserved currents (J^{μ}_{i}) takes the form (10.2). Obviously, in non-Cartesian coordinates, the divergence ∂_{μ} needs to be in covariant form with Christoffel symbols. Since u^{μ} is a covariant four-vector, ϵ, p, n are assumed to be coordinate-dependent scalars. Basically, those are defined in the matter co-moving frame.

These five equations contain five unknowns $(\vec{u}, T, \mu)$, provided thermodynamical equations of state (EoS) $p(T, \mu), \epsilon(T, \mu)$ of matter are known, and thus, they form a *closed set*, which can be solved given the appropriate initial and boundary conditions.

Before the discussion of some simple special cases, let us consider one important general feature of ideal hydrodynamics. The first equation has

[5]In the chapter devoted to holographic models of sQGP, we discuss how one can derive hydrodynamics, order by order.

[6]Greek indices will take values 0,...,3, while the Roman ones take 1...3. The flat Minkowski metric used here is $\mathrm{diag}(1, -1, -1, -1)$, so $u^{\mu}u^{\nu}g_{\mu\nu} = (u)^2 = 1$.

an open index μ, so let us project it to the rest frame by multiplying it by[7] u_μ. Here is what one gets from differentiation:

$$u_\mu \partial_\nu[(\epsilon+p)u^\mu u^\nu - pg^{\mu\nu}] = u_\mu(\epsilon+p)_{,\nu}u^\mu u^\nu + u_\mu(\epsilon+p)u^\mu_{,\nu}u^\nu$$
$$+ u_\mu(\epsilon+p)u^\mu u^\nu_{,\nu} - u_\mu p_{,\nu}g^{\mu\nu}$$

Here, note that:

(i) terms 1 and 3 include the square of $u_\mu u^\mu = 1$;
(ii) term 2 vanishes because $u^\mu_{,\nu}u^\nu = 0$, as follows from differentiation of the square of u;
(iii) term 4, in simplified form, cancels the pressure derivative in term 1;
(iv) the derivative of the energy density can be rewritten using thermodynamic relations:

$$= \epsilon_{,\nu}u^\nu + (\epsilon+p)u^\nu_{,\nu}$$

Considering for simplicity $\mu = 0$ and dealing with only T, one can use $d\epsilon = Tds$ for the first term and $\epsilon + p = Ts$ for the second.

One can put it in the simple form

$$\frac{\partial_\mu \epsilon}{\epsilon+p} = \frac{\partial_\mu s}{s}$$

and canceling T, one ends up with the following nice equation of *entropy conservation*:

$$\partial_\mu(su_\mu) = 0 \tag{10.4}$$

So, in ideal hydrodynamics, all the entropy in the system comes from the initial conditions (and/or the discontinuities, such as shock waves) and is conserved until the end of the evolution. This must be so because, assuming that $L \gg l$, we automatically assumed that the expansion is *adiabatic* in all hydro cells.

10.3 Relativistic Hydrodynamics to the First Order in Gradients

In nonrelativistic theory, the additional terms to the stress tensor linear in velocity gradient were worked out by Navier and Stokes back in the first half of the 19th century.

[7]So far, we mean the flat Minkowski metric, and thus, for now, we may not worry about upper and lower indices. It will, however, soon be necessary, as we will use other coordinates.

In the textbook by Landau and Lifshitz, they simply generalized the Navier–Stokes definitions into relativistic form, writing *dissipative additions* to the stress tensor and the current as follows:

$$T_{\mu\nu}^{\rm dissip} = \eta\left(\nabla_\mu u_\nu + \nabla_\nu u_\mu - \frac{2}{3}\Delta_{\mu\nu}\nabla_\rho u_\rho\right) + \xi(\Delta_{\mu\nu}\nabla_\rho u_\rho) \tag{10.5}$$

$$J_\mu^{\rm dissip} = k\left(\frac{\eta T}{\epsilon + p}\right)^2 \nabla_\mu(\mu_B/T) \tag{10.6}$$

These three coefficients, η, ξ and k, called *shear* viscosity, *bulk* viscosity and *heat conductivity*, respectively, are assumed to be some known functions of matter parameters $T\,\mu$. Therefore, they do not add new unknowns, and the set of equations can still be solved.

Note that the following *projection operators* onto the matter rest frame are used here:

$$\nabla_\mu \equiv \Delta_{\mu\nu}\partial_\nu, \quad \Delta_{\mu\nu} \equiv g_{\mu\nu} + u_\mu u_\nu \tag{10.7}$$

The idea was that only spatial derivatives near the rest frame matter.

With the first-order terms in the stress tensor no longer satisfying entropy conservation, putting them on the r.h.s. of the divergence of the entropy current, one finds sign-definite expressions proportional to each kinetic coefficient. Therefore, if they are positive, *any* flow leads to entropy being increased.

Unfortunately, another general relativistic requirement, that any hydro flow should be causal without superluminal modes, is not in general fulfilled in this formulation.

One may also wonder why the definition of energy density does *not* get appended to some gradient terms. In other words, the motivation for the so-called "Landau condition"

$$u_\mu T^{\mu\nu} = -\epsilon u^\nu \tag{10.8}$$

is unclear.

A *modified* relativistic viscous hydrodynamics, first order in gradients (mRHD1 for short), was proposed by Bemfica *et al.* (2019) and Kovtun (2019). Basically, the Landau condition (10.8) is abolished, and the most general form consistent with relativity is used instead:

$$T^{\mu\nu} = (\epsilon + A_1)u^\mu u^\nu + (p + A_2)\Delta^{\mu\nu} - 2\eta\sigma^{\mu\nu} + u^\mu Q^\nu + u^\nu Q^\mu \tag{10.9}$$

where

$$A_1 = \chi_1 \frac{u^\alpha \nabla_\alpha \epsilon}{\epsilon + p} + \chi_2 \nabla_\alpha u^\alpha, \quad A_2 = \chi_3 \frac{u^\alpha \nabla_\alpha \epsilon}{\epsilon + p} + \chi_4 \nabla_\alpha u^\alpha,$$
$$Q_\mu = \lambda \left(\frac{c_s^2 \Delta_\mu^\nu \nabla_\nu \epsilon}{\epsilon + p} + u^\alpha \nabla_\alpha u^\mu \right) \tag{10.10}$$

This includes more dissipative coefficients, $\chi_1, \ldots, \chi_4$; however, with certain conditions imposed on them, one can prove that in this formulation any flow may be both causal and stable.

The equations themselves are complicated and should be looked at in the original works. We only discuss its one (the simplest but most important) application, sound, in Chapter 12. It shows that in "infrared," two viscosities are sufficient, and only higher terms in the dispersion relation are affected.

More generally, the mRHD1 improvement introduces more dissipative constants, and to get their values for sQGP empirically would be extremely challenging.

In practical applications of hydrodynamics, people used the formulations of hydrodynamic equations with second order in gradients, in particular the one due to Mueller, Israel, and Stewart (MIS). I am not aware that causality has been proven in general, but it worked in applications to heavy ion collisions. Second-order hydrodynamics is significantly more complicated, with new dissipative functions and new equations invented for them. We would not discuss them here; for a review of those, see, for example, Romatschke (2010).

Finally, approximate resummations of all gradients were suggested by Lublinsky and Shuryak (2009): It will be discussed in the chapter on holography.

In concluding this section, let me emphasize again that we are blessed by the fact that dissipative terms in sQGP are additionally small due to the strong coupling regime. It took considerable effort to get quantitative estimates of its shear and bulk viscosities. We are also blessed for the second time that the dissipative terms of sQGP are still large enough to exclude turbulence.

10.4 The Simplest Analytic Solution: The Riemann Wave

Relativistic hydrodynamics is significantly more complicated than the nonrelativistic one. The natural variable now includes the rapidity y instead of the velocity v: It appears inside the hyperbolic functions $\cosh(y), \sinh(y)$ in the equations, which lead to extra nonlinearities.

In a very crude geometric description, the fireball is a 4D domain limited by a 3D surface, consisting of three major parts: (i) the "initial" surface, on which inflows of matter is defined; (ii) a time-like freeze-out surface at which most of particles are born; and finally (iii) a space-like "sides," at which expanding matter is separated from vacuum. Although in practice the problem has been mostly addressed numerically, there are still few instructive analytic solutions.

In a book like this one, those should be introduced in the order of their mathematical complexity; therefore, we begin with the Riemann solution, also known as *the rarefaction fan.* Its first discussion in the context of heavy ion collisions was made by Baym *et al.* (1983); however, it was not really used for decades until Ramamurti and Shuryak (2018) applied it for radial flow at large $p_\perp$, see Section 12.7.

One starts with just *two* coordinates, time and space, and *two* unknown functions. Naively, this setting is already as simple as it can be. Yet, Riemann suggested how one can still simplify it further!

Two unknown functions, namely the energy density $\epsilon(\tau, x)$ and flow rapidity $y(\tau, x)$ *may* in fact be directly related:

$$\epsilon(\tau, x) = F\big(y(\tau, x)\big) \tag{10.11}$$

For 1+1-dimensional relativistic hydrodynamics, we will not change the coordinates, retaining the original t, x. Using standard rapidity notations for four-velocity,

$$u^0 = \cosh(y), \quad u^1 = \sinh(y) \tag{10.12}$$

one has the stress tensor in the form

$$T^{00} = \epsilon \, \cosh^2(y) + p \, \sinh^2(y) \tag{10.13}$$

$$T^{01} = (\epsilon + p) \cosh(y) \sinh(y) \tag{10.14}$$

$$T^{11} = \epsilon \, \sinh^2(y) + p \, \cosh^2(y) \tag{10.15}$$

with the two equations of ideal hydrodynamics being

$$\frac{\partial T^{00}}{\partial t} + \frac{\partial T^{01}}{\partial x} = 0 \tag{10.16}$$

$$\frac{\partial T^{01}}{\partial t} + \frac{\partial T^{11}}{\partial x} = 0 \tag{10.17}$$

For simplicity, we use the EOS $p = c_s^2 \epsilon$ with a constant speed of sound, c_s. If so, the EOS used has no dimensional parameters, and the solution for rapidity therefore must be *self-similar*:

$$y(t,x) = f\left(\frac{t}{x}\right) \tag{10.18}$$

At this point, we have two unknown functions, F and f.

Substituting those into the equations of motion (10.13), one can see that they have similar forms:

$$\frac{F'(f)}{F(f)} + \frac{-2(1+c_s^2)(t\,\cosh(2f) - x\,\sinh(2f))}{x - c_s^2 x + (1+c_s^2)x\,\cosh(2f) - (1+c_s^2)t\,\sinh(2f))} = 0$$

$$\frac{F'(f)}{F(f)} + \frac{2(1+c_s^2)(x\,\cosh(2f) - t\,\sinh(2f))}{t - c_s^2 t - (1+c_s^2)t\,\cosh(2f) + (1+c_s^2)x\,\sinh(2f)} = 0$$

Note here the key consequence of the Riemann assumption: The derivative f' appears in all terms and, if nonzero, does cancel out. As a result, one can find f from the ordinary (not differential) equation, requiring two complicated ratios in the equations above to be equal to each other. These complicated equations lead to (a surprisingly simple) answer:

$$y = f(t/x) = \frac{1}{2}\log\left[\left(\frac{1-c_s}{1+c_s}\right)\left(\frac{t+x}{t-x}\right)\right] \tag{10.19}$$

Note that the logarithm of the second bracket corresponds to the so-called *spatial rapidity*:

$$\eta(t,x) = \frac{1}{2}\log\left(\frac{t+x}{t-x}\right) \tag{10.20}$$

Because the rapidity of the flow y and η both transform additively under longitudinal Lorentz transformation, a linear relation between them could have actually been guessed on general grounds, but not the additional constant depending on the speed of sound.

Substituting it into both EOMs, one finds that both are now indeed the same:

$$\frac{F'(f)}{F(f)} + \frac{1+c_s^2}{c_s} = 0$$

from which

$$F(f) = \exp\left[-\left(\frac{1+c_s^2}{c_s}\right)f\right]$$

times an arbitrary constant, to be determined from the boundary conditions.

In the discussion to follow, we will need the *scalar expansion rate*, so let us give it for the Riemann flow:

$$\partial_\mu u^\mu = \frac{1}{\sqrt{(1-c_s^2)(t^2-x^2)}} \tag{10.21}$$

It is, as should be expected, Lorentz-invariant proper time multiplied by a nontrivial factor.

10.5 Bjorken Flow

The view on strong interaction origins took a decisive turn at the end of the 1960s. Bjorken scaling, the DIS experiment at SLAC confirming it and Feynman's parton model have indicated that at small distances, the "strong forces" become weakened. This alone meant that the Fermi–Landau instant stopping was impossible. Feynman argued that an observer sitting at *any* rapidity y, far from two ends $-Y \gg y \gg Y$, sees basically the same picture of two (nearly infinitely) thin objects passing by and thus should produce the *same* density of particles for all y.

The idea of a rapidity-independent "scaling" distribution of secondaries originated from Feynman's argument and became instantly popular at the start of the 1970s. The existence of a rapidity-independent hydrodynamic solution was perhaps first noticed by Landau, who used the rapidity variable in his classic paper as a somewhat trivial case.[8] The explicit solution of the equations of relativistic hydrodynamics was obtained by Bjorken (1983). It is instructive to first describe it in the original Cartesian coordinates. The setting assumes a collision of two homogeneous infinite sheets of matter, so there will be no dependence on the transverse coordinates x, y, only on the time t and the longitudinal coordinate we now call z. Two 1+1-dimensional equations $\partial_\mu T^{\mu\nu} = 0$ can be rewritten in the following way:

$$\frac{\partial}{\partial t}(s\,u^0) + \frac{\partial}{\partial z}(s\,u^1) = 0 \tag{10.22}$$

$$\frac{\partial}{\partial t}(T\,u^1) + \frac{\partial}{\partial z}(T\,u^0) = 0 \tag{10.23}$$

If $s(T)$ is known, there are two unknowns, e.g. T, u^1 (since $(u^0)^2-(u^1)^2 = 1$), and two equations, as it should be. Their solution can be achieved using

[8]The space-time picture connected with such a scaling regime was in fact widely discussed; see for example, Chiu *et al.* (1975) and Shuryak (1974a).

the 1D "Hubble ansatz" for the four-velocity

$$u^0 = \frac{t}{\tau}, \quad u^1 = \frac{z}{\tau} \tag{10.24}$$

In other words, one assumes that rapidity y is just equal to spatial rapidity (10.20). Another natural assumption is that Lorentz-invariant s, T can only depend on the combination $\tau = \sqrt{t^2 - z^2}$, the Lorentz-invariant proper time. The reader is expected to check that the second equation is identically satisfied and to see that the first one becomes an ordinary differential equation in proper time τ:

$$\frac{ds(\tau)}{d\tau} + \frac{s}{\tau} = 0 \tag{10.25}$$

Rewriting it in the form of $d(s\tau)/d\tau = 0$, one finally has a simple solution:

$$s = \frac{\text{const}}{\tau} \tag{10.26}$$

Exactly as in the Big Bang, for each "observer" (the volume element), the picture is just the same, with the pressure from the left compensated by that from the right. The history is also the same for all volume elements if they are expressed in their own proper time τ.

So far, dissipative phenomena have been ignored. Including the dissipative terms into our equations, one finds the following source for the entropy current:

$$\frac{1}{\epsilon + p}\frac{d\epsilon}{d\tau} = \frac{1}{s}\frac{ds}{d\tau} = -\frac{1}{\tau}\left(1 - \frac{(4/3)\eta + \xi}{(\epsilon + p)\tau}\right) \tag{10.27}$$

with shear and bulk viscosities η, ξ.

Let us also show another approach to the Bjorken flow, which may look more complex at the beginning but is a more natural way to derive further results, such as Gubser's flow. In the above, a relativistic substitute for velocity by the rapidity

$$u^0 = \cosh(y), \quad u^1 = \sinh(y) \tag{10.28}$$

was essential. By analogy, let us introduce the so-called *spatial rapidity*[9]:

$$t = \tau\cosh(\eta) \quad z = \tau\sinh(\eta) \tag{10.29}$$

and use τ, η as two coordinates. We then assume a *separation of variables*, namely that the thermodynamical variables would depend only on τ,

[9] I apologize for notations in which it is denoted by the same Greek letter as viscosity; I hope that no confusion results.

$s(\tau), T(\tau)$, etc., while the rapidity would only depend on η. The *ansatz* above simply reduces the latter dependence to both rapidities being simply equal to each other:

$$y = \eta \tag{10.30}$$

This is rather nice and straightforward; however, let us still remind the reader how tensor equations of motion can be written for an arbitrary coordinate system. All one has to do is substitute the colons (the ordinary partial derivative) with *semicolons*, indicating the *covariant* derivatives. So, our ideal hydro equations now look as follows:

$$T^{mn}{}_{;m} = 0, \qquad j^{m}{}_{;m} = 0, \tag{10.31}$$

For the tensors of ranks 1 and 2 that we only have here, it reads explicitly

$$j^{i}_{;p} = j^{i}_{,p} + \Gamma^{i}_{pk}\, j^{k}, T^{ik}{}_{;p} = T^{ik}{}_{,p} + \Gamma^{i}_{pm} T^{mk} + \Gamma^{k}_{pm} T^{im}, \tag{10.32}$$

where the comma denotes a simple partial derivative, and the Christoffel symbols Γ^{s}_{ij} are given by derivatives of the metric tensor $g^{ab}(x)$:

$$\Gamma^{s}_{ij} = (1/2) g^{ks} \big(g_{ik,j} + g_{jk,i} - g_{ij,k} \big). \tag{10.33}$$

The coordinate transformation is

$$x^{\mu} = (t, x, y, z) \longrightarrow \bar{x}^{m} = (\tau, x, y, \eta)$$

$$t = \tau \cosh \eta \qquad \tau = \sqrt{t^2 - z^2} \tag{10.34}$$

$$z = \tau \sinh \eta \qquad \eta = (1/2) \ln \frac{t+z}{t-z}. \tag{10.35}$$

In the new coordinate system, the velocity field (after inserting $v_z = z/t$) is given by

$$\bar{u}^{m} = \bar{\gamma}(1, \bar{v}_x, \bar{v}_y, 0) \tag{10.36}$$

with $\bar{v}_i \equiv v_i \cosh \eta$, $i = x, y$, and $\bar{\gamma} \equiv 1/\sqrt{1-\bar{v}_x^2-\bar{v}_y^2}$. and the metric in the new coordinate system is therefore

$$\begin{aligned} ds^2 = g_{\mu\nu} dx^{\mu} dx^{\nu} &= dt^2 - dx^2 - dy^2 - dz^2 \\ &= d\tau^2 - dx^2 - dy^2 - \tau^2 d\eta^2 \end{aligned} \tag{10.37}$$

and thus

$$\bar{g}_{mn} = \begin{pmatrix} 1 & 0 & 0 & 0 \\ 0 & -1 & 0 & 0 \\ 0 & 0 & -1 & 0 \\ 0 & 0 & 0 & -\tau^2 \end{pmatrix} \tag{10.38}$$

Out of $2^3 = 8$ Christoffel symbols, only three are non-vanishing:

$$\Gamma^{\eta}_{\eta\tau} = \Gamma^{\eta}_{\tau\eta} = \frac{1}{\tau}, \quad \Gamma^{\tau}_{\eta\eta} = \tau. \tag{10.39}$$

Returning to the example at hand, the Bjorken flow, we do not just use an arbitrary coordinate transformation, but a transformation into the *flow co-moving frame*. Therefore, it connects the Euler and Lagrange approaches to hydrodynamics. In the new coordinate, there is "no flow," so the velocity vector is trivial, $u^{\tau} = 1, u^{\eta} = 0$. Entropy conservation immediately takes the form (10.25).

Extensions of the Bjorken flow used in practice combine longitudinal boost-invariant symmetry with an *arbitrary* transverse expansion. It is convenient (as was done already by Landau) to change longitudinal variables from t, z to more appropriate variables: $\tau = \sqrt{t^2 - z^2}$ and $\eta = \frac{1}{2}\log(\frac{t+z}{t-z})$, which are respectively referred to as the *proper time* and the *spatial rapidity*. Boost invariance then means that the solution for any value of η may be found by boosting the solution at $\eta = 0$ to a frame moving with a velocity of $v = \tanh(\eta)$ in the negative z-direction. With this assumption, the equations of motion become two dimensional and are given at $\eta = 0$ by

$$\begin{aligned}
\partial_\tau(\tau T^{00}) + \partial_x(\tau T^{0x}) + \partial_y(\tau T^{0y}) &= -p \\
\partial_\tau(\tau T^{0x}) + \partial_x(\tau T^{xx}) + \partial_y(\tau T^{xy}) &= 0 \\
\partial_\tau(\tau T^{0y}) + \partial_x(\tau T^{xy}) + \partial_y(\tau T^{yy}) &= 0 \\
\partial_\tau(\tau J^0_B) + \partial_x(\tau J^B_i) + \partial_y(\tau J^y_B) &= 0.
\end{aligned} \tag{10.40}$$

Integrating over the transverse plane, one finds that the net baryon number per unit spatial rapidity, $\int dxdy\,(\tau J^0_B)$, and the transverse momentum per unit rapidity, $\int dxdy\,(\tau T^{0x})$, as well as $\int dxdy\,(\tau T^{0y})$, are conserved. The energy per unit rapidity, $\int dxdy\,(\tau T^{00})$, decreases due to the work done per unit time by the pressure in the longitudinal direction, $\int dxdy\,p$:

$$\partial_\tau(\tau S^0) + \partial_x(\tau S^x) + \partial_y(\tau S^y) = 0. \tag{10.41}$$

Integrating over the transverse plane, we find that

$$\int dx\,dy\,\tau\,s\gamma = \frac{dS_{\text{tot}}}{d\eta} \tag{10.42}$$

is a constant of the motion.

The simplest case is that of *axially symmetric* transverse expansion, as used by Baym *et al.* (1983):

$$\frac{\partial}{\partial t}(rTs\cosh\alpha) + \frac{\partial}{\partial r}(rTs\sinh\alpha) = 0 \tag{10.43}$$

$$\frac{\partial}{\partial t}(T\sinh\alpha) + \frac{\partial}{\partial z}(T\cosh\alpha) = 0 \tag{10.44}$$

appears if one also uses the *transverse rapidity* variable $\alpha = \tanh^{-1} v_r$. One can find in Baym *et al.* (1983) and also in Rischke (1996) nice discussions of the simplest numerical solutions of this equation, as well as of the 1D and 3D spherical ones.

10.6 Landau Flow and Rapidity Distributions

As we already mentioned, one elegant point made by Landau (1953) was entropy conservation, relating Fermi's initial condition with Pomeranchuck's freeze-out conditions.

Landau's solution therefore starts with *instantly* equilibrated matter in a thin (Lorentz-contracted) disk.[10] How quickly and by what mechanism the initial equilibration does happen in real heavy ion collisions will be a subject of a chapter later in the book.

Landau denotes the width of the initial Lorentz-contracted nuclei $\Delta \sim 2R/\gamma_{\rm CM}$ where $\gamma_{\rm CM} = \sqrt{s}/2M$ is the Lorentz factor of the incoming objects in the CM frame. Using variables

$$\tau = \log\left(\frac{t}{\Delta}\right), \quad \eta = \log\left(\frac{t-x}{\Delta}\right)$$

and the ultrarelativistic nature of the flow

$$u^0 \approx u^1 \equiv u \gg 1, \quad u^0 - u^1 \approx \frac{1}{2u}$$

Landau assumes that the flow velocity has the form

$$u^2 = f(\tau,\eta)\frac{t}{t-x} \tag{10.45}$$

where f is a slow (logarithmically) dependent function of the original coordinates. Note that η is basically the spatial rapidity, and his assumption

[10]Landau could not predict the discovery of asymptotic freedom 20 years after his work, which made it clear that QCD is not like QED and that instant equilibration is not feasible, at least not at asymptotically high collision energies. Therefore, Landau's solution is mostly of historical and methodical interest only.

indirectly implies a smooth relation between spatial and momentum rapidities, which, in 1953, was clearly a breakthrough idea. Note further that in such coordinates, the middle of the system, $x \approx 0$ means $\tau \approx \eta$, but the region in which the solution will be found is $t \gg t - x \gg \Delta$.

Ignoring transverse motion and using only two longitudinal ideal hydro equations, he found the solution in the following form:

$$f = \frac{1}{2}\sqrt{\frac{\tau}{\eta}}, \quad \epsilon = \epsilon_0 \exp\left[-\frac{4}{3}(\eta + \tau - \sqrt{\tau\eta})\right] \tag{10.46}$$

which leads to the following distribution of energy and entropy in rapidity:

$$\frac{dE}{d\eta} \sim \epsilon u^2(t - x) \sim \exp\left[-\frac{1}{3}(\sqrt{\tau} - 2\sqrt{\eta})^2\right] \tag{10.47}$$

$$\frac{dS}{d\eta} \sim su(t - x) \sim \exp\left[-\frac{1}{2}(\sqrt{\tau} - \sqrt{\eta})^2\right] \tag{10.48}$$

Unlike energy, which of course is concentrated at large rapidities, the entropy distribution is maximal at $\eta \approx \tau$ or mid-rapidity.

Landau then proceeded to what he called the "conical solution," introducing variable λ related to the angle θ relative to the beam

$$\theta \sim \frac{t - x}{R} \equiv e^{-\lambda}$$

where R is the disk radius. In fact, Landau specified this relation even for large angles via $\lambda = -\log\tan(\theta/2)$; in modern language, it is the *pseudorapidity*.

He also introduces convenient notation for the collision Lorentz factor,

$$\frac{\Delta}{R} = \frac{1}{\gamma_{\mathrm{CM}}} \equiv e^{-L}$$

so that $\eta = L - \lambda$. The transition to conical flow happens at a certain time, after which the angle θ remains constant. The final angular distribution of matter entropy is given by the following curious distribution:

$$dS \sim e^{\sqrt{L^2 - \lambda^2}} d\lambda \tag{10.49}$$

with a maximum at $\lambda = 0$ or large (say $\theta = \pi/2$) angles.

The only experimental data available at the time were emulsion-based, giving mostly an angular distribution of tracks. There were no means to measure momenta; therefore, "pseudorapidity" (as a substitute for rapidity) was used. Naturally, Landau also used this angular variable.

Note that Landau's $L \approx \log(\gamma_{\mathrm{CM}})$ is basically the beam rapidity in the center-of-mass frame. The entropy distribution gives the distribution of particles: It is predicted to be flat around the mid-rapidity, rounding up at the ends. Near mid-rapidity, $\lambda \ll L$, and one can expand Landau's square root, obtaining just a Gaussian:

$$\frac{dS}{d\lambda} \sim e^{-\frac{\lambda^2}{2L}} \tag{10.50}$$

Landau's paper ended with a graph showing the predicted rapidity distribution for a few collision energies without any comparison with any data. Still, there is a remark that his rapidity distribution describes the available cosmic ray data well.[11]

Only in the 1970s could proton accelerators produce high enough energy for *pp* collisions with $O(10)$ secondaries. As already mentioned above, that was the time of Regge phenomenology. With the discovery of QCD, it was the time of gluon ladders, which later morphed into QCD flux tubes.

Hydrodynamics was not mentioned anymore; nonetheless, it was tempting to compare the spectra of secondaries to Landau's predictions. I did so and found (Shuryak, 1972a) that it did describe well the rapidity distributions, provided one makes the following two modifications:

(i) Generalize the Landau solutions to EOS with a different sound velocity $c_s^2 \neq 1/3$.
(ii) Recognize that the "leading nucleons" in *pp* collisions as well as "spectators" in *AA* ones do not come from the "stopped disk" and thus have to be accounted for separately; that was done by Zhirov and Shuryak (1975).

Generalized expressions for the multiplicity and parameter L from Shuryak (1972a) were

$$N \sim s^{\frac{1-c_s^2}{2(1+c_s^2)}}, \quad L = \frac{4}{3}\frac{c_s^2}{1-c_s^4}\log\left(\frac{s}{4M^2}\right) \tag{10.51}$$

The fitted value of the sound velocity for a hadron resonance gas (HRG) used then was

$$c_s^2 = 0.2$$

It worked, maybe by some coincidence.

---

[11]In those days, it was just a number of events recorded by stacks of emulsion plates exposed for months on some mountain tops. Matching tracks between plates and calculating emission angles was done by hand and was very labor-intensive.

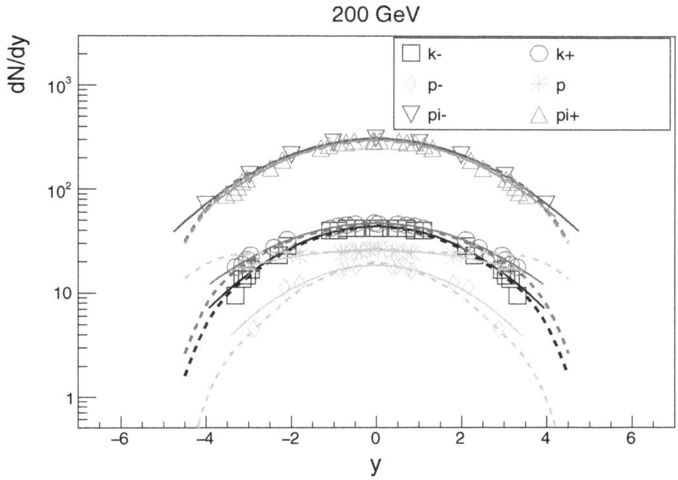

Fig. 10.1. Rapidity distributions of secondaries in the RHIC *AuAu* collisions at 200 GeV/N (points) compared to the CRMC EPOS 1.99 event generator (solid lines) and the Landau expression discussed in the text (dashed lines). The description is good for all secondaries except for protons at large rapidities, which is explained by the contribution of the nucleons from the initial nuclei.

Several decades later, when the heavy ion colliders RHIC and LHC had provided high-quality rapidity distributions, people asked whether they agreed with those predicted by Landau's solution. And again, a very good agreement was found; see, for example, Fig. 10.1 from Rahman *et al.* (2022). The formulae are those given above, with the addition of spectators contributing at larger rapidities. As one can see, the description is, in fact, excellent.[12]

Summarizing, the rapidity distributions for *all* collisions are well described by the (modified) Landau predictions, nearly seven decades after they were made. While it is a remarkable observation, let us think about what it would imply. If one accepts the Landau model as a correct description of what is actually happening, one would also need to accept that:

(i) the matter is instantly stopped, at any collision energy;

[12]The fitted values of the width for the LHC data, however, correspond to $c_s^2 = 0.17 \pm 0.03$, not 1/3. Note that it is only one sigma away from 0.2, the number I derived half a century earlier, back in 1972, from a fit to the hadronic density of states.

(ii) there is no distinction between the hadronic and heavy ion collisions, and the same matter is always produced;

(iii) the EOS of this matter is $c_s^2 \approx 0.2 \neq 1/3, \epsilon \sim p \sim T^6 \neq T^4$, so it is a *resonance gas* rather than a (nearly scale-invariant) QGP!

Needless to say, *none* of these three statements are true, as we learned from theoretical developments and from many other experimental observables, which we detail in the chapters to follow. One obvious lesson from this story is that the agreement of the theory with some experimental observables does not guarantee its validity. One needs to check *all*, or at least as many observables and consequences of the theory as possible. The Landau scenario, its modifications and fits described in this section, were not worked out in vain. Rather, they were quite necessary steps in the learning process. Only in (most) textbooks does this process look straightforward and logical, but it never is in reality. Recognizing the wrong turns in the process is the most difficult — and thus most important — part of it.

10.7 Gubser Flow

This solution given by Gubser (2010) and Gubser and Yarom (2011) retains the rapidity-independence of the Bjorken flow while replacing the translational invariance in the transverse plane by a symmetry under a special conformal transformation. Axial $O(2)$ symmetry is here promoted to higher $O(3)$ symmetry.

However, this nice piece of mathematical physics comes with a certain price. One restriction is that the matter is required to be conformal, with the EOS restricted to $\epsilon = 3p \sim T^4$. Another restriction is that the colliding system has to be of a particular shape, corresponding to the conformal map of the sphere onto the transverse plane. We will see that this last requirement does lead to certain problems, limiting the practical applications of Gubser's solution.

In the case of ideal hydrodynamics, a solution depends on three parameters:

(i) A dimensional q defines the size of the system (and roughly corresponds to the radii of the colliding nuclei).

(ii) A dimensionless f^* characterizes the effective number of degrees of freedom in the matter.

(iii) $\hat{\epsilon}_0$ characterizes the amount of energy (or entropy) the system has.

The original setting uses the coordinates we used above, namely the proper time, spatial rapidity, transverse radius, azimuth angle, $(\bar{\tau}, \eta, \bar{r}, \phi)$ with the metric

$$ds^2 = -d\bar{\tau}^2 + \bar{\tau}^2 d\eta^2 + d\bar{r}^2 + \bar{r}^2 d\phi^2, \tag{10.52}$$

where dimensionless coordinates $\bar{\tau} = q\tau, \bar{r} = qr$, are just the rescaled versions of the actual coordinates.

Looking for solutions independent of both "angles" η, ϕ and using *transverse rapidity* κ defined via

$$u_\mu = (-\cosh\kappa(\tau, r), 0, \sinh\kappa(\tau, r), 0) \tag{10.53}$$

Gubser obtained the following solution:

$$v_\perp(\tau, r) = \tanh\kappa(\tau, r) = \left(\frac{2q^2\tau r}{1 + q^2\tau^2 + q^2 r^2}\right) \tag{10.54}$$

$$\epsilon(\tau, r) = \frac{\hat{\epsilon}_0 (2q)^{8/3}}{\tau^{4/3}\left(1 + 2q^2(\tau^2 + r^2) + q^4(\tau^2 - r^2)^2\right)^{4/3}} \tag{10.55}$$

Figure 10.2 shows the plots of an isotherm (freeze-out surface) on a proper time-transverse radius plane, as well as the transverse velocity. Note that the the part corresponding to a "lid" of the freeze-out surface does correspond to what one gets from numerical solutions, in particular $v_\perp(r) \sim r$. However, the "outside wall" has a completely different shape from what we already discussed in the section on kinetic freeze-out. After $v_\perp$ reaches a maximum, it decreases at large r, rather than staying constant (as we concluded from the Riemann solution and numerics). This happens because Gubser's solution does not describe an explosion into a vacuum but rather into some kind of "atmosphere." Respectively, Gubser's solution can only be used for the time-like part of the freeze-out surface, the "lid."

In the second paper (Gubser and Yarom, 2011), the same solution was rederived by a transformation to the co-moving frame. In order to do so, the metric has to be rescaled:

$$ds^2 = \tau^2 d\hat{s}^2 \tag{10.56}$$

and $d\hat{s}^2$ rewritten via a coordinate transformation from the τ, r to new time and space coordinates ρ, θ given by

$$\sinh\rho = -\frac{1 - q^2\tau^2 + q^2 r^2}{2q\tau} \tag{10.57}$$

$$\tan\theta = \frac{2qr}{1 + q^2\tau^2 - q^2 r^2} \tag{10.58}$$

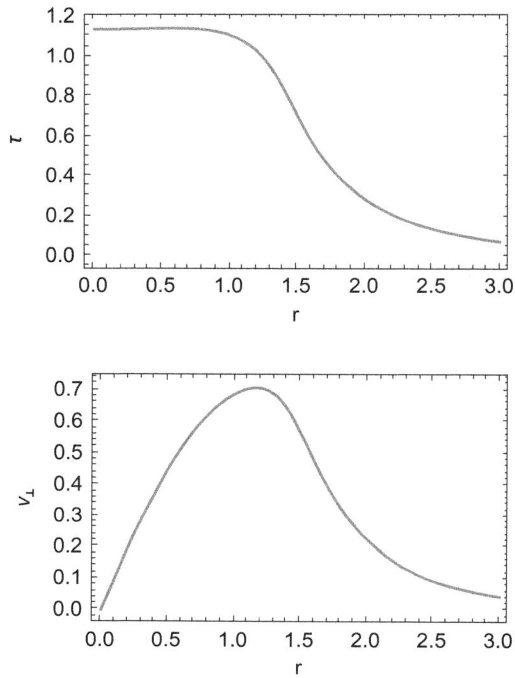

Fig. 10.2. Example of (freeze-out) isotherm on τ, r plot and transverse velocity versus r.

In the new coordinates, the rescaled metric reads

$$d\hat{s}^2 = -d\rho^2 + \cosh^2 \rho \left(d\theta^2 + \sin^2 \theta d\phi^2\right) + d\eta^2 \tag{10.59}$$

The relation between the velocity in the Minkowski space in the (τ, r, ϕ, η) coordinates and the one in the rescaled metric in the $(\rho, \theta, \phi, \eta)$ coordinates corresponds to

$$u_\mu = \tau \frac{\partial \hat{x}^\nu}{\partial \hat{x}^\mu} \hat{u}_\nu, \tag{10.60}$$

In the new coordinates, the fluid is at rest, $u^\rho = 1$, etc.

The energy density transforms as $\epsilon = \tau^{-4}\hat{\epsilon}$. The temperature (in the rescaled frame, $\hat{T} = \tau f_*^{1/4} T$, with $f_* = \epsilon/T^4 = 11$ as in Gubser, 2010) is now dependent only on the new time ρ. The new solution was also generalized to a case with *nonzero viscosity*:

$$\hat{T} = \frac{\hat{T}_0}{(\cosh \rho)^{2/3}} + \frac{H_0 \sinh^3 \rho}{9(\cosh \rho)^{2/3}} \, {}_2F_1\left(\frac{3}{2}, \frac{7}{6}; \frac{5}{2}, -\sinh^2 \rho\right) \tag{10.61}$$

where H_0 is a dimensionless constant made of the shear viscosity and temperature, $\eta = H_0 T^3$, and ${}_2F_1$ is the hypergeometric function. In the inviscid case, the solution is just the first term of expression (10.61), and of course, it also conserves the entropy in this case. The picture of the explosion is obtained by transformation from this expression back to τ, r coordinates.

Small perturbations to the Gubser flow obey linearized equations, which have also been derived by Gubser and Yarom (2011). Those we will discuss later in the chapter on sounds.

The Gubser flow has been used as a theoretical laboratory ever since. A complete Green function has been constructed (Staig and Shuryak, 2011b), leading to pictures of sound circles, which we will discuss. Generalization to perturbations by the quenching jets, with the sounds propagating in the rapidity direction, was done by Shuryak and Staig (2013). For the second order (the Israel–Stuart version) of the hydrodynamics, it has been done by Marrochio *et al.* (2015) and Pang *et al.* (2015). The Boltzmann equation (in tau-approximation) has also been solved in such a setting, see Denicol *et al.* (2014).

There are also a number of phenomenological applications. Without going into those, let me just comment that those are limited by the fact that at large r, the power tail of the solution is completely inadequate for heavy ion collisions. So to speak, the Gubser solution is like an explosion in the atmosphere, while the real ones happen in a vacuum. As a result, in applications, one basically needs to amputate the unphysical regions by hand.

10.8 Particle Spectra and Radial Flow

Reading the textbooks, people always get a bit drowsy: Everything is so smooth, with famous scientists going from one triumph to another in a familiar procession. The laws of physics they discovered are so logical and now nearly self-evident. The textbooks tend to hide the actual drama; they do not spend time describing what "common sense wisdom" was dominant at the time before proven to be wrong by those famous scientists. ("Flat Earth" being perhaps the only exception.) Therefore, at least in this section, let me start with the actual dramatic events.

The original Landau's work on the hydrodynamical model suggested using it for NN collisions, considering heavy ion collisions later as just another application. It was also focused on the quasi-rapidity distribution, the only information of cosmic ray emulsion studies of the 1950s could deliver.

When I started to compare the Landau model with data 20 years later, there appeared data from the high-energy accelerators, and the rapidity distribution — with a modified speed of sound — actually worked. Then came much better data from the CERN Intersection Storage Ring (ISR, the first ever hadronic collider) on spectra of identified secondaries, π, K, p. With my first graduate student, Zhirov, we set out to look for signs of collective *transverse radial flow.* The idea of our paper[13] (Shuryak and Zhirov, 1979) was to compare the spectra of particles of different masses, namely pions, kaons and nucleons.

In *pp* and other hadronic collisions, as we discussed in the introduction, the observed secondaries originate from quarks produced by QCD string breaking. Expressions for pair production in a longitudinal electric field have the feature that the dependence on particle mass and its transverse momentum appear together in the form of the so-called "transverse mass":

$$m_t = \sqrt{m^2 + p_t^2} \tag{10.62}$$

Deviations from universal $m_\perp$-scaling, if observed, would indicate that some mechanism, other than thermal, is in place. So, the collective radial velocity of an explosion can be measured by comparing the spectra of light and heavy particles. For light (pions), the thermal spectrum is exponential in p_t, and boosted by flow, it remains the exponential one with a "blue-shifted" temperature of $T^* = T\gamma_\perp$, including the gamma factor $\gamma_\perp = \cosh(\kappa)$ of the transverse flow. For heavy particles (such as protons), the effect of the Lorentz boost is different. In particular, a very heavy particle would not have any thermal motion, and if captured by the flow, it would have just the momentum $mv_\perp$ with a flow velocity. So, for heavy particles, the momentum distribution should simply follow the distribution of the flow velocity. In reality, protons are in between the two limits.

Unfortunately, the *pp* data from CERN ISR we analyzed showed, in fact, a rather accurate $m_\perp$-scaling:

$$\frac{dN}{dp_t^2} \sim \exp\left(-\frac{m_\perp}{T}\right) \tag{10.63}$$

---

[13]Which, as we later learned, was used in the same year by Siemens and Rasmussen (1979) for low-energy collisions at BEVALAC.

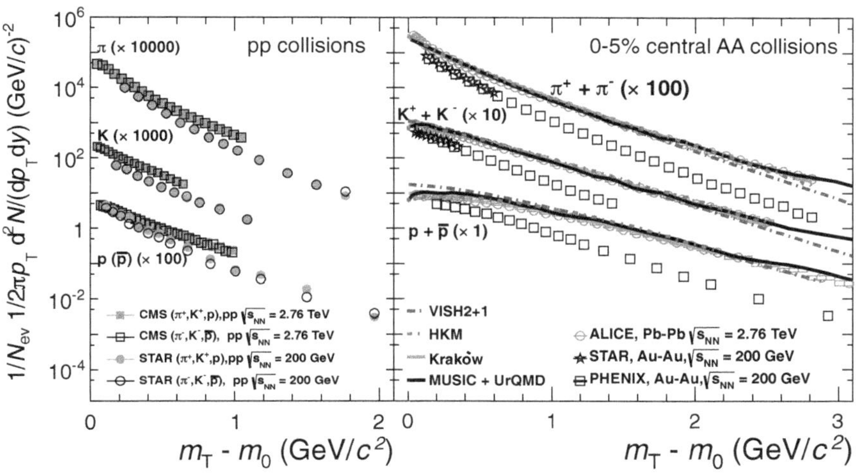

Fig. 10.3. Spectra of secondaries as a function of the transverse mass.

with no sign of any corrections due to transverse collective flow. It turned out that we had to wait[14] about a decade for high-energy heavy ion collisions.

Corresponding spectra in $m_\perp$ are shown in Fig. 10.3. The left plot shows the spectra of π, K, p secondaries in the *pp* collisions, and the right in *AuAu* at the RHIC and *PbPb* at the LHC (CMS). Four decades later, it is still true that *pp* collisions show $\sim\exp(-m_\perp/T)$ spectra with *the same slope* for all three species (although somewhat different at two energies). So, there is no transverse flow in *pp*, neither at the ISR nor at the LHC.

The spectra from heavy ion collisions, shown in the right plot of Fig. 10.3 (from Heinz and Snellings, 2013), are different. The pion spectrum is concave, while the proton spectrum is convex. The lines with various abbreviations correspond to versions of the hydro codes: As one can see, they reproduce shapes of the LHC data quite well.

[14]In 1982, when I gave my first CERN lectures on QGP, it looked quite possible that CERN ISR would be used for that purpose. Unfortunately, CERN leadership at the time decided otherwise: The ISR was physically destroyed "...to store LEP magnets in the building." The hidden irony of that announcement was the fact that the LEP magnets used such a low magnetic field that they were made of concrete, with a small admixture of iron, so there was no problem with their storage even outside any buildings. The actual reason was a focus on a proton–antiproton collider, which was then under construction.

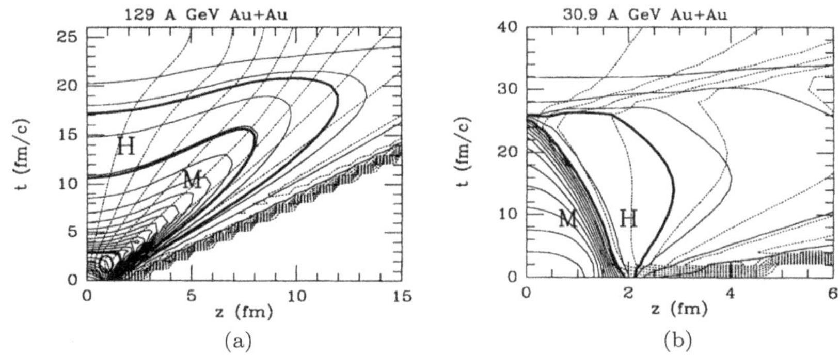

Fig. 10.4. Comparison of the space-time evolution of matter in *AuAu* collisions at beam energy per nucleon of 129 GeV (left) and 30.9 GeV (right) corresponding to the longest fireball lifetime according to hydrodynamical explosion. The lines denote constant energy density, and the Q, M and H regions correspond to QGP, mixed and hadronic phases, respectively. The short-dashed lines are for constant longitudinal rapidity.

Historically, already in the pre-RHIC era, heavy ion collisions at CERN SPS and BNL AGS, in fixed-target mode, had shown the presence of the radial flow. While the mainstream theory of those was various versions of hadronic cascades (RQMD, URQMD), we (Hung and Shuryak, 1995b) had also described these data by hydrodynamical explosion.

We proposed an energy scan of a sort, with the goal of locating the longest-lived fireball. According to this idea, it should correspond to initial conditions corresponding to the *softest point* of the EOS, the minimum of pressure (relative to the energy density) $\min(p(T)/e(T))$. In Fig. 10.4 from Hung and Shuryak (1995b), we show the results of the solution to the hydrodynamical equations: While the former shows a relativistic explosion, the latter is rather a slow "burning." We further discuss the location of the *softest point* in the chapter on femtoscopy and also in Chapter 6 in connection to light nuclei production.

The *hydrodynamical paradigm* states that out of all individual properties of the secondary hadrons, only one — their masses — is important, as one translates from the distribution over collective flow velocities to the observed momenta. All one needs to know is that an object of mass m in a flow with velocity v has a momentum of mv, plus thermal motion, which also only depends on m. Let us check it by a direct comparison of the spectra for a pair of hadrons with the same mass, with otherwise completely different quantum numbers and cross-sections, e.g. p and ϕ.

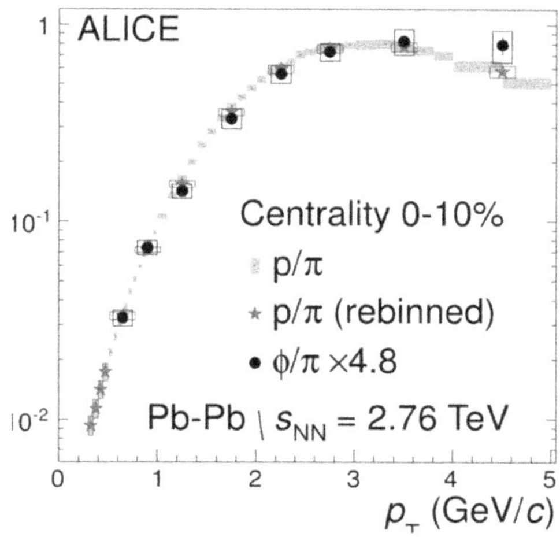

Fig. 10.5. Proton-to-pion and (rescaled) ϕ/π ratios as a function of transverse momentum p_T.

They can hardly be more different, a non-strange baryon versus a $\bar{s}s$ meson, so any "afterburner" code shows their late-time dynamics to be different. And yet, as the data shown in Fig. 10.5 demonstrate, their spectra are practically identical, up to $p_\perp \sim 4$ GeV/c.

Summary and connections: This chapter is among the most important in this book. It describes ideal (local) and first-order relativistic hydrodynamics. We also consider all of the main analytic solutions: Riemann's, Bjorken's, Landau's and Gubser's flows. Using those, one can understand the main qualitative properties of violent relativistic explosions, which heavy ion collisions produce. We discussed specifically only central (axially symmetric) collisions with "radial" (transverse to collision direction) flows, emphasizing the role of $c^2 = dp/d\epsilon$ and "the softest point" of the equation of state. Numerical solutions are needed to get a quantitative description of various phenomena: those are considered in the subsequent chapters.

Chapter 11

ELLIPTIC AND DIRECTED FLOWS

11.1 Geometrical Origins of the Elliptic Flow

We already discussed the geometry of the initial state and pointed out that the region occupied by "participant nucleons" in the transverse plane has a certain shape, as shown in Fig. 7.1. Therefore, secondaries going in different directions in the transverse plane obtain nontrivial distributions in azimuthal angle ϕ.

The observed distributions are usually expanded in the Fourier series in azimuthal angle ϕ. The magnitude of *directed flows* is characterized by parameters v_n, the harmonics

$$\frac{dN}{dp_\perp^2 d\phi dy} / \frac{dN}{dp_\perp^2 dy} \sim \left(1 + 2\sum_n v_n(p_\perp, y)\cos(n\phi)\right) \tag{11.1}$$

where ϕ is the angle between the particle transverse momentum $\vec{p}_\perp$ and the impact parameter $\vec{b}$.[1] At this point, the reader is expected to ask: "How experimentalists know the direction of $\vec{b}$?" Let me not go into technicalities here and just hint that there are many — hundreds or thousands — other observed secondaries, which collectively define its direction, with good enough resolution, even on an event-by-event basis.

We are going to discuss the dependence of v_n on collision parameters, such as collision energy, particle type, centrality class, its $p_\perp$ and rapidity y in the following.

Let us begin with a discussion of the *sign* of the elliptic flow coefficient v_2. The historic developments in the studies of azimuthal distributions

[1]Let me note that the so-called "Lorentz invariant" cross-section is $E_p dN/d^3p = dN/dydp_\perp^2 d\phi$ because $dy = dp_l/E_p$, and it obviously does not change over longitudinal boosts since they modify rapidity y additively.

for relativistic collisions started at Berkeley's BEVALAC, with ion beams of momentum $\sim$1 GeV/c/N. For an extensive summary, see Poskanzer *et al.* (1984). Studies focused on the momentum tensor $\langle p_i p_j \rangle$, which basically tells us about v_1 and v_2 in modern terminology. A new phenomenon called "side splash" was observed if the ions used were heavy enough. Already in these pioneering experiments, it was noted that predictions of cascade codes did not reproduce the phenomenon, while relativistic hydrodynamics (not yet developed to a quantitative level then) leads correct dependencies.

Focusing in this chapter mostly on $v_2(p_\perp)$, note that the pioneering BEVALAC experiments have found *negative* $v_2 < 0$ (i.e. more secondaries out of the collision plane from the tips of the almond). The main reason is that, at such low energies, the "spectator matter" (located "in plane") does not leave an excited matter almond-shaped region of interest rapidly enough; therefore, it leads to *shadowing* of secondaries emitted "in plane."

The next stage of such studies was in the 1990s, using BNL's and CERN "injector synchrotrons" (PS and SPS), with beams of energy $\sim$10–30 GeV/c/N, or $\sqrt{s_{NN}}$ = a few GeV. As the collision energy increases, shadowing by "spectator matter" becomes less important, as it leaves the central fireball more rapidly. Respectively, it was observed that v_2 changed signs and became positive. This is well documented in Fig. 11.1 from the historic paper by Pinkenburg *et al.* (1999). Note that v_2 grows with p_t at 6 GeV/N (squares), yet at the right-hand side reaches the magnitude of only a few percentage points.

At the end of the 1990s, right before the RHIC era, there was naturally a call for predictions. In respect to elliptic flows, the debates that emerged were especially polarized because different models of matter properties suggested quite different predictions.

One point of view, inspired by the parton modes of hard processes and pQCD, considered a weak coupling regime and rare rescatterings. If the produced multiple secondaries were to be partons with only weak interactions with each other, in the first approximation, their angular distribution was expected to be isotropic, $v_2 = 0$. Indeed, a scattering of two hard partons has its angular distribution affected by the direction of their *binary* (parton–parton) impact parameter rather than the global impact parameters of two colliding nuclei $\vec{b}$. The secondary interactions were expected to radiate new soft gluons. If so, there would be more of them on the longer side of the almond $\phi \sim \pi/2$ direction, which has more matter. So, the parton cascades predicted the elliptic flow parameter to have a *negative* $v_2 < 0$.

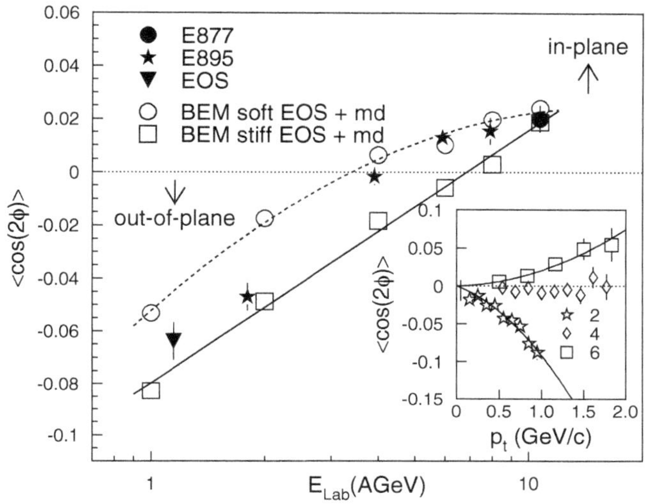

Fig. 11.1. Elliptic flow magnitude in $AuAu$ collisions versus beam energy per nucleon. Filled symbols are experimental, and open ones denote certain theoretical models indicated on the plot. The inset shows the p_t dependence, for 2, 4 and 6 GeV/N beams.

Hydrodynamics, on the other hand, predicts an anisotropic explosion driven by the pressure gradient. The largest gradient is obviously directed along the *shorter* direction of the almond, the in-plane x direction or $\phi = 0, \pi$, so v_2 was predicted to be *positive*.

Another debate was about a large step in energy, from the fixed target AGS/SPS energies to RHIC. The latter was a collider, designed at $\sqrt{s} \approx 200\,\text{GeV/N}$, nearly two orders of magnitude larger than that for a fixed target.[2]

Parton cascades produce v_2 from the softest part of the partonic spectra, the partons are soft enough to be sensitive to the density gradient and thus the geometry of the fireball. With an increase in collision energy, Q_s grows; therefore, this soft part becomes *less* important. Therefore, a prediction was made that v_2 would *decrease* with the collision energy.

Ordinary hadronic cascade codes — RQMD and URQMD — did not predict negative v_2 but predicted its decrease with energy. Basically, it was

[2]Let us recall the kinematics here. For a fixed target, $\sqrt{s} \approx \sqrt{2E_{\text{lab}}m_N}$ So, for $E_{\text{lab}} = 50\,\text{GeV}$, one has $\sqrt{s} \approx 10\,\text{GeV}$. Note also that the next step — from RHIC to LHC — was also approximately a factor of 20.

somehow assumed that mean free paths of secondaries would be of the order of system sizes, leading to badly defined "freeze-out surfaces" and relatively weak collectivity of flows.

The main assumption of the hydrodynamic approach is the opposite: The mean free path is expected to be small, resulting in a collective explosion. The most important argument (Teaney *et al.*, 2001b) was that at the RHIC collider (unlike at the fixed-target AGS/SPS collisions), we were expecting to create a new form of matter, the QGP. Its equation of state $p \approx \epsilon/3$ was $stiffer$ than for the resonance gas $p/\epsilon \approx 0.15 - 0.2$. So, the pressure gradient, i.e. the force, divided by energy density, i.e. matter acceleration, was predicted to grow by about a factor of 2 at RHIC.

With the beginning of the RHIC experiments in 2000, some of the results of which are compared to the SPS data in Fig. 11.2,[3] the answer was readily found: The elliptic flow did increase by a factor 2, as the hydro-QGP scenario predicted.

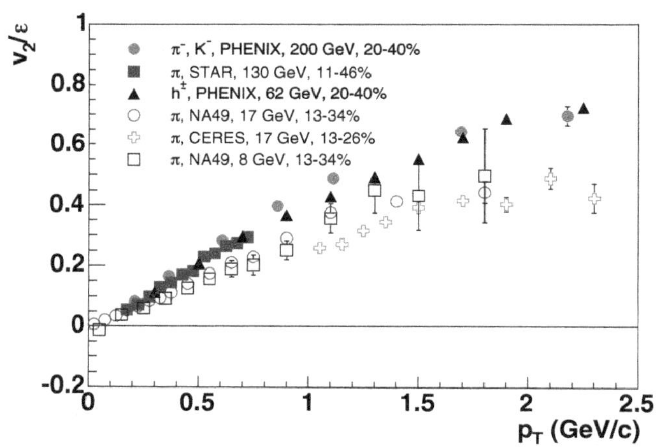

Fig. 11.2. Elliptic flow coefficient normalized to the ellipticity of the initial state $v_2(p_T)/\epsilon_2$ for mid-central collisions at RHIC (filled symbols) and SPS (open symbols).

[3]More accurately, we used a hydro + RQMD model, which correctly describes different kinetic freeze-outs at SPS and RHIC, as well as the correct EOS of "chemically frozen" hadronic gas. If instead a pure hydro had been run to a fixed T_f with equilibrium EOS, there would be no difference in v_2 between SPS and RHIC.

11.2 The $p_\perp$ Dependence of v_2

Experimentally, as we have just shown in the previous figure, v_2 *grows* with $p_\perp$ approximately linearly. This is explained via a hydrodynamical paradigm as follows. Its particle spectra are described by the so-called Cooper–Fry expression:

$$E_p \frac{dN}{d^3p} = \int_\Sigma (d\Sigma_\mu p_\mu) e^{-p^\nu u_\nu(x)/T(x)} \tag{11.2}$$

where the integral is taken over the *kinetic freeze-out surface* Σ, usually approximated by an isotherm, $T(x) = T_f$.

Its main element is a Lorenz invariant Boltzmann factor, in which particle momenta are projected to collective four-velocity $u^\mu(x)$. In the zeroth order of the initial state deformation ϵ_2, the flow is axially symmetric, directed outward,[4] with $u_r \sim r$. In the first approximation in ϵ_2, one includes corrections to flow velocity with a specific dependence on the azimuthal angle

$$u_r^{(1)} \sim \epsilon_2 \cos(2\phi) r$$

Now, if small, one can bring these correction terms down from the exponent in the Cooper–Fry integral. One of its consequences, which is apparent without any calculation, is the *linear* dependence on $p_\perp$:

$$v_2(p_\perp) \sim \epsilon_2 \times p_\perp \tag{11.3}$$

which we have already seen in Fig. 11.2 in the RHIC and SPS data. Basically, it is a consequence of a "splash" picture, in which a collective outward velocity linearly increases, from zero at the center of the exploding system to some maximal value at its edges.

Going way ahead of the logical flow of the presentation, let me nonetheless comment on the next *quadratic* term of the exponent expansion, the one including $(\epsilon_2 \cos(2\phi))^2$. Using $\cos(2\phi)^2 = (\cos(4\phi) + 1)/2$, one observes that it contributes to the *fourth* harmonics of the flow by the amount

$$v_4(p_\perp) \sim \epsilon_2^2 p_\perp^2 \tag{11.4}$$

which can be separated from the "true" $v_4 \sim \epsilon_4 p_\perp$ by its markedly different *quadratic* dependence on $p_\perp$. Indeed, both such terms were not only calculated but also clearly identified in the data.

[4]One may call it a "Hubble-like" flow similar to what one has in the expanding Universe.

Particles with a larger $p_\perp$ originate from the edge of the fireball, where the collective velocity is maximal. Furthermore, they preferentially come from the fraction of this edge, at which the collective velocity is directed toward the observer, so that the spectrum is "blue shifted." The consequence of such "splash" geometry is the $v_2 \sim p_\perp$ prediction.

Now, it is time to show a comparison to hydrodynamical calculations. The observed dependence of the elliptic flow parameter v_2 for pions and protons on the particle transverse momentum is shown in Fig. 11.3 taken from the PHENIX white paper (Adcox *et al.*, 2005).

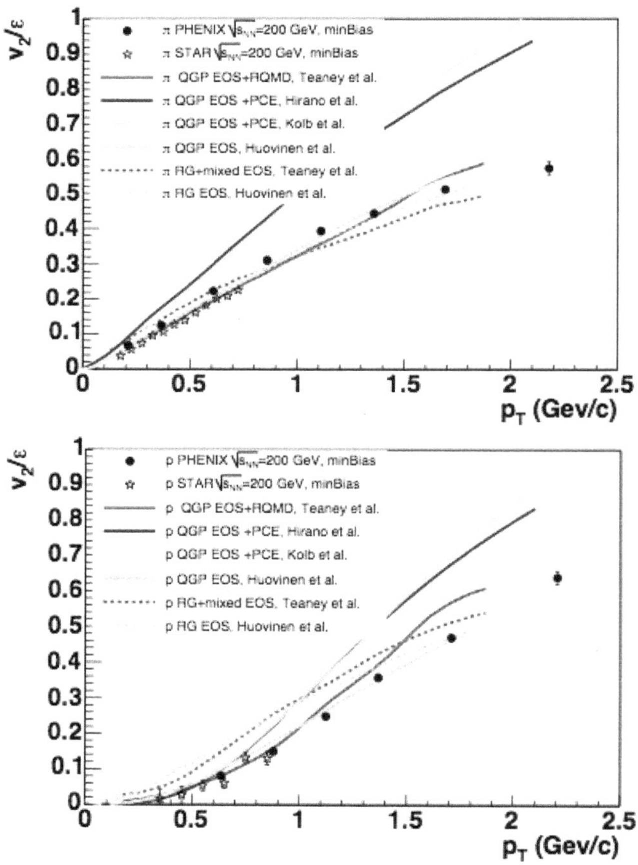

Fig. 11.3. The elliptic flow parameter v_2, the experimental data versus the hydrodynamical predictions, for pions (upper plot) and protons (lower plot).

(Note also that our calculations performed before the data, i.e. the predictions — the red lines online and onscreen — for RHIC were quite accurate, while most of our competition, shown by the blue lines, overshoot the data points.)

Note that, while the agreement is quite good through the whole kinematic region shown in this plot, the theoretical curves stop at $p_\perp = 2\,\text{GeV}$. Why were the calculations stopped there? In fact, even the bravest theorists at the time (myself included) had not dared to venture beyond it since it was commonly expected that above such $p_\perp$, one would enter a jet-dominated regime. We were expecting hydro to work for the "majority of the particles," up to, say, momenta of $p_\perp < 1\,\text{GeV}$.

In fact, the hydro-dominated region extends much further, to $p_\perp \sim 3$–$4\,\text{GeV}$, or a few tens of GeV down in the spectra. Some experiments have put out $p_\perp \sim 3\,\text{GeV}$, while showing events on display: With such a cut, the elliptic deformation is so large, $v_2 \sim 0.2$, that one was able to see the difference between $1 + 2v_2 \sim 1.4$ in the major axes of the ellipse and $1 - 2v_2 \sim 0.6$ with the naked eye on each and every event![5]

11.3 Elliptic Flow Dependence on the Particle Type

A generic feature of hydro effects is that, as the particle mass grows, the flow effects deform the momentum spectra of secondaries more. Basically, the flow shifts low-$p_\perp$ particles upward, as they "move with flow" and acquire additional momentum $M\vec{v}$. This same phenomenon happens with all harmonics of the flow. For elliptic flows, the data points for protons are well below pions and kaons at small p_T, but overcome them near the maximum, see Fig. 11.4 from Krzewicki (2011). (One can also compare the data from RHIC (shaded) with those from the LHC (points) and see that the hydro effects do increase with collision energy. Also, the lines are hydrodynamical calculations, which apparently describe these data quite well.)

Now, one may ask if one can predict the existence/position of those maxima. There are two considerations proposed, which I would call "kinematical" and "dynamical."

The simple kinematical idea is that there is a maximal transverse rapidity of the hydro flow since the edge of the "splash" is terminated by the

[5] I have seen snapshots of those in the talks, but unfortunately, I could not locate them anywhere in published sources.

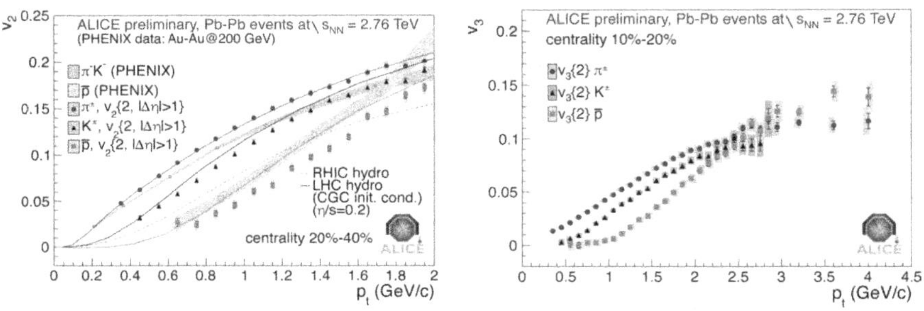

Fig. 11.4. Elliptic and triangular flows of π, K, p.

freeze-out surface. For example, the protons at the maximum of the curve, 4 GeV/c have a transverse rapidity of $\sinh(y_\perp) \approx 4$, which corresponds to $y_\perp \approx 2$. This seems to be somewhat large, but it is possible. The curves for pions and kaons roughly follow the idea.

A dynamical consideration, due to Teaney (2003), also relates the cutoff of hydro to the rescattering of constituents, but in a hydro language related to flow gradients and matter viscosity. The particle distribution inside a hydro cell is, in the first approximation, only thermal, and in the next, it has a correction induced by gradients of flow

$$f = f_{\rm eq}\left(1 + \frac{C}{2T^3}p^\alpha p^\beta \langle \nabla_\alpha u_\beta \rangle\right) \tag{11.5}$$

where the powers of T are written on dimensional grounds. The coefficient C is related to shear viscosity (see the manipulation which led to it in the source paper, if interested):

$$\eta = \frac{C}{15T^3}\int \frac{d^3p}{(2\pi)^3 E_p} p^4 f_{\rm eq} \tag{11.6}$$

For a Boltzmann gas $C = \eta/s$, viscosity is normalized to the entropy density. Teaney's argument is that one should not only include the viscosity in hydro equations but also keep this δf correction in the Cooper–Fry integral. The results, schematically, are

$$v_2(p_\perp) = A\epsilon_2 p_\perp \left[1 - B\frac{\eta}{s}\left(\frac{p_\perp}{T}\right)^2\right]$$

with A, B being some calculated constants. The lesson is as follows: Viscosity causes hydro predictions to develop a maximum and turn around. (Note that, after taking a derivative, the coefficient of the correction term becomes 3. Thus, the maximum is when the viscous correction is about one-third of the main term, so the hydro terms remain perhaps believable there.)

A very nontrivial case is the ϕ meson, which, as we already discussed, has about the same mass as the proton, and indeed, their $p_\perp$ spectra are close. It is also so for the $v_2(p_\perp)$ curve, except at the highest $p_\perp$.

The kinetic argument we just discussed is, however, based on particle rescattering, and the cross-section of that for ϕ is much smaller than for p. that it stops earlier. So, kinetics imply a significantly earlier freeze-out time for ϕ.

An alternative idea (Fries *et al.*, 2003) relates the difference in v_2 for mesons and baryons to a quark recombination mechanism. The authors suggested that in the intermediate region of, say, $p_\perp \in [3, 10]\,\text{GeV}$, the secondaries are formed by recombining quarks from hard pQCD collisions with "soft" ones from the heat bath. The difference between ϕ and the nucleon is then attributed to the fact that the former is a meson and needs one soft quark, while the latter is a baryon and needs to pick up two. This paper proposed the so-called "quark scaling" formula for the elliptic and other flows that suggests that if a hadron is made of n quarks, then its flow harmonics should be proportional to n:

$$v_m^{(h)} \sim n v_m^{(q)}$$

with some universal "quark" $v_m^{(q)}$. The derivation of the "quark scaling" formula (Fries *et al.*, 2003) comes from the expression (82) in that paper, which assumes that the quark distribution function in the phase space is "factorized":

$$w(R;p) = w_i(R;p)[1 + 2v_2^a(p_\perp)\cos(2\phi)]. \tag{11.7}$$

In other words, these authors think that v_2 originates from some anisotropic momentum distribution in *every local hydro cell.* This assumption is orthogonal to the hydro paradigm, which assumes instead that each cell has an equilibrium (isotropic) distribution of particles. It is the *number of cells* moving in different directions which is different and calculated by hydro. This n cannot possibly matter: A hadron can be made of 2, 0 or 22 partons; all you need to know is its mass. Phenomenologically, this relation has been approximately fulfilled by the RHIC data but not the LHC data, and I believe its pursuit is now abandoned.

More consistent models able to describe the particle spectra in the intermediate $p_\perp$ region are still wanted. Perhaps, even a more realistic implementation of the recombination idea would work. The elliptic flow of ϕ can also be related to that of ψ, its heavier brother, with strange quark substituted by charmed ones.

11.4 Elliptic Flow Dependence on Centrality

Note that the "almond-shaped" overlap region can become very thin for peripheral collisions, with the impact parameter $b \to 2R$. Therefore, in this case, the transverse pressure gradients become large, pushing matter to rapidly expand in the *in-collision-plane* direction. The thin almond then relatively quickly becomes thicker, reducing the gradients. What it tells us is that v_2 has only a chance of being generated in a relatively short initial period of time.

This observation is important for defining *how rapid thermalization is* and the onset of transverse pressure. Many who tried some weak coupling cascades for the initial stage were unable to reproduce the centrality dependence of elliptic flow.

On the left-hand side of Fig. 11.5, from Lacey *et al.* (2013), v_2 and v_3 ATLAS data are plotted for p_T = 2–3 GeV as a function of participant

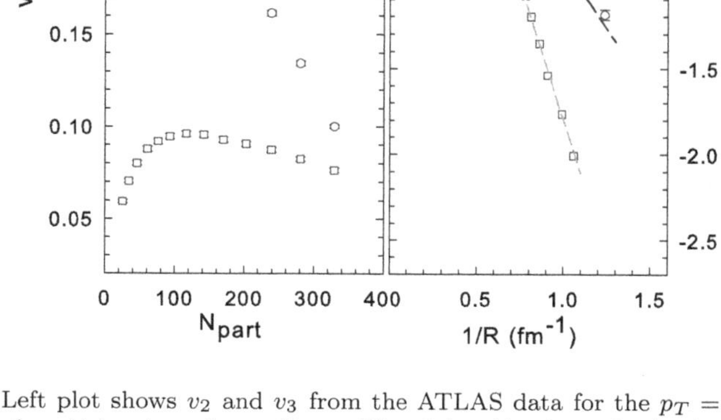

Fig. 11.5. Left plot shows v_2 and v_3 from the ATLAS data for the p_T = 2–3 GeV as a function of participant nucleons N_{part}. The right side shows $\log(v_n/\epsilon_n)$ versus the "system size" variable $\bar{R}$ defined in the text.

nucleons N_{part}. As one can see, these directed flows vanish at both ends of the plot: At large $N_{\text{part}} \approx 400$, it occurs because central collisions have no geometric azimuthal distortions, and the fluctuations are small; at small N_{part}, i.e. highly peripheral collisions, the system is too small to generate any collective effects.

The right-hand side of the plot in Fig. 11.5 shows $\log(v_n/\epsilon_n)$ versus the "system size" variable $\bar{R}$ defined as

$$\frac{1}{\bar{R}} = \frac{1}{\bar{R}_x} + \frac{1}{\bar{R}_y}$$

and thus emphasizing the *smallest* system size for highly deformed systems. As one can see, both sets of data now show perfectly linear dependence. This takes care of the large-size end of the problem: The vanishing of harmonics was in fact induced by the vanishing of ϵ_n.

The so-called "acoustic damping" formula, suggested by Staig and Shuryak (2011a), was motivated by sound amplitude viscous damping. It takes the form

$$\frac{v_n}{\epsilon_n} = \exp\left(-C\frac{\eta}{s}\frac{n^2}{T\bar{R}}\right) \tag{11.8}$$

and includes the ratio of the mean free path to the system size $(\eta/Ts\bar{R})$, also known as the micro-to-macro ratio. As one can see from the plots, it provides a quite accurate description of the data dependence on centrality.

11.5 Can Elliptic Flows Be "Faked"?

The question of to what extent one can substitute hydrodynamics with some kinetic models has been discussed multiple times. It was also suggested that flows (especially in the small systems to be discussed) can be "fake," in the sense that some part of the evolution particles simply fly ballistically without scattering. This issue has been investigated in detail by Romatschke (2015), where a scenario in which quarks and gluons at the QGP phase have no interactions, and they "freely stream" from the point of the initial scattering to the hadronization surface. At this surface, the system switches to the hadronic phase, treated by standard hadronic cascade codes.

In Fig. 11.6 from Romatschke (2015), one finds a comparison of the radial flow profiles of the two cases, with and without interaction at the QGP phase. The profiles are in fact very similar, becoming linear Hubble-like as time goes on. In fact, free streaming generates even a bit stronger flow because it uncouples from the longitudinal direction, whereas the

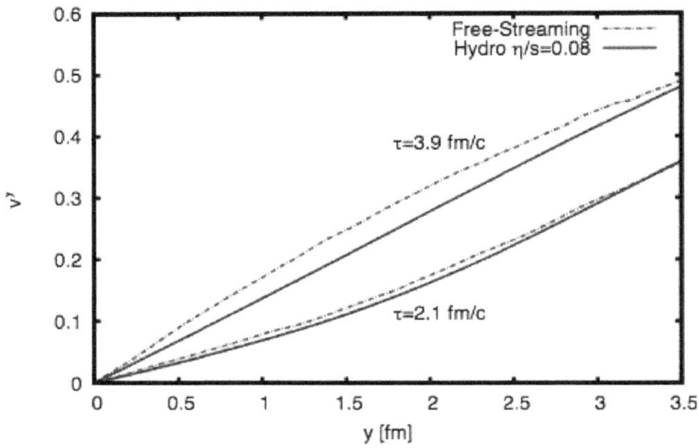

Fig. 11.6. Comparison of the flow profile for hydrodynamics and free streaming.

equilibrated medium does not. Comparing particle spectra and HBT radii, Romatschke shows that this "fake" radial flow is indeed *indistinguishable* from hydrodynamics.

What about flow harmonics? The results for *PbPb* collisions are shown in Fig. 11.7 from the same paper. As one can see, without hydrodynamics, elliptic flows disappear. This is not surprising since if there is no hydrodynamics, then there are no sound waves, and initial bumps simply dissolve without a trace.

In fact, it is an interesting question how *any* v_n can be generated in the free stream. The initial momentum distribution of partons is isotropic, so it must be related to the interaction after hadronization. Romatschke found that before hadronization, they were indeed absent. However, two components of $T_{\mu\nu}$, the flow $\sim u_\mu u_\nu$ part and the dissipative $\Pi_{\mu\nu}$ part, still have nonzero values, which cancel each other in the sum. After hadronization, hadronic interaction kills the second component $\Pi_{\mu\nu} \to 0$ and reveals the effect of the first one.

Not only are the "fake" flow harmonics small, they lack the following two most important features of the "true" hydrodynamical ones: (i) There is no increase with $p_\perp$; and (ii) there is no decrease with the harmonic number $\sim \exp(-n^2)$ induced by viscosity during the time before hadronization.

So, unlike the radial flow, higher harmonics in large (*PbPb*) systems *cannot be faked.*

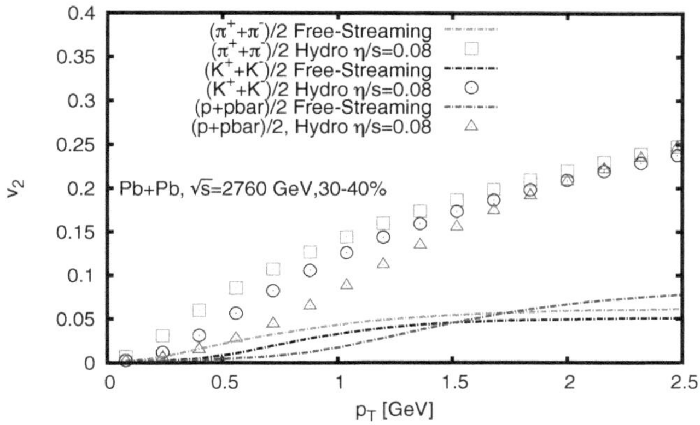

Fig. 11.7. Comparison of the flow harmonic for hydrodynamics and free streaming for *PbPb* collisions.

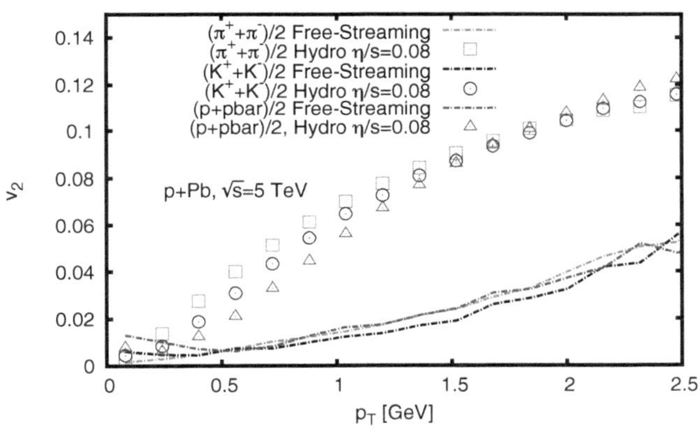

Fig. 11.8. Comparison of the flow harmonic, for hydrodynamics and free streaming in the *pPb* central bin.

What about smaller systems? Romatschke gives the results for pPb at LHC and dAu and He^3Au for RHIC energies. We show the first case in Fig. 11.8. Again, the free streaming fails for v_2 while marginally surviving for v_3.

The conclusion is that flow harmonics with $n = 2\ldots$ cannot be "faked." Even for small systems, this conclusion is robust, even taking into account remaining uncertainties (mostly in the initial stage models and thus ϵ_n values).

11.6 Directed v_1 Flow

Although the "directed" v_1 flow was the first to be actually studied at BEVALAC (and nonrelativistic collisions before that), we consider it at the end of the chapter in a rather brief way. The reason is that it played a relatively modest role in our field, as its main domain is at quite low collision energies. It was addressed in the 1990s by AGS/SPS experiments in a sketchy way, compared to very recent data from RHIC BES II program, which we will discuss.

There are no external forces acting on colliding pair of ions, so the total dipole kick must be zero. It follows that two halves of the system get

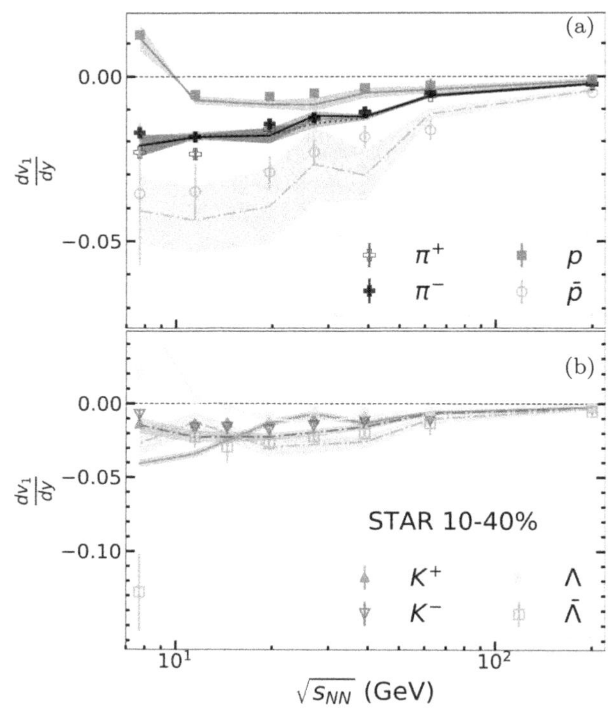

Fig. 11.9. Directed flow slope for various secondaries versus collision energy $\sqrt{s_{NN}}$.

kicks in the opposite directions; therefore, v_1 is an *odd* function of collision rapidity y, fitted near $y = 0$ by a cubic function:

$$v_1(y) = C_1 y + C_2 y^3 \tag{11.9}$$

Since most measurements on colliders are at mid-rapidity, the directed flow is characterized by the first of these coefficients, or the *derivative* at $y = 0$, $dv_1/dy(y = 0)$.

Its energy dependence for different secondaries is shown in Fig. 11.9 from Parida and Chatterjee (2022), the points are from the STAR BESII data and the curves from hydro calculations explained in the paper.

Figure 11.10 from STAR Collaboration (2022c) shows its mass dependence for various baryonic species. The dependence on mass is just linear. Note that light nuclei include the light hypernuclei ${}^3_\Lambda H = pn\Lambda$ and ${}^4_\Lambda H = pnn\Lambda$. The magnitude of v_1 is different for nuclei and hypernuclei, reflecting weaker nuclear forces between Λ hyperons and nucleons.

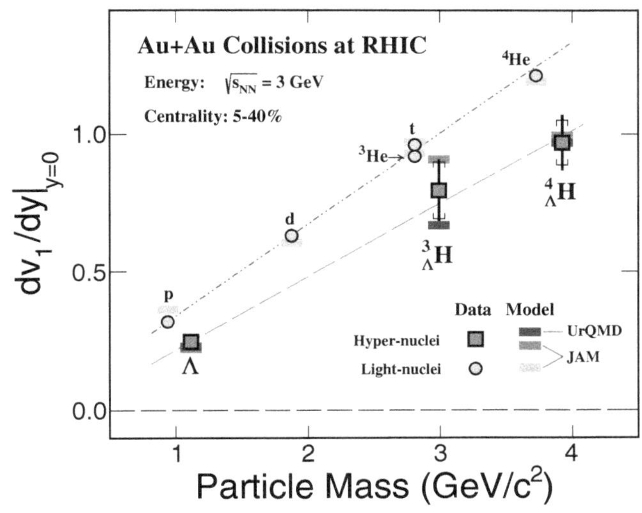

Fig. 11.10. Directed flow slope for various baryonic and light nuclei species versus their masses at $\sqrt{s_{NN}} = 3\,\text{GeV}$. Note that URQMD and Jam are cascade codes, their predictions (shown by short horizontal bars) do account for the observed slope.

Summary and connections: Non-central collisions lead to fireballs lacking axial symmetry, and their initial deformation harmonics ϵ_n eventually produce harmonics of the flow velocity v_n, observed via the spectra of secondaries. The largest $n = 2$ harmonic, or elliptic flow, originates from collision geometry, while $n = 3, 4, \ldots$ ones come from initial state shape fluctuations. Because of the relativistic magnitude of these flows, all of them reach large values at their maxima, located at $p_\perp \sim 3\,\text{GeV}$. Multiple models with free streaming, "parton cascades," and other non-hydro approaches failed to explain the magnitude of v_n at these maxima. Therefore, this observable plays a very important role in heavy ion phenomenology.

Chapter 12

HIGHER FLOW HARMONICS AND SOUNDS

12.1 A Brief Introduction to Sound

Let me start by going back to the simplest nonrelativistic setting. We also consider small perturbations of the fluid at rest, or "hydrodynamic modes."[1] The classic version of the nonrelativistic hydro equations has the form of the matter conservation equation plus the Euler equation of motion

$$\frac{\partial \rho}{\partial t} + \frac{\partial(\rho v^m)}{\partial x^m} = 0 \tag{12.1}$$

$$\frac{\partial v^m}{\partial t} + \left(v^n \frac{\partial}{\partial x^n}\right) v^m = -\frac{\partial p}{\partial x^m} \tag{12.2}$$

Consider fluid at rest with small perturbations. The thermodynamical variables are therefore written as

$$p = p_{(0)} + p_{(1)}, \quad \rho = \rho_{(0)} + \rho_{(1)}$$

and the velocity is only kept to the first order since it is zero for fluid at rest. As a result, the nonlinear term in $\vec{v}$, which creates trouble, is of the second order and ignored: the linearized equations in the terms with $\vec{v}$ have zeroth-order quantities; however, in those with derivatives, we keep the first-order ones. The equations we have simplified to

$$\frac{\partial \rho_{(1)}}{\partial t} + \rho_{(0)} \frac{\partial(v^m)}{\partial x^m} = 0 \tag{12.3}$$

$$\frac{\partial v^m}{\partial t} + \frac{\partial p_{(1)}}{\partial x^m} = 0 \tag{12.4}$$

Further simplification comes if we consider the one-dimensional case (or longitudinal excitation). Then, there are two equations which can be

[1]This is 19th-century textbook material; yet, we do need to include it because hydrodynamics has been totally sacrificed in university physics education.

combined into one second-order equation for the potential, defined via

$$v_m = \frac{\partial \phi}{\partial x^m}:$$

$$\frac{\partial^2 \phi}{\partial t^2} - \left(\frac{\partial p}{\partial \rho}\right)_0 \frac{\partial^2 \phi}{\partial x^m \partial x^m} = 0. \tag{12.5}$$

It is nothing but the standard wave equation which describes the propagation of *sound waves.* In the 1+1 dimensions, these are plane waves $\phi \sim \exp(ik_1x_1 - i\omega t)$, but they can be cylindrical, spherical and so on, depending on the source. Putting such an exponent into the equation, one finds that the wave should move with the velocity[2]

$$c_s = \sqrt{\frac{\partial p}{\partial \rho}}$$

Two comments about this expression: The formula for sound velocity is the same in the relativistic case, when ρ is not the mass density but the total energy density ϵ. So, for ideal QGP, one has $c_s = 1/\sqrt{3}$. However, for matter near T_c, it is much smaller. If the QCD deconfinement transition is the first-order transition, it would have a "mixed phase" in which ϵ can increase without any change in pressure. If so, $c_s = 0$. So, what happens with a sound wave propagating in matter entering the mixed phase? The sound wave "becomes frozen"[3] and stops! It then resumes motion in *both* directions when the matter passes to another (hadronic) phase.

Standard derivation from the Navier–Stokes equation[4] gives the sound dispersion relation, including the lowest-order damping terms

$$\omega = c_s k - i\frac{k^2}{2}\frac{(4/3)\eta + \xi}{sT} + Ck^3 + O(k^4) \tag{12.6}$$

The important feature of this dispersion relation is that at $k \to 0$ (large wavelengths), *all* terms vanish. Indeed, they *must* do so since sound is a

[2]In a simplified form, $\sqrt{p/\rho}$, this formula was first obtained by none other than Newton himself in 1687.

[3]I cannot resist but mention a corresponding story from a non-scientific but still excellent literature. Baron Munchausen, in his famous stories retold to us by Raspe, once told us that he was blowing the trumpet on a day so cold that its sound froze inside the trumpet. Yet, when he came home and the instrument wormed up, it started playing by itself! All I can add to the story is that EOM insisted that the sound should then come *from both ends* of the trumpet.

[4]E.g. Landau–Lifshits Hydrodynamics, Chapter 79.

representative of the family of the so-called *Goldstone modes*, appearing when some continuous symmetry is spontaneously broken. In this case, the symmetry breaking is due to an arbitrary selection of a frame in which matter is at rest. Since at $k \to 0$ the mode becomes the symmetry transformation. Therefore it is not allowed to have any energy or interaction in this limit.

Now, about the second term, note that it has an extra power of k, which means that for a small k, it is extra small. That is why large-wavelength sound can propagate large distances,[5] and its amplitude is damped according to

$$T_1^{\mu\nu}(t) = \exp\left(-\frac{k^2 t}{2}\frac{(4/3)\eta + \xi}{sT}\right) T_1^{\mu\nu}(0) \tag{12.7}$$

Let us now include two more terms in the sound dispersion relation

$$\omega = c_s k - \frac{i}{2}k^2\Gamma_2 + C_3 k^3 + i\Gamma_4 k^4 + O(k^5) \tag{12.8}$$

In general, assuming a small-amplitude wave, one linearizes hydrodynamic equations, and the dispersion relation is a root of the determinant of the system of equations.

It is instructive to do it for mRHD1, expanding it in powers of k, one gets the subsequent ordinary equations from which these coefficients can be found. The leading order $c_s k$ remains the same, and the dissipative term is

$$\Gamma_2 = (4\eta - \chi_1 + \chi_2 + 3\chi_3 - 3\chi_4)/(8\epsilon_0) \tag{12.9}$$

Once the bulk viscosity is identified with a combination of χ_i (Bemfica *et al.*, 2019),

$$\zeta = \chi_3 - \chi_4 + \chi_2/3 - \chi_1/3 \tag{12.10}$$

it gives the lowest-order damping of the sound waves in the same form as the Navier–Stokes formulation with two viscosities, η and ζ. However, higher-order terms are modified: I found (unpublished) that in mRHD1:

$$\begin{aligned} C_3 = &\frac{1}{128\epsilon_0^2}(\sqrt{3}(-9\zeta^2 + 12\zeta(-2\eta + \lambda) \\ &+ 4(-4\eta^2 + 4\eta\lambda + (\lambda - \chi_1 + \chi_2)(\chi_1 - 3\chi_3)))) \end{aligned} \tag{12.11}$$

[5] For example, you hear the lecture about 10 m away, which is many orders of magnitude larger than the molecular mean free path in the air.

$$
\begin{aligned}
\Gamma_4 = &\frac{3}{128\epsilon_0^3}(9\zeta^2\lambda + 16\eta^2\lambda \\
&+ 3\zeta(8\eta\lambda - \lambda^2 + \lambda(\chi_1 - 3\chi_3) - (\chi_1 - \chi_2)(\chi_1 - 3\chi_3)) \\
&- 4\eta(\lambda^2 - \lambda(\chi_1 - 3\chi_3) + (\chi_1 - \chi_2)(\chi_1 - 3\chi_3)) \\
&- (\lambda + \chi_1)(\lambda - \chi_1 + \chi_2)(\chi_1 - 3\chi_3))
\end{aligned} \tag{12.12}
$$

This naturally brings up several questions. First, how do you calculate all these dissipative coefficients? There is no good answer to it. In the work of Bemfica *et al.* (2019), one can find the evaluation of χ_i in kinetic theory; however, this is only supposed to work in a regime of weak coupling. In strong coupling, one can derive those holographically, see Chapter 21.

Can one perhaps measure them experimentally? In particular, would the corrections to the dispersion relations of sound be important in some settings? The coefficient C_3, especially its sign, is crucial to acoustic cascade, which we will discuss in Chapter 13. The modification of the damping Γ_4 can, in principle, be accessed via higher harmonics of azimuthal flows, which we discuss in the following.

Now, are there other hydrodynamical modes? What happens if we take the direction of $\vec{v}$ to be orthogonal to $\vec{k}$, the direction of propagation? As one can see from the first equation, it will not have the second term, and thus, for a plane wave, it will give $\omega = 0$. The transverse modes are associated with rotational motion in fluids, for which one cannot assume that velocity is a gradient. With viscosity included, there appears another term, and two transverse modes appear but with purely *diffusive* (non-propagating) modes. With two sounds — left- and right-moving ones — it makes four modes, which is expected since there are four fields involved.

12.2 Sounds on Top of the Little Bang and v_3 Harmonics

Before we get into the technical side, let us discuss the problem at hand qualitatively in simple physics terms. We already argued above that the individual sources — "bumps" in the initial state, as we call them — are deviations from overall smooth initial/nuclear shapes. Ignoring the correlations of nucleons inside nuclei, one may assume these bumps to be uncorrelated, so one should consider the sounds they generate individually, ignoring their interference in the ensemble average. The picture suggested is thus similar to a number of stones thrown into the pond, which produce multiple expanding circles. Those visibly interfere,

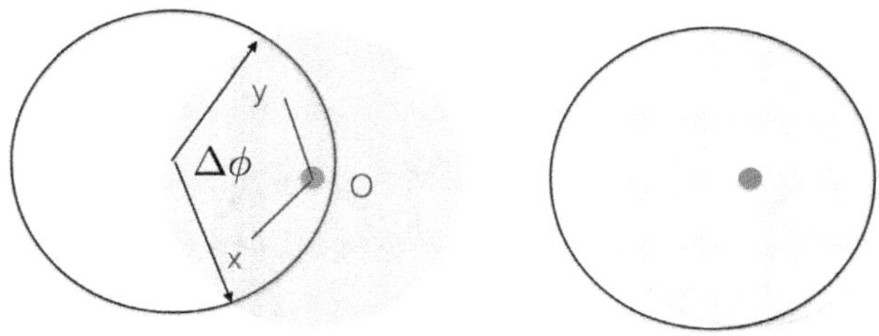

Fig. 12.1. The perturbation is shown by a small blue circle at point O: Its time evolution to points x and y is described by the Green function of linearized hydrodynamics, shown by two lines. The perturbed region, shown by the gray circle, is inside the sound horizon. The sound wave effect is maximal at the intersection points of this area with the fireball boundary: $\Delta\phi$ indicates the angle at which there is a peak in the two-body correlation function. Shifting the location of the perturbation, from (a) to (b), results in a rather small shift in $\Delta\phi$.

but they do not really interact with each other since their amplitudes are small.

Consider first the left-hand side of Fig. 12.1, in which we show the transverse plane view of the fireball at a freeze-out time. The black dot is the original source of the sound wave, from which a spherical wave (distorted by flow) originated. The region where it has passed is shown as a gray disk. Here, the dissipated part of the sound wave energy was deposited. The circle itself represents the *sound horizon*. Parts of the fireball outside cannot possibly be affected. The main density perturbations carried by the wave are located near the circle.

We already studied the radial and elliptic flows and understand that due to the radial flow of a Hubble form $v_r \sim r$, the *rim* of the fireball is the place of maximal outward radial flow. Thus, it is especially important for particle spectra at relatively large $p_\perp > 2\,\text{GeV}$. The sound perturbation carries extra energy, and one finds from the calculation that the main contributions are from two points in Fig. 12.1, where *the "sound circle" and the fireball rim intersect.*

Note that the freeze-out time is approximately twice the fireball size: $\tau_f \approx 2R$. The sound horizon is this time multiplied by the speed of sound, which is a bit smaller than that in asymptotically hot QGP, $c_s = 1/\sqrt{3} = 0.577$. As a result, the sound circle radius $R_h = c_s\tau_f$ happens to be numerically close to the radius of the fireball itself. The directions to the two intersection points in Fig. 12.1 are therefore at $\phi = \pm 60° \approx \pm 1$ radian.

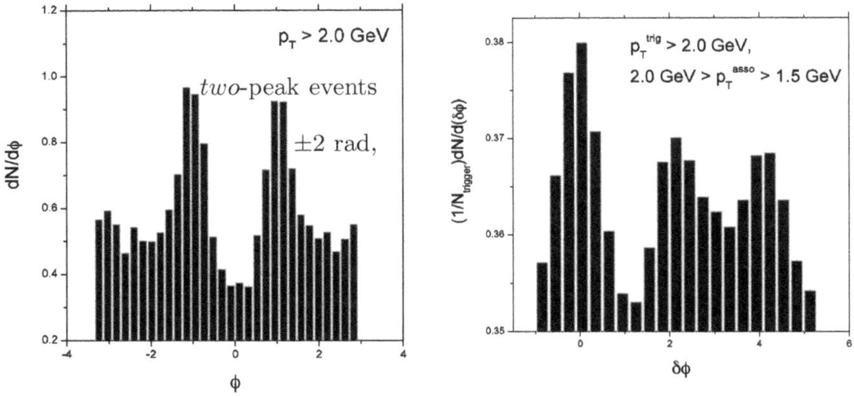

Fig. 12.2. Modification of the single-particle distribution in the azimuthal angle ϕ (left) and the corresponding modification of the two-particle correlation (right) in the relative angle $\delta\phi = \phi_1 - \phi_2$.

The corresponding modification of the single-body angular distributions, calculated hydrodynamically, is shown in Fig. 12.2 (left), from Andrade *et al.* (2010). Those two peaks are due to excesses of particles at the corresponding two azimuthal locations, the intersection of the sound front and the fireball surface. Note that the peaks are located at angles $\pm 60° \approx \pm 1$ radian, perfectly matching the location of the intersection points explained above.

This geometric picture explains why the particular harmonics, v_3 is in fact about the largest in amplitude. Indeed, proceeding to *two-particle* distribution, as a function of the *relative* angle $\delta\phi = \phi_2 - \phi_1$, one finds *three* peaks, one at $\delta\phi = 0$ and the other two at $\delta\phi \approx 120°$ or ± 2 rad. The first one appears if both observed particles, 1 and 2, come from *the same* peak in 12.2 (left). The two others appear when the observed particles belong to *different* peaks. Note further that the area of the first peak is indeed about equal to that of the two other peaks combined.

Now that we qualitatively understand how *two* peaks in the angular distribution generate *three* peaks in the two-particle correlation function, let us return to Fig. 12.1 and discuss what happens when the "bump" (sound source) at the initial time is located closer to the center, as shown in the right-hand side of the figure, rather than on the surface. In this case, two intersection angles move toward $\pm 90° = \pm\pi/2$ rad. However, more importantly, the contribution of such sources to ϵ_n is suppressed by an additional power of the radius r^n in its definition, giving advantage to the "bumps" on the fireball edge. It is further clear from symmetry that a bump located at the center cannot lead to deviations from the azimuthal symmetry of the particle distribution and thus generates no sound angular harmonics.

12.3 Sound Harmonics on Top of the Gubser Flow

A particular model of the initial state expressing locality and statistical independence of "bumps" has been suggested by Ollitrault, in which the correlator of fluctuations is given by the Poisson local expression

$$\langle \delta n(x) \delta n(y) \rangle = \bar{n}(x) \delta^2(x-y) \tag{12.13}$$

where $\bar{n}(x)$ is the average matter distribution. The immediate consequence of this model is that, for the central collisions (on which we focus now), ϵ_m are the same for all $m < m_{\max} = O(10)$ (until the bump size gets resolved).

In order to calculate perturbations at later times, one needs to apply the Green functions *twice*, describing perturbations propagating from the original source O to the observation points x and y, as shown in Fig. 12.1(a). This was first done by Staig and Shuryak (2011b) analytically for the Gubser flow.

Since we already discussed Gubser's solution, let me just recall that it can be best expressed in the co-moving coordinates $(\rho, \theta, \phi, \eta)$ related to the usual ones via

$$\sinh \rho = -\frac{1 - q^2\tau^2 + q^2 r^2}{2q\tau} \tag{12.14}$$

$$\tan \theta = \frac{2qr}{1 + q^2\tau^2 - q^2 r^2} \tag{12.15}$$

In such coordinates, the internal quantities, such as the temperature, depend on the "time" ρ:

$$\hat{T} = \frac{\hat{T}_0}{(\cosh \rho)^{2/3}} + \frac{H_0 \sinh^3 \rho}{9(\cosh \rho)^{2/3}} \, {}_2F_1\left(\frac{3}{2}, \frac{7}{6}; \frac{5}{2}, -\sinh^2 \rho\right) \tag{12.16}$$

where H_0 is a dimensionless constant obtained from shear viscosity and temperature, $\eta = H_0 T^3$ and ${}_2F_1$ is the hypergeometric function. In the inviscid case, the solution is just the first term of expression (12.16), and of course, it also conserves the entropy in this case. The freeze-out surfaces are usually approximated by the isotherms, which are also simply $\rho = \text{const}$

coordinate surfaces. We will show how they look later, when we use this solution in the discussion of small systems.

Now, we are ready to consider small perturbations on top of the Gubser flow. Those obey linearized equations, which have also been derived by Gubser and Yarom (2011). We begin with the zero viscosity case, so that the background temperature (now referred to as T_0) will be given by only the first term in (10.61). The perturbations over the previous solution are defined by

$$\hat{T} = \hat{T}_0(1+\delta), \tag{12.17}$$

$$u_\mu = u_{0\,\mu} + u_{1\mu} \tag{12.18}$$

with

$$\hat{u}_{0\,\mu} = (-1,0,0,0) \tag{12.19}$$

$$\hat{u}_{1\mu} = (0, u_\theta(\rho,\theta,\phi), u_\phi(\rho,\theta,\phi), 0) \tag{12.20}$$

$$\delta = \delta(\rho,\theta,\phi) \tag{12.21}$$

Plugging expressions (12.17) and (12.18) into the hydrodynamic equations and only keeping linear terms in the perturbation, one can get a system of coupled first-order differential equations. Furthermore, if one ignores the viscosity terms, one may exclude velocity and get the following (second-order) closed equation for the temperature perturbation:

$$\frac{\partial^2\delta}{\partial\rho^2} - \frac{1}{3\cosh^2\rho}\left(\frac{\partial^2\delta}{\partial\theta^2} + \frac{1}{\tan\theta}\frac{\partial\delta}{\partial\theta} + \frac{1}{\sin^2\theta}\frac{\partial^2\delta}{\partial\phi^2}\right) + \frac{4}{3}\tanh\rho\frac{\partial\delta}{\partial\rho} = 0 \tag{12.22}$$

Following are a few comments on this equation. First of all, looking at only the first two terms with second derivatives over ρ and θ, we recognize the sound wave equation, with the square of the speed of sound $c_s^2 = 1/3$ of conformal (scale-invariant) QGP, now modified by a time-dependent coefficient reflecting the overall expansion of the zeroth-order basic solution. Second, looking closer at the part of the equation within the round brackets, one can recognize the angular part of the Laplacian (in θ, ϕ) in 3D spherical coordinates. This is because the Gubser flow is a conformal map from the $O(3)$ spherically symmetric solution, and one can think of deriving this sound *before* mapping. Finally, since the initial perturbations are assumed to be rapidity-independent, we also ignored this coordinate here, although there is no problem including it as well, as there is no dependence on the spatial rapidity η anywhere in the coefficients.

The remarkable property of this equation is that all four coordinates can be separated $\delta(\rho, \theta, \phi) = R(\rho)\Theta(\theta)\Phi(\phi)$ with three equations for them

$$R''(\rho) + \frac{4}{3}\tanh\rho R'(\rho) + \frac{\lambda}{3\cosh^2\rho}R(\rho) = 0 \tag{12.23}$$

$$\Theta''(\theta) + \frac{1}{\tan\theta}\Theta'(\theta) + \left(\lambda - \frac{m^2}{\sin^2\theta}\right)\Theta(\theta) = 0 \tag{12.24}$$

$$\Phi'(\phi) + m^2\Phi(\phi) = 0 \tag{12.25}$$

Furthermore, all three are analytically solvable, with the results

$$\begin{aligned} R(\rho) &= \frac{C_1}{(\cosh\rho)^{2/3}} P^{2/3}_{-\frac{1}{2}+\frac{1}{6}\sqrt{12\lambda+1}}(\tanh\rho) \\ &\quad + \frac{C_2}{(\cosh\rho)^{2/3}} Q^{2/3}_{-\frac{1}{2}+\frac{1}{6}\sqrt{12\lambda+1}}(\tanh\rho) \\ \Theta(\theta) &= C_3 P_l^m(\cos\theta) + C_4 Q_l^m(\cos\theta) \\ \Phi(\phi) &= C_5 e^{im\phi} + C_6 e^{-im\phi} \end{aligned} \tag{12.26}$$

where $\lambda = l(l+1)$ and P and Q are the associated Legendre polynomials. The part of the solution depending on θ and ϕ can be combined in order to form spherical harmonics $Y_{lm}(\theta, \phi)$, such that its generic solution can be written as[6]

$$\delta(\rho, \theta, \phi) = \sum C_l R_l(\rho) Y_{lm}(\theta, \phi) \tag{12.27}$$

with some arbitrary coefficients C_l. The solution sourced by a delta function — the Green function of the second-order equation — can then be found (Staig and Shuryak, 2011b) and, with it, a solution to any other initial conditions. We made a movie, which I cannot append here, which shows a distorted and expanding "sound circle."

The azimuthal dependence of the secondary particles can be calculated from the Cooper–Fry integral, from which the two-body correlation function follows. It is shown in Fig. 12.3(a), from Staig and Shuryak (2011b). As we already explained in the previous section, it features a peak at zero $\delta\phi = 0$ (generated when both observed particles come from the same peak) and two peaks displaced by $\Delta\phi = \pm 2$ rads (two particles come from

[6]A generic case of the separation of three variables would require splitting of the operator into three parts and the definition of three eigenvalues, which combine together to give zero. However, in this particular case, $l(l+1)$ is the eigenvalue of the whole angular part of the Laplacian, including both the θ, ϕ variables.

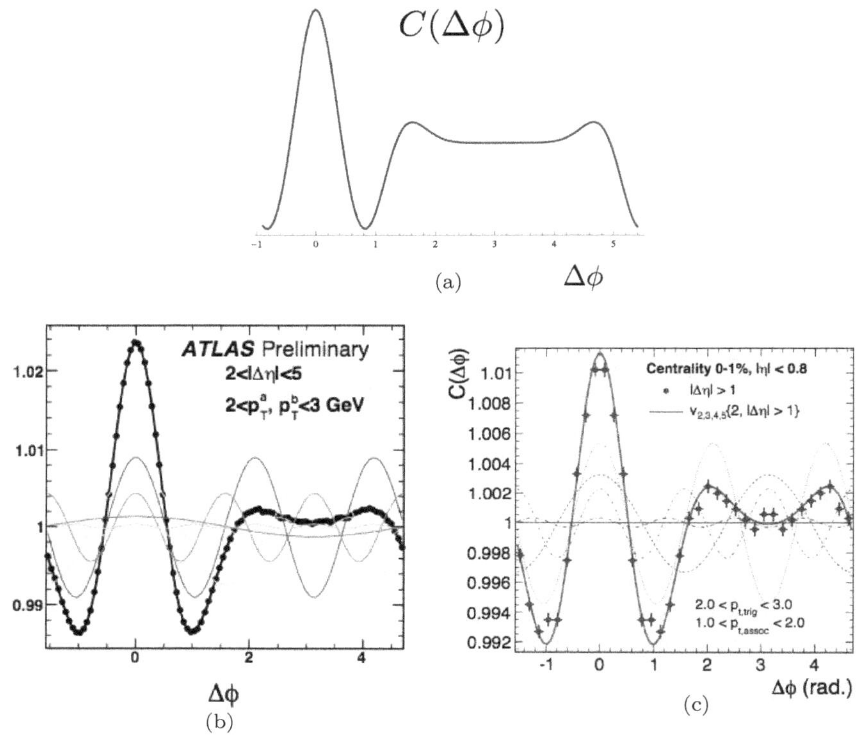

Fig. 12.3. (a) Calculated two-pion distribution as a function of the azimuthal angle difference $\Delta\phi$ for the viscosity-to-entropy ratios $\eta/s = 0.134$. (b) Experimental data from ATLAS and (c) from ALICE on the two-particle correlation function, both for ultra-central collisions. Thin lines show the corresponding decomposition of these correlation functions into angular harmonics. Note that the largest of them is the $m = 3$ "triangular flow."

two different locations). This calculation has been made and presented at the first day of the Annecy Quark Matter conference in 2009.[7] Two experimental correlation functions from ATLAS (Jia, 2011) (for the "super-central bin" with the fraction of the total cross-section 0–1%) and ALICE (Aamodt *et al.*, 2011b) are also shown in Fig. 12.3(b). The observed shape of the azimuthal correlation corresponds to the predicted one strikingly well.

[7] It was presented *before* the experimental data were shown, about an hour later, making it a true prediction.

12.4 Are Acoustic Maxima/Minima Observed in Harmonic Amplitudes?

Let us begin with a question of why one would need to use the Fourier harmonics of the correlation function itself, instead of just a function of the relative angle (as discussed above). Consider the simplest case of central collisions, which are on average axially symmetric. The answer to it lies in the fact that experiments provide not only the two-particle correlation function discussed but also more complicated correlators of $4, 6, 8, \ldots$ particles, from which one can also deduce the magnitude of higher flows v_n. (We define those later, when discussing the issue of the so-called *collectivity* of the flow.)

The correlation function calculated by Staig and Shuryak (2011b) can, of course, be expanded in harmonics, and one can plot the harmonic power spectrum of the squares of the coefficients. The results of this first calculation are shown in Fig. 12.4(a). It had shown the existence of strong oscillations, as seen from the plot, with the first peak close to $n = 3$ and the minimum around $n = 7$. The nature of the oscillations is the same as in the Big Bang: Any oscillator has energy constantly shifting, for example from kinetic to potential energy, and at some time moments, the signal based on, say, the velocity can be missing.

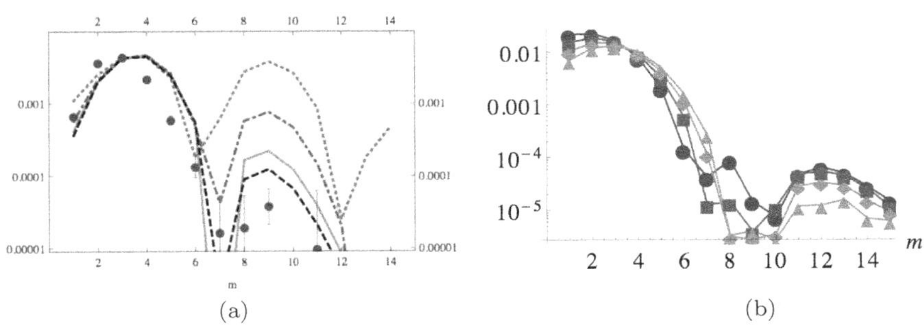

Fig. 12.4. (a) The mean squared amplitude of harmonics $\langle v_n^2 \rangle$. The lines are from analytic linearized hydrodynamic calculations of the correlation function harmonics, v_m^2, based on a Green function from a point source for four values of viscosity $4\pi\eta/s = 0, 1, 1.68, 2$ (from top to bottom at the right). The closed circles are the ATLAS data for the ultra-central bin. (b) The same quantity at $p_\perp = 1\,\text{GeV}$, with different curves demonstrating the dependence of the minimum on the initial distance r_0 of the perturbation. Calculations for $r_0 = 7.5, 8.0, 8.5, 9\,\text{fm}$ are shown by the blue circle, red square, brown diamond and green triangle symbols, respectively.

A subsequent study by Gorda and Romatschke (2014) of the analytic linearized perturbation on top of the Gubser flow gave more information about the minima, see Fig. 12.4(b). While the minima are clearly there for one point-like source, its location depends on the radial location of the original source r_0. Gorda and Romatschke (2014) further studied the question and found that for $r_0 \sim 5 - 6\,\text{fm}$, the minimum shifts to $m = 10$. The question of whether such minima can survive realistic ensemble average and perhaps ever be observed experimentally was debated.

The power spectrum of microwave background perturbation in the Universe is known to display several "acoustic minima," when the phase of sound oscillations at the freeze-out takes certain "disfavored" values. Staig and Shuryak (2011b) predicted the same phenomenon to happen in the "Little Bang," with the location of the first minimum predicted to be $n = 7$. This prediction was met with skepticism, as numerical hydro simulations have not shown that. Yet, a decade later, this prediction was confirmed by the ALICE data on the harmonics of the flow, see Fig. 12.5 from Acharya *et al.* (2020). Note that the highest harmonics v_6, v_7, v_8, v_9 do indeed show a minimum at around $n = 8$.

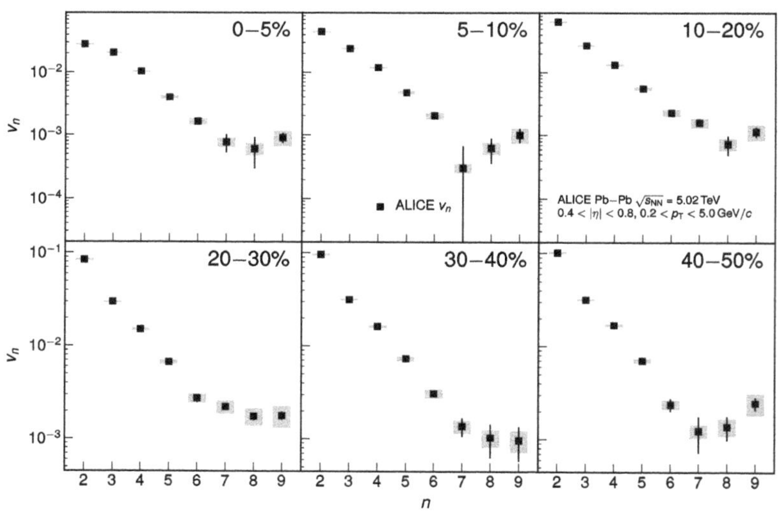

Fig. 12.5. Alice data on flow harmonics magnitude for different centrality classes of PbPb collisions.

Another debated question was whether there is a *maximum* in angular harmonic power spectra. The question more specifically addresses the relative magnitude of the $m = 2$ and $m = 3$ harmonics.[8] The experiments show that in some bins, the "inversed" ($v_3 > v_2$) order is seen; see, for example, the so-called ultra-central ATLAS data shown in Fig. 12.4(a), as well as the CMS central bin data in Fig. 12.4(b). (The latter include slightly larger impact parameters and thus feed more geometry-related contributions to v_2.) The "inverse magnitude" $v_3 > v_2$ is also predicted by both analytic calculations just discussed (Staig and Shuryak, 2011b; Gorda and Romatschke, 2014) at large r_0, while all sophisticated event-by-event studies led to the opposite conclusion, i.e. $v_3 < v_2$ for all bins.

12.5 Event-by-Event Hydrodynamics and Acoustic Damping

We have discussed so far the *analytical* approach to the issue of higher harmonics of the flow based on linearized equations and their Green function. The sound was thus generated from one delta-function-like local source. While I do find it more instructive, it is not the only approach to this problem. In fact, the mainstream approach in the field is the so-called *event-by-event hydrodynamics*, which generates an ensemble of initial conditions and then solves viscous hydro equations for each of them. To give proper credit to the people doing it, let me present some of the very successful examples of such works.

Following suggestions by Ollitrault, various models of the ensemble of the initial states are translated into a set of complex numbers:

$$\epsilon_n \exp(\mathrm{in}\Phi_n) = -\frac{\int d^2 r_\perp r_\perp^n e^{\mathrm{in}\phi} \rho(\vec{r}_\perp)}{\int d^2 r_\perp r_\perp^n \rho(\vec{r}_\perp)} \tag{12.28}$$

The modulus ϵ_n is used to represent the magnitude of intital deformation, and Φ_n shows its direction in the transverse phase, as ϕ is the azimuthal angle. As hydrodynamical calculations show, the observed v_n (harmonic of

[8]The harmonics $m = 1$ is known to be especially small due to the vanishing dipole. So, it should be removed from consideration. We now discuss only central collisions, in which all effects come from fluctuations, not the geometry.

the flow velocity) is simply proportional to ϵ_n.[9] This relation holds *per event*, so the ensemble of v_n is, up to a constant, the same as the ensemble of ϵ_n.

In Fig. 12.6, one can see the results of an event-by-even calculation based on the hydro code MUSIC (Gale *et al.*, 2013). It has a hadronic afterburner and includes the decays of all resonances up to 2 GeV in mass. One may say that this model does excellent work and fixes the value of the viscosity with a reasonable accuracy. So, what else remains to be discussed?

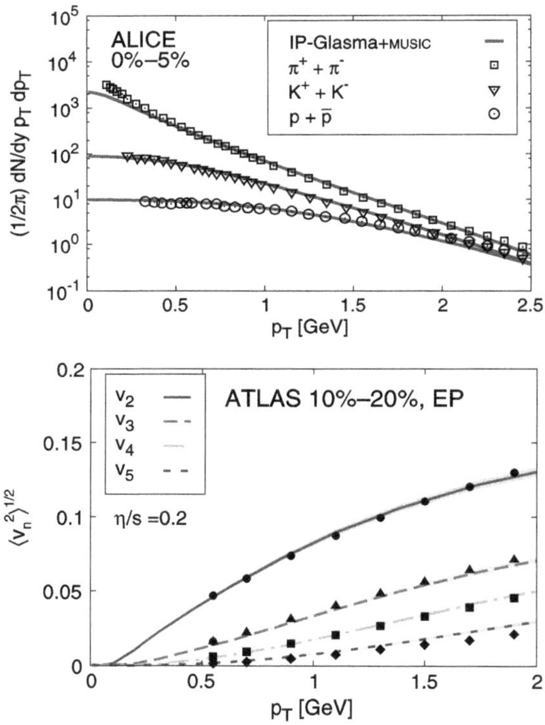

Fig. 12.6. (left) Calculated spectra of identified secondaries compared to the ALICE central *PbPb* collision data. (right) Root-mean-square anisotropic flows $\langle v_n^2 \rangle^{1/2}$ as a function of p_T (GeV), compared to the data of the ATLAS collaboration. Experimental error bars are smaller than the size of the points.

[9]A deviation from this occurs due to composite harmonics, which we will discuss in the following section.

Well, one still needs to understand certain phenomena and dependencies. We already discussed the acoustic maxima/minima. The next question is whether these calculations do or do not agree with the overall damping of the modes, the acoustic systematics. We already briefly discussed the "acoustic damping" formula, suggested by Staig and myself (Staig and Shuryak, 2011a). The expression (11.8) was discussed in the previous chapter in connection with the proportionality of $\log(v_n/\epsilon_n)$ to the mean gradient $1/\bar{R}$.

Let us now discuss the dependence of two other variables: n and the viscosity. Let us first look at the middle plot (b) of Fig. 12.8 (from Lacey *et al.*, 2013), based on the experimental data from the ATLAS collaboration. As one can see, the linear dependence $\log(v_n/\epsilon_n) \sim n^2$ seems to be holding well. The left plot (a) corresponds to hydrodynamical calculations with a different viscosity, also plotted versus n^2. The same dependence $\log(v_n/\epsilon_n) \sim n^2$ is clearly seen in the calculations as well.

Furthermore, by extracting the exponential slopes from Fig. 12.8(a) and plotting them in Fig. 12.8(c), one finds a linear dependence as well. However, as seen from the plot, at zero input viscosity, one finds a nonzero slope,[10] although a smaller one. Gorda and Romatschke (2014) have extensively studied this: They also see an $\sim n^2$ dependence of the damping while also observing "an offset" in the exponent, not only being proportional to the viscosity value used.

Other sophisticated event-by-event hydrodynamic calculations of the harmonics, such as those done by MUSIC (Rose *et al.*, 2014), are also well described by an $\exp(-n^2 * \text{const})$ dependence, as expected from the acoustic damping discussed above, see Fig. 12.7(a).

Completing the subject of higher harmonics of flow, let me end with the following questions: *Can one similarly define a complete set of independent harmonics for a generic non-Gubser setting? And, even more importantly, can those be observed?*

Let me specify the first question a bit more. Of course, one can define linearized hydrodynamics on top of the (numerically found) smooth (average fireball) flow. Let us, as before, ignore for now the rapidity variable and focus on the transverse plane r, ϕ.

There is no doubt that one can define the $e^{im\phi}$ azimuthal harmonics. There is no doubt that any Hermitian differential operator does have a

[10] I do not know the origin of this effect, but I suspect it is due to "numerical viscosity," not explicitly incorporated but still present in the code due to the finiteness of the cells.

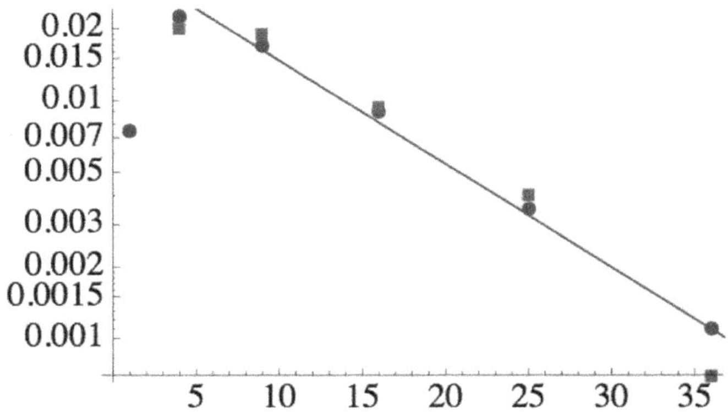

(a)

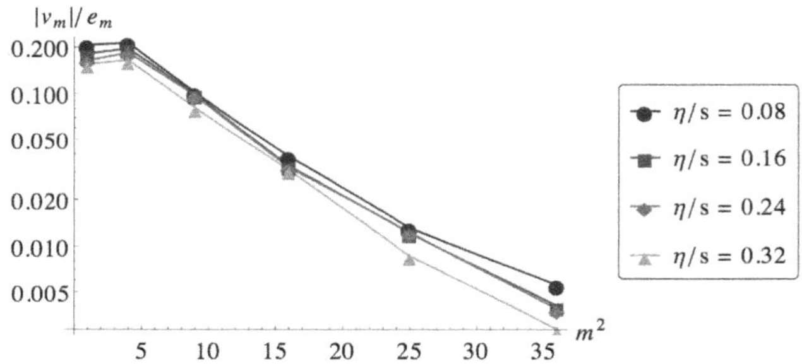

(b)

Fig. 12.7. (a) $v_n\{2\}$ plotted versus n^2. Blue closed circles correspond to viscous event-by-event hydrodynamics in the "IP Glasma+Music" model, with the viscosity value $\eta/s = 0.14$. The straight line, shown to guide the eye, demonstrates that the "acoustic systematics" does in fact describe the results of this heavy calculation quite accurately. The CMS data for the 0–1% centrality bin, shown by the red squares, in fact display larger deviations, perhaps oscillatory ones. (b) The harmonics induced by a fixed perturbation at $r_0 = 4\,\text{fm}$ for variable viscosity.

certain orthogonal basis with real eigenvalues. So, the problem must have a solution. The problems are practical.

One problem is that outside of the Gubser flow, one cannot combine the eigenfunction of r dependence with m into one common quantum number l. So, there would be a *double* sum over harmonics. If this is done, there appears to be another problem: The fluctuating ensemble of initial conditions should be projected to those harmonics. While it may resemble the current definition of ϵ_m, with r^m in it, it will not be exactly this function. One would then need to recalculate all ϵ_m.

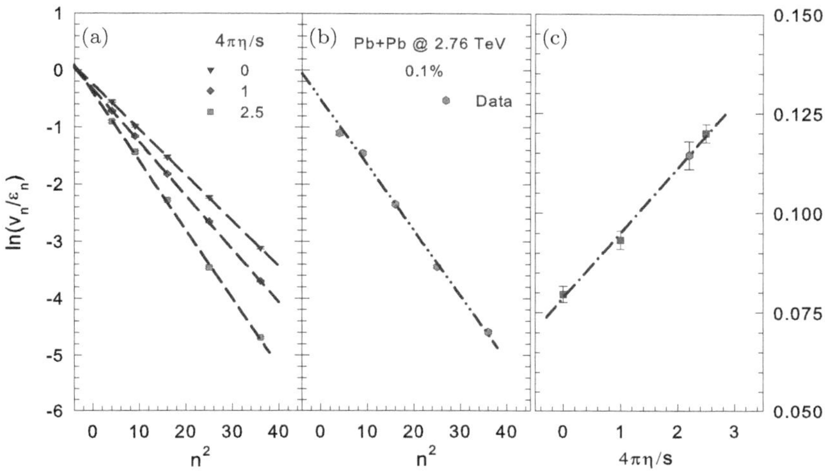

Fig. 12.8. (a) The $\ln(v_n/\epsilon_n)$ versus n^2 from viscous hydrodynamical calculations for three values of specific shear viscosity, as indicated. (b) $\ln(v_n/\epsilon_n)$ versus n^2 for $Pb+Pb$ data. The p_T-integrated v_n results in (a) and (b) are for ATLAS 0.1% central $Pb+Pb$ collisions at $\sqrt{s_{NN}} = 2.76\,\text{TeV}$; the curves are linear fits. (c) $\ln(v_n/\epsilon_n)$ versus $4\pi\eta/s$ extracted from the curves shown in (a) and (b).

(From the experimental side, a step in this direction has been taken by Mazeliauskas and Teaney, 2015, who defined the so called "subleading harmonics" of flow, which can be extracted from the experimental data. The main difference between the leading and subleading n flows is indeed in the radial dependence.)

It has been shown that the linear relation $v_n \sim \epsilon_n$ assumed in the linearized theory above is not always fulfilled because there are nonlinear terms. For example, the harmonic v_4 receives a contribution proportional to ϵ_2^2, v_5 from $\epsilon_3\epsilon_2$, etc.

Let me also mention that, in practice, people use two sets of definitions. One, called the *moments*, we have used all along:

$$\epsilon_n e^{in\Phi_n} \equiv -\langle r^n e^{in\phi}\rangle / \langle r^n \rangle$$

Another set is formed by the so-called *cumulants*. As an example, let us define it for $n = 4$ as

$$C_4 e^{i4\Phi_4} \equiv -\left[\langle r^4 e^{i4\phi}\rangle - 3\langle r^2 e^{i2\phi}\rangle^2\right] / \langle r^4 \rangle \tag{12.29}$$

The factorized elliptic distribution $\sim \exp\left(-\frac{x^2}{2\sigma_x^2} - \frac{y^2}{2\sigma_y^2}\right)$ has a nonzero ϵ_4 but zero S_4.

Further studies of such effects can be found in the work of Teaney and Yan (2012), who use cumulants as their basis. Nonlinear effects include a particularly curious case: v_1 harmonics generated by a nonlinear $\epsilon_2 \cdot \epsilon_3$ interaction. I would not discuss them in detail and only note that these "nonlinear" effects do *not* originate from the nonlinear terms in the hydrodynamic equation but mostly from an expansion of the Cooper–Frye exponent $\exp(p^\mu u_\mu/T)$, containing flow velocity in powers of these perturbations. Obviously, they contain a higher power of p_t and can therefore be separated from linear effects.

12.6 Nonlinear Effects Lead to "Composite" Harmonics

In our discussion above, we assumed that perturbations of hydrodynamical flow are small and are therefore based on the calculations on the linearized hydrodynamics, on top of the basic explosion, e.g. on top of Gubser's flow (Staig and Shuryak, 2011b). However, this assumption is only correct at a late enough time, while the initial state is quite inhomegeneous, and the nonlinear terms in the equation may not be small.

It is, of course, possible to identify interactions between sound harmonics. Two papers, by Qiu and Heinz (2011) and Gardim *et al.* (2012), specifically addressed harmonic numbers for which nonlinear effects were expected to be the strongest. Those are $n = 4 = 2 + 2$ and $n = 5 = 3 + 2$, since the second and third harmonics are the largest. They have indeed found that the calculated v_4 and v_5 do not have strong correlations with the magnitude and directions of the initial deformations ϵ_4, ϵ_5.

Instead of a simple proportionality between ϵ_n and v_n, one then uses a relation with a quadratic term, e.g.

$$v_4 e^{i4\phi_4} = \kappa_4 \epsilon_4 e^{i4\Phi_4} + \kappa_4' \epsilon_2^2 e^{i4\Phi_2} \tag{12.30}$$

with two (event-independent) constants κ_4 and κ_4' calculated using hydrodynamics.

A systematic way to separate the linear and nonlinear harmonics was worked out by Teaney and Yan (2012). All explicit formulae can be found in that paper; let me explain the main idea in words. The Boltzmann factor of matter with flow four-velocity u_μ has the form $\exp(u_\mu p^\mu)$. Azimuthal harmonics are part of u_μ, and expanding the exponent to linear order, one finds these harmonics in the numerator. Since the coefficient is momentum, these linear harmonics all have *linear* growth with $p_\perp$ (for pions). It is indeed what we have seen, for example, in an elliptic flow, $v_3(p_\perp) \sim p_\perp$.

Now, expanding the Boltzmann factor to the *second* order, one finds the products of two angular harmonics $\exp(i(n_1+n_2)\phi)$, producing "composite" terms with $n = n_1 + n_2$. An important point made by Teaney and Yan (2012), is that those can be identified in the data via *quadratic* dependence:

$$v_n^{\text{composite}}(p_\perp) \sim p_\perp^2 \tag{12.31}$$

This idea works quite well, and fitting the observed v_4 and v_5 to a superposition of terms linear and quadratic in $p_\perp$, one can indeed identify the magnitude of the nonlinear effects in flow harmonic generation. For an extensive discussion of the energy dependence of the composite harmonics by the STAR collaboration, one should consult a more recent work (STAR Collaboration, 2022a).

12.7 Flow Magnitude at the Fireball's Edge

While most of the previous work focused on a part of the freeze-out surface at the latest proper time, here we focus on the complementary part, to be referred to as "the outer edge," where the highest transverse rapidity of flow $\kappa \sim 1.4$ is achieved. Following Ramamurti and Shuryak (2018), we first describe this surface analytically, using the Riemann rarefaction wave, and then numerically, using the numerical hydrodynamic code MUSIC.

For central collisions, we observe a good description of the spectra for π, K, N in central *PbPb* LHC collisions in this extended region. We further suggest that "the outer edge" has very small azimuthal asymmetry even for non-central collisions, smaller than those predicted by standard hydrodynamics.

Let me start with a question: *Do we know up to what transverse momenta (transverse rapidities $y_\perp$) hydro/thermodynamics described the observed spectra?* The first pre-RHIC papers reached up to $p_\perp \sim 1\,\text{GeV}$ only, as it was expected then that above it, "minijets" would dominate. This turned out to be wrong: Due to jet quenching, their contribution actually becomes visible only at much higher $p_\perp$. Therefore, the hydro description has a chance to be extended by a few more decades in spectra.

On purely phenomenological grounds, one can locate a transition between the exponential-like versus the power-like (hard jets) regimes, for example, by plotting the logarithmic derivative (Ramamurti and Shuryak, 2018). A sharp transition between those is indeed well seen for pions at

$y_\perp \approx 4.2$ or $p_\perp \sim 5\,\text{GeV}$. This suggests that the hydro/thermodynamical description may perhaps be extended further down the spectra.[11]

As already discussed in Section 9.3, the fireball has an "outer edge" limited by the freeze-out surface, and it is where the maximal transverse rapidity of the collective flow is reached. It was argued by Ramamurti and Shuryak (2018) that hydro flow near the edge can be described analytically using the Riemann solution, also known as *the rarefaction fan.* Its first discussion in the context of heavy ion collisions was made by Baym *et al.* (1983) but never used since.[12]

The main simplification leading to the solution we discuss is due to Riemann, who assumed that in the rarefaction fan region, the two unknown functions, i.e. the energy density $\epsilon(\tau, x)$ and flow rapidity $y(\tau, x)$, are *not* independent but, in fact, are directly related to each other:

$$\epsilon(\tau, x) = f\big(y(\tau, x)\big). \tag{12.32}$$

Therefore, the isotherms (close to freeze-out surfaces) are also simultaneously constant transverse rapidity $y(\tau, x) =$ const., the *same lines* in this region!

The resulting spectra for this model are shown in Fig. 12.9 for the pions and protons. The previous studies focused on the "lid" contribution, shown by the dash-dotted lines, which provides a good description of the shape of the distribution up to $p_\perp \sim 3\,\text{GeV}$. Note further that the excess seen in the data at small $p_\perp$ is due to the so-called "feed-down" contribution, resulting from the decay of multiple mesonic and baryonic resonances. It is well known and calculated in statistical hadronization models, such as those by Andronic *et al.* (2018).

The new contribution of the "outer edge" that we introduced now, shown by the solid lines, is small in the integral, of the order of a percent. However, it becomes dominant at higher transverse momenta, improving agreement for pions for $p_\perp = 2$–5 GeV and for protons for $p_\perp = 3$–6 GeV. The ALICE data are from Abelev *et al.* (2014). It significantly extends the description of the spectra in terms of absolute probability by 2–3 orders

[11] If one questions the validity of thermal equilibrium to an extent able to correctly predict the tails of thermal spectra, we recall that at chemical freeze-out, the Boltzmann exponent does describe correctly the production of light nuclei up to at least He^4, with mass about 4 GeV.

[12] At that time, the magnitude of the flow was expected to be very small. For example, in that paper, the authors mention pions with $p_\perp \sim 30\,\text{MeV}$ affected by the radial flow. This is more than a hundred times(!) smaller than what we are discussing now.

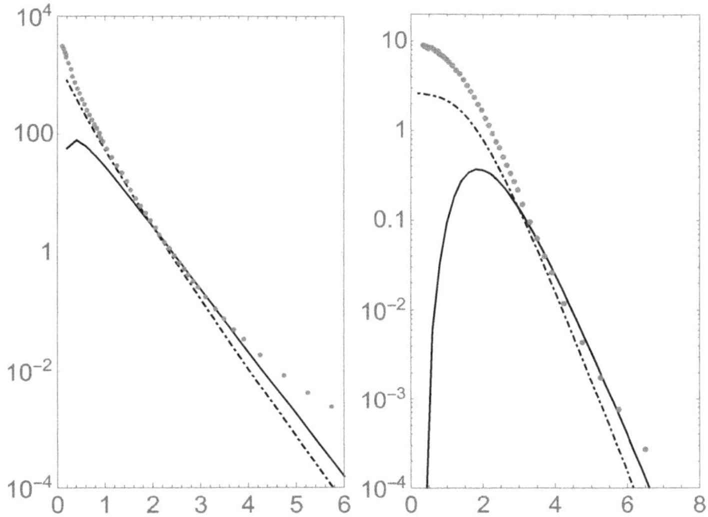

Fig. 12.9. The transverse momentum spectrum $dN/dydp_\perp^2$ $((\mathrm{GeV/c})^{-2})$ versus transverse momentum $p_\perp$ (GeV/c) for pions (left) and protons (right). The points correspond to the 0–5% centrality data from ALICE; the dash-dotted lines show the "lid" contribution, and the solid lines show that of the "outer edge," with the additional factor $P_{\mathrm{edge}} = 1/4$.

of magnitude. This is important because it now extends right up to the momenta of $p_\perp > 5$–6 GeV, where the contribution from hard partonic reactions eliminates the "nobody's land" region in between.

We end this section with an open question: *Can one extend these ideas to non-central collisions and elliptic flows?* So far, nobody has succeeded in doing so. The freeze-out surfaces can, of course, be calculated, but one has to correctly implement the relatively large hadronic mean free paths.

12.8 Sounds in the Loops

We know from other fields of physics, such as that of liquid He^4, that many effects can indeed be understood by phonon kinetics. In the chapters on holography, we will see that sounds in matter are dual to gravitons in the bulk. In particular, sound emission from quenched jets can be seen as a dual to gravitational radiation.

Hydrodynamics is an effective theory in its simplest ideal form (zero mean free paths and viscosities) described by local thermal and velocity fields. The hydrodynamical modes include sounds, longitudinal pressure waves, related to breaking of translational invariance by matter. Thus, their interactions fall into a certain pattern familiar from the physics of pions.

Therefore, one may try to build effective Lagrangian for sound quanta, similar to the effective chiral Lagrangians. We have already seen from the "composite harmonics" phenomenon that sounds do interact with each other due to nonlinear terms in hydro equations. It is also clear that they would contribute to certain correlation functions via "sound propagators." So, the key elements of a diagrammatic description of higher-order effects are available. Two or more of those would then create rescatterings as well as certain "loop" effects, such as how the fundamental fields of any QFT do.

Progress along these lines was reviewed by Kovtun (2012). One particular object we care about is a correlator of two stress tensors. First, via the Kubo formulae, it describes contributions to kinetic coefficients. Second, it is used to calculate the production rate of gravity waves in a cosmological setting (see the following chapter).

Skipping the derivation, I present one particular result by Kovtun *et al.*, namely a correction to shear viscosity due to a loop diagram with two sound propagators:

$$\Delta\eta_{\rm loop} = \frac{17}{120\pi^2}\frac{T(\epsilon+p)p_{\rm max}}{\eta} \tag{12.33}$$

Two curious features need to be commented upon. The original matter viscosity η appears here in the *denominator*! Indeed, if it is small, sounds live long and can propagate far, inducing a large correction (at small η) to the loop correlator.

Second, because this effective theory is non-renormalizable, this loop diverges in the UV, i.e. there appears $p_{\rm max}$ which is the highest momentum retained in the momentum integral.

Summary and connections: This chapter started with nonrelativistic linearized hydrodynamics, reminding the reader what the sound and diffusion modes are. We then argued that "higher flow harmonics, $n = 3, 4, \ldots$, are basically sounds, propagating on top of an exploding fireball. If so, their damping should follow "acoustic systematics," and indeed it does. This statement had given us a quantitative handle on QGP viscosity (see more discussion in Chapter 22). Furthermore, there are "phase factors," describing a particular phase of the oscillation at the freeze-out. A maximum at $n = 3$ and a minimum at $n = 7$–8 were predicted and, recently, observed. In the following chapter, we take a detour from heavy ions to cosmology, describing the "sounds of the Big Bang."

Unlike quarks and gluons, which do not make good quasiparticles in strongly coupled QGP, collective modes such as sounds are, in fact, weakly interacting and long-lived. In a way, sounds are also "penetrating probes." It was not surprising that the lowest harmonics survived from the initial to the final state, giving us the best handle on the initial state we have. In this chapter, we have seen multiple manifestations of them, as azimuthal harmonics v_n, now measured up to $n \sim 10$. We have seen the appearance of "composite harmonics" due to their nonlinear interactions.

Chapter 13

SOUNDS IN THE BIG BANG

13.1 Sounds in the Big Bang Seen in Cosmic Microwave Background

While this book is not about cosmology, it would be wrong not to discuss phenomena very similar in their physical origins to what we discussed in the previous chapter in relation to high harmonics of the flow in the "Little Bang."

Let me briefly outline the setting. While in the optical photon domain, the sky is black (except stars and galaxies, of course), at microwave frequencies it is "white," homogeneously filled by electromagnetic radiation. First discovered as "ubiquitous noise," it has a thermal spectrum with a temperature of only $T_{\rm CMB} = 2.725$ K. This spectrum was famously discovered in 1965 by Arno Penzias and Robert Wilson (awarded the Nobel Prize in 1978). While unexpected, it was in fact already predicted by Gamow[1] and collaborators as early as the 1940s based on ideas about the "hot Big Bang" scenario of the Universe evolution.

The initial conditions of the Big Bang led to a very homogenous Universe, expanding according to the Friedmann equations following Einstein's GR. Yet, nothing is perfect: Under continuous scrutiny by subsequent experiments, small $\Delta T/T \sim 10^{-5}$ perturbations were found.[2]

[1]Gamow's adviser in his university years was none other than Alexander Friedman, the father of the expanding Universe solution to Einstein's equations.

[2]How it happened is explained in the 2006 Nobel Lecture of J. C. Mather. It is in fact a modern-day miracle story of a graduate student whose PhD balloon project failed (and was completed by another student) but who had the guts to propose it to NASA in a satellite version, which succeeded and paved the way to this remarkable discovery and his Nobel Prize. The main idea was to keep a liquid helium source exactly at 2.725 K and measure only a small *difference* between the effective temperatures of the waves coming from variable direction in the sky and this source.

They must be sourced by some "initial events" at some random places or waves: Their exact origin remains unknown.[3]

The early Universe was dominated by quanta with masses $m_i \ll T$ (quarks and gluons but also leptons, neutrinos, etc.), and thus, EOS is $\epsilon \sim p \sim T^4$. Obviously, perturbations would propagate with the same speed of sound $c_s = 1/\sqrt{3}$ as in QGP, making "sound circles" (actually, spheres) with a radius of $r(t) = \int_0^t c_s(t')dt'$.

As the temperature dropped, the last relevant example of such a species is the *photons*. When it reaches $T_{\text{neutralization}}$ (of the order of an electron volt), the primordial plasma was neutralized, and the mean free path of photons increases rapidly by many orders of magnitude. The Universe then becomes "optically thin," which is why we can see those photons today (degraded in energy from optical to microwave frequencies).

The decoupling of photons from baryonic matter implies that photons' blackbody pressure is no longer there, and the sound velocity drops down to that in the baryonic component alone, which is orders of magnitude smaller. In practical terms, one may say that the sounds stop instantaneously. If there would be only a single source, we would see a single circle in the sky, with some extra density of matter. Yet, of course, there are multiple sources with random amplitudes, whose shapes are unknown, etc.

The observed deviations from constant T are treated statistically. The so-called "power spectrum" of it across the sky is decomposed into a sum of 2D angular harmonics:

$$\frac{\delta T}{T} = \sum_{l,m} c_{l,m} Y_{l,m}(\theta, \phi)$$

and the square of the modulus of the coefficients $|c_{l,m}|^2$ is plotted against l in Fig. 13.1 (the data are from the Planck mission (Ade *et al.*, 2014a)).

The first feature of the power spectrum plot is that there is a *maximum* at $l \sim 300$, or the angle of $\delta\phi \sim 360°/l \sim 1°$, of the order of one angular degree.[4]

Another place in which "circles in the sky" reveal themselves is in the correlation function of galaxies. Plotting the two-point correlator of their

[3]Cosmologists attribute those to quantum zero-point oscillations of scalar "inflation" fields in the inflation models. If so, they come from an era when the temperature was $T_{\text{inflation}} \sim 10^{16}$ GeV. It can, however, also result from phase transitions, such as electroweak and/or QCD phase transitions, which we will discuss soon.

[4]How did the ancient Sumerians justify the division of a circle into 360 parts? It is a good round number, close but not equal to the number of days in a year.

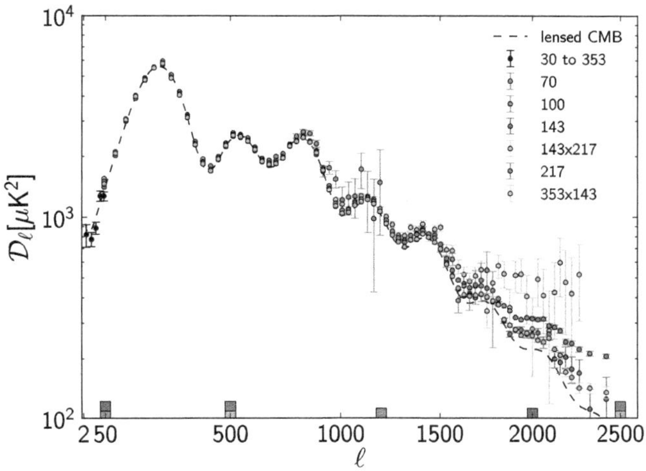

Fig. 13.1. Power spectrum of cosmic microwave background radiation measured by the Planck collaboration. The dashed line at the lower right side indicate the correction taking away "gravitational lensing" by the foreground matter, which makes some locations appear brighter than they actually are.

density, one finds a peak at a distance of[5]

$$r(t_{\text{freezout}}) \sim 100\,\text{Megaparsec (Mps)}$$

which appears because even a perturbation on top of the smooth background, as small as 10^{-5}, is able to seed the formation of galaxies better. Note that by simply relating the angular size of the *acoustic maximum* to its absolute scale, one may conclude that the size of the Universe today must be about two orders of magnitude larger, ~10 Gigaparsec, which is clearly an important conclusion for cosmology!

The second feature of the power spectrum in this semi-logarithmic plot is the clear *exponential damping* of higher harmonics. This is the effect of viscous damping of the acoustic harmonics, similar to what we already discussed for the Little Bang. Although the mechanism of this damping is nontrivial, we do not have time for its discussion.

The main reason we discuss this plot is the third feature, namely the oscillations of the magnitude. Their explanation goes as follows. All hydrodynamic harmonics become excited at the initial time $t = 0$ by the

[5]For non-astronomers, "parsec" is a distance from which Earth's orbit is seen at an angle of one angular second; it is ≈3.2 light years or $3 \cdot 10^{16}$ m.

Big Bang, and all become frozen at the same time (of plasma neutralization) as well. The oscillation *phases* at the freeze-out moment depend on the harmonic number l: For larger l, they oscillate more rapidly, and the phase is larger. For some phases, all oscillator energy is "potential energy" (the photon density, or T^3); for some (we do not observe) in kinetic energy. So, the former harmonics show observed maxima, the latter minima. The binary correlator is proportional to $\cos^2(\phi^l_{\text{freezeout}\}})$.

One more estimate comparing the Little and Big Bangs may be instructive. If the measured highest harmonics number is, say, $l = 3000$, the sky contains the number of bins (and thus amplitudes) of $\sim(3000)^3 \sim 10^7$. It is a large number, but not very large. Let me also mention that cosmology has only *one* event! Heavy ion collision experiments have given us $\sim 10^{10}$ events already.

Before leaving the subject of CMB and large-scale perturbations, it would be fair to acknowledge several unexplained observations. The amplitudes of lower harmonics $l = 2 \ldots 20$ do show certain strange features and correlations (see, for example, a summary of the Planck collaboration (Ade *et al.*, 2014b)). The planes in which quadrupole and octupole are maximal are quite accurately aligned (why?). Odd harmonics are larger than even ones in a systematic way. Nearly all of them have magnitudes deviating from smooth fits to the spectrum, some by a significant factor.

Let me also add that it was observed by Shamir (2020) that the directions of the *spins* of galaxies are not entirely random, despite their very large number, $\sim 10^{11}$, in the Universe and quite complicated formation processes. One finds statistically significant nonzero dipole and quadrupole harmonics of this "averaged spin field" over the Universe, suggesting that they originated from some primordial vorticity field. Nobody has a clue what has created all of that.

Finally, although a statistical basis of the following claim remains under dispute, let me still mention that Gurzadyan and Penrose (2011) have announced that they were able to observe several *co-centric sound circles* in the CMB data, at some locations on the sky, with two and even three circles. If this is indeed so, it means at certain locations, there were "repeated perturbations" which somehow were able to emit sound waves *more than once*! If confirmed,[6] it would hardly be incompatible with our standard cosmological model, assuming that the Big Bang was a single event.

[6]Penrose, who became a Nobel Prize winner in the year 2020, he devoted a lot of time to these findings in his Nobel lecture, rather than to his famous achievements for which the prize was for.

13.2 Sound Cascades after Cosmological Phase Transitions

In this section, we consider cosmological stages at temperatures many orders of magnitude higher than those discussed in the preceding section ($T_{\text{decoupling}} \sim 1$ eV). We know that (at least) two phase transitions need to happen: (i) deconfinement in QCD, and (ii) the electroweak (EW) transition, in which the Higgs field gets its VEV, leading to the breaking of SU(2) symmetry.

Let us briefly recall the numbers related to these phase transitions. Step one is to evaluate the redshifts of the transitions, which can be done by comparing the transition temperatures $T_{\text{QCD}} \approx 156$ MeV and $T_{\text{EW}} \approx 160$ GeV with the temperature of the cosmic microwave background $T_{\text{CMB}} = 2.73\,\text{K}$. This leads to

$$z_{\text{QCD}} \approx 7. \times 10^{11}, \quad z_{\text{EW}} \sim 5 \times 10^{14}\,. \tag{13.1}$$

During the radiation-dominated era, to which both the QCD and EW eras belong, the solution to the Friedmann equations leads to a well-known relation between time and temperature[7]:

$$t = \left(\frac{90}{32\pi^3 N_{\text{DOF}}(t)}\right)^{1/2} \frac{M_P}{T^2}\,, \tag{13.2}$$

where M_P is the Planck mass and $N_{\text{DOF}}(t)$ is the effective number of bosonic degrees of freedom with masses lower[8] than T.

Plugging in the corresponding T, one finds the time of the QCD and EW phase transition to be

$$t_{\text{QCD}} = 4 \times 10^{-5}\,\text{s}, \quad t_{\text{EW}} \sim 10^{-11}\,\text{s} \tag{13.3}$$

Multiplying those by the speed of light, one arrives at the approximate sizes of the Universe and finds that at the QCD transition, it should be measured in km while at the EW one, it should be measured in mm. What will be more important is obtaining the respective redshift factors. Since temperature scales with it, comparing it with the CMB, one finds that the t_{QCD} scale today should correspond to about $3 \times 10^7\,\text{s} = 1$ year and the EW to $5 \times 10^4\,\text{s} = 15$ hours.

Phase transitions are expected to be associated with somewhat increased fluctuations. If the transitions are first order (as it was naively expected

[7]Note that we use not gravitational but particle physics units, in which $c = 1$, but the Newton constant is $G_N = 1/M_P^2$.

[8]For example, for a QCD transition, those are photons, neutrinos, quarks except c, b, t and leptons except τ.

for both QCD and EW phase transitions two decades ago), one expects inhomogeneities $O(1)$ (as one can see in a boiling tea pot). If they are second order, there appears to be an "order parameter" with an infinite correlation length and thus "critical opalescence."[9] Unfortunately, both in the QCD and Standard Model[10] with physical masses (quarks and Higgs, respectively), they are just relatively strong *crossovers*. One intriguing source of sounds can be the "sphaleron explosions," which we discussed in Section 23.5.

The question then is what happens with the sound waves *after* they get excited. Since the damping starts in sound dispersion (12.6) at an order of k^2, long-wavelength sounds at $k \to 0$ have very long lifetimes.[11] Out-of-equilibrium phenomena may produce much higher amplitudes of hydrodynamical perturbations at small k, provided the interaction of sound waves enters the so-called *inverse acoustic cascade*. The sound momenta and frequencies are, however, limited from below, $k > k_{\text{cutoff}}$, and thus, the sound intensities n_k are limited as well. The most obvious infrared cutoff of the wavelength is the horizon size of the Universe, $\lambda_{\text{cutoff}} \sim$ a few km.

To explain the idea (Kalaydzhyan and Shuryak, 2015), let me use an analogy with Earth's atmosphere. The loudest sounds on Earth have nothing to do with thermal equilibrium sounds but rather with thunderstorms and hurricanes. Those appear as a result of *inverse acoustic cascades* in Earth's atmosphere, in which millions of small vortices of size, say, 0.1 m, are successively combined until they reach sizes of 10^3 km (increasing by an impressive seven orders of magnitude) at the cutoff set by Earth's size.

An explanation of why it happens is provided by the theory of "acoustic turbulence," which, for various types of waves, was developed by Zakharov *et al.* (1990). The study of turbulence, either driven or free, began with the hydrodynamics of fluids. Kolmogorov proposed the famous stationary power solutions based on the notion that *energy flux through the wave spectrum* must be the same for all scales. For weak turbulence, governed by the Boltzmann equation, similar power solutions were found. Furthermore, they

---

[9]In Chapter 6, we will discuss current searches for the QCD critical point at a finite baryon chemical potential. However, in the Universe, it is very close to zero; therefore, we are quite sure it is not the case there.

[10]Since we do not understand the nature of dark matter, this statement may change in the future. Also, some modifications to the Standard model can do that.

[11]Note that when we speak to each other, one can hear from many meters away, while the "microscale" of the air — the interatomic distances — are of the order of μm. The reason is the same.

were shown to be *attractors*, meaning that, with time, all turbulent regimes converge to them.

The turbulent regime was shown to depend on the *sign* of the $O(k^3)$ term in the dispersion relation $\omega(k)$. If it is positive, a sound wave can decay into two nearly co-moving waves. Naively, this leads to a reduction in momentum, but in fact, the cascade proceeds in the direction of the inverse $2 \to 1$ reaction, which is called the *direct cascade* toward large k. Yet, when the *sign* of the $O(k^3)$ term is *negative*, there are no decays but $2 \to 2$ scattering, and this creates the *inverse* cascade. We do not know what the sign is for hadronic matter, so we need to get lucky here. Let me not go into details here and just note that sound–sound scattering amplitude is not simple, and (similar to the gluon–gluon case) it is dominated by small angle scatterings producing very large cross-sections. For the inverse cascade, we are interested in the momentum density

$$n_k \sim k^{-s}$$

with a negative power of s, which is found from the "constancy of the energy flux" equation and, for our case, given by $s = 10/3$. Now, recall that the scale range, from $\lambda \sim 1/T$ to the horizon, is 18 orders of magnitude, and in power s, it will be so huge that the cascade would shift into a somewhat different regime called "strong turbulence." Although it leads to a more complex equation for s, the index is believed to be only increasing, to say $s \approx 4$–5.

A turbulent inverse cascade may also be created by other waves: vortices or magnetic hydrodynamical phenomena. If the latter is the case, it would be in principle possible to see it in the domains of intergalactic magnetic fields (not yet detected), see Neronov *et al.* (2021).

In summary, it has been proposed by Kalaydzhyan and Shuryak (2015) that after the QCD or EW phase transitions, one might get a *cosmic thunderstorm* of huge amplitude, reaching the causal cutoff, the horizon of that era.[12]

13.3 Gravity Waves as "Penetrating Probes" of the Big Bang

Now, suppose we get lucky and the conditions for inverse cascades happen to be satisfied: How would we learn about that? Even at the largest

[12]For non-cosmologists, the horizon scales of EW, QCD and CMB times are projected to be completely different from the scales of today.

wavelength, all sounds will eventually dissipate, and there was nobody there to listen...

Here, we made use of the same idea as we used in heavy ion collisions: One has to look for "penetrating probes" which interact with matter sufficiently weakly. Photons and dileptons worked for our "Little Bang," but they would not do for the Big Bang. Even neutrinos have too small mean free paths. The only ones left are the *gravity waves* (GWs).

The cosmological horizon provides a natural infrared cutoff on the gravitational radiation wavelength. In the radiation-dominated era, it is inversely proportional to time, so the estimates above give a cutoff on the periods of the gravitational waves in the present time. GWs from the EW era are expected to be searched for by future space GW observatories, such as eLISA. A discussion of their potential sensitivity can be found elsewhere. The observational tools for the GW at the period scale of *years* are based on the long-term monitoring of the millisecond pulsar phases, with subsequent correlations between all of them. The basic idea is that when a GW is falling on Earth and, say, stretches distances in a certain direction, then in the orthogonal direction, one expects distances to be contracted. The binary correlation function for the pulsar time delay is an expected function of the angle θ between them in the sky.

There are existing collaborations — North American NanoHertz Observatory for Gravitational Radiation, the European Pulsar Timing Array (EPTA), and the Parkes Pulsar Timing Array — which actively pursue both the search for new millisecond pulsars and collecting the timing data for some known pulsars. It is believed that about 200 known millisecond pulsars constitute only about 1% of their total numbers in our galaxy.

The NANOGrav collaboration announced, in 2021 (Arzoumanian *et al.*, 2021), the first detection of gravitational waves with a year-long period through the pulsar correlation method. The frequency of this signal is actually $f \sim 3 \cdot 10^{-9}$ Hz, and the magnitude (in cosmological units of the fraction of the Universe energy density) is $h^2\Omega_{\rm GW} \sim 10^{-9}$. The proof that the wave is a tensor wave (negative correlations at $\theta \sim 90°$) needs to be established, but as of now, its statistical significance is not yet sufficient. Optimistically, it is perhaps the first observable consequence of a cosmological QCD phase transition.

Summary and connections: This chapter is rather different from all others, as it is devoted to cosmology. Instead of QGP, it describes phenomena near the time when the temperature falls to a fraction of 1 eV and the primordial plasma (of photons, electrons and ions) neutralizes. The speed of sound rapidly drops at this stage, leaving "frozen sounds" visible in the sky, as perturbations of the microwave background radiation. We discussed strong similarities between the resulting observations and those in heavy ion collisions, such as "acoustic maxima/minima." We went on to speculate further about the relations between acoustic cascades at the time of cosmological QCD and EW phase transitions and primordial long-time gravitational waves. Indeed, gravity waves are the only "penetrating probes" reaching us in the early stages, and they can be directly produced from a pair of sound waves.

Chapter 14

FEMTOSCOPY

14.1 The Basic Theory of HBT Correlations

The main idea of *two-quanta* (also called *intensity*) interferometry originated in radioastronomy (Hanbury Brown and Twiss, 1954) and is therefore often referred to as the "HBT method." It has since been used for multiple purposes.[1]

Its first appearance in hadronic physics was a decade later, when Goldhaber *et al.* (1960) ascribed a specific observed correlation between identical pions, produced in the reaction $\bar{p}p \to$ pions, and their Bose–Einstein statistics.[2] To use it as a research tool aimed at measuring fireball sizes was first advocated by Kopylov and Podgoretsky (1973). This idea, i.e. to use HBT not only as a telescope but also as a microscope, so to speak, is reflected in the modern name of the method: *femtoscopy*.

Although known for decades, it was only occasionally used for the analysis of hadronic collisions. This may perhaps be attributed to the following two reasons: (i) The method needs very high statistics of pion pairs; (ii) there is very limited interest in the space-time visualization of these processes in that community. As heavy ion collisions became more readily available, this naturally attracted more attention.

The main idea is illustrated in Fig. 14.1. Consider two point sources, S1 and S2, emitting an infinitely long train of monochromatic radiation.

---

[1]There should be plenty of applications since the simplest version of HBT correlations is even used for demonstrative purposes in graduate-level labs. Let me just mention here my own little example of using the HBT correlations of two X-rays in the synchrotron radiation to measure the (sometimes very small, sub-micron) beam size of modern electron storage rings. Three decades later, it was indeed tested on the radiation ring at Argonne National Laboratory, and it really worked!

[2]It is apparent from the paper that they did not know about the two-photon HBT correlations in radioastronomy mentioned above.

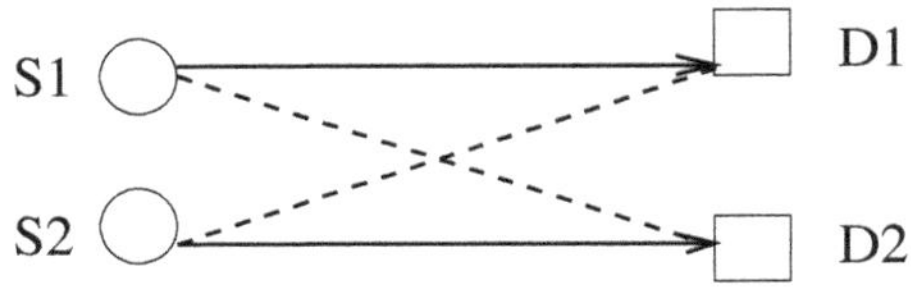

Fig. 14.1. Two quanta are produced by sources $S1, S2$ and detected by two detectors $D1, D2$. The process can proceed via two different possible paths, indicated by the solid and dashed lines, which interfere with the quantum-mechanical amplitude.

Two detectors, D1 and D2, measure the *intensity* (not the amplitude!) of radiation at two distant points. Their signals are correlated with each other. What actually interferes are two *diagrams*, shown in this figure by solid and dashed lines, respectively. For identical particles, it is impossible to tell which source emits which particle! The amplitude in the coincidence experiment is the *sum* of two terms:

$$M \sim e^{i(k_1 R_{11'} + k_2 R_{22'})} + e^{i(k_1 R_{12'} + k_2 R_{21'})} \tag{14.1}$$

We used a *plus* sign between the two amplitudes, assuming that the particles are bosons.[3]

The probability is a modulus square of the amplitude, which is equal to

$$|M|^2 \sim 1 + \cos[\Delta\Phi] \tag{14.2}$$

where $\Delta\Phi$ is the difference of phases in the previous expression. When the distance from the sources to the detectors is much larger than their separation, one may rewrite the argument of the cosine term as $kL\theta$, where L is the "baseline," i.e. the distance between the detectors, and θ is the angular size of the source.

Substituting here the typical order of magnitude of the radioastronomy wavelength $1/k \sim 0.1$ m and the distance between two radiotelescopes $L \sim 10^4$ km, one finds that the angular resolution of the method can reach $\theta \sim 10^{-8}$, which is better than what optical telescopes can deliver. That is why they were able to measure stars' radii, which famously started with Sirius.[4]

[3]For fermions, it would be a *minus* sign, as Pauli's exclusion principle demands; therefore, instead of a peak, one finds a dip in the two-body distribution. Note, however, that for the usually observed *unpolarized* fermions, the effect is a dip only down to one-half of the original level, not zero. In fact, the method has been actually used for protons in heavy ion collisions, but we will not discuss it.

[4]One of the brightest stars in the sky, so the statistics were a problem even then.

(One can, in principle, also use HBT for optical wavelengths, reaching even smaller angles; however, in this case, the base L needs to be much shorter because of lack of coherence. The signal of a radio telescope can be electronically recorded and phases correlated — with appropriate time delay — even *a posteriori.*)

In a very similar way, in the two-body spectrum of, say, a pair of identical π^- pions, one cannot say where in the fireball each pion comes from. If R is the source size, the condition for HBT correlation to be seen is

$$|\vec{k}_1 - \vec{k}_2||\vec{x}_1 - \vec{x}_2| < 1 \tag{14.3}$$

Aiming at large fireballs with $R \sim 6$ fm, one thus needs two pion momenta to be equal within $1/R \sim 30$ MeV or so, while their thermal momenta at freeze-out are about one order of magnitude larger. The probability of finding such a situation for a random pair of pions is small, say $(\Delta k/k)^3 \sim 10^{-3}$ or less. Nonetheless, since a fireball produces approximately a thousand π^-, there are $\sim 10^6$ pairs to make a histogram, times, of course, millions or even billions of events. So, statistics are no longer a problem for heavy ion collisions. As we will discuss, one can even cut into pairs of different total momentum and create a kind of movie of fireball expansion, or select different directions of the pair and get pictures of the fireball from different sides.

Another aspect of the problem deals with the time structure of the signal, which we have so far ignored. It is clear that even if the momenta of both quanta are the same and the condition given above is satisfied, their interference is impossible if these two quanta do not overlap in time. Using the language of four-momenta, one finds an additional term in the argument of the cos term, with times and frequencies substituting for distances and wave vectors. The interference is absent if the two frequencies measured are too different, therefore it is required that

$$|\omega_1 - \omega_2||t_1 - t_2| < 1 \tag{14.4}$$

where τ is the time between the emission moments of two quanta, which is about the lifetime of the source. Basically, these two "second-order coherence" conditions determine where the HBT effect is seen in the two-body spectra.

Experimentally, the existence of the Bose correlation was first identified by the comparison of identical (say $\pi^-\pi^-$) pair distributions to those of non-identical (e.g. $\pi^+\pi^-$) pairs. A modification of this idea currently used is based on taking π^+ and π^- from different (or "mixed") events.

In this case, one can get rid of $\pi^+\pi^-$ resonances (such as σ, ρ mesons) and also of other interactions between π^+ and π^-.

It should be emphasized that the two-quanta interference effects to be considered in the following are very different from more familiar interference effects at the one-quantum level. In particular, the "thermal" sources of radiation considered are not assumed to give "coherent" radiation in the ordinary sense but show the effect in coincidence-type measurements with two (or more) intensity-sensitive detectors.[5]

14.2 The Correlation Function in the Random Source Model

These ideas can be put into a more general form using a *random source model* and the correlators of sources developed by Shuryak (1973b). It begins with an introduction of some (scalar) currents or sources emitting pions at some 4D point $j(x)$. Let us refer to the positions of two sources as x_1 and x_2, which can also be expressed via the mean position $X = (x_1 + x_2)/2$ and the separation $R = x_1 - x_2$. One can imagine a multi-pion production amplitude as (a set of) Feynman diagrams, in which the positions of each emission vertex is integrated over d^4x. Let us now select only the positions of our two detected identical pions x_1 and x_2 and integrate over *all others.* It would give the *correlator* of the two sources:

$$\langle j^*(x_1) j(x_2)\rangle = I(X, R) \tag{14.5}$$

Here, the angular bracket refers to all unused (maybe even unobserved) particles as a kind of "heat bath." If one introduces the Fourier transform of the correlator

$$\tilde{I}(P, Q) = \int d^4X d^4R e^{iXP+iQR} I(X, R) \tag{14.6}$$

one can see that $\tilde{I}(P, Q = 0)$ is just the average single-body spectrum. We often also use the mixed representation, also known as the Wigner or *source* function,

$$S(P, R) = \int d^4X e^{iXP} I(X, R) \tag{14.7}$$

which tells us where the particles with momentum P come from.

[5]The "coherent" sources such as lasers, on the contrary, do not produce HBT correlation effects, which need extermination of the cross terms due to the chaoticity of phases, see the following.

The main result of Shuryak (1973b) is that, writing down the amplitude as the sum of two diagrams indicated in Fig. 14.1, making a square, and averaging over random sources, one finds that the two-body spectrum contains the product of two such intensities plus/minus the interference effect if particles are bosons/fermions:

$$\frac{d^6N}{dp_1dp_2} \sim \left(\tilde{I}(p_1,0)\tilde{I}(p_2,0) \pm \left|\tilde{I}\left(\frac{p_1+p_2}{2}, p_1-p_2\right)\right|^2\right) \tag{14.8}$$

Note that if the four-momenta of both particles are identical, the second term is the same as the first. So, at this point, the effect is expected to be twice as large for bosons and zero for fermions.

The shape of the measured correlation function is related to the Fourier transform of the source space-time distribution, which is the information we are looking for. So, naively, we may think that by simply measuring the correlator, we would be able to perform a Fourier transform back and recover the source. It does not, however, work quite like that because the measured spectra are six dimensional, and for the source, we need an eight-dimensional transform. Two missing dimensions appear because the observed particles are on shells.

Lacking direct inversion, one can do what people always do in such cases: parameterizing the answer and extract parameters by comparing it with data. The simplest parametrization for which the Fourier transform is easy to do is a Gaussian, which is commonly used in practice:

$$S(R,P) \sim \exp\left(-\sum_{\mu\nu} B_{\mu\nu}(P)\Delta x_\mu \Delta x_\nu\right) \tag{14.9}$$

where a bar denotes the effective source center for a given mean momentum of the pair P and $\delta x_\mu(P) = R_\mu - \bar{R}_\mu$ is the deviation. Now, denoting the spatial $\int d^4RS\ldots$ average over the source by angular brackets, one finds that the quadratic form of the coordinates is the inverse of B:

$$\langle \Delta x_\mu \Delta x_\nu \rangle = B^{-1}_{\mu\nu}(R) \tag{14.10}$$

Note also that by performing the Fourier transform of the source function and plugging it back into the main expression for the correlator, one finds a convenient expression:

$$C(Q,P) = 1 + \exp\left[-\sum_{\mu\nu} q_\mu q_\nu \langle \Delta x_\mu \Delta x_\nu \rangle_{\vec{P}}\right] \tag{14.11}$$

Here, the coordinate quadratic form measured still depends on the vector of the total momentum. Note that all the information about the center position $(\vec{x}_1 + \vec{x}_2)/2$ is lost in the correlation function, and only *relative* positions or times are thus measured. Furthermore, standard Cartesian parameterization for the tensor is based on three coordinates: *long* (longitudinal or z direction), which is along the beam); *out* or y, which is along the vector $\vec{P}_\perp$; and *side* or x, which is orthogonal to both. For azimuthally symmetric central collisions, the non-zero components of the quadratic form are the following four radii (all are functions of $\vec{P}$):

$$R^2_{\rm side} = \langle \Delta y^2 \rangle, \qquad R^2_{\rm out} = \langle (\Delta x - \beta_\perp \Delta t)^2 \rangle, \qquad R^2_{\rm long} = \langle (\Delta z - \beta_l \Delta t)^2 \rangle,$$

$$R^2_{\rm ol} = \langle (\Delta z - \beta_l \Delta t)(\Delta x - \beta_\perp \Delta t) \rangle$$

where $\vec{\beta} = \vec{P}/P_0$ is the mean velocity of the pair.

These four parameters are usually reported by experimental groups based on their measured two-particle momentum distributions. For non-central collisions, the source is elliptic; thus, there are more non-zero components which depend on the absolute orientation of the out and side directions with respect to the impact parameter plane.

14.3 Femtoscopy in Heavy Ion Collisions, the "Coherence Factor" and the Final State Interaction

We discuss these parameters separately in the following. Now, let me just note that their product $R_{\rm long} R_{\rm out} R_{\rm side} \equiv V_{\rm emission}$ gives us the emission volume of the freeze-out surface (projected to the fixed detector frame). One of the first dynamical issues discussed in the femtoscopy context, already in my first work (Shuryak, 1973b), was the question of whether or not this volume is proportional to the total multiplicity

$$V_{\rm emission} = N_{\rm tot} V_0$$

with some fixed volume per particle. Recall that was the main assumption of the statistical model applied to $\bar{p}p \to n$ pions. Indeed, the femtoscopy volume was found to scale like that.

Jumping by 50 years, from 1972 to the present day, let me show the ALICE summary plot (Fig. 20.2, left; from Grosse-Oetringhaus, 2014), showing the dependence of the invariant two-pion and three-pion radii on the multiplicity. Using these data, one can check if the freeze-out size scales as $\langle N_{\rm tot} \rangle^{1/3}$. The answer is that, indeed, it does, although for all heavy ion data (black points) and the "small systems" of pp, pPb (colored points), the volume-per-secondary V_0 is *not* the same.

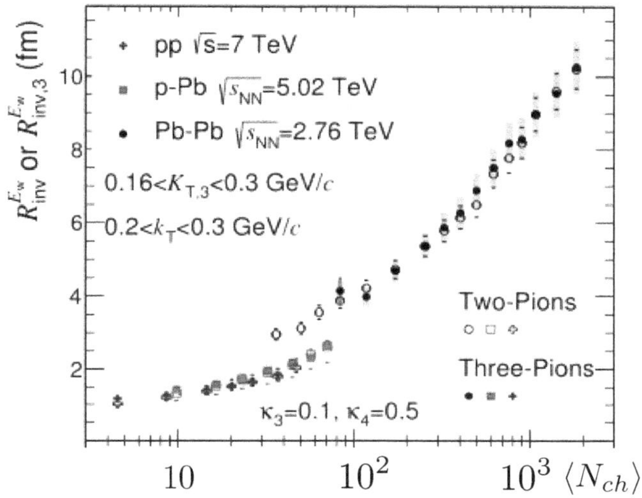

Fig. 14.2. ALICE data on the femtoscopy invariant radius as a function of multiplicity for $pp, pPb, PbPb$ collisions.

Let us now switch to another issue, that of "coherency," which was historically much discussed in the early days of femtoscopy. The "random source" model assumes that the phases of the sources are random and that the two quanta observed are mutually incoherent. Is it really true for the pion productions that we are discussing?

For stars, the excited atoms which emit two observed photons are millions of kilometers away from each other at different locations in the star photosphere, and obviously, their phases cannot be correlated.

The number of emitted pions per high-energy heavy ion collision is now counted in the thousands, and the size of the fireball is assumed to be large compared to any correlation length (mean free path) of the excited matter. It seems quite clear that two observed pions are likely to come from different places in the fireball, and their mutual incoherence (randomness of the phases) should be the case.

The validity of this assumption is, however, not obvious for hadronic reactions, such as minimum bias pp collisions, in which there is a rapidity of only a few pions per unit and the transverse size of the system is small. This is especially true for low-energy reactions, such as the reactions $\bar{p}p \to$ several pions (which we discussed in connection with the early applications of the statistical model). Let us recall again an assumption, popular in the 1960s when these data were first taken that the matrix element for these processes is simply some constant. If this were true, all pions would be produced in the same quantum state from a point-like source with the same (zero) angular momentum, and the production amplitude would be a product of the same functions, and the pion symmetrization would not lead to any additional correlations. Their experimental observation in the original paper (Goldhaber *et al.*, 1960), has proven this idea to be wrong. The modulus of the production amplitude is, perhaps approximately, constant over the phase space, but certainly not its phase.

Phase coherency can also, in principle, occur in another physical situation: in the Bose–Einstein condensation of the pions. We already discussed the pion chemical potential at freeze-out and subsequently concluded that it does *not* happen.

At this point, the reader may ask, why do I emphasize those trivial points. The reason is that when people have located peaks in the two-pion distributions, they found that their heights were *not* a factor of two, as predicted by Bose statistics, but lower than that. So, in practice, one does not use (14.11) directly but the parameterization with the so-called "coherency factor" $\lambda < 1$, such as

$$C(q, K) = 1 + \lambda \exp\left(-\sum_{\mu\nu} q_\mu q_\nu \langle \Delta x_\mu \Delta x_\nu \rangle\right) \tag{14.12}$$

However, the historic name of this factor is misleading. In fact, it stands *not* for the coherency of the pion sources, which do not exist, but due to a feed-down from the *long-lived resonances.*

A certain fraction of pions originates from the decay of $\omega, K_s, \ldots$, from which some fraction f_{llr} of the observed pions originate. The interference is observed only if both are "prompt" pions, so $\lambda = (1 - f_{\text{llr}})^2$. Of course, if the resolution of the detector were ideal, one would see the second component in the correlator $\delta C_{\text{llr}} \sim (1 - \lambda)(\text{narrow peak})$ so that if two pions are completely identical, one would still have $C(q = 0, K) = 2$, as the Bose statistics demands. However, even if we are able to resolve the contribution of the long-lived resonances, we are not going to include their (well-known) lifetimes in the "lifetime of the source."

Our last subject in this section is the "final state interaction" between the observed particles. The pions are not photons, and they interact on the way from the source to the detector.

In particular, there is the Coulomb interaction and the low-energy strong interaction. The inclusion of these effects is made by substituting the wave function of two interacting particles,

$$\langle \Delta k | \Delta x \rangle = \psi_{\Delta k}(\Delta x),$$

instead of the plane waves (Gyulassy *et al.*, 1979). Identical pions or protons have both repulsive Coulomb forces as well as those due to strong interaction. At short distances, protons have a repulsive "core," while pions do not. At large distance, the Coulomb forces — the long-range ones because the photons are massless — dominate. Fortunately, the Coulomb problem is analytically solved (see QM textbooks), and the corresponding wave function is well known.

Particles with close momenta travel for a long time in close proximity; therefore, even a small force may produce a large effect. This indeed happens if the Coulomb parameter is no longer small. As we will see from the following expression, the correction is large when

$$\gamma_c \equiv m_\pi \frac{e^2}{q} \sim \frac{1}{137} \frac{1}{|\vec{v}_1 - \vec{v}_2|} > \frac{1}{2\pi} \tag{14.13}$$

The correlation term should be corrected by the so-called "Gamow factor,"[6] which takes care of electric repulsion and has the form

$$G = \frac{2\pi\gamma_c}{\exp(2\pi\gamma_c) - 1} \tag{14.14}$$

which is 1 at small γ_c but exponentially small at large γ_c due to Coulomb repulsion of the identical pions.

It is not a problem to take into account strong $\pi^-\pi^-$ interaction because, at a small relative momentum q, it is negligible due to the Goldstone nature of the pions. For the protons, the scattering length is not small, and one does account for it in practice.

Finally, another type of final-state interaction is due to the known interaction of the observed pair with other particles. In particular, for heavy ion collisions, one may correct for the electric field outside the nuclei.

14.4 Three-Pion Correlations

High statistics of heavy ion events, with large numbers of pions in them, have enabled us to look at higher-order Bose effects, including those in $\pi^-\pi^-\pi^-$ and $\pi^+\pi^+\pi^+$ systems.

[6] It is the same one which appeared in the alpha radioactive decay and is the ratio of the wave function at zero for Coulombic repulsion relative to free propagation. Note that this factor has poles, corresponding to the bound states of the pionic atoms.

There are a number of advantages compared to two-pion systems. When all three quanta have exactly the same momenta, there is a peak in the three-particle distribution, which is higher. Also, the possible influence of resonances is greatly reduced. On the other hand, the three-body Coulomb wave function is not analytically known and needs to be accounted for numerically.

Triple correlation can be written in terms of the Fourier transform of the source in the same form as for the double one, $C_2 = 1+|F_{Q_{12}}|^2$, namely

$$C_3 = 1 + |F_{Q_{12}}|^2 + |F_{Q_{23}}|^2 + |F_{Q_{31}}|^2 + 2|F_{Q_{12}}||F_{Q_{23}}||F_{Q_{31}}| \tag{14.15}$$

where $\vec{Q}_{ij} = \vec{k}_i - \vec{k}_j$. If $Q \to 0$ and $F(0) = 1$, then $C_2 = 2$ and $C_3 = 6$. The "coherence parameter" λ is not inserted into this formula, it is part of the first power for all binary terms and the power of 3/2 in the last term.

Among its first applications in heavy ion physics was that of the NA44 collaboration (Bearden *et al.*, 2001) at SPS. Their combined three-particle correlator $C_3(Q_3)$, expressed in terms of $Q_3^2 \equiv Q_{12}^2 + Q_{23}^2 + Q_{13}^2$, is shown in Fig. 14.3. One can see that at the maximum $Q_3 = 0$, the value is not 6 but about 3. This is expected due to the pions from long-lived resonance decays. The fitted parameter $\lambda_3 = 0.57 \pm 0.04$ is quite consistent with that from the two-pion data, $\lambda_2 = 0.52 \pm 0.04$.

In addition, looking at Fig. 20.2, which depicts the modern ALICE data on the radii we discussed before, we can note that it includes the data from both the two-pion and three-pion femtoscopy and that those agree with each other quite well.

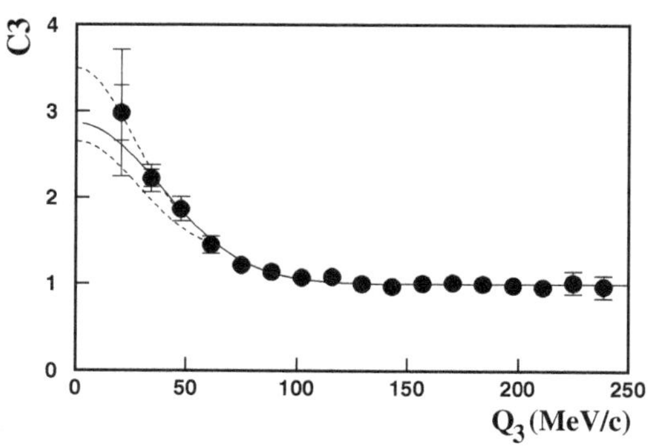

Fig. 14.3. One of the first observations of the three-pion correlation function in *PbPb* collisions at SPS.

14.5 Expanding Sources and the Regions of Homogeneity

Watching the k_T dependence is like looking at an exploding fireball through filters that allow light of different colors. Since we know from particle composition that there is thermal equilibrium at chemical freeze-out, we basically have thermal matter boosted by the flow. Femtoscopy data at different k_T basically tell us what the sizes/shapes of the parts of the fireball boosted to the particular flow velocity. Moreover, since it is now done as a function of azimuthal angle ϕ, one can understand its dependence on the specific direction.

The idea of pairs with different k_T having different sources is explained in Fig. 14.4.

A more technical definition of *the region of homogeneity* was introduced by Makhlin and Sinyukov (1987). If the source from which pions originate is expanding, pions with different momenta come from its different parts. The maximum would come from a spatial point at which the mean *velocity* of the two pions $\vec{\beta}$ is exactly equal to the collective velocity $\vec{v}_{\text{coll}}$. The points from a region around it can only contribute if the magnitude of random thermal motion is strong enough, so that the gradient of the flow field can be compensated by it:

$$(\delta x_\mu \partial_\mu) v_{\text{coll}} < v_{\text{therm}} \tag{14.16}$$

For example, if we speak about the longitudinal radius, the right-hand side of this condition tells us that $R_l \sim 1/\sqrt{(M_\perp T)}$, where $M_\perp^2 = p_t^2 + m^2$ is the transverse mass. We have just predicted that the pions with a larger p_t show a smaller longitudinal radius, which has indeed been observed. Another consequence of this argument, also confirmed experimentally, is that the R_l HBT radii for $\pi^\pm\pi^\pm$ and $K^\pm K^\pm$ identical pairs with *the same* $M_\perp$ should be the same.

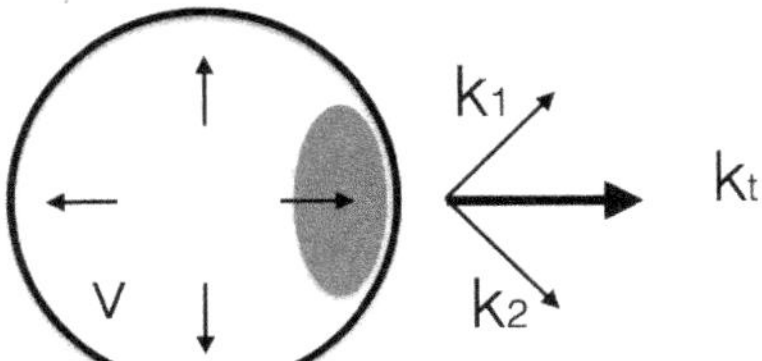

Fig. 14.4. Sketch of the radial flow (arrows directed radially from the fireball center) explaining how it influences the HBT radii. At small k_t, the whole fireball (the large circle) contributes, whereas at larger k_t, one sees only the part of the fireball which is co-moving in the same direction as the observed pair. This region, shown by a shaded ellipse, has a smaller radius and an anisotropic shape, even for central collisions.

14.6 The Femtoscopy of Anisotropic Fireballs

The ellipticity of the freeze-out surface was nicely demonstrated by femtoscopy data, such as from STAR (Adams *et al.*, 2004), which we follow in this section.

A standard expression used for the two-pion correlator is

$$\begin{aligned} C(\vec{q}, \Phi) &\sim (1-\lambda) + \lambda K(q) \\ &\times [1 + \exp(-q_0^2 R_0^2(\Phi) - q_s^2 R_s^2(\Phi) - q_l^2 R_l^2(\Phi) - q_s q_0 R_{\rm os}^2(\Phi))] \end{aligned}$$

where the four-vector q is the momentum difference of the pair, $\Phi = \phi_{\rm pair} - \Psi_{\rm event}$, λ is the "coherence parameter" discussed above, and the four radii are the fit parameters. Note that they all depend on the azimuthal angle of the vector $\vec{k}_1 + \vec{k}_2$ relative to the event reaction plane. The factor $K(q)$ is the Coulomb correction factor discussed in the preceding section.

The fit to the STAR data is shown in Fig. 14.5. The left plot compares three centralities of the collision: It shows that the most central collisions have the largest radii, ~6 fm, while the most peripheral bins have sizes about twice as small. All radii show modulations of $\sim\cos(2\Phi)$, smaller for the central and more pronounced in the peripheral bin. From these modulations, one can deduce the ellipticity of the system at the freeze-out and compare it to that of the initial state (from the Glauber model). What is found is that the final one is smaller than the initial one by a factor of 2.5–3. This, along with the phases of the oscillations, confirms that expansion occurs mostly in the in-plane $\Phi = 0$ direction, as the original hydro pressure gradient argument suggested.

The right plot is for one centrality but different total momentum of the pair k_T. As one can see, the radii strongly *decrease* as k_T increases. The reason for this effect is schematically illustrated in Fig. 14.4. At small $k_\perp$m the detector sees hadrons emitted from the whole fireball; however, the larger $k_\perp$ is, the brighter the *blue-shifted*, small (shaded) part in which the radial flow is maximal in the direction of $\vec{k_t}$. This follows from the maximization of the Doppler-shifted thermal spectrum $\sim\exp\left(p^\mu u_\mu / T_{\rm freezeout}\right)$. The other half of the fireball, moving in the opposite direction, is red-shifted and basically invisible.

14.7 Conditions for the Longest Emission Time

The Beam Energy Scan program at RHIC also provided femtoscopy data, which detailed the energy dependence of the fireball size/lifetime. As we

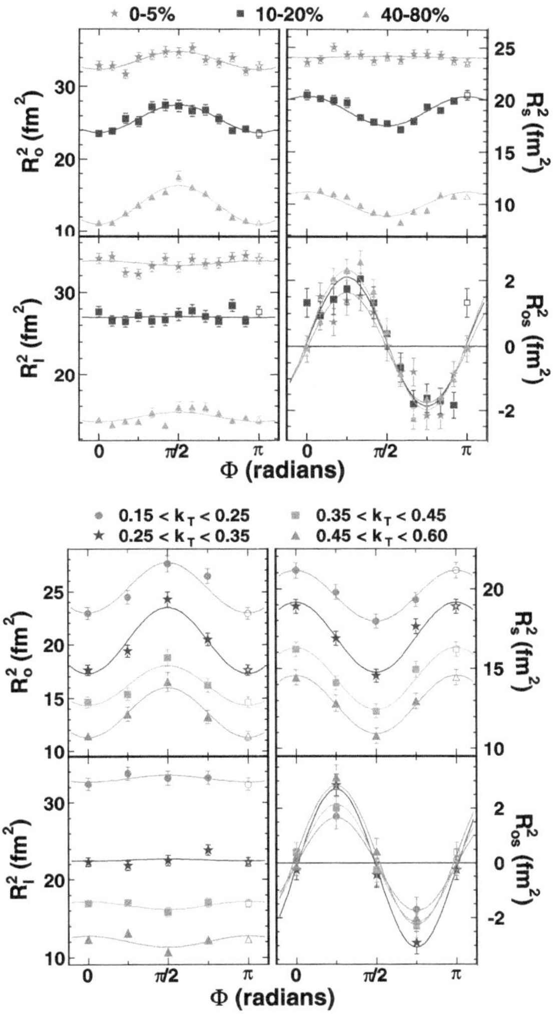

Fig. 14.5. The squares of radii versus the angle Φ relative to the even reaction plane. The left plot is for three centrality classes, and the right plot is for a fixed centrality of 20–30% but for four bins in the pair total momentum of $k_T = |\vec{k}_1 + \vec{k}_2|$.

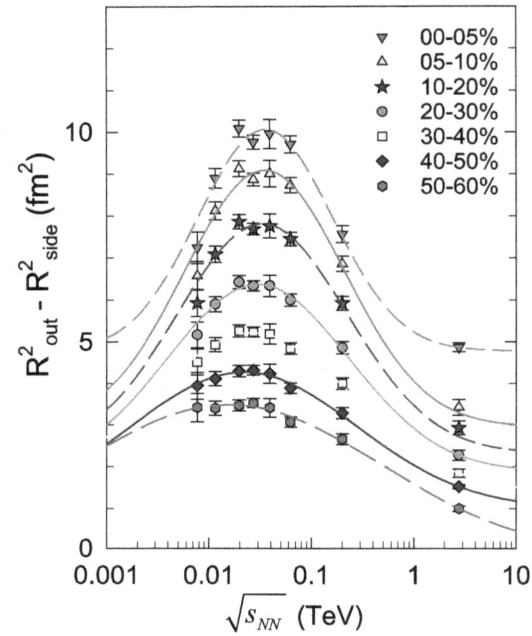

Fig. 14.6. Compilation of the femtoscopy data for $R^2_{\text{out}} - R^2_{\text{side}}$. The solid and dashed curves represent Gaussian fits to the combined data sets.

discussed above, the *emission duration* τ^2 is defined via a difference between the squares of the "outward" and "side" radii.

In Fig. 14.6, we show a compilation of data for $R^2_{\text{out}} - R^2_{\text{side}}$ from Lacey (2015) at different energies and centralities. This combination is related to *emission duration*[7] from the kinetic freeze-out surface.

As one can see, there is a broad maximum at a certain collision energy of $\sqrt{s} \approx 30$ GeV. As one can see, the effect is more pronounced for larger systems (with centralities of 00–05%), as compared to peripheral collisions.

The physical origin of this maximum is the *softest point* of the EOS, which we already discussed in the chapter on hydrodynamics. At a collision energy of $\sqrt{s_{NN}} \sim$ 20–30 GeV, the combination of radii has a pronounced maximum. At approximately these energies, the initial value of p/ϵ (or the

[7]This should not be confused with the absolute lifetime of the fireball. Emission durations can be rather short and yet have a long lifetime. This duration is related to the *curvature* of the freeze-out surface.

sound speed $dp/d\epsilon$) reaches a minimum, slowing the initial velocity of the explosion.

Summary and connections: Second-order interferometry, or the HBT method, which is also known in our field as *femtoscopy*, allows us to "see" the shape of the kinetic freeze-out surface. It was specifically used to confirm azimuthal modulations for non-central collisions. Another important set of observations is the measurement of "emission duration," which has a significant maximum at certain collision energies.

Chapter 15

VORTICITY OF THE FIREBALL AND HADRONIC POLARIZATIONS

15.1 Equilibrium and Kinetic Spin Polarization

Let us begin by recalling the most surprising property of sQGP: Its unexpectedly rapid equilibration, which we believe is explained by its strong coupling regime. It is then natural to ask if rapid equilibration of momenta extends to the spin degrees of freedom of secondaries. The way to check it would be to create *vorticity* in a system and then see if it induces *spin polarization* of various secondaries. In fact, a hydrodynamical picture of the flow in heavy ion collisions does provide for nonzero vorticity, with different directions and magnitudes.

Let me remind the reader that in thermal equilibrium with a rotating fluid (for details, see, for example, Becattini *et al.*, 2013), the spin distribution follows from a standard textbook statement that for a rotating body, the density matrix has an additional factor

$$\hat{\rho} \sim \exp\left(\frac{\vec{\omega}\hat{\vec{J}}}{T}\right) \tag{15.1}$$

where $\vec{\omega}$ is the angular velocity of the rotating frame and $\hat{\vec{J}}$ is the operator of angular momentum of a particle in consideration. Indeed, the term in exponent is just an addition to the Hamiltonian calculated in a rotating frame.

In nonrelativistic hydrodynamical flow, vorticity is given locally by the curl of the velocity:

$$\vec{\omega} = \frac{1}{2}[\vec{\partial} \times \vec{v}] \tag{15.2}$$

A more natural definition used in relativistic hydro theory is the so-called *thermal vorticity*:

$$\bar{\omega}_{\mu\nu} = -\frac{1}{2}(\partial_\mu \beta_\nu - \partial_\nu \beta_\mu), \quad \beta_\mu = \frac{1}{T} u_\mu \tag{15.3}$$

entering the so-called "Cooper–Frie formula for the spin," with integrals over the freeze-out surface:

$$S^\mu(p) = -\frac{\epsilon^{\mu\rho\sigma\tau}}{8m} p_\tau \frac{\int d\Sigma_\lambda p^\lambda n_F (1-n_F)\bar{\omega}_{\mu\nu}}{\int d\Sigma_\lambda p^\lambda n_F} \tag{15.4}$$

Let me also mention that if a nonzero magnetic field is also present, then this formula is modified as

$$\bar{\omega}_{\mu\nu} \to \bar{\omega}_{\mu\nu} + \frac{\mu}{S} F_{\mu\nu} \tag{15.5}$$

where μ/S is the ratio of magnetic moment to spin.

Now, checking if spins are or are not equilibrated is, in principle, simple: One should measure the polarization of secondaries with different spins ($J = 1/2$ baryons, $J = 1$ vector mesons, and perhaps also $J = 3/2$ baryons, such as Δ^{++}), extract the magnitude of $\vec{\omega}$ and see if it is universal.

Of course, it is by no means obvious that the spin polarization time τ_{spin} can be short enough to equilibrate the spins. Again, in principle, when comparing momentum and spin equilibration times, one should simply compare total scattering cross-sections to those with spin flip.

The first setting, in which the issue was discussed by Kapusta *et al.* (2020b), is a weakly coupled QGP. Aiming at lambda hyperon polarization, we focus on the strange quark with a mass of $m_s \sim 100$ MeV (note that we are at $T > T_c$, so there are no chiral condensates or constituent quarks). The smallness of the effect arises from the fact that the squares of helicity flip amplitudes must be proportional to m_s^2, while those for non-flip scattering have invariants such as $(p_s + p_i)^2$, which are larger. Thus, pQCD tells us that spin relaxation time should be $\sim(6T/m_s)^2$ times longer than momentum relaxation. The reactions included are

$$s + q \to s + q, \quad s + \bar{q} \to s + \bar{q}, \quad s + g \to s + g$$

The explicit expressions for flip and non-flip cross-sections can be found in the work of Kapusta *et al.* (2020b). The conclusion is that spin relaxation time changes between 10^3 fm/c at zero momentum and 10^5 fm/c at $\vec{p}_s = 1$ GeV/c. This is obviously much longer than fireball lifetimes, so one has to conclude that perturbative effects in QGP cannot equilibrate spins.

In discussing the issue of nonperturbative spin kinetics, we follow Kapusta *et al.* (2020a), who discuss instanton-induced spin-flip amplitudes described by the 't Hooft Lagrangian. The setting now is in the hadronic phase below the phase transition, $T \approx T_c$; therefore. quarks are "constituent quarks" with corresponding effective masses, e.g. $M_s \sim 500\,\text{MeV}$. Again, omitting details, we only mention the main physics here, which is the fact that instanton density decreases strongly with increasing temperature.[1] The relaxation time at $T = 150$ MeV was found to be of the order of 1 fm/c, again jumping to 10^3 fm/c at $T = 200$ MeV.

The conclusion drawn from these kinetic calculations is that spin equilibration is indeed possible, yet it is most likely to happen in a hadronic phase, near and below the critical QCD temperature $T_c \approx 155$ MeV.

15.2 Polarization of Λ or $\bar{\Lambda}$ in Experiments

The first secondaries whose polarization were measured are Λ and $\bar{\Lambda}$ hyperons. Those are "self-analyzing" because in weak decays of strange quarks, the P-parity is strongly violated. In the observed decays, for example via $\Lambda \to p\pi^-$, the outgoing pion momenta are strongly correlated with the direction of the hyperon's spin. The original data from the STAR collaboration (Adamczyk *et al.*, 2017) are illustrated in Fig. 15.1.

The rotation of the final state as a whole simply follows from angular momentum conservation. What is, however, non-trivial is how the final angular momentum is distributed between secondaries with different rapidities. In hydrodynamic language, rotational flow is induced by shear viscosity, which is small in the sQGP phase and thus expected to be quite small for high-energy collisions. At low collision energies, with greater amounts of viscous matter and longer fireball lifetimes, one expects stronger effects.

In heavy ion collisions, the natural direction of $\vec{\omega}$ is normal to the collision plane (which needs to be experimentally defined on an event-by-event basis).

A summary of the original STAR data (Adamczyk *et al.*, 2017) is presented in Fig. 15.1. The first thing to note is that the polarization is found to grow toward small collision energies, as expected. To convert these data into absolute vorticity requires knowledge of the freeze-out

[1] We know that from the semiclassical theory and also from the topological susceptibility $\chi_{\text{top}}(T)$ measured on the lattice.

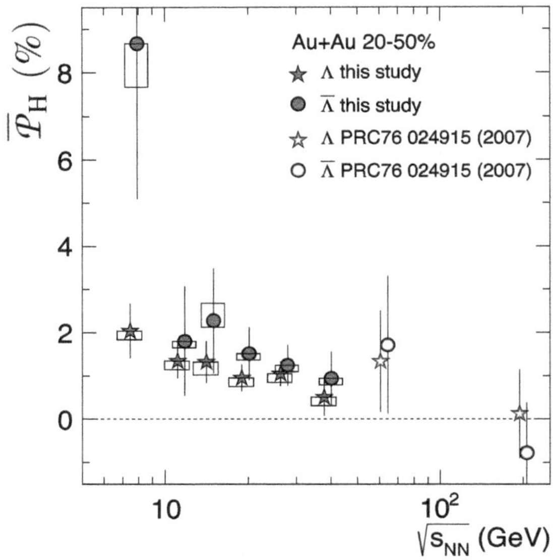

Fig. 15.1. The average polarization P_H (where $H = \Lambda$ or $\bar{\Lambda}$) from 20–50% central $Au + Au$ collisions, plotted as a function of the collision energy.

temperature, a fraction of Λ originating with "feeddown" from heavier resonances, and some knowledge of how much these secondary Λ retain the polarization of their parents.

Without providing precise numbers, let me just quote the order-of-magnitude value of the rotation frequency:

$$\omega \sim 10^{22} s^{-1} \sim \frac{1}{30} \frac{c}{\text{fm}} \tag{15.6}$$

During a fireball lifetime of $\sim$10 fm/c it only rotates by one-third of a radian. A more quantitative comparison with hydro simulations will allow us to quantify the freeze-out temperature of the spin-flip reactions.

15.3 Polarization of Other Secondaries and Possible Magnetic Effects

Resonances which decay via strong interaction do not have a P-parity violation, so the first-order correlations of the $(\vec{S}\vec{p})$ type (used for Λ) are absent. Yet, their polarized spin density matrix affects the decays and thus can, in principle, be measured.

For vector mesons, the diagonal matrix elements of the spin density matrix, $\rho_{11}, \rho_{00}, \rho_{-1-1}$, may deviate from 1/3 (the unpolarized value), which can be observed. The first two vector mesons for which "spin alignment" was observed are K^* and ϕ. The natural direction of increased eigenvalue is the one normal to the reaction plane. Existing data as a function of collision energy are shown in Fig. 15.2, adapted from Abdallah *et al.* (2022a). The alignment for the ϕ meson at lower energies is clearly very large, while for K^* it is consistent with zero. Note, however, that the lifetime of ϕ is long and the rescattering probability is small, while K^* scatters on pions and comes from the vicinity of the freeze-out surface. So, the latter should be subject to much stronger depolarization. Why strange quarks in ϕ have such a high preference for the zero projection of the spin on $\vec{J}$ ($\rho_{00} > 1/3$) is one of the puzzles in this field.

Another intriguing feature of the STAR data is the systematic difference between Λ and $\bar{\Lambda}$ polarizations. Unfortunately, it is not yet a statistically significant effect. If it were confirmed, it would indicate the presence of a significant magnetic field, $\vec{B}$, in the medium. The magnetic moments of Λ and $\bar{\Lambda}$ are, of course, of the opposite sign, so their interactions with $\vec{B}$ would be added/subtracted from the vorticity effect. The magnetic

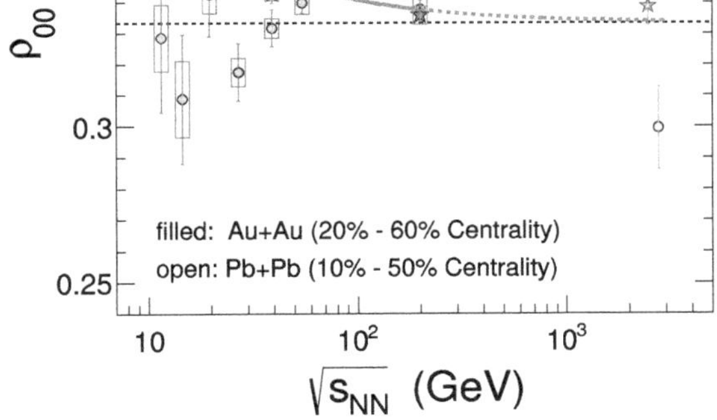

Fig. 15.2. The measured matrix element ρ_{00} as a function of beam energy for the ϕ and $K^?0$ vector mesons within the kinematical windows indicated.

moments are

$$\mu_{\bar{\Lambda}} = -\mu_{\Lambda} = 0.61\ \mu_N \approx \mu_s \tag{15.7}$$

There is no doubt that there exists an initial strong pulse of magnetic field induced by spectator protons. The partial capture of its flux through the electric currents in sQGP has been discussed in the literature, such as by Tuchin (2013); yet, there is no consensus on what the possible magnitude of the field at freeze-out can be. A quantified $\Lambda - \bar{\Lambda}$ polarization difference would imply the existence of a magnetic field at the time of freeze-out, which would be a nontrivial discovery.

In my comment (Shuryak, 2016), I proposed to look for a polarization of hadrons with magnetic moments *larger* than that of Λ hyperon. The most suitable pair seems to be Δ^{++} resonance, which has a magnetic moment one order of magnitude larger than Λ, namely

$$\mu_{\Delta^{++}} \approx 3\mu_u = 5.6\,\mu_N \tag{15.8}$$

It also possesses a Δ^0 isotopic partner with *zero* magnetic moment, which is useful for a background measurement.

15.4 Longitudinal Polarization versus Vorticity

The discussion above focused on spin directed along the most natural direction, which is that of the total angular momentum of the fireball, $\vec{J}$. Yet, a sufficiently complicated flow can generate local vorticity in different directions. In particular, an elliptic flow creates a quadrupole component of vorticity; see Fig. 15.3 (right), adopted from Becattini and Lisa (2020). Unexpectedly, the quadrupole pattern observed by STAR (Fig. 15.3, left) does not agree with the calculation, neither in magnitude nor in *sign*.

Although this embarrassing contradiction puts into question even the simplest scenario in which spin is locally equilibrated, it leads us to several new ideas. Alternatives can either involve some kinetic studies *out of equilibrium* for spins or the introduction of other operators to which the spin can be coupled. According to a more recent review (Becattini, 2022), there are such operators containing (in nonrelativistic form) the gradient of $1/T$ or μ_b/T (rather than the curl of velocity in the vorticity). So, an unmoving fluid with a nonzero temperature gradient can create spin polarization in a moving hadron as well!

At the time of this writing, the issue has not yet been satisfactorily resolved, so more work is needed. In particular, the novel effects related to

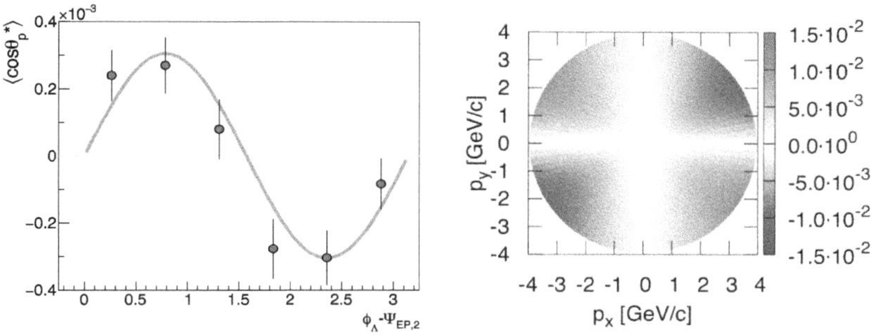

Fig. 15.3. Left: Polarization of Λ versus the emission angle with respect to the event plane based on the STAR data at 200 GeV/N. A guiding line is included for visual reference. Right: The predicted theoretical pattern of vorticity in the transverse plane for the same colliding system and centrality.

operators other than vorticity must be more commonly present in fluids than in QGP, and it would probably be easier to see what their effects can actually be.

In theory, I am not yet sure whether the derived "spin hydrodynamics" equations and their subsequent use in some works make sense. Another application worth studying in this context, which is of a huge scale and pure classical context, is the mean spin field in the Universe revealed in the studies of galactic spin correlations (Shamir, 2020) (as mentioned in the section on the Big Bang sounds).

Summary and connections: This chapter described the transfer of the angular momentum stored in matter (vorticity) to the spin polarization of secondaries. The mechanism of such transfer should rely on *spin–orbit* forces in the system, which are, however, far from being well understood. The initial data on the polarization of Λ hyperons suggest that this transfer is strong enough for polarization to reach its equilibrium thermal value. The difference between the effects for Λ and $\bar{\Lambda}$, if reliably observed, would be a measure of the magnetic fields captured in matter. Similar long-range correlations between the rotation axes of galaxies in cosmology have also recently been found.

Chapter 16

MATTER AT HIGH BARYON DENSITY

16.1 Quark Matter in Stars?

Historically, discussions of cold and very dense matter were among the first applications of the view that strong interaction gets weak at small distances. Let me recall that these ideas first came in the form of Bjorken scaling and the parton model around 1970, soon explained by the discovery of asymptotic freedom and QCD in 1973. It then became apparent that at very large densities, matter should be close to the Fermi gas of quarks.

Note that the name of the main conference of the field, *Quark Matter*, was fixed at that stage and never changed, although its actual focus was on high-energy (heavy ion) collisions, which inevitably produce entropy and therefore gluons and quarks and thus hot quark-gluon plasma.

While the main focus of the community during the past two decades, since RHIC's first run in 2000 and the LHC in 2010, was mainly on the highest collision energies and temperatures, basically with negligible baryon chemical potential, lately the attitude of the RHIC program has begun to change toward lower collision energies. Furthermore, a number of nuclear centers are building low-energy colliders (in GSI, Darmstadt in Germany and JINR, Dubna in Russia), with dedicated fixed-target facilities planned in China and Japan.

One should not be surprised by such a reversed trend. From an experimental point of view, while the intermediate energy region of $\sqrt{s} = 1 - 20$ GeV has been studied in the 1990s, at CERN SPS and BNL AGS, neither the strength of experimental groups nor the quality of the detectors of that period could match what is available today at RHIC or can be built now.

From a theoretical point of view, the domain of $\mu_B \sim (0.5 - 1)$ GeV still remains *terra incognita*. Many different phases of dense matter were proposed, but which of them are actually realized and what type of phase transitions separate them remains unknown. At the moment, RHIC Beam Energy Scan (BES) is the major program to look for transitions, now nearly half done and less than half of that has been analyzed.

A traditional application — neutron stars (NSs) — has been recently revived. First came the discovery of very heavy $M \approx 2M_{\text{Sun}}$ pulsars, which put serious constraints on the EOS. Then came gravity wave detection, with the historic (so far the only one) "multi-messenger" detection of a two neutron star merger in 2017. Those put experimental limits on the "deformability" of NSs and thus on matter EOS. Many groups performed simulations of NS mergers and searched for optimal observables. Needless to say, we are all waiting to see if more events will come our way. As we will see, it would also be important that gravity wave detectors are being upgraded into the kHz frequency range, which is key to observing the last moments of the mergers.

Turning now to physics, we need to remember the basics. We are not going to discuss finite nuclei or nuclear matter in any detail, as there are many reviews and books about it. Yet, some elementary introduction is needed to motivate the transition from it to the quark matter phases.

Nuclear matter is a dilute Fermi liquid: A volume per nucleon in it is about $V/N \approx 6\,\text{fm}^3$, which is significantly larger than a nucleon's volume. As any liquid, it has a liquid–gas phase transition, which is the first-order line ending in some endpoint at $T_c \sim 10$ MeV. Near it, the so-called $multifragmentation$ phenomenon occurs, which is related to the critical opalescence phenomenon observed at all *second*-order transitions.

Following Landau's Fermi liquid theory, the phenomenology of nuclear matter is based on a set of certain parameters defined at the Fermi surface. In modern terminology, this theory is also derived from the renormalization group equation as a limit when the parameter Δk (a width of a shell around the Fermi surface, at which effective theory is defined) tends to zero. For estimates, we use the standard phenomenological equation of state near the minimum:

$$\epsilon(n) = m_N n + \frac{3}{2}Tn + \frac{1}{18}Kn_0\left(1 - \frac{n}{n0}\right)^2 \tag{16.1}$$

where ϵ, n are the energy and the baryon charge density, respectively, T is the temperature and the so-called "nuclear compression modulus" $K \approx 200$ MeV. The equilibrium density of the nuclear matter value is $n_0 = 0.16\,\text{fm}^{-3}$.

One way to approach the issue of the QCD phase transition line is to look at the value of chiral condensate modification by a nonzero density. Its modification can be evaluated as

$$\frac{\langle\bar{q}q\rangle^*}{\langle\bar{q}q\rangle} = 1 - \frac{\Sigma_{\pi N} n}{f_\pi^2 m_\pi^2} + \cdots \approx 1 - 0.36\frac{n}{n_0} \tag{16.2}$$

where $\Sigma_{\pi N}$ is a parameter of πN scattering known as the nucleon sigma term. Extrapolating this linear dependence, one may think that the chiral condensate should vanish[1] at $n \sim 3n_0$. By extrapolating multiple sophisticated versions of nuclear matter theory, one finds that their predictive power ends roughly at the same density.

16.2 From Nuclear to Quark Matter at Zero Temperature

We begin with a simple argument, suggesting why nuclear matter should turn into the quark matter at high density. The nuclear matter remains nonrelativistic basically until the Fermi momentum p_F reaches m_N, so

$$\epsilon_F \sim p_F^2/2m_N \sim n^{2/3} \tag{16.3}$$

while the quark matter is always relativistic:

$$\epsilon_F = p_F \sim n^{1/3} \tag{16.4}$$

Therefore, at a large enough n, the latter has a smaller energy per baryon and should eventually win.

The minimization of the energy per baryon is correct only at $T = 0$. The general (although admittedly less intuitive) way to compare any two phases is to see which has the *lowest* grand canonical (Gibbs) potential $\Omega = -pV$. In other words, the winning phase should have the *largest pressure* at a given T, μ_B. The mechanical, thermal and chemical coexistence conditions include the equality of pressures

$$p_{\text{nuclear/matter}}(\mu_c, T) = p_{\text{quark/matter}}(\mu_c, T) \tag{16.5}$$

temperatures and chemical potentials.

The simplest and most often used model for quark matter is that of an ideal relativistic Fermi gas, complemented by the so-called bag constant.

[1]It was suggested that at this point, other condensates may appear, such as Migdal's pion condensate or Kaplan–Nelson–Brown kaon condensate. Their inhomogeneous versions are still discussed in the literature, but we will not discuss them here, as support for them has dwindled over time.

For N_f massless quark flavors, its pressure is

$$p = N_f \frac{\mu^4}{4\pi^2} - B \tag{16.6}$$

where μ per quark is the same as Fermi momentum. Such quark matter would be "self-contained" at the minimum of E/A or at $p = 0$, or at

$$\mu_0 = (B/4\pi^2 N_f)^{1/4}. \tag{16.7}$$

If this is lower than the chemical potential per quark in nuclear matter, $(m_N - E_{\rm binding})/3$, the curves would never cross and quark matter would be the only truly stable phase, with nuclear matter being just metastable. The window for this to happen corresponds to $56\,{\rm MeV/fm}^3 < B < 90\,{\rm MeV/fm}^3$, with the lower limit coming from the condition that non-strange quark matter is still above the nuclear one, and thus, nuclear physics, as we know it, remains. While scenarios with stable "quark matter drops" or "strangelets" were and still are discussed, we do not think it is a realistic option.

Bag-type models, popular in the early days of QCD, do not give a consistent description of hadronic structure. Out of many ways to demonstrate it, let us just mention that it predicted multiquark states made of light quarks, such as pentaquarks and dibaryons, which do not exist.[2]

Returning to (16.7), let us comment that, while we still do not know the value of B, i.e. the difference in energy density between the perturbative and physical vacuum, we do know that the bag model underestimated it by a large factor. The actual magnitude of "vacuum pressure" of $B \sim 1\ {\rm GeV/fm}^3$, as can be deduced from lattice data. If so, the value of μ_0 corresponding to the quark Fermi gas increases by a factor of approximately three, opening a significant gap between our descriptions of the nuclear and ideal quark phases. What happens inside this gap remains unknown at this point and is the subject of intense debates (and empirical parameterizations: see the following).

Naive expectations based on experience with conventional forms of matter would suggest that compression of nuclear matter in liquid phase may lead to a *solid* or crystalline phase.

Indeed, sophisticated theoretical tools (also based on condensed matter physics) tell us that, at the Fermi sphere surface, a number of instabilities can occur. They can be followed using the corresponding renormalization

[2]Do not confuse those with, say, the four- and five-quark states with at least one heavy quark, which do exist and have been recently observed.

group (RG) equations. The instability, which grows more rapidly, would be the winner.

One of the first is the *Peierls instability*, induced by particle–hole attraction near the Fermi surface. In 1+1-dimensional models, it generates a crystalline phase. It is more difficult to deal with it in $d = 2, 3$ spatial dimensions because the Fermi sphere surface in these case becomes divided into a certain number of "*patches*," defining discrete directions of modulations.[3] (Those, of course, modulate both the baryon density and the chiral condensate.) The modulation wavelength corresponds to a momentum of $\approx 2p_f$. For a discussion of chiral crystals or chiral wave phases; see, for example, Rapp *et al.* (2001) and Pisarski *et al.* (2021). While crystalline nuclear matter can well be the answer, theoretical models so far lack the accuracy to tell the difference between its multiple options, so we will not discuss them in any detail.

Let me instead preview this chapter's content based on certain limits of the two basic parameters of the problem: the number of colors N_c and the number of light flavors N_f.

For a large number of colors $N_c \to \infty$, one may think that gluon dynamics dominates all quark effects. In particular, it suggests that deconfinement and chiral transition critical lines $T_c(\mu_B)$ remain nearly horizontal on the phase diagram up to large μ_B. This picture leads to "quarkyonic matter" (McLerran and Pisarski, 2007), a quark Fermi sphere with baryon-like excitations on the surface. How a quark Fermi sphere can be consistent with confinement remains, however, unclear.

The opposite limit of "smallest number of colors" $N_c = 2$ baryons are diquarks, which are bosons. Naively, a cold gas of those should be Bose condensed. For $N_c = 3$, we see strong diquark correlations in baryons, opening the discussion of the *color superconductor phases*, to be discussed in the following.

While at this time it is not possible to say which of these possibilities wins at intermediate densities, the particle–particle and hole–hole pairings have to win at very high densities. This option is discussed in the following section.

Before we do so, let me mention another limit, invented by Veneziano:

$$x_V \equiv \lim_{N_f, N_c \to \infty} \left(\frac{N_f}{N_c} \right) = \text{const} \tag{16.8}$$

[3]Such patches naturally generate crystals of many types, with multiple phase transitions. For instance, *water ice* has about two dozen of them.

which we also discuss in holographic models. The "physical" value, at the end of calculations, should be set at $x_V = 1$. Let me illustrate it with the (extremely simplistic) idea that the "key observable" of hot QGP is the energy, proportional to the number of effective degrees of freedom defined at large N_c as

$$N_{\rm DoF} \equiv 2N_c^2 + \frac{7}{8}2 \cdot 2 \cdot N_f N_c = 2N_c^2 \left(1 + \frac{7}{4}x_V\right) \tag{16.9}$$

The next issue is the shape of the phase boundary on the T–μ_B plane. One idea (perhaps simplistic) is that a certain physical quantity $X(T, \mu_B)$ is most important for the phase diagram, which then approximately follows the line

$$X(T, \mu_B) = \text{const}$$

For example, one may think of energy density as such a quantity.

Let me however propose another option using as X the *electric Debye screening mass*. The motivation is that the instanton suppression factor depends on it *exponentially* (Shuryak, 1982b):

$$dn_{\rm inst}(\rho, T, \mu_B) = dn_{\rm inst}(\rho, 0, 0) \exp\left[-\frac{2\pi^2\rho^2}{g^2}\kappa_D^2(T, \mu_B)\right]$$

and instantons generate nonperturbative effects important for the phase transition. According to one-loop perturbative calculations (Shuryak, 1978b), for $N_c = N_f = 3$, it is

$$\kappa_D^2(T, \mu_B) = \frac{3}{2}g^2 \left(\frac{\mu_q^2}{\pi^2} + T^2\right) \tag{16.10}$$

If so, the critical line would correspond to a constant value of the last bracketed term; therefore, the critical value of the baryon number chemical potential at zero T is predicted to be

$$\mu_B^c(T = 0) = 3\mu_q^c(T = 0) = 3\pi T_c(\mu_b = 0) \approx 1500\,\text{MeV} \tag{16.11}$$

16.3 Color Superconductivity and Diquarks

This topic has been described in more detail in my previous book (Shuryak, 2004); therefore, here, we only mention the main results.

It is impossible to begin without recalling the following two basic ideas about superconductivity: (i) that the so-called Cooper pair with zero momentum is an exceptional channel; and (ii) that the interaction in it is analogous to the 1D case, when any attraction, no matter how weak, creates a bound state.

The kinematics of scattering of two quasi-particles ($|p| > p_F$), or two holes ($|p| < p_F$), or a particle–hole with the initial momenta $\vec{p}_1, \vec{p}_2$ near the Fermi surface is displayed in Fig. 16.1. Two spheres shown by thick solid circles have the radius of Fermi momentum p_F, and a couple of concentric nearby spheres shown in thin lines indicate the range occupied by thermal excitations, particles and/or holes, respectively. If the total momentum $\vec{p}_1 + \vec{p}_2$ is nonzero, as in Fig. 16.1(a), all possible phase spaces for the final momentum $\vec{p}_3$ consists of two small rhomboid intersections of those spherical shells in 2D or a thin circle obtained by the rotation of those intersections around the total direction of the momentum in 3D. As $T \to 0$, its available volume is $\sim T^2$.

The situation is different only if the total momentum is zero, $\vec{p}_1 + \vec{p}_2 = 0$. As shown in Fig. 16.1(b), in this case, the whole spherical region near the Fermi sphere is available, and as $T \to 0$. its volume is $\sim T^1$, or much larger. Recall that for ordinary superconducting metals, the temperature-to-Fermi-energy ratio is very small, $T/\epsilon_F \sim 10^{-4}$, so this difference is really huge.

The second point we would like to make is that in the latter situation, the direction of $\vec{p}_1$ is irrelevant (as soon as $\vec{p}_2 = -\vec{p}_1$), and in low-energy s-wave scattering, the area of the Fermi sphere just enters as a common factor. It does not really matter whether we consider the 2D or 3D problem: Only the 1D component of the momentum, the *radial* one, really matters and

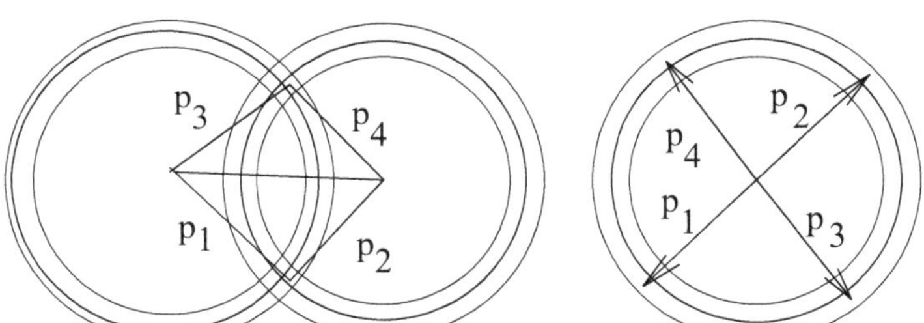

Fig. 16.1. The left side of the figure shows the kinematics of scattering $\vec{p}_1 + \vec{p}_2 = \vec{p}_3 + \vec{p}_4$. When all four momenta are forced to be close to the surface of the Fermi sphere (between two thin circles around the thick circle), the final momentum has to be inside small diamond-shaped overlap regions. The exception, however, is the case when the total momentum is zero: This case is shown in the right figure. In this case, the whole region near the Fermi surface is open for the final momenta.

enters the equations. Therefore, all the equations for superconductivity (such as the famous BCS *gap equation*) are basically the same for any dimension.

If so, one should now recall a fact we all learned in quantum mechanics courses: In 1D, unlike higher dimensions, any attractive potential creates at least one bound state. In field theory language, this means that one should calculate a "fish" one-loop diagram and find that it is logarithmically divergent.

A more modern way of doing the same thing is to use the renormalization group approach. Using the expansion of scattering amplitude for a pair with momenta $(\mathbf{p}, -\mathbf{p})$ to another pair with momenta $(\mathbf{k}, -\mathbf{k})$ on a Fermi sphere in harmonics,

$$f(\theta) = \sum_{l=0}^{\infty}(2l+1) f_l P_l(\cos\theta)$$

or, inversely, $f_l = \frac{1}{2}\int_0^{\pi} d\theta \, \sin\theta \, P_l(\cos\theta) f(\theta)$. The partial-wave amplitudes f_l RG evolve independently,

$$\frac{df_l}{dt} = -\frac{\mu^2}{2\pi^2} f_l^2 \tag{16.12}$$

leading to the solution

$$f_l(t) = \frac{f_l(0)}{1 + \frac{\mu^2}{2\pi^2} f_l(0) t}$$

Note that if at $t = 0$, all $f_l > 0$, the interaction is repulsive in all channels, then the four-fermion interaction vanishes at the Fermi surface. However, if one of $f_l(0)$ is negative, it will develop a singularity (analogous to the Landau pole in QED RG) at $t = -\frac{2\pi^2}{\mu^2 f(0)}$, which is known as the Cooper pairing. The pole is reached first by the channel having the largest negative $f_l(0)$. This singularity indicates the BCS instability of the Fermi surface with respect to any attractive interaction. The BCS gap is proportional to the energy scale at which the Landau pole is reached, which is

$$\Delta \sim \exp\left(-\frac{2\pi^2}{\mu^2 f(0)}\right) \tag{16.13}$$

For asymptotically dense cold quark matter, the interaction between quarks is assumed to be perturbative gluon exchange. Since its electric part is screened more than its magnetic part, the latter dominates. The derivation of the superconducting gap and other features in this asymptotic limit has been done by Son (1999). His result for the gap has rather unusual behavior:

$$\Delta \sim \frac{\mu}{g(\mu)^5} \exp\left[-\frac{3\pi^2}{\sqrt{2}g(\mu)}\right] \tag{16.14}$$

in which the coupling is not squared. Although the coupling at large chemical potential is weak, $g(\mu) \sim 1/\sqrt{\log(\mu/\Lambda_{QCD})}$, the suppression due to the exponent is weaker than the power in front, so the gap (weakly) grows with μ.

At intermediate densities $n \sim 10n_0$, quarks are not very light and ultrarelativistic, and one may think that their electric attraction dominates. Such a mechanism was considered as early as the 1970s (Barrois, 1977; Bailin and Love, 1984). The expected gaps Δ (and T_c) predicted are ~1 MeV, which is of the same scale as nucleon pairing in nuclear matter.

Historically, the second was the *instanton mechanism* of quark pairing, in the context of the simplest two-flavor[4] problem (Rapp *et al.*, 1998; Alford *et al.*, 1998). These two papers, submitted to arXiv on the same day, have created an avalanche of hundreds of papers on the subject. The expected gap was much larger $\Delta \sim 100$ MeV and $T_c \sim 50$ MeV. What is even more important is that it was based on the same instanton-induced t' Hooft interaction which induces the chiral symmetry breaking in vacuum and "diquarks" inside the nucleons.

The symmetries of the 2SC phase are similar to the electroweak part of the Standard Model, with the condensed scalar isoscalar *ud* diquark operating as a Higgs. The colored condensate breaks the color group, making five out of eight gluons massive.

The three-flavor-like phase, CSC3 proposed by Alford *et al.* (1999), was based on one gluon exchange interaction, but in fact it is favored by instantons as well (Rapp *et al.*, 2000). Its unusual features include *color-flavor locking* and the *coexistence* of both types of condensates: $\langle qq \rangle$ and $\langle \bar{q}q \rangle$. It combines the features of the Higgs phase (eight massive gluons) and the usual hadronic phase (eight massless "pions").

Furthermore, in the *two-color* QCD, there is the so-called Pauli–Gursey symmetry, which relates these two condensates. It also makes relations between the masses of baryons (=diquarks) and mesons. The pion is degenerate with the lightest scalar diquark.

[4]The reader who has come this far in this book surely knows by now that the 't Hooft interaction between two quarks of the same flavor does not exist.

So, at high densities, the chiral condensate $\langle \bar{q}q \rangle$ simply rotates into a superconductor one $\langle qq \rangle$, while the gap remains the same (Rapp *et al.*, 1998). At low densities and temperatures, there is "diquark matter," and at high densities and low temperatures, a "superfluid"[5] with $\langle ud \rangle \neq 0$. The additional advantage of two-color QCD is that there is no sign problem at nonzero μ, allowing first principle lattice simulations. Let me mention one of the latest papers by Boz *et al.* (2020), who studied the $N_c = N_f = 2$ QCD. They found that in a rather large interval of quark $\mu = 400$–$700\,\text{MeV}$, the diquark condensate is nonzero for μ-independent

$$T_{\text{qq}} \approx 100\,\text{MeV}$$

The deconfinement temperature (from the inflection point of the Polyakov line) is also μ-independent but twice as large:

$$T_{\text{deconf}} \approx 200\,\text{MeV}$$

So, for intermediate $T_{\text{qq}} < T < T_{\text{deconf}}$, it is some quantum liquid of uncondensed (bosonic) baryons. Remarkably, the density is not far from that of an ideal quark Fermi gas:

$$n_{\text{ideal}} = N_f N_c \left(\mu T^2 + \frac{\mu^3}{\pi^2} \right) \tag{16.15}$$

so in this sense, it follows a "quarkyonic" idea.

Can one use a "small N_c approximation" and try to extrapolate these results to real-world QCD, for which such lattice simulations at nonzero μ are so far impossible? Well, the strength of both Coulomb and instanton-induced attraction is different in quark–antiquark and quark–quark channels via a universal factor $1/(N_c - 1)$. For the case we just discussed, $N_c = 2$, it is 1. Thus, two-color QCD has meson–diquark symmetry. For our world, $N_c = 3$, this factor is $1/2$. Crudely, one may assume that the diquark condensation temperature would therefore be about half:

$$T_{\text{qq}}(N_c = 3) \approx \frac{1}{2} T_{\text{qq}}(N_c = 2) \approx 50\,\text{MeV} \tag{16.16}$$

The same value was obtained from the solution of the BCS-like gap equation for color superconductivity.[6] What deconfinement and chiral transitions look like, we do not yet know.

---

[5] We call it a color superconductor when the diquark is colored. In this case, it is colored neutral. From the point of view of electromagnetism, however, it is still a superconductor because *ud* diquark has a nonzero charge.

[6] It is not impossible that heavy ion collisions in the few-GeV/N energy range will freeze out in the vicinity of this line.

16.4 Neutron Stars and Their Merger

The creation and life of NSs belong to books on astrophysics, and we cannot go into them in any significant depth. Let me just state that they are born in core-collapse supernovae. For decades, studies of binary NSs faced significant constraints due to possible deviations from General Relativity. Recently, due to historic observations by the Ligo–Virgo collaboration, the first NS–NS merger was seen. We would not even attempt their discussion here.

The properties of a stable single NS can be obtained from a solution to the Tollman–Oppenhimer–Volkov (TOV) equation, with the hydrostatic one including the effects of General Relativity, for $p(r)$, starting at the center with a certain initial pressure of p_0 and continuing to $p(R) = 0$, which gives the star's radius R. The mass of the star is then given by[7]

$$M(R) = \int_0^R \rho(r) 4\pi r^2 dr$$

By changing p_0, one gets the so-called mass–radius curves.

All this can be done provided the EOS $p(\epsilon)$ is known. Assuming certain EOSs, one can get mass–radius curves, providing the *maximal possible mass* $M_{\max}$ for it. By comparing with the maximal mass seen by astronomers, one can exclude certain EOSs. There is vast literature which compares these curves to data, so let me just say that the current *heavy-weight champion* among them has $M_{\max} \approx 2M_{\text{Sun}}$.

Measurements of the masses can be very accurately done if an NS is in a binary, as the periods of the motion depend on it. Measurements of the radius is much more tricky. Even for a few of them, for which we know the distance and can measure the temperature and total luminosity, one cannot reliably assume the radiation to be blackbody since that depends on the *albedo* of the surface, and its conditions remain unclear.

Let me mention in connection to this one cute method to get information about R. As we all were told at school, Galileo observed how black spots on the Sun move to its edge and figured out that those are not due to some independent objects (exoplanets) but just spots on the Sun itself, as they all rotate with the same frequency. Each black spot is invisible when it is on the invisible side and has easily calculable effects on the luminosity when it is on this side.

[7]Note that, contrary to expectation, this integral is correct despite the fact that it does not include a covariant volume element depending on the internal metrics g_{rr}.

Since NS are very relativistic (their radii R and mass-in- gravitational-units are comparable), the light deflection angles are not small (as for the Sun). As a result, light goes *round the corner* of the star, so one may see some fraction of its back side. Since some stars have "hot spots," the total radiation as a function of time has shapes calculable by GR and depending on the NSs' radii.)

The very first NS merger observed via emitted gravity waves put measurements of the radius on much simpler grounds because those revealed at what distance between the NS centers their surfaces touched each other. We now know that $R \approx 12$ km.

Furthermore, these data provided the first information about NS deformability by tidal forces. There are also many details of the wave amplitude shape depending on the EOS; however, without doing such calculations myself, I would not describe these fascinating predictions. (Unfortunately, the most nontrivial part of the signal happens to be at frequencies higher than the current window of the LIGO detector.)

Let me just summarize what is learned from empirical information about dense-matter EoS. Its qualitative form can be discussed in terms of the *maximal value of the speed of sound* $\max(c_s^2)$ allowed. In Fig. 16.2 adapted from Annala *et al.* (2019), the allowed EoSs are shown, color coded according to their maximal value of the slope:

$$c_s^2 \equiv \frac{dp}{d\epsilon}$$

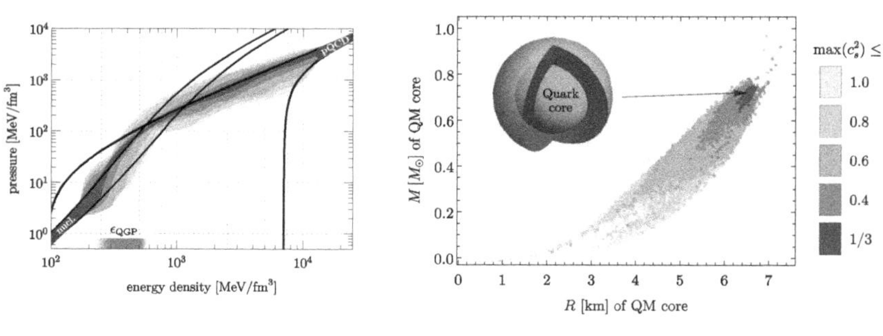

Fig. 16.2. (Left) The band of allowed NS-matter EoSs constructed with the speed-of-sound interpolation method with color-coded upper limits for c_s^2. The symbol ϵ_{QGP} refers to the energy densities, where the deconfinement transition takes place in high-temperature QGP, while the black solid lines represent extrapolations of the theoretical low- and high-density EoSs beyond their regimes of validity. (Right) The masses and radii of quark matter cores, of maximal mass neutron stars for the ensemble of equation of states color coded according to $\max(c_s^2)$.

Although such interpolations are smooth by construction, one can observe a change in slope at an energy density of $\sim$1 GeV/fm^3, perhaps indicative of the phase transition. Note that it is in the ballpark of energy density at which the transition happens at zero baryon density but at a high temperature.

The right plot shows how masses and radii of quark matter cores depend on the selected EOS. If the allowed $\max(c_s^2) = 1$ (the speed of light), the quark core is small[8]; if it is 1/3, the radius may reach 6 km, or half of the total radius.

A large $\max(c_s^2)$ implies significant repulsion on the nuclear matter side of the transition and/or repulsion among diquarks. The realization of that came as a surprise to the nuclear physics community.

The problem is related to features of "repulsive nuclear core potential" in NN scattering. Its radius is well constrained, but the magnitude is not. For normal nuclei, it is enough to know that the wave function inside the core is negligible.

Let me, in retrospect, provide some possible explanations, relating them to the dibaryon issue. Naive (bag-type) models view dibaryons as six quarks packed inside the volume of a nucleon. Obviously, the mass is larger than $2M_N$, as the argument goes, because of the Pauli principle. Yet, when one calculates the related Fermi-statistics effect, one finds that it is very modest. Indeed, light quarks have three colors, two spins and two flavors, i.e. 12 states, and we now discuss only six. The total zero color constraint exists, although it is not very restrictive either.

Current views, based partly on Regge phenomenology, tell us that a nucleon is made of a quark + diquark, connected by a flux tube. The diquark correlation is explained as the instanton effect, with a large binding of $B_{ud} \sim 200$ MeV. So, a dibaryon has just three diquarks with a huge binding of $3B \sim 900$ MeV as compared to only the constituent quark masses $6M_{\text{eff}} \approx 2M_\Delta \approx 2.4$ GeV.

Experiments tell us it cannot possibly be the case. Attempts to put two nucleons in a single volume are met with a repulsive core of height in the tune of hundreds of MeVs. Apparently, diquarks cannot coexist peacefully, but repel each other. Why?[9]

---

[8]The total mass is around 2 M_{Sun}, and the total radius is around 12 km.

[9]In fact, the instanton mechanism does provide an explanation. An instanton is not an empty bag. A single instanton has only *one* zero mode per flavor: The color and spin of quarks on it are fixed by the instanton gauge field. So, only two (u, d) quarks, out of six,

Of course, we are interested in the changes in the EOS related to the deconfinement transition to quark matter. A quite visible change in slope on the left of Fig. 16.2 suggests the place where it most likely is located. However, this figure corresponds to single-static nonrotating NSs. Actual simulations of the time dependence of the NS mergers reveal much more interesting possible scenarios of the future observations.

In Fig. 16.3 from Bauswein *et al.* (2019), one can see the results of calculations for two EOSs: a pure nuclear-type and another with phase transitions to quark matter.

Frequency, in general, increases as the merger process progresses. The existence of visible peaks at certain frequencies corresponds to intermediate configurations of the system which persist for some time. Frequency basically gives the ratio of total (approximately conserved) angular momentum J to the time-dependent momentum of inertia $I(t)$. The system with quark matter is more dense/compact, so its I is smaller and its frequency is higher. Thus, the peaks are shifting to the right. Unfortunately, the values of interesting frequencies are 3–4 kHz, which is above the window of frequencies at which the LIGO detector can currently operate. So, one needs to wait for better gravity wave detectors to compare plots like this one with the data.

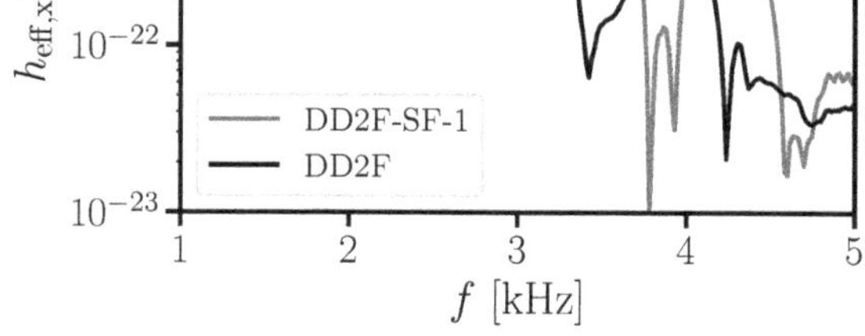

Fig. 16.3. Gravitational wave signals calculated for the merger of two NSs (with a mass of 2.7 solar masses), as a function of the observed frequency f. The black and green lines correspond to the EOS without and with the phase transition to cold quark matter.

can get the t'Hooft attractive force. The Pauli principle's restrictions are much stronger in this case. The diquarks therefore do repel each other.

Summary and connections: The subject of this chapter deviated from our main topic of heavy ion collisions and is related to neutron stars, which are studied by astrophysicists. Our focus is on the transition from neutron to quark matter at high densities and zero (or low) temperatures. Lacking a lattice gauge theory, one can only use the phenomenology of EOS from the properties of NSs and their mergers. There is evidence for rapid changes in the speed of sound, taken as indications for the QCD phase transition, perhaps even of the first order. Better gravity wave detectors observing few-kHz signals would be able to get a much better handle on it and on the fraction of cold quark matter in such stars.

Returning now to the NS merger and corresponding theoretical calculations, let us focus on the latest stages of the process. In Fig. 16.3 from the same paper, the frequency spectrum of the gravitational waves for EOS with and without phase transition is shown. The peaks in frequency are related to the existence of certain quasi-stationary intermediate conditions in the merged NSs, before most of them go inside the horizon into the final black hole. As one can see from the plot, these peaks are quite sensitive to the EOS. Unfortunately, they correspond to frequencies of 3–4 kHz, and at the moment, there are no gravitational wave detectors which can detect such a frequency range. (The LIGO signal is traced for frequencies from approximately 40–500 Hz).

Chapter summary: Unlike hot hadronic matter, for which we have heavy ion collision experiments and powerful lattice thermodynamics, for cold dense matter, one can only rely on astronomical data about NSs. Even though many ideas to extend lattice methods to complex weights have been attempted, none of them have been practical so far. As a result, there is considerable uncertainty about high-density phases as well as the order of the phase transitions. Their approximate location is, however, known from empirical EOS describing the NSs. The first NS–NS merger observed in 2017 was a historic event, important to nuclear astrophysics. Unfortunately, it did not clarify much about matter conditions in the NS cores.

Chapter 17

MODELS OF THE INITIAL STAGE

The reader who has come this far in this book is perhaps ready for a more substantial discussion of the *initial* stage of the collision. Indeed, when we first passed it, we only had an introductory discussion of the collision geometry, including the initial fluctuations due to the random locations of the nucleons in the nuclei at the collision moment.

We begin this chapter with pQCD-based views on the partonic structure of the initial stage and the equilibration process via the partonic cascades. We then proceed to the experimental data on the distributions of the so-called global observables, such as total transverse energy and multiplicity. The following section is devoted to the issue of *chemical quark equilibration*, which is related to electromagnetic probes. Then, we discuss the *anisotropic stage* of the collision, for which a specialized tool — *anisotropic hydrodynamics* — has been developed.

Needless to say, the initial, or pre-equilibrium, stage is the one on which we are much less certain, with a wide range of opinions present in the literature and being intensely debated. This is, of course, due to the fact that it is rather nontrivial to obtain direct experimental information about it.

Perhaps, the most important conceptual controversy in the field remains the conflicting conclusions emerging from weakly coupled and strongly coupled scenarios of the initial state and equilibration.

Significant progress in the theory of a weakly coupled initial state has been made in the concept of *turbulent cascades* with stationary and time-dependent self-similar solutions. Both classical glue simulations and parton cascades came up with out-of-equilibrium attractors possessing power spectra with certain indices, which are qualitatively different from those in equilibrium. From a practical perspective, these studies suggest that the stress tensor remains anisotropic for a long time. However, more recent works indicate that the nontrivial attractor is only approached if the coupling is unrealistically small.

Strongly coupled approaches, especially based on AdS/CFT and related models, view equilibration as a process dual to the *gravitational collapse*, resulting in black hole production in the bulk of AdS_5. As soon as some trapped surface (a black hole) is present, the equilibration is very rapid: Any kind of "debris" simply falls into it. Mathematically, the non-hydro modes have imaginary parts comparable to the real ones, which are numerically quite large (21.29). In this scenario, there are no cascades or even quasiparticles, and the only propagating modes are sounds (dual to gravitons).

Whether or not the stress tensor remains anisotropic beyond the short initial period is still an open question. Theoretical efforts to combine hydrodynamics with out-of-equilibrium parameterization of the stress tensor were discussed earlier: The situation at any realistic anisotropy is thus under theoretical control.

In order to decide on the equilibration time and anisotropy, one needs to develop experimental observables sensitive to the early stage. (My specific proposal — the dilepton polarization — is discussed in Section 17.5.1.)

17.1 Partonic Substructure of the Initial State, Color Glass Condensate and GLASMA

Nucleons are not, of course, some point-like fundamental particles. They have a rather complex internal structure and are made of quarks and glue (strings, string junctions, etc.).

Captured at some arbitrary moments of collision, the nucleons, and thus the corresponding cross-sections, are also fluctuating. The locations of "partons" — quarks and gluons — in the transverse plane are prescribed by different models of the initial state. Accounting for that one generates different — more spiky — initial states, with many more angular harmonics.

The theoretical study of parton equilibration in the weak coupling domain has a long history. The "bottom-up" approach (Baier *et al.*, 2001) was based on soft gluons radiated by scattered hard partons. The name reflects that in this scenario thermal occupation starts at the IR end. (Note that it is opposite to the "top-down" equilibration in holographic models we discuss in the following.) The main predictions of that model were how the equilibration time and the initial temperature scaled with the magnitude of the coupling:

$$\tau_{\rm eq} \sim 1/(\alpha_s^{13/5} Q_s), \quad T_i \sim \alpha_s^{2/5} Q_s \tag{17.1}$$

Some details were changed later, when Weibel, Nielsen–Olesen and other instabilities were incorporated. Its core assumption is the dominance of small-angle scattering of the gluons, justified by large impact parameters of the order of the inverse (perturbative) Debye mass. Perturbative means $M_D \approx gT \ll T$ or small $g \ll 1$. For the boundary of this regime — the coupling values of $g = 1, \alpha_s = 1/4\pi$ — the equilibration time is predicted to be very long $\tau_{\rm eq} Q_s \sim 700$, much exceeding the duration of the collisions.

Perhaps, the most popular model of the initial state among the theorists is the *GLASMA model* (McLerran and Venugopalan, 1994). According to it, the initial state is GLASMA, with a random (Gaussian-distributed) gluonic field characterized by "cells" of the size dictated by a single "saturation scale" parameter Q_s. The pQCD "phase diagram" of this model has already been shown in Fig. 2.1 When the density of gluons becomes large enough and nonlinear effects become important, the GLASMA is in its "dense regime." Its boundary, shown in Fig. 2.1, is defined by the saturation momentum $Q_s(x)$, separating it from the dilute partonic phase. $Q_s(x)$ is expected to grow with collision energy (smaller x) and a higher atomic number A. At the highest LHC energies and atomic numbers, its value is as large as $Q_s^2 \sim 10\,\text{GeV}^2$, believed to be in the perturbative domain. However, another boundary, of the "confining regime" at the bottom of the figure, is indicated by an extremely small $Q^2 < 0.03\,\text{GeV}^2$. This is unacceptable: The boundary of pQCD is in fact at least a factor 30 or more higher. There are no gluons with a virtuality below 1 GeV.[1] Modern lattice simulations show that the gluon effective mass in QGP is of the order of 1 GeV at $T = T_c$ and grows further at $T > T_c$.

The impact-parameter (IP) GLASMA model uses location-dependent $Q_s(x, y)$, related to the gluon density by

$$\frac{dN_g}{d^2x} \sim Q_s^2(\vec{x}) \tag{17.2}$$

The typical values of Q_s at the RHIC (LHC) are about 1.5(2.0) GeV, so there are many such "cells" per $1\,\text{fm}^2$. The fields themselves, both electric and magnetic, are directed longitudinally. Therefore, sometimes these "cells" are referred to as "GLASMA flux tubes."

[1] Recall that the Nambu–Jona–Lasinio model of 1961 already had strong nonperturbative forces at $Q < 1$ GeV, creating chiral symmetry breaking.

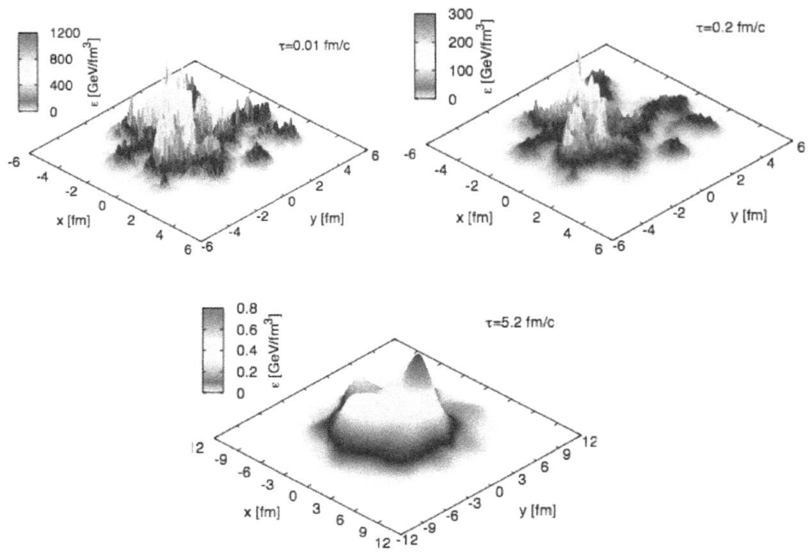

Fig. 17.1. Transverse energy profile from the IP GLASMA model for a semi-peripheral (b = 8 fm) $Au + Au$ collision at s = 200 A GeV at times τ = 0.01, 0.2 and 5.2 fm/c. From τ= 0.01 fm/c to 0.2 fm/c, the fireball evolves out of equilibrium according to the Glasma model.

Figure 17.1 from Schenke and Venugopalan (2014) illustrates what is done in the IP GLASMA approach. Independent fluctuations of color directions in different cells seed later hydrodynamical harmonic flows. At a certain proper time, 0.2 fm/c in this example, the GLASMA evolution is stopped, and the stress tensor is matched with that of ideal fluid. For technical reasons, the value of the viscous tensor is set to zero. How important is the selected time of 0.2 fm/c? Note that the second picture is hardly different from the first, except that the overall scale of the energy density is reduced. Indeed, 0.2 fm is a small distance relative to the nuclear size, and all one finds at this time is dilution due to longitudinal stretching. By starting hydrodynamics right from the second picture, Schenke and collaborators implicitly assume that hydrodynamic cells can indeed be as small as 0.2 fm and that their code can cope with huge gradients between the cells. (Typically, hydrodynamics starts at a later time, 0.6 fm/c or more.)

Following the evolution to the bottom figure, at time 5.2 fm/c, one finds it to appear rather different. The original bumps have disappeared, and instead, a new one at a different location has formed. Indeed, we know from hydro that sound perturbations cannot stand still, but rather move with a speed of sound. At intersections of "sound circles" from the primary bumps, random enhancements of density are observed. Yet, since the bumps are statistically uncorrelated, those should get averaged out, at least in two-particle correlations, and only correlations from the same circle will remain.

How many harmonics are needed to describe pictures such as those shown in Fig. 17.1? Taking 0.2 fm as the resolution and 4 fm as the fireball size, one finds that the upper picture requires about $20 \times 20 = 400$ pixels to be represented by certain stress tensor components. Yet, at the freeze-out, there are only a few angular harmonics observed! So, 99% of the information shown in those pictures does not survive; it predictably disappears via viscous damping.

The transition from the initial out-of-equilibrium partonic state to hydrodynamics uses certain versions of the *kinetic* description. We focus on one particular issue only: the amount of *entropy production*. More precisely, we discuss the difference between the entropy produced *per wounded nucleon pair* in an AA collision and that in the pp case.

17.2 The Lund Model

The name "Lund model" refers to a more traditional view of the initial state, which it describes via a set of independent QCD flux tubes stretched between the color charges of outgoing quarks. Such a state with many longitudinally extended flux tubes is loosely called "a spaghetti" state.

The QCD flux tubes should not be confused with the GLASMA flux tubes just described above: They are quantized and carry integer *electric* fluxes. They exist in the QCD vacuum, which does not have magnetic confinement. So, they never have magnetic fluxes. Unlike "GLASMA tubes," they are expected to exist for a relatively long time before breaking and thus become extended not only in rapidity but in real longitudinal extension as well. The Lund model view is the basis for event generators, such as Pythia, which is rather popular among experimentalists.

Unlike GLASMA, which defines the cell size via Q_s and the total parton density at beam rapidities, in the Lund model, string multiplicity is defined by the number of "color exchanges" that occur as two beams pass each other. While exchanges with large momentum transfers can be evaluated by pQCD, most of those do not belong in this category and are "soft color exchanges." With some fitted — rather than theoretically based — cross-sections and cutoffs of the processes, this string multiplicity is determined empirically.

As we show in the following, the number of strings can be deduced from distributions of "global observables," such as particle multiplicities and transverse energy in the events.

One may ask whether one can deduce the number of independent cells/strings from the magnitude of fluctuations, e.g. distributions over the strengths of higher-order flow harmonics. Unfortunately, the answer to that is *negative*: The directed flows do not have enough resolution to tell us about the "color cell" sizes. High harmonics $n > 8$ are in fact damped by viscosity effects, and the wavelengths of the observed harmonics $n < 8$ are way too large to resolve the nucleon substructure. As we discussed above, flow harmonics are well understood via simple models with randomly placed nucleons.

17.3 Multiplicity and $E_\perp$ Distributions Induced by Initial State Fluctuations

By *global observables*, we mean distributions of the total multiplicity of secondaries, or total transverse energy

$$E_\perp = \sum_i \sqrt{(p_i^\perp)^2 + m_i^2}$$

of the observed events. These distributions are, however, discussed in conjunction with the definition of the centrality classes and are therefore tied to observations of the "wounded" and "spectator" nucleons.

Historically, the RHIC data on the ratio of the $E_\perp$ distributions for AA and pp collisions were described using a parameterization:

$$\frac{dE_\perp^{AA}}{d\eta} \Big/ \frac{dE_\perp^{pp}}{d\eta} = (1-x)\frac{N_p}{2} + xN_{\text{coll}} \tag{17.3}$$

in which the main "soft" term is simply linearly proportional to the number of participants N_p, while the second term (admixture of which is weighted by the parameter x) was interpreted as the number of hard "binary collisions" N_{coll} scaling as a higher power of N_p.

The motivation of such nonlinear fits by the admixture of the "hard" component, however, became questionable as the discovery of strong jet quenching pushed the "hard" (power-like) component down the particle spectra so much that they could hardly be used to explain the nonlinear terms in those fits.

One possible interpretation may be as follows. At an early time, when transverse energy is generated, there are "minijet" collisions with nonlinear contributions. Yet, the subsequent jet quenching indeed kills them, so their descendants are not observed in the spectra.

Another possible interpretation for the multiplicity and $E_\perp$ distributions was proposed by Mitchell *et al.* (2016). The notion of a "participant nucleon" is in their theory substituted by a "participant quark," with their number denoted by N_{qp}. The model — an incarnation of the additive quark model of the 1960s — views a nucleon as a set of three constituent quarks, which interact separately with a certain cross-section. Strikingly, the number of "quark participants" N_{qp} — within a 1% accuracy(!) turns out to be proportional to the r.h.s. of Eq. (17.3) *without* any need for a nonlinear term. The observed $E_\perp$ turned out to be perfectly linear in N_{qp}, see Fig. 17.2(b) from Mitchell *et al.* (2016) and Adler *et al.* (2014). Therefore, in this model, one can ascribe the multiplicity to that Lund-model picture, in which each participant quark is connected by the QCD string to some other participant quark, with those strings becoming ancestors for the observed distribution of secondaries. We will return to "wounded quarks" concept in the discussion of the Pomeron in Section 26.2.

Needless to say, the additive quark model does not agree with the CGC/GLASMA picture, in which cells, or flux tubes, of the size $1/Q_s \times 1/Q_s$ are statistically independent sources, defined locally in the transverse plane. Presumably, the CGC/GLASMA picture should become valid at a certain high-density regime, while the low-density regime is described by the Lund-type models based on the flux tubes and, perhaps, constituent quarks at their ends. However, in the data on global observables, we do not indicate any transitions between the two regimes, and their distributions are basically the same at all energies.

Let us seek further guidance from phenomenology. Note that if the number of independently decaying "clusters" is N, the width of the observed distributions should scale as $O(N^{-1/2})$. The multiplicity distributions have long tails toward large values, which are usually fitted by the negative binomial or similar distributions with two parameters, or some similar convolutions of two random processes with different parameters. Its second moment should tell us how many "progenitors" (clusters, clans, ancestors and cells) the system goes through.

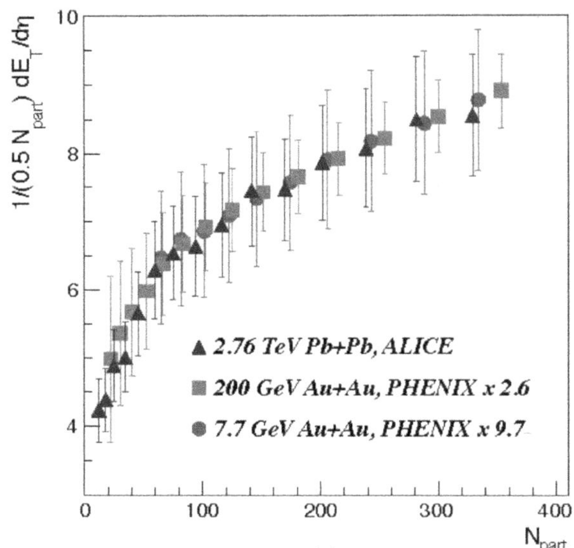

(a)

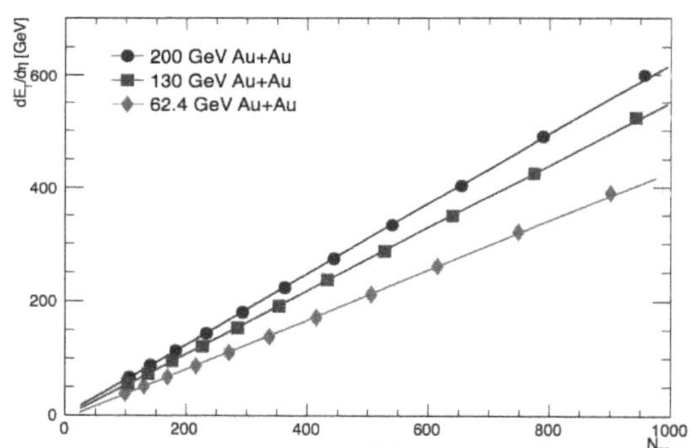

(b)

Fig. 17.2. Distributions over participant nucleons (a) and participant quarks (b). Note that the latter is perfectly linear, while the former is not.

Let us consider the following three schematic models: (i) the usual Glauber model in which N is the number of participant nucleons N_p; (ii) the model based on the number of participant constituent quarks $N_{\rm pq}$; (iii) and the CGC/GLASMA model, and calculate the number of independent cells and the resulting fluctuations.

In the last case,

$$N_{\rm GLASMA} \sim (\pi R^2) Q_s^2 \tag{17.4}$$

For central AA collisions, the area is geometrical, equal to 100 fm^2 and

$$N_p \sim 400, N_{\rm pq} \sim 1000, N_{\rm GLASMA} \sim 10^4 \tag{17.5}$$

For central pA collisions, the area is given by the NN cross-section $\sigma \sim 100\,\text{mb} = 10\,\text{fm}^2$. So, one gets a very different number of "clusters":

$$N_p \sim 16, N_{\rm pq} \sim 40, N_{\rm GLASMA} \sim 10^3 \tag{17.6}$$

Therefore, these models predict vastly different fluctuations.

A brief summary of such fits to data: The "participant quark" model describes $AuAu$ and dAu data extremely well, while for pp collisions, it underpredicts the tail of the distribution. Even six participant quarks — the maximal of the model — is not enough; there seem to be more "clusters" than that. The lesson is perhaps that the highest-multiplicity pp collisions are indeed the first case when "soft" models become insufficient. Models based on pQCD gluons as the main component of the hadronic wave function multiplied by their hard scatterings perform better on these "tails." In particular, these are so for Pythia (pQCD + strings) event generators, which do describe the high-multiplicity tail of pp collisions well.

17.4 When and How the QGP Gets Its Quarks?

Perturbative cross-sections suggest that *initial chemical equilibration* via quark–antiquark pair production is relatively slow and should occur much later than thermal equilibration of the glue. Without delving into the details, let us follow the original perturbative reasoning (Shuryak, 1992), known as the "hot glue" scenario.

The lowest-order cross-section of the gluon–gluon scattering is

$$\frac{d\sigma}{dt} = \left(\frac{\pi\alpha_s^2}{s^2}\right)\left(\frac{9}{2}\right)\left(3 - \frac{ut}{s^2} - \frac{us}{t^2} - \frac{st}{u^2}\right) \tag{17.7}$$

with four terms in the bracket corresponding to four diagrams and the local four-gluon vertex and gluon exchanges in the s, t, u channels. Forward scattering $t \to 0$ gives a divergent cross-section. In QGP matter, it gets regulated by the gluon effective mass M_g in all denominators.

Transverse energy is given by the *transport* cross-section, with an additional factor $\sin(\theta)^2$ in it, so its divergence is only logarithmic, and

$$\sigma^T_{gg\to gg} \approx \frac{9\pi\alpha_s^2}{2s}\log\left(\frac{s}{M_g^2}\right) \tag{17.8}$$

The cross-section of gluon conversion into quarks

$$\sigma_{gg\to\bar{q}q} \approx \frac{\pi\alpha_s^2}{3s}\log\left(\frac{s}{M_q^2}\right) \tag{17.9}$$

where M_q is the effective in-matter quark mass. Since the numerical factor here is much smaller than for gg scattering, one may expect the so-called "hot glue" scenario (Shuryak, 1992; Shuryak and Xiong, 1993), in which gluons get quickly thermally equilibrated, yet chemical equilibration is lagging, with quark/antiquark density remaining significantly smaller than in fully equilibrated QGP. Introducing an effective quark fugacity, $\xi_q < 1$ at early stages, one observes that our main tool — the dilepton production — scales quadratically in it

$$q + \bar{q} \to \gamma^* \to l_+ + l_- \tag{17.10}$$

Going a bit ahead of the logic of this presentation, let me say that an empirical comparison of the electromagnatic processes with data does not seem to need an extra small quark density. If indeed chemical quark equilibration is happening very quickly, then it must be produced by processes other than perturbative diagrams. (Recall that we have already met similar situation in discussion of spin equilibration.)

17.5 Anisotropic Hydrodynamics

The partonic initial state leads to the initial out-of-equilibrium stage of the collision, which is highly anisotropic in momentum distribution. However, during the collisions, partons are naturally separated in time according to different rapidities and create "floating matter," in which the spread of longitudinal momenta is *smaller* than the transverse one. At the later hydrodynamical stage, the viscosity effects reduce such anisotropy. Knowledge of the viscosity value allows one to calculate the anisotropy at this stage, provided its initial value is known.

What happens in between is still a matter of debate. Weak coupling theory, such as partonic cascades, predicts large anisotropy of the momentum distribution to persist for a long time, while strong coupling (e.g. holographic) approaches lead to rapid convergence to anisotropies consistent with hydrodynamics. For a discussion; see, for example, Martinez and Strickland (2010) and subsequent papers.

The issue of anisotropy has two practical aspects. The experimental one, to which we return in Section 17.5.1, is a question of how one can experimentally monitor the anisotropy of matter at various stages of evolution. The theoretical question is whether one can extend the hydrodynamical description for strongly anisotropic matter. There have been significant developments along the lines of the so-called *anisotropic hydrodynamics*, abbreviated by "*aHydro*". The idea (Martinez and Strickland, 2010; Florkowski and Ryblewski, 2011) is to introduce the asymmetry parameter into the particle distribution and then, from the Boltzmann equation, derive an equation of motion for it, complementing the equations of hydrodynamics with a new one describing the separate development of transverse and longitudinal momenta.

The solutions of various versions of anisotropic hydrodynamics were compared to the exact solution of the Boltzmann equation itself, derived for a Gubser geometrical setting by Denicol *et al.* (2014). This paper contains many instructive plots, from which I selected the normalized temperature shown in Fig. 17.3 and the sheer stress Π_{ξ}^{ξ} shown in Fig. 17.4. In both cases, the pairs of points correspond to small and very high viscosity values, separated by two orders of magnitude. They roughly represent the strongly and weakly coupled regimes.

Gubser's variable ρ is the "time" coordinate. In all four plots, one can see that all curves coincide in the interval $-2 < \rho < 2$, but deviate from each other both at large negative values, corresponding to the very early stages and, for large positive ones, describing very late times. In fact, all practical applications of hydrodynamics were indeed made within a finite interval of ρ, with other regions being "before formation" and "after freeze-out."

Solutions for two — hugely different — viscosities show a very similar trend. Israel–Stuart hydrodynamics seems to follow the solution of the Boltzmann equation quite well. Even the free streaming regime is not very far from either hydro or exact Boltzmann: This would be surprising for the reader if we did not already know that the radial flow, unlike higher harmonics, can indeed be "faked." If these authors were to calculate the elliptic and higher flows, the results would be quite different. One would expect that for $4\pi\eta/s = 100$, higher harmonics would be completely obliterated. The plots for the shear stress show different behaviors for small and large viscosities, but again, all curves coincide inside the "hydro window" of $-2 < \rho < 2$. Even going well outside that domain, we never see discrepancies between them by more than 20%.

The overall conclusion one can draw from those works is quite simple: All versions of hydrodynamics used in practice are sufficiently accurate for realistic small QGP viscosity $4\pi\eta/s \approx 2$, at least during the times when these models are actually applied.

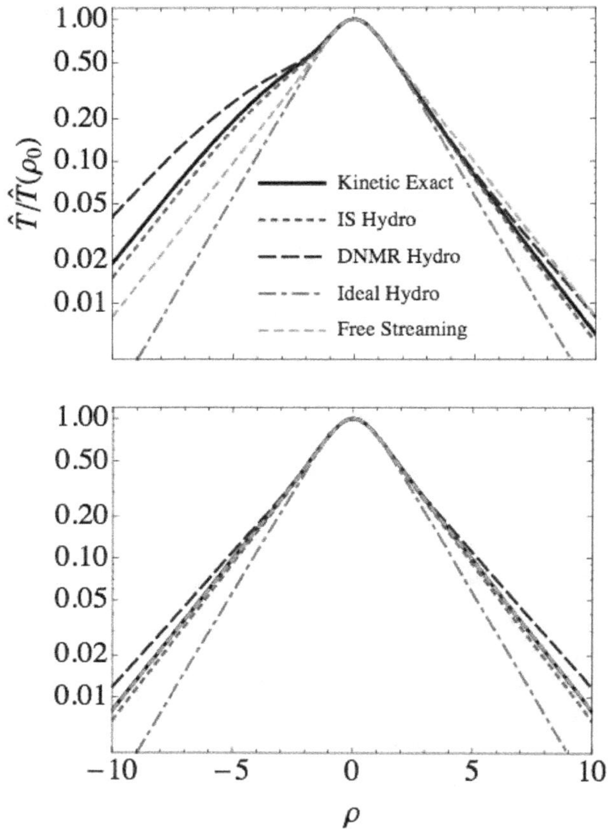

Fig. 17.3. The normalized temperature for $4\pi\eta/s = 1, 100$, upper and lower plots, respectively. The meaning of different curves is explained in the upper plot.

For how long does the matter remain anisotropic? Let us formulate the question a little more precisely. High-energy heavy ion collisions produce fireballs which explode in a very anisotropic manner, with longitudinal rapidities reaching several units, while the transverse rapidity hardly exceeds a unit value. The question in the title of this section does not refer to this obvious effect but to *local anisotropy* inside the hydro cells. At late times — due to the small viscosity value for sQGP — it must be relatively small. A large anisotropy, corresponding to a large $O(1)$ difference between the longitudinal and transverse components of the matter stress tensor, is expected to be present at the early collision stages.

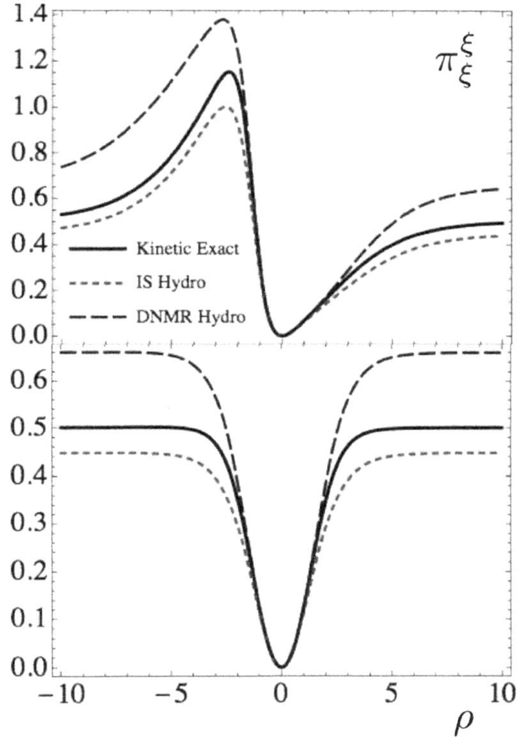

Fig. 17.4. The shear stress π_ξ^ξ versus coordinate ρ, for viscosity $4\pi\eta/s = 1, 100$, the upper and lower plots, respectively.

In strict terms, we do not yet have a quantitative answer to this question. It crucially depends on whether the out-of-equilibrium matter is weakly or strongly coupled. Instead of going into this debate, we will just cover interesting theoretical developments in anisotropic hydrodynamics in the following section, which would perhaps be needed *if* there is an extended anisotropic stage of the collisions, and then proceed to a proposal to verify it experimentally using the dilepton polarization.

17.5.1 *Dilepton polarization and the (early-time) pressure anisotropy*

It is well known that when spin-1/2 particles (such as quarks) annihilate and produce lepton pairs, the cross-section is not isotropic but depends on the lepton emission angle:

$$\frac{d\sigma}{d\Omega_k} \sim (1 + a \cdot \cos^2\theta_k) \tag{17.11}$$

where the subscript corresponds to a momentum k of, say, the positively charged lepton and θ_k is its direction relative to the beam. At large momentum transfer, the dominant is the perturbative Drell–Yan process, and the anisotropy parameter a is $a = 1$.

At intermediate dilepton masses, they originate from the initial stages of matter equilibration. I suggested (Shuryak, 2012a) that the parameter a can be used to control the anisotropy of the early stage of the collision. In particular, if matter is anisotropic, i.e. the longitudinal pressure is small relative to the transverse pressure, $p_l < p_t$, then the annihilating quarks mostly move *transversely* to the beam, which leads to *negative* $a < 0$. Such a regime is expected due to a "self-sorting" process in which partons with different rapidities automatically become spatially separated after the collision.

For illustration, one can use a simple one-parameter angular distribution of quarks over their momenta p in the form of

$$W \sim \exp[-\alpha \cos^2\theta_p] \tag{17.12}$$

and calculate $a(\alpha)$ resulting from it. It does show that a may reach negative values as low as -0.2 at the initial anisotropic stage. During emission from later stages, when local equilibration is approximately reached, this parameter vanishes, $a \to 0$.

17.6 Sensitivity of Heavy Ion Collisions to the Nuclear Structure of the Ions Involved

The so-called "isobar run" at the RHIC was designed to compare a number of observables for collisions of ${}^{96}_{40}Zr$ ions with those of ${}^{96}_{44}Ru$. Since they have the same $A = 96$ and only four protons changed into neutrons, it was assumed that these differences would be small, being mostly due to electromagnetic (Z-dependent effects), such as the chiral magnetic effect (see Section 23.6).

The STAR data (Abdallah *et al.*, 2022b) provide the *difference* between those two collision pairs, reaching so far an unprecedented level of accuracy, $O(1/2\%)$. These data give us an opportunity to test better current models of heavy ion collisions.

The differences between these two nuclei in various observables, such as multiplicity distribution and the magnitudes of elliptic and triangular flows,

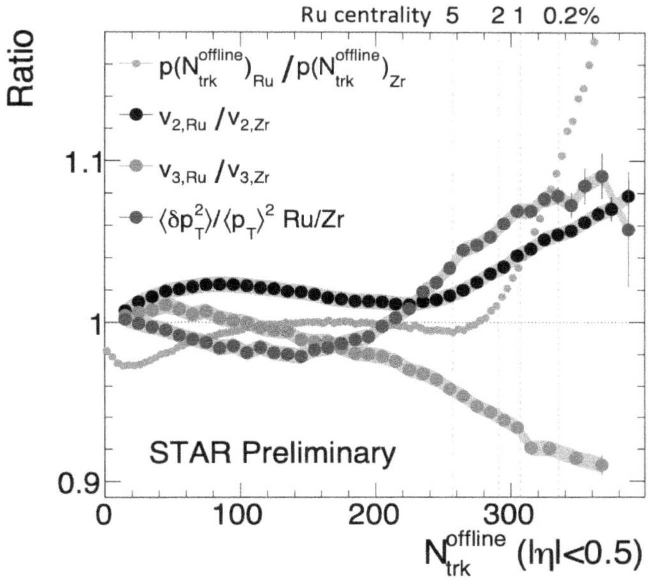

Fig. 17.5. Four observables defined in the plot, all in the form of ratios for *RuRu* to *ZrZr* collisions from STAR as a function of centrality. The lower scale is the number of tracks per unit rapidity, and the upper scale with thin dashed lines is the percentage of total cross-section.

were found to be not so small: of the order of a few percents and up to 10% in some limits, see Fig. 17.5. Further studies have indicated that the origin of that is not so much due to electromagnetic effects as to the different properties of related "valence" protons and neutrons in these nuclei.

So, *how different* may the two nuclides used, $^{96}_{40}Zr$ and $^{96}_{44}Ru$, be? Naively, since four of the neutrons are turned into protons, the resulting effects should be proportional to a small factor of 4/96. However, it is not that simple: The shell model states in which these protons/neutrons are placed are, in fact, rather different.

In particular, a simple "liquid drop" model would suggest that, due to Coulombic repulsion pushing *protons* to larger radii, there would be more protons than neutrons at the nuclear surface. Yet, as we will soon see, the inverse is true. For these nuclei, the effect of Coulomb repulsion is, in fact, smaller than shell-level effects.

A compilation of the ratios for the four observables measured by the STAR collaboration (from Zhang and Jia, 2022) is shown in Fig. 17.5. Apparently, for noncentral collisions (the left side of the plot, >5%), deviations are at the 2% level, but for central collisions (right side of the plot), they reach the level of $O(10\%)$ or more. Thus, naive original expectations were incorrect, and these two nuclei turned out to be in fact very different!

In order to understand *why* $^{96}_{40}Zr$ and $^{96}_{44}Ru$ are so different and be able to quantitatively explain these results, one needs to turn to the nuclear structure. In particular, we discuss:

(i) the wave functions of "valence" quasiparticles, p "holes" and n "particles" in the nuclear shell model, see Appendix.
(ii) the structure of "excitation trees" for both nuclei, which can be understood as "vibrators" and "rotators," with certain parameters to be used in the description of shape fluctuations. Rotational bands indicate the deformation of Ru but not Zr.
(iii) the range of energies ΔE, out of which one needs to consider excitations to reproduce the shape of the virtual state in which one finds nuclei at the collision moment. We interpret it via the "preheating temperature" parameter $T_\perp$. To introduce the width of this range is *unavoidable* because nuclei are not rigid objects. The moments of inertia and non-diagonal matrix elements among the first couple of levels and the first 20 or so are *not the same*. So, to quantify fluctuations in nuclear shapes, one needs to define which set of excitations is involved.

The single-nucleon states, calculated in collective nuclear potentials, defined by the so-called shell model, are shown in a textbook as in Fig. 17.6.

The states are supposed to be filled from bottom up, according to the sequence of single-nucleon energies. As it follows, 50 is a "magic number," and the double-magic ^{100}Sn should be a nice spherical nucleus with filled shells. The nuclei we are interested in differ from it by (2 or 6) neutrons in the $2d_{5/2}$ state and (6 or 10) proton holes in the $1g_{9/2}$ states. The key point is that those states have radically different radial distributions. They differ not only in orbital momentum (2 versus 4), but even in the principal quantum number (2 versus 1).

Let us calculate the corresponding wave functions following my note (Shuryak, 2022b). Using nuclear potential,

$$V(r) := -\frac{V_0}{1 + \text{Exp}[(x - R)/a]} \tag{17.13}$$

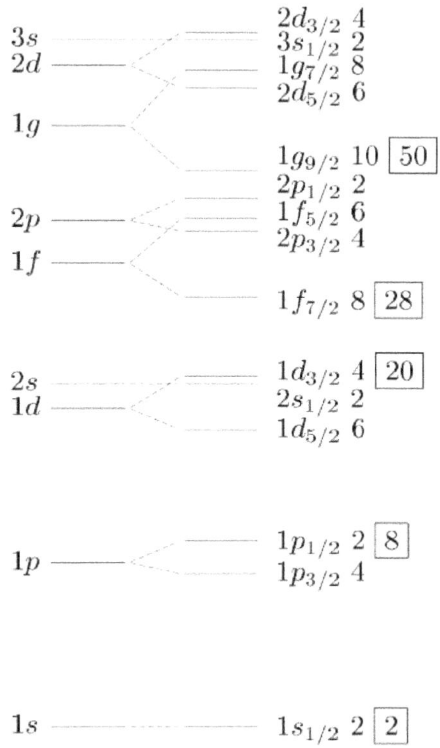

Fig. 17.6. Shell model levels from textbooks.

with $R = (1.25/0.197) * 96^{1/3}, V0 = 0.057; a = 0.65/0.197$ (all energies in GeV, distances in inverse GeV), we calculated the corresponding wave functions, see Fig. 17.7. Indeed, they have very different shapes. Note that the former has a node, located exactly where the latter has a maximum.

The radial dependence of densities can be taken in the form of

$$\begin{aligned} Ru &= d_{5050}(r) - 6\psi_p^2(r) + 2\psi_n^2(r) \\ Zr &= d_{5050(r)} - 10\psi_p^2(r) + 6\psi_n^2(r) \end{aligned} \tag{17.14}$$

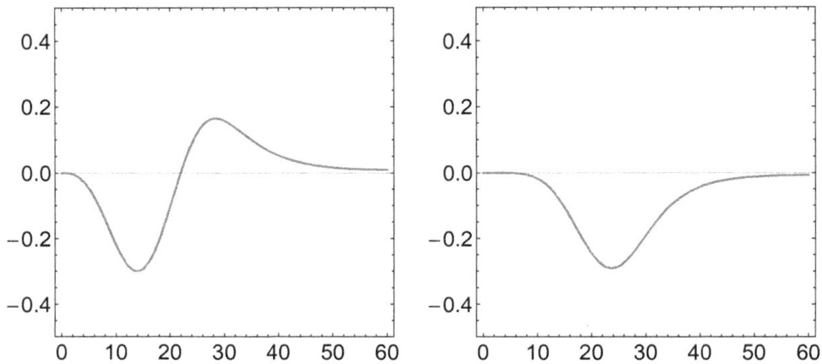

Fig. 17.7. (Unnormalized) wave functions $2d$ and $1g$ as a function of radial coordinate $r\,(\mathrm{GeV}^{-1})$.

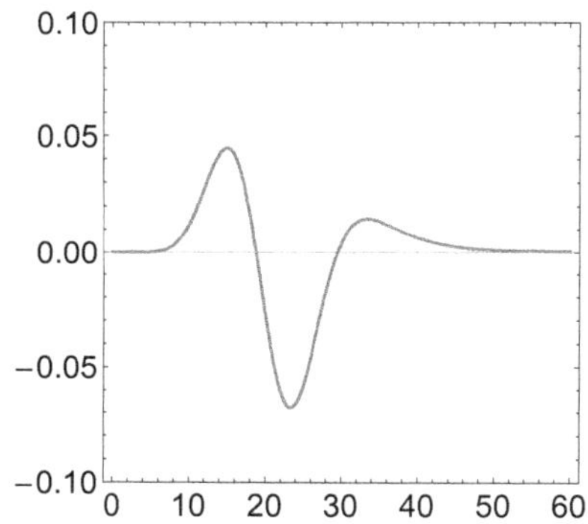

Fig. 17.8. The density ratio Ru/Zr as a function of the distance (GeV^{-1}).

where the first term is a parameterization for the double-magic 50–50 nucleus. Their difference is shown in Fig. 17.8. Note a certain excess[2] of neutrons at large r.

Furthermore, the nuclei in question have a sufficiently large number of quasiparticles, on top of the double magic 50–50 one, so that there is a question about their *shapes*. The generic idea that some nuclei are spherical and some are deformed goes back to the 1950s, and by now, it has taken its proper place in textbooks. What is important here is that it is *not*

[2]It is qualitatively similar to a "neutron halo" obtained in the literature based on "density function theory."

formulated in terms of the ground state wave function $|0\rangle$ (which for even–even nuclei is always "spherical" 0^+).

To understand the shapes of nuclei, one should look not only at their ground states but also the quantum numbers of *excitations*. Certain subsets of excited states $|n\rangle$ are interpreted in terms of particular models, such as "rotator" and "vibrator." Their properties are the momentum of inertia Θ in the former case, vibrational frequency in the latter and transition matrix elements. So, excitation energies E_n, the diagonal and nondiagonal matrix elements of operators (such as magnetic dipole or electric quadrupole moments) should all be taken into consideration while deciding on "nuclear shapes" at the collision moment. It is necessary to define "collective variables" describing these shapes.

These collective variables are variables of model Hamiltonians containing certain "potentials":

$$\langle \beta_2, \gamma, \ldots | \hat{H} | \beta_2, \gamma, \ldots, \rangle = E(\beta_2, \gamma, \ldots) \tag{17.15}$$

calculated to explain the quantum dynamics of the nuclei. For example, if the collective potential has a minimum at zero values of deformations $\beta_2 = \beta_3 = 0$, the nuclei are "spherical," otherwise they are "deformed." These potentials lead to the prediction of certain families of excited states.

Furthermore, there are quantum oscillations around the minimum, which we describe as "phonon" states. If these potentials can have two or more minima, one may define several "vibrators" and look for their excitations among the experimentally observed states. Sometimes, the potentials are flat in a wide range of variables, and the nucleus is declared "soft" with an indefinite shape. We will return to specific examples of those relevant for our "isobar" nuclei in the following, from the nuclear structure literature.

Is the structure of excited states relevant for heavy ion collisions, or does one only need to understand the ground state? These questions have been debated in the field for many years.

My first encounter with these issues was just before the first run of the RHIC, considering in Shuryak (2000) what would happen if we collided strongly deformed nuclei, such as ^{238}U.[3] The classical notion of random directions of nuclear deformation axes would lead to a variety of situations ("tip-to-tip," etc.), and my then primitive simulations addressed the

[3]Believed to be an ellipsoid with about 30% deformation, similar to a ball used in American football.

question of whether one would be able to distinguish them experimentally. After sending the paper to PRC, I got a referee report proclaiming that the study is wrong and very misleading. The referee argued that since the ground state is $J^P = 0^+$, it is spherically symmetric, and thus, the idea of an intrinsic nuclear shape is nonsense. My defense was based on the argument that, if the collision energy is not too small, the relevant intermediate state does not correspond to the ground state alone, but it is a certain "wave packet" made out of a number of excited states. I eventually succeeded, and my paper got published.[4]

If so, which set of excitations would one need to use specifically? An answer suggested in Shuryak (2022b) is a direct — albeit still model-dependent — way to use the potentials $E(\beta_2, \gamma, \ldots)$ from nuclear structure calculations to define the initial state in high-energy heavy ion collisions.

Generically, an act of measurement fixes nucleon transverse coordinates within a certain uncertainty $\Delta\vec{x}_i$, resulting in an uncertainty in the total energy ΔE. Therefore, not just the ground state but excited ones, from a strip $E_n < \Delta E$, would contribute to the density matrix of the virtual state.

Now, what should the probabilities P_n in that expression be? Here, we would like to evoke a standard statistical argument. Suppose ΔE is large enough to encompass a large number of states, which contribute about equally, so that the most important factor in the sum over states would be simply the density of states itself, or its entropy

$$N(E) \sim \exp[S(E)] \tag{17.16}$$

A standard expansion of it, with $\Delta S/\Delta E = 1/T$, generates a Boltzmann distribution over the state's energies $\exp(-E/T)$. In other terms, one may think that nuclei at the collision moment can be viewed as "preheated" ones. If so, the density matrices over relevant collective variables can be evaluated as thermal ones.

Here comes the main question: What can this temperature of preheating be? Is it that we can simply suggest, "in anticipation" of the QGP production in the collision, that the preheating temperature be what we usually call T_0 in hydrodynamical applications, namely hundreds of MeVs? The answer is *no*, it is not. The reason is that equilibration of all degrees of freedom to a common T_0 requires a certain amount of time, which is commonly assumed to be about $\tau \sim 1/2$ fm/c *after* the collision. At the

[4]Yet, this discussion had moved the publication date to the next millennium!

collision moment, one should discuss transverse and longitudinal degrees of freedom separately.

The accuracy of localization in the *transverse plane* $\vec{x}_\perp$ for each nucleon is given by a typical impact parameter in NN respective collisions. An estimate of it is

$$\Delta x_\perp \sim \sqrt{\frac{\sigma_{NN}}{\pi}} \sim 1\,\text{fm} \tag{17.17}$$

The uncertainty relation then tells us that each nucleon gets a kick of magnitude $\Delta p_\perp \sim \hbar/\Delta x_\perp \sim 0.2$ GeV. This corresponds to the nucleon kinetic energy

$$\Delta E_\perp \sim \frac{\Delta p_\perp^2}{2M_N} \sim 20\,\text{MeV} \tag{17.18}$$

and we suggest the *transverse temperature* $T_\perp$ should be of this order.

(The exchanges of *longitudinal* momenta in NN collisions are much larger, but they are not relevant for the distribution in the transverse plane we are discussing. Both small $T_\perp$ and huge T_{long} will eventually equilibrate into common T_0, but we do know that does not happen at the collision moment. If they would, the state at the collision moment would have a very high T and would need a description in terms of quarks and gluons, *a la* the homogeneous CGC gluon state without nucleon correlations. We do know that it is *not* so, or else fluctuations of higher angular harmonics would be much smaller than what is actually observed.)

Uncertainty in energy means that we will not deal with the ground state of the nucleus but with some density matrix made out of excited states with $E_n < \Delta E_\perp$. An idea of how it will look can be formed by assessing another density matrix corresponding to *Euclidean* time duration $\beta \sim \hbar/\Delta E_\perp$. A periodic motion with such "Matsubara" time corresponds to a density matrix of the system at a certain effective "transverse temperature":

$$T_\perp \sim \Delta E_\perp \sim 20\,\text{MeV} \tag{17.19}$$

In other terms, we suggest that "in anticipation of a collision," the nuclei are "preheated" to such a temperature.

We now return to particular nuclei ${}^{96}_{40}Zr$ and ${}^{96}_{44}Ru$ and note that even standard shell model calculations show that there should be a significant difference between them (see Appendix). The pp and nn pairs become strongly correlated with the Cooper pairing. Since there are several such pairs, their states are not simple, and this is what nuclear structure professionals calculate.

The spectroscopy of the excited states of the two nuclides in question provides key information about their structure. Before we go into specifics, let me note that experimentally, it is followed up to around nucleon separation energy, or $O(10\,\mathrm{MeV})$. Since our estimated $T_\perp$ is unfortunately higher, we will not yet have a full set of excitations needed to calculate the thermal density matrix from them. Yet, we do have enough excitations to understand what the main excitation types of both nuclei are, whether they are "rotors" or "oscillators" and with what parameters.

17.6.1 $^{96}_{40}Zr$: *Configurations and "excitation trees"*

One family of (collectivized) particle–hole bound states is known as *nuclear phonons.* In the first approximation, their effective Hamiltonian is that of a harmonic oscillator, and the lowest states are approximately equidistant. The quantum numbers of a "phonon" depend on those of particle-holes, and those of n-phonon states can be deduced from those using standard rules of summed angular momenta. For example, most typical quadrupole oscillation phonons have $J^P = 2^+$, two-phonon states around twice excitations are with $J^P = 4^+, 2^+, 0^+$, etc. A more accurate description is provided by anharmonic oscillators for the "interacting bosons model" (IBM); for a recent discussion of Zr isotopes in it and general references, see Gavrielov *et al.* (2022).

One important concept is that nuclei can be thought of in terms of several *coexisting configurations.* Furthermore, each configuration has its own *excitation tree* (also called a "band"). Since transitions between states are mainly confined inside each, these trees are relatively distinct experimentally (see the following).

In the particular case of $^{96}_{40}Zr$, configuration A corresponds to a closed proton sub-shell and only nn pairs, while configuration B contains two-proton excitation (from below to above sub-shell gap) with a $2p-2h$ state, etc. Each of them have its own effective Hamiltonians $H_A, H_B, \ldots$, with relatively small but nonzero mixing terms $H_{AB}, \ldots$ (which we will further ignore). The IBM approach is to formulate Hamiltonians not in terms of quasiparticle pairs but in terms of scalar and quadrupole "phonons." Their numbers are defined as

$$\hat{n}_s + \hat{n}_d = s^+ s + \sum_\mu d^+_\mu d_\mu \tag{17.20}$$

(A microscopic derivation of IBM relates the number of phonons to the number of quasiparticle pairs, but we will not discuss that.)

The other important concept is that of *dynamical symmetries*. Unlike the usual symmetries, it does not imply certain operators to commute with the Hamiltonian but that a number of operators form a closed algebra, and thus states can be calculated algebraically using representations of the corresponding groups. Since there are $1 + 5 = 6$ phonon operators, the largest group is $U(6)$, which, in a particular case, can be reduced to its subgroups ($U(5), SU(3), SO(6)$, etc.). Related to that is the concept of *collective motion paradigms*, which correspond to such dynamical symmetries. The simplest is spherical vibrations $[U(5)]$, axially symmetric $[SU(3)]$, γ-soft deformed rotor $[SO(6)]$, etc. A geometrical interpretation of the states obtained can be visualized by *coherent states* with certain parameters, such as quadrupole shape parameters (β, γ) related to the following creation operator:

$$b^{+} = \frac{1}{\sqrt{1+\beta^{2}}}\left(s^{+} + \beta\cos(\gamma)d_{0}^{+} + \beta\sin(\gamma)(d_{2}^{+} + d_{-2}^{+})/\sqrt{2}\right) \tag{17.21}$$

The IBM Hamiltonians are made of quadratic parts in s, d operators and quartic ones, typically in the form of quadrupole–quadrupole, with quadrupole quadratic in s, d. The Hamiltonian averaged over these states defines the "energy profile":

$$E(\beta, \gamma) = \langle \beta, \gamma | \hat{H} | \beta, \gamma \rangle \tag{17.22}$$

describing quantum motion in terms of the corresponding collective variables.

In the chart of nuclides (Z, N), there exist multiple domains in which excitation trees have the same symmetry, and the effective Hamiltonian just displays a smooth change in parameters. They are separated by lines of *"mini phase transitions."* We put these words within parenthesis for a few reasons. First of all, these transitions happen for each of the "excitation trees" individually. Second, they indicate excitations of just several (not macroscopically large) number of pairs; therefore, they would only be observed by high-accuracy data. Finally, since (Z, N) change in a discrete manner (by two protons or neutrons, for even–even nuclei), there are no true critical points or singularities, but just jumps from one phase to another.

Let us show how it looks in practice for the particular nucleus in question. The experimental and calculated parts of the spectra, from Gavrielov *et al.* (2022), are shown in Fig. 17.9. Focusing on the configuration B excitation tree (black, right), one observes a typical set of states of a (slightly anharmonic) oscillator, with 2^{+} phonon state, $4^{+}, 2^{+}, 0^{+}$ two

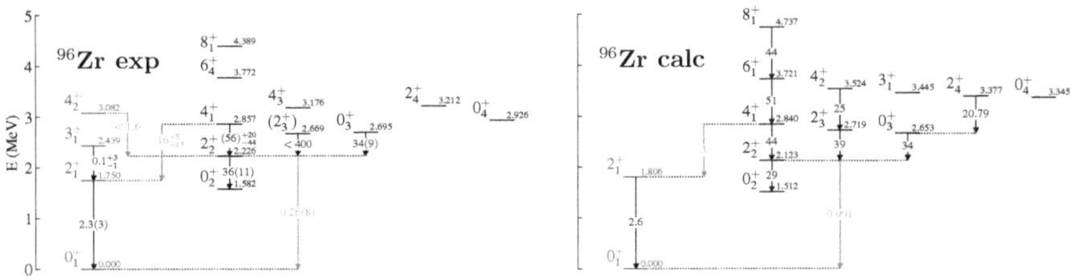

Fig. 17.9. Blue (left) and black (right) are states corresponding to “excitation trees” growing from configurations A and B, respectively.

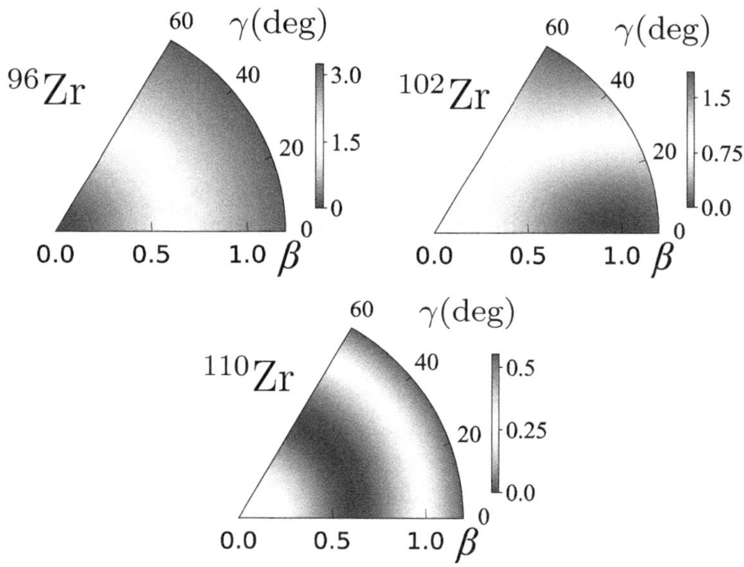

Fig. 17.10. Contour plots of $E(\beta, \gamma)$ for three Zr isotopes, displaying different symmetries.

phonons, up to three phonons states. The ratio of their energies to that of a single phonon are indeed close to 2, 3, etc., confirming the vibrational interpretation of the tree. three phonons, etc. The corresponding picture of $E(\beta, \gamma)$ is given in Fig. 17.10, for three Zr isotopes. As one can see, they correspond to qualitatively different “phases” of configuration B. The one we focus on, ^{96}Zr, has a potential with a single minimum at the origin, corresponding to a basic spherical shape. Its potential seems to be independent of the angle γ.

However, even the isotope ^{102}Zr (with three extra n pairs) shows a completely different potential. Now, the minimum is at large β and zero γ. Adding four more neutron pairs to ^{110}Zr, we again find that another "mini phase transition" line was crossed since the shape of the effective potential gets qualitatively different once again.

Now, we return to our main problem: the evaluation of the density matrix. If the collective motion is described by a harmonic oscillator, the probability of finding a (configuration B) nucleus with particular β, γ is then Gaussian. Furthermore, when $T \gg \omega \approx 2.226 - 1.582 = 0.644$ MeV, the thermal density matrix should be given just by the classical Boltzmann factor:

$$P(\beta, \gamma) \sim \exp\left[-\frac{E(\beta, \gamma)}{T}\right] \tag{17.23}$$

17.6.2 $^{96}_{44}Ru$: *Deformations and rotations*

We now focus on the second nuclide used in the STAR experiment. Reducing the $^{96}_{44}Ru$ problem to four pairs, one nn pair and three pp ones, may appear to be a simpler problem, yet there are four pairs of θ, ϕ variables. Doing quantum mechanics in $8 - 2 = 6$ dimensions (global orientation obviously cannot matter) is still not easy.

Fortunately, a lot of information is available about the excitations, see Fig. 17.11. A clear separation into excitation trees or five "bands" is shown.

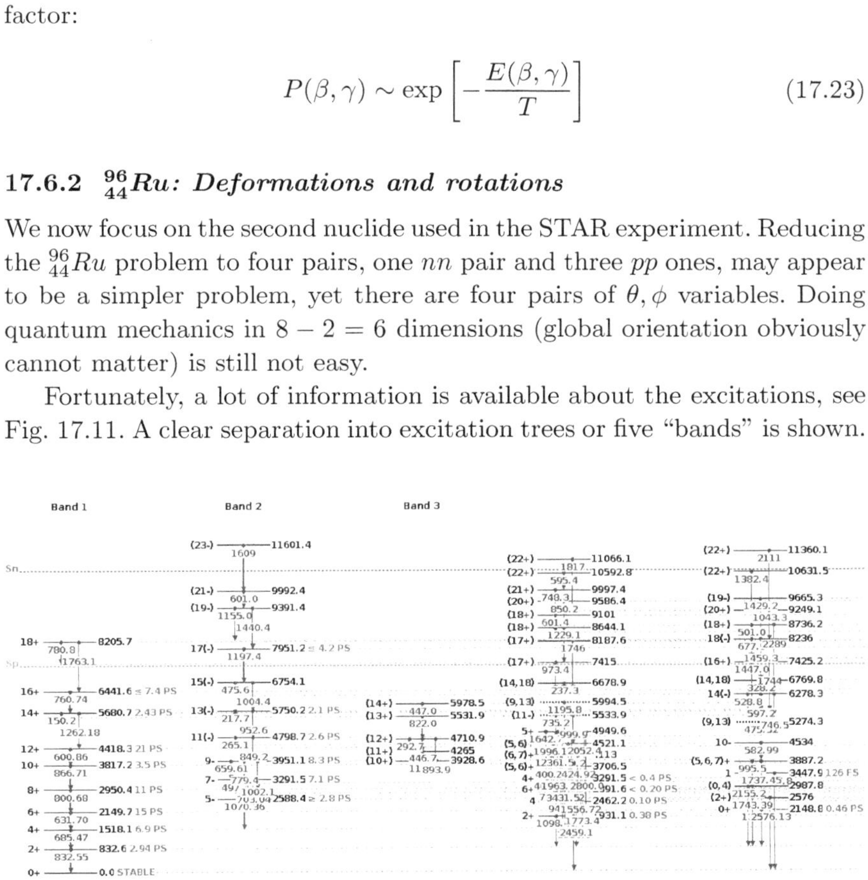

Fig. 17.11. Excitation levels of ^{96}Ru (replotted from the BNL webpage on nuclear excitations).

The first one is a set of states with $J^P = 0^+, 2^+ \cdots + 18^+$, a typical *rotational band.* Since a spherical nucleus cannot be rotated, we learned that this band corresponds to a deformed but axially symmetric configuration.

Two ways of how information on the band can be used. We define the J-dependent moment of inertia and rotational frequency by

$$I_J = \frac{J(J+1)}{2E_J}, \quad \omega_J = \frac{E_{J+2} - E_J}{2} \tag{17.24}$$

and get for the former (GeV^{-1})

$$I_{J=1...8} = 3603.17, 6587.18, 9768.8, 12201.7, 14408.5,$$
$$17653.8, 18485.9, 21112.8$$

Here, we see that nuclei are "flexible" (not rigid), with momentum of inertia (and thus deformation) growing with J. It remains significantly smaller than the moment of inertia for "solid state sphere rotation," which for a sphere is

$$I_{\text{solid}} = (2/5) * M_{\text{tot}} R^2 \approx 32400\,\text{GeV}^{-1}.$$

Therefore, only a part of nuclear matter is actually rotating (which has been known since the 1950s). Again, defining deformation at the collision moment, one has to specify how many states are included in the wave package, or how much preheated the nuclei actually are.

On the other hand, consider one compact Cooper pair sitting at the equator. It will add to the moment of inertia an amount of

$$I_{\text{pair}} = 2MR^2 \approx 1700\,\text{GeV}^{-1}$$

which is smaller than the observed I_J values. But, of course, there are four Cooper pairs sitting somewhere on a sphere, and the observed values can correspond to some particular arrangements of those. Clearly, as J grows, the pairs become unpaired by centrifugal force and become "normal," thus growing I_J.

Looking at ω_J from the rotational band, one finds that it is nearly constant. This indicates that all excitations rotate with about the same rotational frequency, and all increases in J are due to increases in the moment of inertia. The "depairing" of Cooper pairs is not a sharp transition, as observed in heavier nuclei, but a gradual depairing of quasiparticles.

Let us now discuss the second band (a tree of similar states). All of them are $P = -$, so clearly they are not axially symmetric. The root of this tree is 5^- state, which obviously cannot be described using the usual IBM building blocks, 0^+ and 2^+ phonons: Some Cooper pair should be unpaired for that.

Further excitations in this tree also indicates rotations. (The addition of quadrupole phonons cannot describe it since it would generate many more states which are not there.)

Now, we have learned an important lesson: The superposition of excitations from both first trees would generate parity-odd terms in the density matrix, e.g. 3^- or pear-like shapes. If so, one may expect larger triangular flows for these nuclei, as was indeed observed.

Summary and connections: As the collision energy grows, the initial state is expected to change from a set of QCD flux tubes, described by the Lund model, to GLASMA, with a highly excited (even classical) gauge field as its energy density grows. The collective interaction of multiple flux tubes is further discussed in Chapters 26 and 29. The distribution over transverse energy and wounded nucleons seems to be best described by the "additive quark model" for constituent quarks. The issue of initial chemical equilibration remains open. The dependence on the initial shape, at unprecedented $<1\%$ levels, has been provided by the "isobar run" at the RHIC, and we discussed how the initial nuclear shape is related to the wave functions of "valence nuclei" and related issues of nuclear structure.

Chapter 18

THE PENETRATING PROBES

18.1 Using Temperature as a Clock

The main reason electromagnetically produced secondaries — photons and dileptons — occupy a special place in heavy ion physics is that they are "penetrating probes." They are rather weakly coupled to matter, with their rescattering cross-sections small enough to be neglected. Therefore, they provide, so to speak, an X-ray image of the fireball, integrated over its whole history. Another useful analogy is the solar neutrinos, which reach us carrying the details about all the nuclear reactions happening in the Sun's interior.

Of course, quanta emitted from inside the fireball come on top of photons and leptons from decays of secondary hadrons, such as $\pi^0 \to \gamma\gamma$. Fortunately, those can be subtracted, as both secondary spectra and decay modes are well known.

The first ideas about electromagnetic probes were proposed by Feinberg, soon after Landau's hydrodynamical paper of 1953. His first view was that photons would be described by the usual blackbody radiation formulae.[1] Two decades later, Feinberg (1976) still mentioned that it would be the case in high-energy collisions, admitting that at then available energies, photons and leptons would be produced at some (then uncalculable) rates.

Yet, a naive thermal description of photons and leptons is incorrect, as they never come to thermal equilibrium. This is exactly because nothing keeps them inside the fireball. However, Feinberg got one principle point right: They are enhanced by a factor proportional to the fireball lifetime. Indeed, unlike hadrons, which are emitted from the freeze-out surface, the yield of penetrating probes is given by the four-volume integral of the rates.

[1] I recall that he sadly told that Landau did not approve of it, but he did not explain why.

In the first papers on the subject based on the QGP description of matter, photon and lepton production rates were calculated at the lowest QCD orders by Shuryak (1978a, 1980). Those signals were proposed to be one of the key signatures of QGP.

Especially important are "dileptons," lepton pairs e^+e^- or $\mu^+\mu^-$, originating from a virtual photon. While being one more power of small electromagnetic coupling down in rate compared to photon emission and being quite difficult to measure experimentally, dileptons provide more valuable information about matter conditions. While photons provide only the distribution over their transverse momenta $p_\perp$, dileptons give us both the distribution over $p_\perp$ and the invariant mass $M^2 = (p_\mu^+ + p_\mu^-)^2$. Unlike $p_\perp$, an invariant mass distribution is not affected by matter flow and is therefore a much more reliable way to measure the matter temperature.

Another valuable use of the dilepton mass spectrum is that it tells us about the fate of vector mesons in matter. It is quite clear, on general grounds, that the dileptons produced in the QGP phase should be different from those originating from hadronic matter. Unlike those, QGP-originated invariant mass spectra should not have peaks corresponding to familiar vector mesons, $\rho, \omega, \phi, J/\psi \ldots$, but instead be just smooth decreasing functions of the dilepton invariant mass $M^2 = (p_{l+} + p_{l-})^2$.

The evaluation of the dilepton/photon yields is done via the convolution of two factors:

- the production rate $R(T)$ per d^4x as a function of T;
- the so-called *temperature profile of the collision* introduced by Shuryak (1978a, 1980):

$$\text{Yield} = \int d^4x R(T(x)) = \int dT \Phi(T) R(T) \tag{18.1}$$

$$\Phi(T) = \int d^4x \delta(T(x) - T) \tag{18.2}$$

where the integral is done over the four-volume of expanding excited matter, between the 3D surfaces of matter initiation and its freeze-out. The space-time dependence of the temperature $T(x)$ is determined, for example, from hydrodynamics or kinetic models. The profile $\Phi(T)$ depends on collision, such as on its centrality and initial state deformations. The rate function $R(T)$ is, on the other hand, *universal*; it depends on matter parameters, such as T, only (assuming matter is equilibrated, see the following).

All dilepton/photon calculations, from the 1970s to now, follow this setting. The key original observation is that the temperature profile $\Phi(T)$

is dominated by the latter stages of the collisions with small T. It falls as a power, roughly as

$$\Phi(T) \sim \left(\frac{1}{T}\right)^7$$

The rate, on the other hand, includes the Boltzmann factor

$$R(T) \sim \exp\left(-\frac{M}{T}\right)$$

and peaks at the *highest* T. The product of the two factors has a sharp maximum, roughly at $T_{\max} \approx M/7$. Therefore, the dilepton mass M relatively accurately defines *when* it was produced, serving as a kind of clock (Shuryak, 1978a, 1980).

Let me present an example from modern event-by-event hydrodynamics used by Paquet (2017) and shown in Fig. 18.1. Let us start with plot (b), which shows matter distribution in the transverse plane at an early time, $\tau_0 = 0.4$ fm, corresponding to the beginning of the hydrodynamical

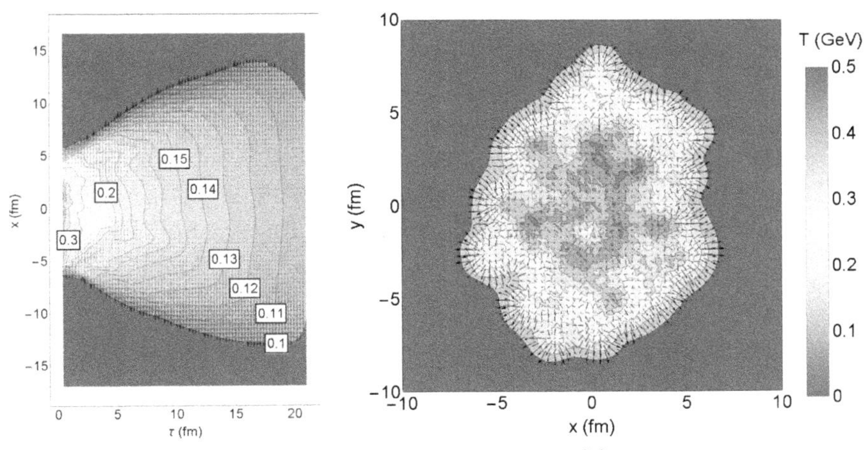

(a) (b)

Fig. 18.1. (a) Temperature profile (in GeV) at mid-rapidity from a hydrodynamic simulation for a central Au–Au collision with $\sqrt{s_{NN}} = 200$ GeV. The abscissa is the proper time τ, and the ordinate is a cut in the transverse plane. Contours are shown every 50 MeV for temperatures above 200 MeV and every 10 MeV below. The arrows show the relative size of the flow velocity $\vec{u}$ along the transverse axis. (b) Temperature and flow profile in the transverse plane at $\tau_0 = 0.4$ fm. The same color coding is used for both figures, but the arrows representing the flow velocity are scaled differently for clarity.

description. As one can see, at some locations, the temperature reaches $T \sim 500\,\mathrm{MeV} \approx 3T_c$ (dark red) and includes a number of "bumps," which we discussed in the chapter on sounds. One can see that the initial hot region occupies only a very small part of space-time.

As seen from plot (a), the critical temperature $T \approx 155$ MeV is reached through most of the fireball at time $\tau \sim 10$ fm/c, and from that time on, the matter is in hadronic phase. The last isotherm $T \approx 100$ MeV corresponds to the final kinetic freeze-out, from which the individual observed hadrons originate, which happens at $\tau \sim 20$ fm/c. So, it seems natural to divide the collision into two eras. For theoretical reasons, however, it is sometimes convenient to divide it into three periods: QGP, near-T_c and hadronic. Central heavy ion collisions at the RHIC show that those three take roughly the same duration of $\Delta\tau \approx 7$ fm/c.

So, in summary, the observation of the dileptons and photons provides a means of testing the time development of the collisions, telling us about matter *at certain temperature slices* of the process.

Finally, in this section, let us discuss one more reason *why* they have additional importance: the issue of *gluon/quark chemical equilibration.* We do know from the nucleon structure functions that the partons of the proton with the CM energy $E \sim 1$ GeV are 90% gluons. This composition is very different from that of equilibrium QGP, in which gluons only make up about 30% of the degrees of freedom. The question then is how quickly the gluons can be converted into quarks.

The production of photons and dileptons is tied to quarks only since gluons (and magnetic monopoles abundant near T_c) have no QED electric charge.

18.2 The Main Experimental Results on Photons/Dileptons

More details about the theory of penetrating probes are provided later. Now, let us turn to the history of the *dilepton/photon experiments.* Before we turn to specifics, one needs to emphasize that these experiments are generally much more difficult as compared to measurements of hadronic observables. In addition to a large background from electromagnetic decays after freeze-outs, which needs to be subtracted, they are related to smaller cross-sections and are therefore significantly limited by statistics. This puts severe demands on both the detector acceptance and the accelerator luminosity. Historically, the dilepton experiments have had the unfortunate tendency to be conducted "too late," when the particular experimental program is nearing its end.

At Berkeley's BEVALAC, home of the first relativistic heavy ion collisions, the DLS dilepton spectrometer was built and recorded data in the 1980s. Yet, its detector efficiencies were quite complicated, and they were studied and finalized years after its last run. A striking dilepton signal was found, but with large uncertainties. The Brookhaven fixed-target AGS program of the 1990s had no dilepton experiments.

The CERN SPS-based heavy ion program conducted subsequent experiments, which were able to identify properties of dileptons that originated from a hadronic fireball. Experimental studies of dilepton production in heavy ion collisions have been performed at CERN SPS in three experiments:

(i) CERES (NA45), which studied the low mass (M = 0–1 GeV) e^+e^- pairs;
(ii) HELIOS-3, which was focused on medium mass $\mu^+\mu^-$ (M = 1–2 GeV);
(iii) NA38/50/60 concentrated on higher mass $\mu^+\mu^-$. All three observed quite significant enhancement over "standard sources," ranging from factor 5 at CERES (in some kinematic regions) to about 3 at NA38/50 at M = 2–3 GeV.

The NA50 experiment evolved into its later incarnation, NA60. The question of whether the observed leptons do or do not come from charm quark decays was answered as follows: NA60 had new silicon-based tracking detector able to study charm production background directly. The issue of charm decay contamination of the lepton pairs has been resolved: (Contrary to general belief) most of the leptons were due to electromagnetic radiation from matter, not charm decays.

As already mentioned, the main difficulty in dilepton studies is the large combinatorial background from decays of secondary hadrons, in particular the so-called "Dalitz decay" of η, ω mesons. It took considerable time and effort to separate them. The dilepton production in the light system pp (or $p-Be$) has been measured, with their spectra well explained by a "cocktail" of known resonance decays. Only then were people able to proceed to heavy ion collisions of interest, which contained a fireball of hadronic matter, subtract the "cocktail" contribution, and get the radiation from matter.

The dimuon NA60 experiment, led by Professor Specht, set the gold standard in this field, and thus, it is fair to present some of them here. Let me start with Fig. 18.2 (right) adapted from Specht (2010), which shows the

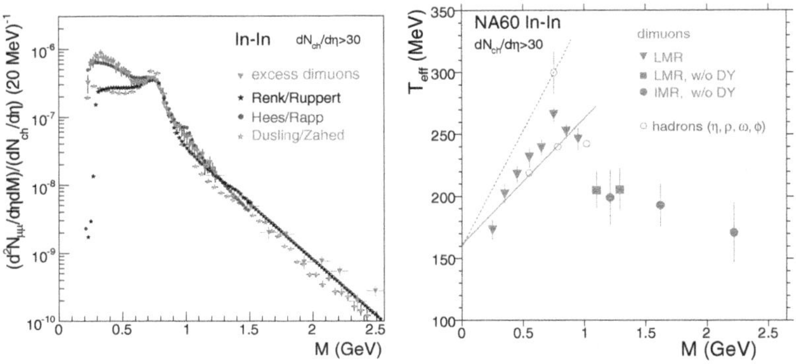

Fig. 18.2. Left panel: Acceptance-corrected invariant mass spectrum of the excess dimuons, integrated over p_T, compared with three different sets of thermal-model results in absolute terms. Right panel: Inverse slope parameter T_{eff} of the acceptance-corrected m_T spectra versus dimuon mass. Hadron results are shown for comparison.

effective slope of the p_T distribution as a function of the dilepton mass M. One can see two completely different trends, for light dileptons with $M < 1$ GeV and for the intermediate mass dileptons (IMD) with

$$M_\phi \approx 1\,\text{GeV} < M < M_\psi \approx 3\,\text{GeV}$$

The light pairs show the growing trend $T_{\text{eff}} \sim p_T$, typical for radial flows. The fact that its value is compatible with that of the hadrons (open circles) tells us that these light dileptons are produced *late*, when the radial flow is already developed.

The intermediate mass dileptons show, on the contrary, a mass-independent T_{eff}. The reason for that is that they are produced *early*, before the radial flow is developed. The corresponding T_{eff} is thus the initial temperature of the fireball, which is what we wanted to measure!

The left panel of Fig. 18.2 shows the invariant mass spectrum of the excess dimuons, integrated over p_T, compared with three different sets of thermal-model results. One can see a bump near $M \approx 0.8$ GeV, corresponding to the ρ mass, described by all theories, and an excess to the left of it. The Hees–Rapp model, which apparently describes it, is based on a self-consistent treatment of the resonances (Chanfray *et al.*, 1996) in which the ρ width in matter is greatly enhanced. Another plot from this paper, Fig. 18.3, shows the (approximate) in-matter spectral density of the ρ current in more detail. The red curve, called "vacuum rho," shows what the signal would be for unmodified resonance. The green Brown–Rho curve

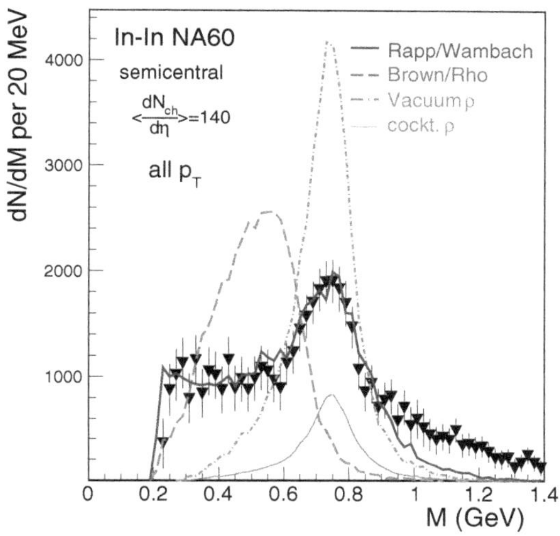

Fig. 18.3. Excess dimuons before acceptance correction, reflecting the in-medium ρ spectral function averaged over space-time and momenta. The theoretical predictions also shown are renormalized to the data for $M < 0.9$ GeV. Beyond 0.9 GeV, other physical processes take over.

is a result of the downward mass shift of ρ, and the blue Rapp–Wambach curve corresponds to ρ in-matter widening. Apparently, the last model does the best job here.

At the RHIC, the dilepton measurements were done by the dedicated PHENIX detector. Later, STAR also joined after electromagnetic calorimeters were added on top of the tracking TPC detector. The main results of both are enhanced (over the "cocktail") production of dileptons of small invariant masses, although for the most central collisions, their exact magnitude is still somewhat disputed between STAR and PHENIX.

(b) IMD are well measured, but the contribution from charm/bottom decays remains uncertain at RHIC. PHENIX hoped to solve this problem with a better vertex detector, while STAR expects to use $e - \mu$ correlations based on a new muon subsystem;
(c) direct photons have spectra consistent with standard rates and hydrodynamics in shape, but not in absolute magnitude.
(d) The dilepton production timing/origins were put into debate again by observations of an unexpectedly strong elliptic flow v_2 of the direct photons. Dilepton data from PHENIX with a "hadron-blind detector"

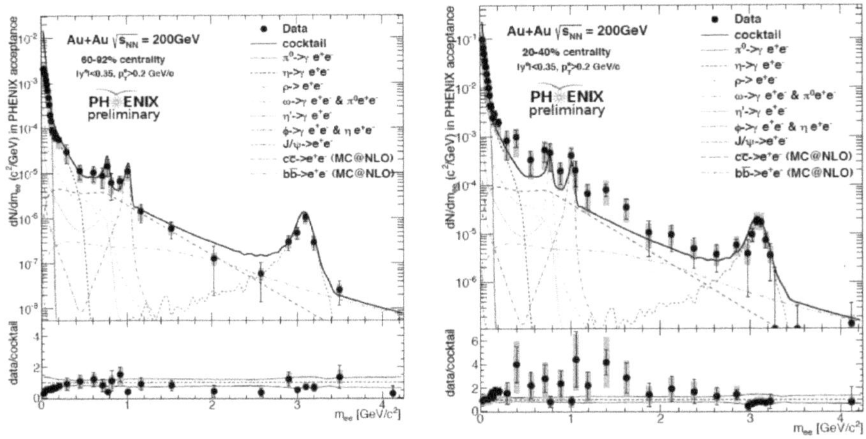

Fig. 18.4. Invariant mass spectrum of dielectron pairs in $Au + Au$ collisions at $\sqrt{s_{NN}} =$ 200 GeV in two centrality bins. The data points show statistical (vertical bars) and systematic (shades) uncertainties separately. The shaded band represents the systematic uncertainty of the cocktail.

is shown in Fig. 18.4 adapted from Tserruya (2013). The left panel shows the plot for peripheral collisions. The curves of different colors represent various processes contributing to the "cocktail" of late-time processes. Their sum — the thin black line — represents the data quite well. The right panel corresponds to a more central bin, 20–40% of the cross-section, and in this case, the data (points) are systematically above the "cocktail." Again, a clear excess is present, both below and above the ρ–ω peak.

At the LHC, the dilepton measurements are (to my knowledge) not yet fully developed. At the time of this writing, we have the ALICE results (De Falco, 2015), which are still dominated by the hadronic cocktail and do not indicate a statistically significant signal from the matter radiation.

Photon measurements by ALICE were performed, and in particular, they confirmed the strong (and puzzling) value of the direct photon elliptic flow, consistent with that originally observed by PHENIX. Figure 18.11 shows a comparison of theoretical models (curves): The poor agreement is known as the "direct photon puzzle."

18.3 The Basic Sources of Dileptons/Photons and Their Rates

The sources of dileptons are split into the following categories:

(i) instantaneous $\bar{q}q$ annihilation, known as the Drell–Yan partonic process;
(ii) $\bar{q}q$ annihilation at the pre-equilibrium stage, after the nuclei pass each other;
(iii) $\bar{q}q$ annihilation in the equilibrated sQGP;
(iv) meson–meson annihilation at the (kinetically but not chemically) equilibrated hadronic stage;
(v) the so-called "cocktail" contribution, consisting of leptonic decays of unstable secondaries. Electromagnetic and electroweak decays all occur long *after* freeze-out, so this component can be calculated from spectra and statistically subtracted. We will not discuss this component any further.

The corresponding windows in the dilepton mass are: $M > 4$ GeV for (i), $1 < M < 3$ GeV for (iii) and $M < 1$ GeV for (iv). So, the early-stage dileptons mostly fall into the 3–4 GeV window.[2]

The simplest (I would call them the zeroth approximation) models for dilepton production rate are based on two well-tested processes:

- $\pi\pi$ annihilation in the hadronic phase (small T)
- $\bar{q}q$ annihilation in QGP (large T). This process is similar to the well-known Drell–Yan process, only with the weight coming from the thermal ensemble rather than from the nucleon structure functions.

These two basic processes can both be accounted for by the "standard rate" formula (Shuryak, 1978a):

$$\frac{dR}{d^4q} = \frac{\alpha^2}{48\pi^4} F e^{-\left(\frac{q_0}{T}\right)} \tag{18.3}$$

where the rate R is counted per three-volume per unit time, and q is the four-momentum of the virtual photon ($q^2 = M^2_{e+e-} = M^2$). The remaining factor F for $\pi\pi$ annihilation is the usual pion form factor, for example,

[2]This window of masses also contains prominent peaks due to $J/\psi, \psi', \psi''$ decays. Those, according to the definition, belong to the "cocktail" and therefore should be subtracted.

written in the standard vector-dominance form. For the QGP case, the factor F is simply a constant, up to small corrections, containing the sum over the squares of the electric charges of all quarks:

$$F_Q \stackrel{\text{def}}{=} 12 \sum_q e_q^2 \left(1 + \frac{2m_q^2}{M^2}\right)\left(1 - \frac{4m_q^2}{M^2}\right)^{\frac{1}{2}}, \quad \text{(QGP)} \tag{18.4}$$

$$F_H \stackrel{\text{def}}{=} \frac{m_\rho^4}{[(m_\rho^2 - M^2)^2 + m_\rho^2 \Gamma_\rho^2]}, \quad \text{(Hadronic)} \tag{18.5}$$

18.4 Perturbative Resummations of the Dilepton/Photon Rates

The need for more careful resummation can be explained as follows. In our schematic treatment of the perturbative cross-sections above, one can see that if one does the integral over momentum transfer as

$$\int \frac{dt}{(-t + M_g^2)^2} \approx \frac{1}{M_g^2}$$

and use the perturbative gluon effective mass $M_g^2 \sim g^2T^2$, one finds that a power of the small parameter $\alpha_s = g^2/4\pi$ is canceled out. It shows that the normal counting of the coupling powers of Feynman diagrams gets upset by "soft" scattering with momentum transfer $t \sim g^2T^2$.

It is consistently done in a number of papers. Here, we follow Ghiglieri and Moore (2014) for low-mass $M \sim gT$ dileptons: References to previous literature on the next-to-leading order (NLO) results can be found in this paper. The main settings for all of them are as follows. The square of the amplitude of the lowest order is expressed via the (imaginary part of) the quark loop diagram, see the upper part of Fig. 18.5. The NLO quark loop diagrams are those shown in the lower part of this figure, which have an

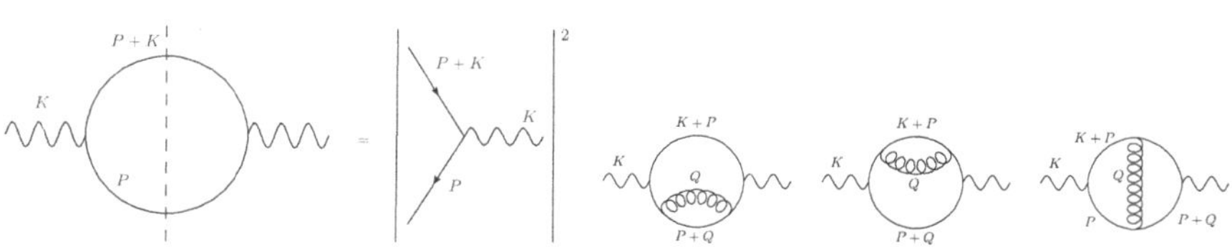

Fig. 18.5. Leading order diagrams (upper) and next-to-leading ones (lower) for dilepton production.

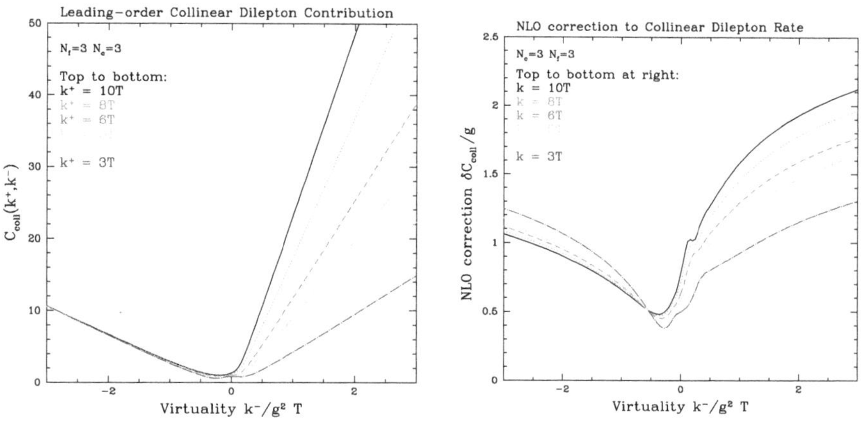

Fig. 18.6. The leading order result (left) and the next order correction (right): g times the correction should be added to the leading order on the left.

additional gluon. Cutting this diagram into different ways, one obtains all next-order processes, with some including the extra gluon produced and some having gluons inside virtual corrections.

The results are given in Fig. 18.6. The result depends on the virtuality. The LO result (left) is in this quantity $\sim O(20\text{–}30)$, while the (coupling times) correction is $\sim O(1)$. So, the perturbative conclusion is that the next order effect always *increases* the rate, but not by much.

Unfortunately, one needs to discard these nice conclusions because the assumptions of the perturbative setting used,

$$\frac{M^2}{T^2} \sim g^2 \ll 1$$

are not in fact valid in experimental settings. As the lattice data we discussed before show, their left hand side is in fact significantly larger than one, unless at extremely high T.

18.5 Dileptons from the Hadronic Phase and Nuclear Matter

So far, we have mentioned only one type of process, the $\pi^+\pi^- \to e^+e^-$, proceeding via the vector $\rho, \omega, \phi, J/\psi\ldots$ resonances. The main issue here is whether one can observe the modification of the spectral shapes of these mesons due to expected in-matter mass shifts caused by

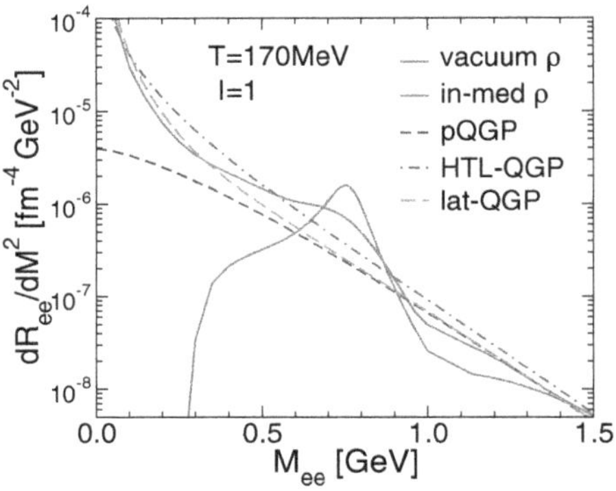

Fig. 18.7. Comparison of dilepton production rates calculated in different approximations (see text).

vacuum parameter changes, in particular the modification of the chiral condensate.

There exists a very important distinction between these cases. ρ is wide enough, i.e. it has a short lifetime of ~1 fm/c, to decay inside the fireball. As we already discussed, in QGP conditions, it seems to be completely wiped out.

J/ψ and, in most cases, ω, ϕ live long enough to decay outside of matter and thus are included in the "cocktail," which is to be subtracted out from the data, as not belonging to the discussion of the excited matter.

Yet, if kinematics is correctly selected, ϕ can decay in either the fireball or even inside normal nuclear matter in pA collisions. This last case has been studied at the 12 GeV KEK experiment E325 (Muto *et al.*, 2007), which reported a significant modification of the spectral shape. These results, shown in Fig. 18.8, compare invariant mass dilepton spectra in pC and pCu collisions for different velocities of the pair. The upper-right plot demonstrates a clear left shoulder going well beyond the usual spectral shape (shown by the histogram). A new dilepton experiment is under construction at KEK to extend these measurements.

Theoretical discussions of ϕ mass modification included two parts. One is the modification of ϕ width, induced by a modification of the K meson mass (Lissauer and Shuryak, 1991). Another is the generic reduction in the masses of mesons due to a reduction in the chiral condensate value (Brown and Rho, 1991).

The in-matter modifications of those mesons are due to resummation of certain series of diagrams. In Fig. 18.9, we show some examples of the

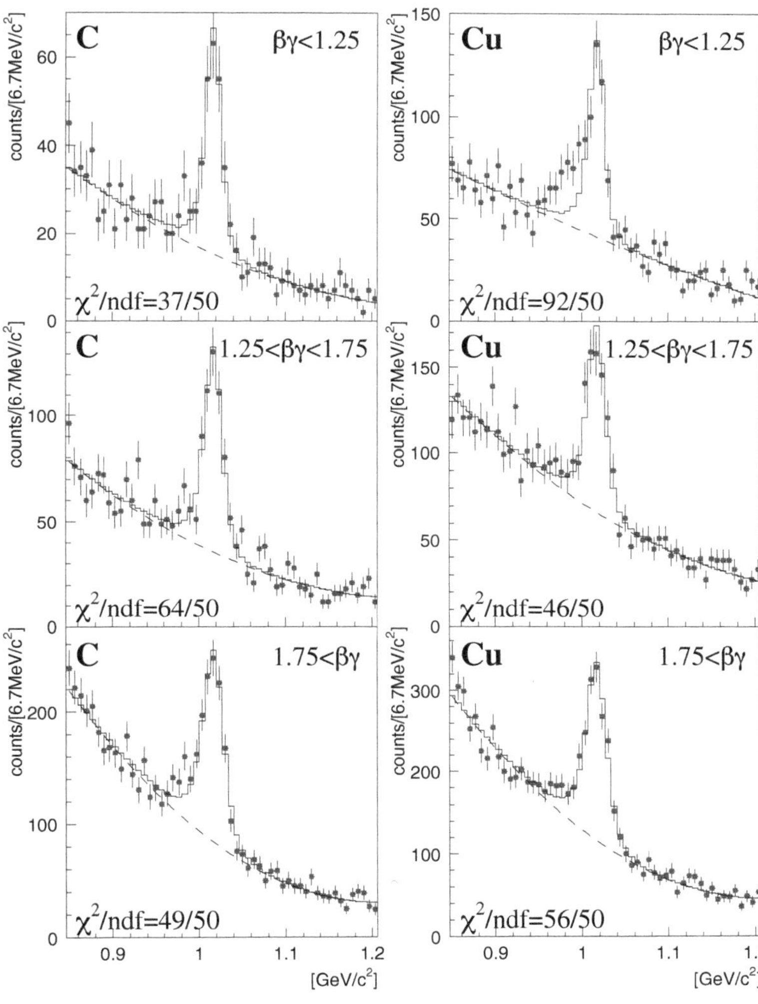

Fig. 18.8. Spectral shape of the ϕ peak in pCu collisions, as a function of the velocity-times-gamma factor $\beta\gamma$. The solid lines show the fit result assuming normal shape and background.

relevant processes. Figure 18.9(a) shows a process in which the vector ρ is not on a shell but is a virtual one, created by a collision of other particles. Figure 18.9(b) shows the radiative de-excitation of baryonic resonances with the emission of dileptons: This process is important, as emphasized by (Chanfray *et al.*, 1996). Figure 18.9(c) is a critical process pointed out by

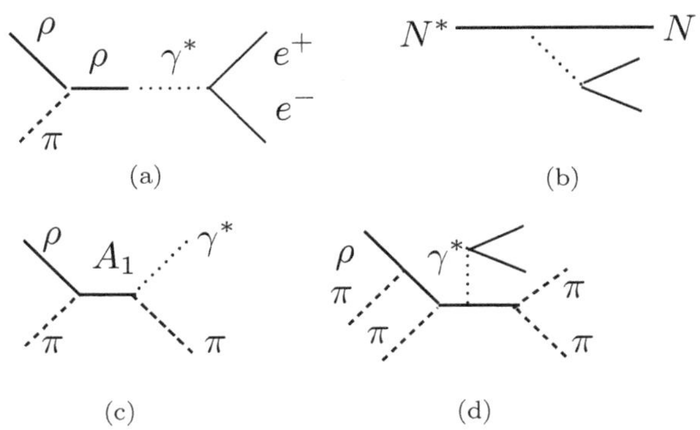

Fig. 18.9. Diagrams for photon production via mesons.

(Xiong *et al.*, 1992): The axial $a_1(1230)$ resonance, while quite heavy, has a very large coupling to the pion-photon channel.

The processes (a) and (c) are binary reactions, and (b) is a resonance decay, so their rates/couplings/form factors can be deduced directly from vacuum experiments. Out of these couplings/form factors, one can construct more complicated diagrams, such as that shown in Fig. 18.9(d), in which $\rho\pi\pi \to e^+e^-\pi\pi$.

It is of course desirable to get a consistent set of diagrams, deriving them systematically from certain effective Lagrangians with appropriate symmetries. Such diagrams are effective chiral Lagrangians describing the interactions of soft pions (for a review, see, for example, Gasser and Leutwyler, 1984), and Lagrangians for vector mesons possessing the so-called "hidden gauge symmetry" (for a review, see Harada and Sasaki, 2006). Both of them were later naturally rederived from holographic models, in which vector and pseudoscalar mesons appear as small oscillations of branes. How they systematically follow from the Born–Infeld action of the brane has been reviewed, for example, by Sakai and Sugimoto (2005). Furthermore, holography lets one calculate the correlators of the electromagnetic currents directly, without cutting for specific processes.

Let us now return to the issue of *in-matter modification of mesons*. One can relate the modification of mesons (e.g. of ρ) to the $\pi\rho$ and $N\rho$ forward scattering amplitudes, or momentum-dependent optical "potentials," which predict a relatively modest downward and broadening shift in m_ρ.

When matter is no longer dilute and modifications are no longer small, one should perform consistent resummations and try to minimize the free energy. This leads to a set of *coupled gap equations*. As a good example of this kind of work, let me mention the Rapp–Wambach approach (Chanfray *et al.*, 1996), in which a very strong broadening of the ρ meson was predicted based on the properties of ρ–N interaction and N^* resonances. One can also understand this broadening as due to mixing between ρ with excitations of the lowest baryon resonances, the most important of which is[3] $N^*(1520)N^{-1}$. We had already shown the predictions of the corresponding models with the NA60 data.

Let us now see how these models compare to each other. In Fig. 18.7, adapter from Rapp (2013), one can see how dilepton production rates differ if one considers in-vacuum rho versus the state-of-the-art many-body calculation, which shows very little trace of the ρ peak. Presumably, this fact is a manifestation of the "hadron–parton" duality.

Whatever the reasons may be, one finds that such theoretical evaluations of the rates of dilepton production, both in QGP and dense hadronic matter close to T_c, lead to values that are about the same as those of the zeroth-approximation "partonic" rate. If so, this rate can be used as a "standard candle" to calibrate the temperature profile of the exploding matter.

The general properties of the two-current correlators following from OPE can be written as a set of Weinberg-type sum rules (Kapusta and Shuryak, 1994), which relate certain moments of the difference between vector and axial spectral densities to particular chirally-odd order parameters, such as $\langle \bar{q}q \rangle$ and f_π. As all of them should vanish at $T = T_c$, these sum rules demand that both spectral densities are the same at $T > T_c$.

The so-called intermediate mass (1.5–3 GeV) dileptons (IMDs) are a region where a significant enhancement of dilepton production has been observed, by the NA38/50 experiment. One possible interpretation (put forward by the experimentalists themselves) is the enhancement of open charm production. In order to explain the data, it should, however, be quite large, up to a factor of 3 for the most central collisions. It has been, however, shown (Rapp and Shuryak, 2000) that there is no need for this anomalous charm production because the thermal radiation of dileptons can naturally describe it, both for the invariant mass and the transverse momentum distributions, see Fig. 18.10.

At low-energy high-baryon-density matter, it is expected to find the critical point, followed by the first-order transition. Recent studies of how

---

[3]This notation means the baryonic resonance *minus* a nucleon.

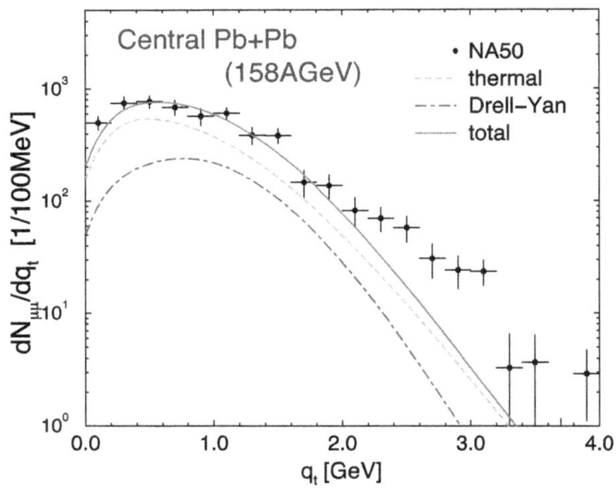

Fig. 18.10. A comparison between the NA50 IMD data and the thermal QGP calculation.

dilepton production can be affected by such scenarios can be found in Seck *et al.* (2022).

18.6 The Direct Photons

We discussed above that the dileptons can be produced by both a partonic prompt Drell–Yan process and in-matter thermal annihilation. The production of direct (i.e. not from hadronic decays) photons can also be made by a number of direct and thermal sources. The partonic production of photons is well known and studied in pp collisions, so we will not describe those in detail.

The rate of photon production in matter, per unit time and volume, of a real photon of momentum p is

$$\frac{EdN}{d\vec{p}dtd\vec{x}} = -\frac{1}{(2\pi)^3}\, n_B(E)\, \mathrm{Im}\, \Pi^{R}{}_{\mu}{}^{\mu}(E,\vec{p}) \,, \tag{18.6}$$

where $\Pi^{R}{}_{\mu}{}^{\mu}(E,\vec{p})$ is the retarded photon polarization tensor. The prefactor $n_B(E)$ provides the expected exponential damping $\exp(-E/T)$ when $E \gg T$.

In the original paper suggesting thermal photons from QGP (Shuryak, 1978a), only two lowest-order reactions have been considered: (i) the "Compton" scattering $gq \to \gamma q$ and (ii) annihilation $\bar{q}q \to \gamma g$.

The corresponding rates for hard photons are

$$\mathrm{Im}\,\Pi^{R}(E,\vec{p}) \sim e^2 g^2 T^2 \left(\ln\left(\frac{ET}{m_q^2}\right) + C\right) \tag{18.7}$$

However, it is clear that in general, the photon production, even at large photon energy, $E \gg T$, is more difficult to evaluate because of possible *bremsstrahlung*-like phenomena. In particular, a process of the next order in g, such as $qq \to qq\gamma$, can be as large as the leading one of the momentum transfer is soft $q \sim gT$. So, a consistent resummation of hard thermal loops is needed. In brief, the next-order diagrams with *bremsstrahlung* lead to rates of the order

$$\mathrm{Im}\,\Pi^{R}(E,\vec{p})\Big|_{\mathrm{brems}} \sim e^2\; g^2T^2 \mathrm{Im}\,\Pi^{R}(E,\vec{p})\Big|_{\mathrm{annil}} \sim e^2\; g^2 TE \tag{18.8}$$

The $q\bar{q}$ annihilation with scattering (a "3 to 2" process crossed from *bremsstrahlung*) actually grows with the energy E and dominates when $E/T \gg 1$. Phenomenological applications of these results have been carried out, and the two-loop processes have been included in hydrodynamic evolution codes to predict the rate of real photon production. It is found that the two-loop processes (especially the annihilation with scattering) lead to some numerical increase in the rate.

The issue of photon production from hadronic gas has also been rather extensively discussed in the literature. Nadeau *et al.* (1992) included all diagrams with ρ, π mesons and found that it is about the same as the lowest-order QGP. Xiong, Brown and myself (Xiong *et al.*, 1992) have argued that one process,

$$\pi\rho \to a_1 \to \gamma^*\pi$$

contributes about a factor of three more to the rate. As a result, one again sees an approximate "duality," like for dileptons, namely that *the photon production rates in hadronic and QGP phases at close T seem to be very similar.* This fact makes it more difficult to claim "QGP radiation" as a signal but makes theoretical predictions less uncertain.

Early WA80/98 photon data and direct + thermal photon calculations were generally in agreement. However, despite strong efforts, large background from π^0 decays only allowed the upper limits in the region crucial for thermal QGP production. Data from the PHENIX collaboration at RHIC are shown in Fig. 18.11. The left plot shows the $p_\perp$ spectrum: Its shape corresponds to the calculated thermal rates, but its magnitude is higher by about a factor of 3. Perhaps, not all diagrams/sources of photon

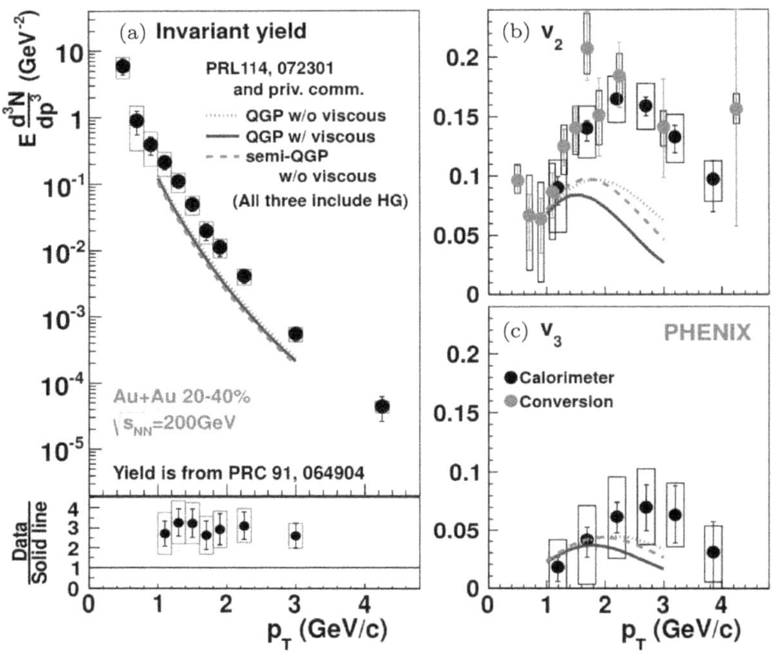

Fig. 18.11. Illustrations of the "direct photon puzzle" (from Drees, PHENIX presentation at QM2015). The yield (a), v_2 (b) and v_3 of the direct photons. Points are data, and curves are from theory based on the hydrodynamical model, the reference indicated on the figure.

production are included. The two plots on the right show what became known as the "photon puzzle": Photons seem to participate in elliptic and triangular flows even stronger than those predicted by hydrodynamical models. We will discuss these predictions in the following section: Here, I just say that there is no commonly accepted explanation. Since at $p_\perp = 2 - 4$ GeV, these photons come from the "outer edge" of the fireball, it is a part of a wider issue of "nobody's land" between hydro-dominated and jet-dominated regions, in which there is no commonly accepted theory for hadrons spectra as well.

There is yet one way to beat the π^0 decay background: the two-photon interferometry; see for example Srivastava and Kapusta (1994). Indeed, photons created from the decay cannot interfere, as they appear far from each other, and only photons directly produced in the fireball can. Unfortunately, this interesting idea is very hard to put into actual measurements, as the direct photon rate is orders of magnitude below the background.

18.7 More Complicated Sources for Photons/Dileptons: Multi-Gluon or Phonon + Magnetic Field Processes

This scenario has been recently challenged: Higher-order processes with *virtual quark loops* can produce electromagnetic effects as well, even without on-shell quarks. First, contrary to general expectations, the quark loop effect in GLASMA has been suggested to be significant (Chiu *et al.*, 2013). The enhancement is due to multi-gluon, virtual quark loop and dilepton processes, such as

$$ggg \to (\text{quark loop}) \to \gamma^* \to l_+ + l_-. \tag{18.9}$$

The magnitude of the correction to the production rates due to these processes still needs to be quantified.

An explicit calculation of the rate of the two-gluon to two-photon transition $gg \to \gamma\gamma$ was done by Basar *et al.* (2014). One of the photons is assumed to be the ambient magnetic field at the time of collision, while both gluons are combined into a colorless stress tensor:

$$T^\mu_\mu + \vec{B} \to (\text{quark loop}) \to \gamma^*(= \text{dileptons}) \tag{18.10}$$

The terminology introduced in this paper is as follows. The process in which gluons appear as the average matter stress tensor $\langle T_{\mu\nu}\rangle$, producing photons (real or virtual) due to a time-dependent magnetic field, is called *magneto-thermo-luminescence* (MTL). The average stress tensor is nearly constant over the whole fireball; therefore, its Fourier decomposition has very small momenta, $p \sim 1/R$.

Individual events, however, also possess fluctuations of the matter stress tensor $\delta T_{\mu\nu}$, with a chaotic spatial distribution and thus with the Fourier transforms with non-negligible momenta. Since those are referred to as "sounds," their interaction with the ambient magnetic field is called *magneto-sono-luminescence* (MSL). If observed, the MSL process tests both the amplitudes of the short-wavelength sounds and also the magnitude of the magnetic field. We already discussed many uses of sounds above; let us here only comment on the magnetic field. It is easy to evaluate its early values, resulting from Maxwell equations and the currents due to the spectators in peripheral collisions. Since sQGP is believed to be a good conductor, these fields are expected to create currents that capture a fraction of the field inside the plasma (Tuchin, 2013), perhaps lasting for many fm/c. Magnetic field evolution is important to know for other applications as well, e.g. chiral magnetic effects.

So far, luminosity and acceptance limitations have led experiments to focus on the most luminous central collisions. In those, however, the ambient magnetic field is absent. Now, it is perhaps time to look at dileptons in semi-peripheral collisions as well. To differentiate the effect of the ambient magnetic field from others, RHIC considers runs of isotopes with similar A but different Z and N values.

18.8 Elliptic Flows of Dileptons and Photons

We have argued repeatedly that the elliptic (and higher harmonics of) flow is due to hydrodynamical expansion of the fireball, mostly along the direction of the largest pressure gradient. The observed hadronic data are well explained by hydrodynamics.

The question to be discussed here is whether the same is true for the elliptic flows of *dileptons and photons*, which is supposed to test the *time dependence* of elliptic flow development.

In Fig. 18.12(a) adapted from Paquet (2017), we show the time evolution of the elliptic photon flow, and in Fig. 18.12(b), the corresponding evolution of the radial flow. One can deduce from these calculations that the elliptic photon flow rapidly grows and develops, mainly at intermediate times of $\tau = [1 \ldots 5]$ fm, to its surprisingly large values. Its p_T dependence is linear, as for hadrons, again in agreement with hydrodynamics.

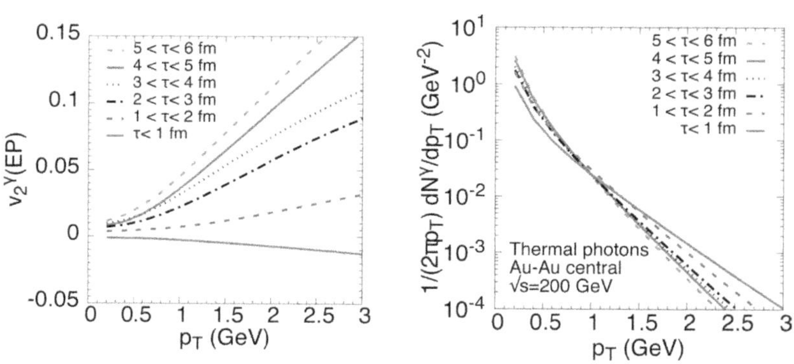

Fig. 18.12. (a) Thermal photon $v_2(EP)$ and (b) thermal photon spectra at different times, for the central event illustrated in Fig. 26.4.

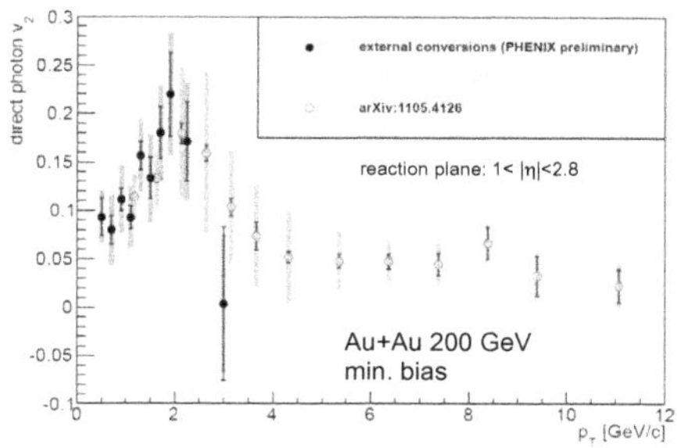

Fig. 18.13. Direct photon $v_2(p_T)$ published result (open circles) in minimum bias $Au+Au$ collisions at $\sqrt{s_{NN}} = 200$ GeV, compared to the result (solid points) obtained in an independent analysis based on external conversions. The vertical error bars represent the statistical uncertainties, and the shaded boxes indicate the systematic uncertainties.

The spectra, shown in Fig. 18.12(b), show, however, the slopes going in the opposite direction, indicating that the *decreasing* temperature is the most important effect. The radial flow builds up much more gradually, reaching the observed values at times $\tau \sim 15$ fm/c (not shown in the plot).

Now we proceed to the experimental data. PHENIX has observed a surprisingly large photon v_2 using two different experimental methods, see Fig. 18.13.

Summary and connections: The electromagnetic processes that produce photons/dileptons, are of special interest because they are "penetrating probes," which occur over the entire lifespan of a fireball. Their spectra, therefore, give us a thermal history of the explosion. Among the most studied issues is the "melting of the ρ meson." The theory of the rates is generally consistent with observations of the photon/dilepton spectra, except for the observed elliptic flow parameter v_2, which is *larger* than predicted. Perhaps, this indicates that more weight should be given to electromagnetic processes at later stages of collisions.

Chapter 19

HEAVY QUARKS AND "JETS" INTERACTING WITH MATTER

Heavy c, b quarks are produced at the very beginning of the collision, then propagate through hot/dense matter and finally appear in the detectors in the form of hadrons. Those can either have "heavy flavor" (e.g. $D = c\bar{q}$, $B = b\bar{q}$ mesons and similar baryons) or "hidden flavor" (such as $J/\psi = \bar{c}c$, $\Upsilon = \bar{b}b$). We already discussed charm fugacity and charmed hadron production in the chapter on chemical freeze-out.

"Jets" are remnants of gluons or quarks produced with large transverse momenta of $p_\perp \sim 20$–$200\,\mathrm{GeV}$, significantly exceeding matter temperatures. In a vacuum, these colored objects flying away from the fireball produce "*bremstrahlung*-like" cones of other gluons. Behind each are trailing, confining flux tubes: Their breaking into mesons leads to a high population of secondaries inside a certain cone around the direction of the jet. Its angular radius is relatively small, typically $r_{\mathrm{jet}} = 0.1$–0.2 radians. This feature is used in jet reconstruction algorithms.

Propagating through matter, they get modified or "quenched." Rescattering leads to a certain growth of momenta orthogonal to the jet, with a corresponding decrease in the longitudinal component.

19.1 Do Heavy Quarks Flow?

Before we turn to theory, let us assess the empirical information on c, b quark production and their subsequent interaction with ambient matter.

The simplest approach here would be a direct comparison of spectra (of certain heavy-flavored hadrons) in pp and AA collisions. In the former case, there is no matter surrounding the locus of the hard production process; in the latter, there is such matter.

The pp data tell us about the primary momentum spectrum, with which heavy quarks are born. Those can vary from semi-relativistic to very large ones. Going through matter, they suffer certain friction and slow down. On the other hand, matter itself is not at rest: It is developing a certain radial flow velocity. So, slow heavy quarks may even be subject to "acceleration by friction." Ultimately, if the time until their decoupling is sufficiently long, one may expect heavy quarks to "move with the flows," like gluons and light quarks do.

One further expects that the answer to this question will depend on the rescattering cross-sections of particular hadrons in which the heavy quarks will ultimately appear. The smallest sizes (and thus cross-sections) are for the lowest quarkonia ($J/\psi, \Upsilon$), while heavy-light hadrons ($D, \Lambda_c, B, \Lambda_b$) are larger and interact stronger.

Let us begin with radial flows and then proceed to elliptic flows. In Fig. 19.1 adapted from Andronic *et al.* (2021b), we show a comparison of the ALICE p_t spectrum of D^0 (points) with hydrodynamical predictions (shadowed lines). The experimental spectrum clearly exceeds predictions for large p_t due to the absence of charmed jets in the hydro code. Yet, when compared to spectra in pp collisions, one finds that this very spectrum is

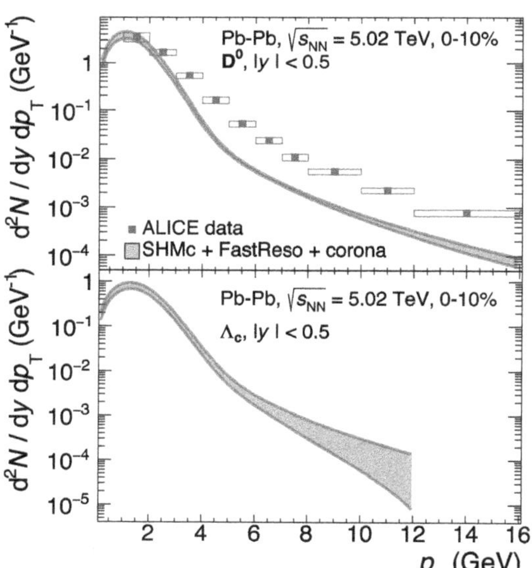

Fig. 19.1. p_t spectra of D^0, Λ_c in hydrodynamics versus the ALICE data.

quite significantly suppressed, by a factor of $R_{AA} \approx 0.2$ (not shown). So, we have evidence for "charm stopping," which, however, is not as strong as to force charm quarks to "move with the flow."

Elliptic flows are even more instructive. Indeed, in elementary *pp* collisions, elliptic flow is absent. If there is no matter, all directions are the same, and there could not be any angular correlation with other secondaries. If a fireball is present and has a certain ellipticity in the transverse plane, then the path length would be direction-dependent and friction would also generate a certain ellipticity in the heavy hadrons.

An answer to the question in the title of this section can be readily obtained from Fig. 19.2 (from CMS (Sirunyan *et al.*, 2021), which also includes the charmonium data from ALICE). Indeed, while for J/ψ, elliptic flow is well established with impressively small error bars, the elliptic flow coefficient v_2 for Upsilons is consistent with zero. So, the conclusion is that heavy c and b quarks do *not* behave in the same manner: b quarks largely ignore matter, while c quarks have an elliptic flow coefficient of v_2 close to that of light quarks!

If so, why is that the case? One option for an answer can be due to the heavy mass of b quarks: It is more difficult for QGP to drag along Upsilon (whose mass is about 10 GeV) as compared to J/ψ, which is three times lighter.

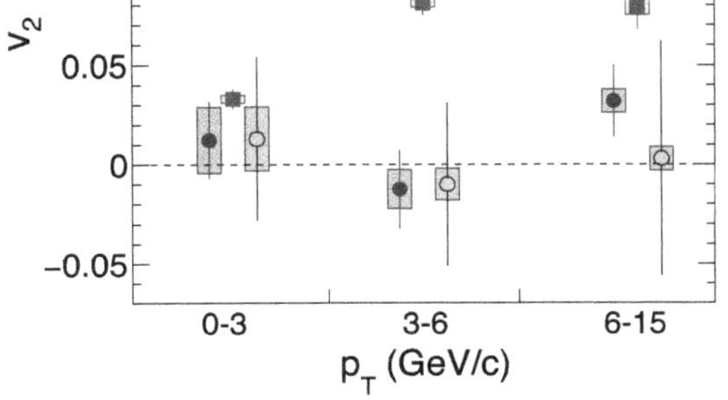

Fig. 19.2. Comparison of the elliptic flow coefficients for Upsilon and J/ψ mesons in $p_\perp$ bins indicated at the lower part of the plot.

Another important factor can be the different production histories of different quarkonia. Most of J/ψ is believed (at the LHC) to be produced from recombination at chemical freeze-out, while due to the smaller density of b quarks, Υ can only be produced as such in hard processes. In other words, it may be much easier to drag along a *single* c quark into the flow, than a $\bar{b}b$ bound pair Υ.

Finally, one can point out that c and b quarks — produced via hard partonic processes — are distributed in the transverse plane, unlike the light quarks/gluons. They are produced closer to the fireball center and therefore have fewer chances to be picked up by flow.

19.2 Heavy Quark Diffusion

After our brief introduction of the main experimental facts, let us discuss the theory. General considerations show that a heavy quark with a mass of $M \gg T$ and a thermal momentum of $\sim \sqrt{MT}$ cannot change much by a *single* scattering with the light quarks or gluons with momenta $\sim T$. In order to quench (stop) them, multiple interactions with matter are needed. Standard theoretical tools describing them include the Langevin or Focker–Planck equations. The former is a *stochastic* equation:

$$\frac{d\vec{p}}{dt} = -\eta\vec{p} + \vec{\xi}(t), \quad \langle\vec{\xi}(t)\vec{\xi}(t')\rangle = \kappa\delta(t-t') \tag{19.1}$$

with the kinetic coefficient η, and the second constant κ must be related to it via the fluctuation-dissipation theorem (the Einstein relation). Here are two useful forms of this relation:

$$\eta \approx \frac{\kappa}{2MT}, \quad D_s \approx \frac{2T^2}{\kappa} \tag{19.2}$$

including the *spatial diffusing constant* D_s.

The perturbative evaluation of η from lowest-order scatterings gives way too small a result. Higher-order corrections seem to be larger than those of the leading order; however, no convergence of the series is at hand. Therefore, we will not go into this theory here.

Nonperturbative evaluations begin to be available when using lattice, in particular. Straightforward current–current correlators can be measured, but in Euclidean time. Transforming those to Minkowskian time, with the aim of identifying the *transport peak* at small momenta, is a very hard task. Instead, one can focus on the force–force correlator and measure κ instead. In Fig. 19.3, we show a compilation of κ/T^3 versus T/T_c in gluonic plasma.

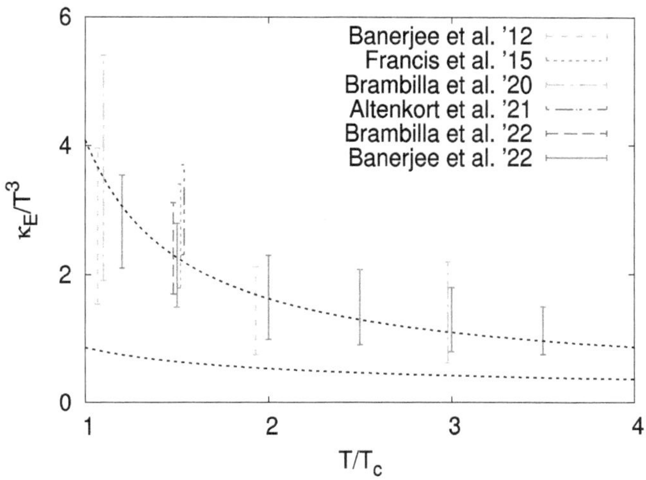

Fig. 19.3. Lattice evaluation of κ/T^3 temperature dependence.

Dynamical quarks are expected to increase it substantially, but there are no data yet (to my knowledge). Similar to other kinetic coefficients, such as viscosity η/s and jet quenching parameter $\hat{q}/T^3$, there are indications of a peak at T_c.

Let us now discuss quarkonia dynamics in more detail. It includes:

(i) prompt creation at the initial stage, with subsequent excitation and dissolution;
(ii) diffusive propagation in the flowing medium;
(iii) charmonium recombination, from independent heavy quarks at a bound state and ultimately to a thermally equilibrated charmonium.[1]

Since the initial production of $\bar{Q}Q$ is local, the bound states are produced according to the magnitude of their wave function at the origin $|\psi_n(0)^2|$, with the ground state being dominant. While the probability of producing bound states is small $\sim 10^{-2}$, historically, the main attention has been focused on their "successive melting" (Matsui and Satz, 1986) in QGP. Even earlier, I calculated ionization via the gluon absorption process in

[1]We have discussed the statistical theory of charmed hadron yields in the pertinent chapter on chemical equilibration.

Shuryak (1980). All these early works assumed that quarkonia, once broken by matter, would never return.

However, as noted by Young and Shuryak (2009), in strongly coupled regimes, diffusion is suppressed; therefore, sQGP would instead help preserve the lowest charmonium states. Indeed, if Q and $\bar{Q}$ cannot move away from each other, at the end, those would get projected back to the lowest S-states with the largest wave function at the origin. Yes, perhaps there are no identifiable charmonium states during the fireball lifetime, so what? We only see them at the end.

An important role of charmonium recombination has been pointed out by Braun-Munzinger and Stachel (2000) and studied in cascade-like models, for example by Grandchamp and Rapp (2002). The fundamental question here is, of course, whether the interaction with matter is weak or strong. A "theoretical quarkonium" with a parametrically large mass of $M_Q \gg \Lambda_{\rm QCD}, 2\pi T$ has size so small that all matter quasiparticles are outside it, and smallness is ensured by the multipole expansion. If so, and if the collisional width of quarkonium states remains small compared to their spacing, the basis of bound states retains its importance. If the states overlap significantly, they lose their physical importance.

Since heavy quarks move and change momenta relatively slowly, at the classical level, the problem is also amenable to the Langevin or Focker–Planck equation, e.g.

$$\frac{\partial P}{\partial t} = \frac{\partial}{\partial t} D\left(\frac{\partial P}{\partial \vec{r}} + \frac{P}{T}\frac{\partial V_{Q\bar{Q}}}{\partial \vec{r}}\right) \tag{19.3}$$

For a single heavy quark, this equation has only the first term in the right hand side. However, for quarkonium, one has to add a new second term with the force from $\bar{Q}Q$ interaction $V_{Q\bar{Q}}(r)$. The dissociation of quarkonia is classically described by diffusion *against the binding force.*

One more modification is the addition of certain "quantum effective potential" $\sim\hbar^2/Mr^2$, which prevents the classical "falling into small r" phenomenon. While (i) describes a flux from small-size to larger-size states, eventually to "full ionization," (ii) describes the flux directed downward in energy toward the chemical equilibrium.

The equilibration is seen by tending at large times to the *thermal fixed point*

$$P \sim \exp(-V_{Q\bar{Q}}/T),$$

at which the right-hand side vanishes. This happens if one starts at small distances, as in (i), or large distances, as in recombination setting (ii).

Also, don't forget that the equilibrium depends on charm fugacity (chemical potential), as the heavy quarks are not in chemical equilibrium.

Let me comment on a curious (but so far not really used) idea of mine, that of "classical stationary states." Indeed, the Focker–Planck equation has solutions in which the left-hand side is zero and the last bracket on the r.h.s. is *nonzero* but time-independent. Its value can be positive or negative, describing states with stationary flows. Such solutions are well known in the theory of turbulence, and some of those can also turn out to be "attractors."

At the full quantum level, there are approaches based on Schwinger–Keldysh double-time formalism or the so-called "quantum open systems" (see, for example, a review by Rothkopf, 2020) based on the *Lindblad equation* for the density matrix:

$$\frac{d}{dt}\rho_{Q\bar{Q}} = -i\left[H_{Q\bar{Q}}\rho_{Q\bar{Q}}\right] + \sum_k \gamma_k (L_k \rho_{Q\bar{Q}} L_k^+ + \rho_{Q\bar{Q}} L_k^+ L_k/2 - L_k^+ L_k \rho_{Q\bar{Q}}/2) \tag{19.4}$$

which preserves the hermiticity of $\rho_{Q\bar{Q}}$, the probability interpretation $\mathrm{Tr}[\rho_{Q\bar{Q}}] = 1$ and positivity $\langle n|\rho_{Q\bar{Q}}|n\rangle > 0$ for all n during the time evolution. The Lindblad operators L_k describe the interaction with matter in terms of "internal" variables (e.g. charm quark coordinates/momenta). The eigenstates of the "Lindbladian operator" describe stationary yet out-of-equilibrium states. Only the lowest of those corresponds to thermal equilibrium. This approach is a fully quantum generalization of the Langevin or Focker–Planck equations and is potentially of great interest. Yet, so far, most of this theory is based on perturbative scattering, and we will not go into details.

Completing this section, one may ask whether the particular states observed do or do not follow the thermal distribution with $T = T_{\mathrm{ch}}$. Basically, we discussed the yields of $J/\psi, \psi', \psi''$, etc., already in Fig. 8.11 in the chapter on chemical freeze-out, which answers this question affirmatively for charmonia. There are also the ALICE data for three upsilons, and it is my understanding that their population is *not* thermal. This means that relaxation times for b quarks are not small relative to fireball lifetimes.

A short summary: While charm quarks basically behave in the same way as light quarks (apart from the density given by fugacity), the b quark is different. It neither catches with the flows nor does it arrive at the freeze-out in a chemically equilibrated way.

19.3 Jet Quenching

Let me start by briefly reminding the reader of the history of QCD jet physics. The so-called "Bjorken scaling" property of deep inelastic scattering (DIS) was confirmed by an SLAC experiment at the end of the 1960s. This led Feynman to the *parton model*, introducing the notion of quasi-free and point-like hadron constituents. Among its first consequences was the successful prediction of processes other than DIS, in particular the Drell–Yan process, a lepton pair production from $\bar{q}q \to e^+e^-$. Since the 1970s were also the time of active development or building of the first e^+e^- colliders, it was natural to think that the inverse reaction to the Drell–Yan process would produce a quark–antiquark pair. In the weak coupling regime[2] — justified if the energy/virtuality $\sqrt{s}$ is large enough — one would expect to see a "di-jet," two oppositely directed narrow cones of hadrons. Those indeed were observed and studied in detail eventually at the largest e^+e^- collider, CERN LEP. After two-jet events were well identified, three-jet events were also detected, in which a quark–antiquark pair was supplemented by a gluon jet.

Jets, produced in hard collisions of partons, originate at a high momentum transfer scale of $Q \sim$ 20–200 GeV $\gg$ 1 GeV and combine both a perturbative cascade from this scale down and flux tube breaking, all the way to low-mass hadrons. The theory of hard collisions includes the *hard block* described by perturbative QCD, combined with *structure and fragmentation functions* and with nonperturbative objects treated phenomenologically.

"Jet quenching" in AA heavy ion collisions includes modification of the jet shape due to interaction of the leading parton and subleading cascade with ambient matter. The original proposal to use jet quenching as an observable, telling us about the dynamics of fireball matter, was put forward by Bjorken (1982). Its first usage was the so-called "fireball tomography," revealing its size and shape.

The fixed-target heavy ion collisions at SPS/AGS in the 1990s did not have sufficient energy; therefore, the well-formed jets and their quenching were first observed at RHIC. In the early 2000s, observation of the jets as such, identified by specific algorithms, was not developed. It was, however, sufficient to observe the highest-$p_\perp$ part of the spectra.

---

[2]We will discuss the process $e^+e^- \to$ *hadrons* in strong coupling holographically in Section 21.4.

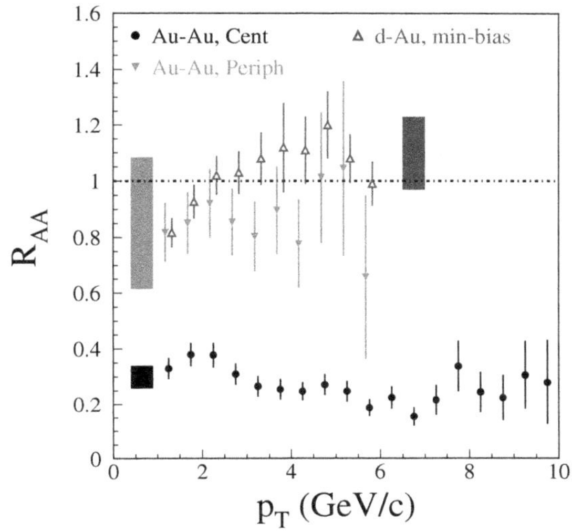

Fig. 19.4. The ratio R_{AA} of the yield of high $p_\perp$ hadrons in dAu and $AuAu$ collisions to the expected one according to parton model scaling. A strong deviation from one of these ratios indicates the jet quenching phenomenon.

The corresponding ratio, $R_{AA}(p_\perp)$, was defined as the AA spectra normalized (properly at an appropriate centrality to the expected number of hard collisions) by the spectrum in pp collisions. Figure 19.4 from the historic PHENIX white paper (Adcox *et al.*, 2005) summarizes its experimental values.

Note that the open triangles are for $d - Au$ collisions: Within the error bars, those are consistent with "no quenching," $R_{AA}(p_\perp) = 1$, which means their spectra in this case are basically *the same* as in pp, with no modifications. The red closed triangles denote peripheral $AuAu$ collisions: (inside still relatively large error bars due to small statistics at such $p_\perp$). They indicate weak quenching, reducing $R_{AA}(p_\perp)$ by $\approx$0.1–0.2. The main finding (shown by the closed black points at the bottom) is for central $AuAu$ collisions. In this case, the high-$p_\perp$ spectrum is reduced by a factor of 3–5. Furthermore, it was rather striking that this value appeared independent of the $p_\perp$ value!

Strong quenching at such $p_\perp$ was later confirmed at the LHC. To put their comparison into proper perspective, note first that for a similar reduction of momenta, the ratio R_{AA} depends strongly on the steepness of the spectrum. Power parameterization

$$\frac{dN}{dp_\perp^2} \sim \frac{1}{p_\perp^n}$$

gives at RHIC rather high power $n \sim 8$, while it is smaller at LHC.[3] As n grows and the spectrum becomes steeper, even a modest energy loss would lead to a large change in R_{AA}.

A better characterization of quenching is the relative energy loss $\Delta p_\perp / p_\perp$, which for a 100 GeV jet is typically of the order of 10%. The exact numbers for jet R_{AA} depend on the jet definition, such as the angular cone radius used in the jet finding algorithm.

Note also that the largest $p_\perp$ on the previous plot is only 10 GeV, the highest value statistically allowed at the time. At the LHC, jet studies were extended to much higher $p_\perp$, up to a few hundred GeV. More importantly, the development of jet finding algorithms allowed for the extension of R_{AA} measurements to full jets, rather than just one high-$p_\perp$ hadron. The modification due to quenching is still present, although of a different magnitude.

The "tomography" application was not forgotten. Since fireballs have nontrivial initial shapes, jet quenching features nontrivial azimuthal anisotropy, depending on the collision centrality. Mid-central and peripheral collisions produce a strongly deformed almond-like fireball, so that jets have rather different path lengths in matter in the x direction (along the impact parameter) and y normal to it. Since the directions of the reaction plane are known, on an event-by-event basis, those can be compared. Traditionally, these results are presented in the form of the same "ellipticity" parameter $v_2(p_\perp)$, as used for the elliptic flow

$$R_{AA}(p_\perp, \phi) \sim 1 + 2v_2(p_\perp)\cos[2(\phi - \Phi_{\text{event}})] \tag{19.5}$$

The measurements of v_2 at RHIC at large $p_\perp$ resulted in an apparent puzzle: The experimental v_2 value was clearly *way too large* for the models used at the time (Shuryak, 2002). It was first pointed out that varying matter absorption from small to infinitely large.[4] The measured v_2 was found

[3]The power n is different at RHIC and LHC because, at the same $p_\perp$, the jets correspond to very different fraction of the total momentum and thus different regions of the parton distribution functions.

[4]In the latter case, the only jets going out are those from the fireball surface, and for an almond of known deformation, one can easily calculate the corresponding "geometrical limit" value $v_2^*(b)$, for a given impact parameter.

to *exceed* even this upper geometrical limit. We return to the proposed resolution of this puzzle later in the chapter.

The other observable characterizing jet quenching is the *di-jet asymmetry parameter* defined by

$$A_j = \frac{p_{\perp,1} - p_{\perp,2}}{p_{\perp,1} + p_{\perp,2}} \tag{19.6}$$

where jet 1 is used as a trigger and jet 2 is the largest energy jet in the same event besides jet 1. The measurements of di-jet events with $p_\perp \sim 100$ GeV became possible when the LHC detectors CMS and ATLAS were involved. Its measurements for pp collisions gave a distribution that strongly peaked near $A_j = 0$, corresponding to perfect back-to-back identical jets. Yet, for central PbPb collisions, the measured distribution in A_j is wide, with a maximum at around 0.3. There were events in which a $p_\perp \sim 100$ GeV jet was not accompanied by any visible jet on the opposite side! Yet, it is not due to some unobserved particle (neutrino) because the total balance of the transverse momentum between the two hemispheres was still maintained. So, the momentum as large as 100 GeV has apparently been "eaten up" by the fireball (dissipated into only soft hadrons).

19.4 Radiative Theory of Energy Loss in QED and QCD

Let us now outline the main ideas of the theory of jet quenching. The original proposal by Bjorken included only elastic scatterings of the jet parton on those in matter. However, it was soon realized that it is subleading to the *radiative energy loss*, in which the jet parton emits forward-going gluons. Let us begin this chapter by explaining the key kinematical differences between radiative energy loss in two similar theories, QED and QCD.

In both cases, a charge with large momentum passing through the medium can scatter, deviating slightly from the straight line. Therefore, a soft photon or a gluon is emitted, which typically carry a small fraction of the energy ϵ:

$$x \equiv \frac{\epsilon}{E} \ll 1 \tag{19.7}$$

For simplicity, let us view the medium as a randomly placed (or correlated) static charge, generating static fields. The high-energy charge, passing through these fields, suffers transverse "kicks." As a result, its transverse momentum $q_\perp$ grows. If fields are uncorrelated, the growth in the transverse momentum is "diffusive" as a function of distance traveled:

$$\langle q_\perp^2(L)\rangle \sim \mu^2 N_{\text{kicks}} \sim \mu^2 \frac{L}{\lambda} \tag{19.8}$$

where μ is a typical kick and λ is the distance between the collisions, known as the *mean free path*. As the transverse momentum becomes larger and larger, the longitudinal velocity becomes a little less than the speed of light:

$$1 - v_{||} \approx \frac{\langle q_\perp^2\rangle}{2E} \tag{19.9}$$

Here, the similarity between QED and QCD ends. In QED, the charge lags *behind its own field*, which is massless and still moves at the speed of light. Eventually, the field and the charge acquire a large enough phase difference to become two incoherent, separate entities.

Let us now proceed to soft gluon radiation in QCD. There is a drastic difference: Unlike the photon, the soft radiated gluon also has a nonzero color charge and thus suffers transverse kicks. It also deviates from its straight path. Moreover, since its energy is smaller than that of the parent color charge, $\epsilon \ll E$, its deviations are larger. As a result, it is the soft gluon which "lags behind"[5] its parent charge!

In both cases, the combination of the squared of the kick strength normalized to mean free path,

$$\hat{q} = \frac{\mu^2}{\lambda} \tag{19.10}$$

incorporates all the matter properties one needs to know. In fact, this quantity is a very important kinetic coefficient of matter. To measure its value and dependence on the matter temperature $\hat{q}(T)$ is among the main goals of jet quenching studies. To relate it to kinetic theory, let us recall that in weak coupling, $1/\lambda = n\sigma$, the product of the density of scatterers and the collision cross-section. Furthermore, $\mu^2\sigma$ is proportional to the so-called *transport cross-section*, in which angular distribution is weighted with the square of momentum transfer. We will return to this point later in the section.

For understanding purposes, it is also useful to think for a moment about a setting opposite to random kicks, namely a motion in a constant

[5]In the synchrotron radiation setting (Shuryak and Zahed, 2003c), the curvature radius of the soft gluon is smaller than that of the large energy charge, and the curvature of the latter can be ignored.

external field, electric or magnetic. In this case, the charge classically moves along a curved path, either parabolic or circular. The accompanied photons, however, have a tendency to continue along the straight line tangent to this line. The resulting radiation we call *bremsstrahlung* or synchrotron radiation.

To turn these ideas into a quantitative theory, one has to answer the next important question: Do the emissions due to subsequent "kicks" interfere with the radiation process, or do the kicks occur rarely and one can simply add up their effects:

$$\frac{dE}{dL} = \text{const}, \quad \Delta E \sim L \tag{19.11}$$

As everybody knows, the latter regime normally takes place when particles fly through ordinary matter in the detectors. Indeed, scatterers (atomic nuclei) are separated by large distances at the atomic scale. Radiative energy loss, in this case, is given by the celebrated Bethe–Heitler formula multiplied by the number of scatterings.

However, for very large energy $E \to \infty$, it is not true since the "formation length" of soft photons eventually reaches the atomic scale. This regime leads to the Landau–Pomeranchuck–Migdal (LPM) effect[6] by Anthony *et al.* (1995).

QCD radiation, including the LPM effect, has been derived by Baier, Dokshitzer, Mueller, Peigne and Schiff (BDMPS). The crucial point is that the process is *not* linear in distance but quadratic:

$$\frac{dE}{dL} \sim L, \quad \Delta E \sim L^2 \tag{19.12}$$

The applicability of the pQCD issue: While each kick can be relatively soft, their combined action typically puts the transverse momentum in a range high enough that it becomes applicable.

Let us now derive the main BDMPS results following Baier *et al.* (1998). The probability for a charge to have a transverse momentum of $q_\perp$ is written as a loss and gain terms:

$$\lambda(z)\frac{\partial f(q_\perp^2, z)}{\partial z} = -f(q_\perp^2, z) + \int \frac{1}{\sigma}\frac{\mathrm{d}\sigma}{\mathrm{d}^2\vec{q}_\perp'}(\vec{q}_\perp', z) f((\vec{q}_\perp - \vec{q}_\perp')^2, z)\mathrm{d}^2\vec{q}_\perp' \tag{19.13}$$

[6]Proposed in the 1950s and experimentally confirmed at SLAC much later in 1995. Sadly, none of the authors lived to see this confirmation. Migdal died in 1991 in Princeton, just a few years earlier.

where $\lambda(z)$ is the mean free path of the jet particle and σ is the cross-section. Taking the Fourier transforms of f and of the potential

$$\tilde{f}(b^2, z) = \int \mathrm{d}^2\vec{q}_\perp e^{-i\vec{b}\cdot\vec{q}_\perp} f(q_\perp^2, z) \tag{19.14}$$

$$V = \frac{1}{\sigma}\frac{\mathrm{d}\sigma}{\mathrm{d}^2\vec{q}_\perp}, \quad \tilde{V}(b^2, z) = \int \mathrm{d}^2\vec{q}_\perp e^{-i\vec{b}\cdot\vec{q}_\perp} \frac{1}{\sigma}\frac{\mathrm{d}\sigma}{\mathrm{d}^2\vec{q}_\perp}(\vec{q}_\perp, z) \tag{19.15}$$

we can diagonalize the master evolution equation by taking the Fourier transform of the right-hand side of Eq. (19.13),

$$\int \mathrm{d}^2\vec{q}_\perp' \int \mathrm{d}^2b' e^{i(\vec{q}_\perp' - \vec{q}_\perp)\cdot\vec{b}'} \tilde{f}(b^2, z) \int \mathrm{d}^2b \tilde{V}(b^2, z) e^{-i\vec{b}\cdot\vec{q}_\perp'} \tag{19.16}$$

to get

$$\lambda(z)\frac{\partial\tilde{f}(b^2, z)}{\partial z} = [1 - \tilde{V}(b^2, z)]\tilde{f}(b^2, z) \tag{19.17}$$

As a model for the potential,

$$V(q_\perp^2) = \frac{1}{\sigma}\frac{\mathrm{d}\sigma}{\mathrm{d}^2\vec{q}_\perp}(\vec{q}_\perp) = \frac{\mu^2}{\pi(q_\perp^2 + \mu^2(z))^2} \tag{19.18}$$

we find that, at small b, the evolution equation for the jet has the form

$$\frac{\partial\tilde{f}(b^2, z)}{\partial z} = -\frac{b^2}{4}\hat{q}(z)\tilde{f}(b^2, z) \tag{19.19}$$

The parameter $\hat{q}(z) \equiv \mu^2(z)/\lambda(z)$ is the main property of the matter — a kind of kinetic coefficient — that determines all features of the jet quenching. To derive the explicit form for this, we take the Fourier transform of the potential:

$$\begin{aligned}\tilde{V}(b^2, z) &= \int \mathrm{d}^2\vec{q}_\perp e^{-i\vec{b}\cdot\vec{q}_\perp}\left(\frac{\mu^2(z)}{\pi}\right)\frac{1}{(q_\perp^2 + \mu^2(z))^2} \\ &= 2\mu^2(z)\left(\frac{1}{9}b_1^3F_2\left(2; \frac{5}{2}, \frac{5}{2}; \frac{b^2\mu(z)^2}{4}\right.\right. \\ &\quad \left.\left. + \frac{\pi(I_0(b\mu(z)) - \mu(z)bI_1(b\mu(z)))}{4\mu^3(z)}\right)\right.\end{aligned} \tag{19.20}$$

Expanding around $b = 0$,

$$\tilde{V}(b^2, z) = \frac{\pi}{2\mu(z)} - \frac{1}{8}\pi\mu(z)b^2 + \mathcal{O}\left(b^3\right), \tag{19.21}$$

Then,

$$\frac{[1-\tilde{V}(b^2,z)]}{\lambda(z)}\tilde{f}(b^2,z) \equiv -\frac{b^2}{4}\hat{q}(z)\tilde{f}(b^2,z) \tag{19.22}$$

where the transport coefficient $\hat{q}$,

$$\hat{q}(z) \equiv \frac{1}{\lambda(z)}\left[\frac{4}{b^2}(1-\tilde{V}(b^2,z))\right] = \frac{\Delta p_\perp^2(z)}{\lambda(z)} \tag{19.23}$$

is defined as the average of the square of the transverse momentum acquired per unit length. This expression is not convergent at $b \to 0$ (see the expression in parentheses in Eq. (19.22)), but in logarithmic approximation,

$$\hat{q}(z) \approx \frac{1}{\lambda(z)}\int_0^{1/b^2} \mathrm{d}^2\vec{q}_\perp \vec{q}_\perp^{\,2} V(\vec{q}_\perp^{\,2},z) = \rho(z)\int_0^{1/b^2} \mathrm{d}^2\vec{q}_\perp \vec{q}_\perp^{\,2}\frac{\mathrm{d}\sigma}{\mathrm{d}\vec{q}_\perp^{\,2}}(\vec{q}_\perp^{\,2},z), \tag{19.24}$$

where $\rho(z)$ is the density of scatterers. This shows that this formalism is equivalent to the transport cross-section method of finding the transverse kick; e.g. for a non-relativistic particle traveling in the z direction through the field of a single scatterer, we have

$$\Delta p_\perp = \int_{-\infty}^{\infty} \frac{b\ \mathrm{d}z}{(b^2+z^2)^{3/2}} = \frac{2}{b} \to \Delta p_\perp^2 = \frac{4}{b^2}. \tag{19.25}$$

So far, we have discussed a setting in which a fully dressed charge enters the matter. Although such a setting is natural for a charged particle entering the detector, it is actually not suitable for the jet quenching problem. As it was famously explained by Feinberg (1980), partons after hard collisions are born "semi-naked," with only part of their equilibrium field, up to a distance of $1/Q$, the inverse of the hard momentum scale. The decaying process of the "semi-naked" charge into a fully dressed hadron is, in fact, the reason why we see jets. In QCD, the perturbative part of this process is described by the so-called Dokshitzer–Gribov–Lipatov–Altarelli–Parisi (DGLAP) evolution equation. The nonperturbative part is described by a decay of confining flux tubes, extended between the produced jets. Together, those are believed to well describe the observed *fragmentation functions* of jets.

Therefore, for jets produced in matter, there are two processes occurring simultaneously and moving in opposite directions in terms of momentum scale: (i) DGLAP going from the hard scale $Q^2 \sim E^2$ downward, and (ii) the BDMPS process of multiple kicks, in which $q_\perp^2$ grows due to kicks. We will not go into their simultaneous description, as the literature includes many other approaches to radiative jet quenching. Instead, we end this section by mentioning two issues which were much discussed when the radiative energy loss theory was confronted with the data.

The first of them is known as the "Casimir ratio" problem. If the kicks are not too soft, it is expected that will be described by perturbative cross-sections, as discussed above in this section. If so, the ratio of dE/dL for quark and gluon jets should be the ratio of universal color factors in the cross-section

$$\frac{dE_q/dL}{dE_g/dL} = \frac{4}{9} \tag{19.26}$$

While it is hard to separate quarks and gluon jets individually, it can be done statistically. However, if jet energy is small compared to $\sqrt{s}$, most jets are gluonic ones. The tagged jets with c, b quarks are definitely quark jets. But the measured ratio is stubbornly closer to one than to 4/9.

There are further problems with heavy quark jets when their energy is not too high. As pointed out by Dokshitzer and Kharzeev (2001), the so-called "dead cone effect" should additionally suppress the radiative energy loss for c, b quarks. The data, however, do not show such suppression. The resolution to this puzzle, suggested by Wicks *et al.* (2007) and subsequent literature, has to do with complementing the quenching mechanism with

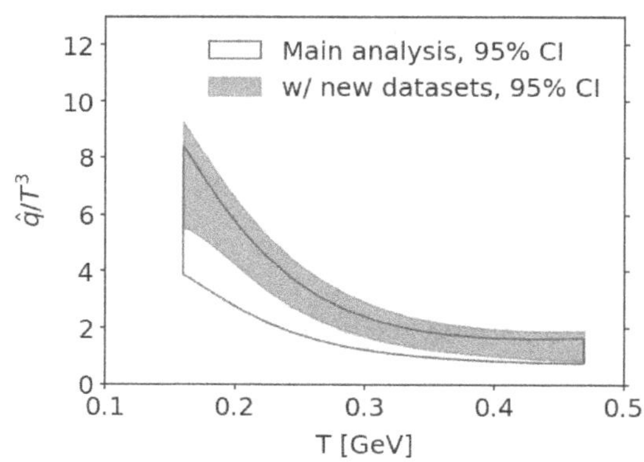

Fig. 19.5. A comparison of the main analysis of the data for $\hat{q}(T)/T^3$ (red open band) and that obtained by including the gamma-jet fragmentation function and hadron–hadron correlations (blue strip).

certain elastic processes. To my knowledge, there is no widely accepted explanation.

Finally, a comment about the magnitude of the ratio $\hat{q}(T)/T^3$. As it is proportional to the transport cross-section, it is expected to be small in a weak coupling regime. The actual comprehensive analysis (Xie *et al.*, 2022) based on several observables led to the conclusions plotted in Fig. 19.5, which shows that it is not at all small. Its temperature dependence is discussed in the following section.

19.5 Angular Anisotropy of Jet Quenching and Magnetic Monopoles

We now return to the puzzle of "too large v_2," exceeding predictions for *any* value of the quenching parameter in the existing models. This disagreement amounts to about a factor of 2, way above any experimental uncertainties.

Another hint that something unusual is going on came from a comparison between quenching at the RHIC and LHC. The densest matter — the QGP — was assumed to be causing the quenching. Therefore, it was universally assumed that the dimensionless ratio $\hat{q}(T)/T^3 \approx$ const, and thus, the quenching parameter scales as the entropy density or multiplicity. The multiplicity at the LHC is roughly twice that at the RHIC, so it was expected that $\hat{q}(T)$ there would also be twice as large. However, measurements have shown that $\Delta E/E$ is not that different, which indicates that $\hat{q}(T)/T^3$ *decreases* with increasing T, perhaps due to the running coupling. This observation implies that $\hat{q}(T)/T^3$ should have a maximum at some temperature.

A pure phenomenological analysis of the asymmetry puzzle, performed by Liao and Shuryak (2009), answered this question. If one thinks of $\hat{q}(T)/T^3$ as some unknown *function* of T, it is natural to slice the fireball into bins according to certain isotherms and calculate the v_2, which would be generated if *only this bin* was responsible for quenching. The result of such a study is shown in Fig. 19.6 adapted from Liao and Shuryak (2009). As one can see in Fig. 19.6(a), the most dense matter at the right of the plot does not generate large anisotropy. The reason is that it is located close to the center of the almond fireball and has close to a spherical shape. A very dilute matter, as seen on the left side of the plot, is also rather

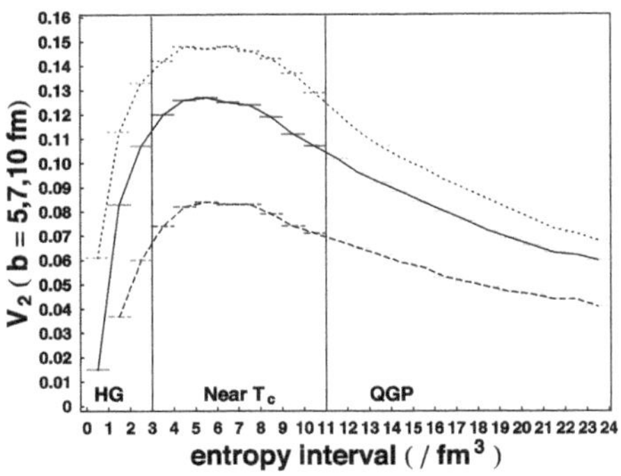

(a)

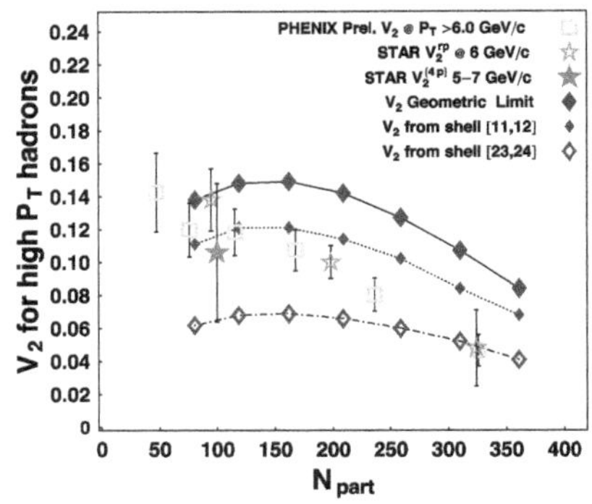

(b)

Fig. 19.6. (a) The v_2 corresponding to each entropy bin versus the entropy density (fm^{-3}). Three sets of points correspond to the impact parameters $b = 5, 7, 10$ fm, bottom to top. (b) v_2 as a function of the centrality measure, N_{part}, compared with available RHIC data, stars for STAR and green squares for PHENIX.

spherical. The angular anisotropy, small in both limits, has a maximum for a particular temperature in between. To our surprise, this maximal value was in fact quite close to what was observed in experiments: see Fig. 19.6(b)! The conclusion reached by Liao and Shuryak (2009) was that most of the jet quenching must be produced by matter in the entropy density range $s = 3$–$10\ \text{fm}^{-3}$. It turns out that it corresponded to a small range of temperatures close to the critical one, T_c.

If $\hat{q}(T)/T^3$ does indeed peak near T_c, why would that be the case? Quark and gluon densities, divided by T^3, are expected to vanish at T_c due to the approach of electric confinement. The only quasiparticles not subject to it and known to have a peak at or above T_c are the magnetic monopoles.

The idea that monopoles can be scatterers can be easily implemented in the BDMPS formalism. The generic form of $\mathrm{d}\sigma/\mathrm{d}q_\perp^2$ is the same:

$$\frac{\mathrm{d}\sigma}{\mathrm{d}q_\perp^2} = \frac{C}{(q_\perp^2 + \mu^2)^2}, \tag{19.27}$$

As we already mentioned above, for quarks, $C_F = 4/3$ and for gluons, $C_A = 3$. For magnetic monopoles, it is

$$\frac{\mathrm{d}\sigma_{gm}}{\mathrm{d}q_\perp^2} = \frac{3\pi F^2(q_\perp^2)}{(q_\perp^2 + \mu_M^2)^2}, \tag{19.28}$$

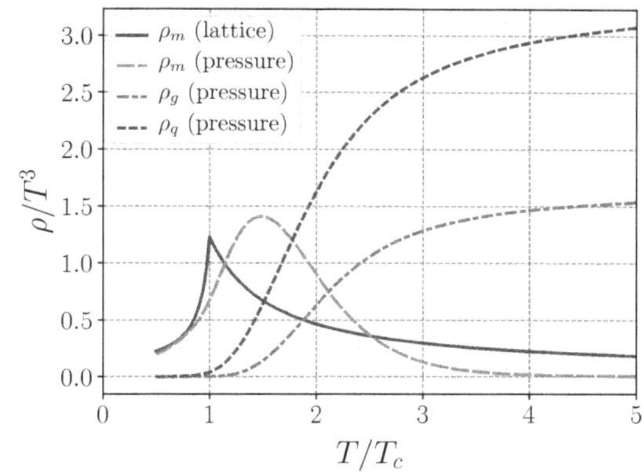

Fig. 19.7. (Color online) Electric and magnetic quasiparticle densities used. The (blue) solid line shows the magnetic monopole density as directly observed on the lattice. The (red) long dashed line is the monopole density extracted from the thermodynamics (pressure), along with the densities of quarks (purple, short dashed) and gluons (green, dot dashed).

with $F(q_\perp^2)$, the monopole form factor. For point-like monopoles, we have $F(q_\perp) = 1$. There are no factors of α_s in the monopole cross-sections due to the Dirac condition, which makes their magnitudes larger than the electric-scatterer cross-sections.

The BDMPS-like energy loss of a parton as it traverses the medium is given by an integral over the line along which the parton travels, with quark, gluon and monopole densities and corresponding contributions to the $\hat{q}$.

The results for R_{AA} and v_2 of fragmented jets are seen in Figs. 19.8 and 19.9, respectively. The collider data are from Abelev *et al.* (2013b) and Chatrchyan *et al.* (2012b) for LHC 2.76 *Pb-Pb* TeV R_{AA}, from Adare *et al.* (2008, 2013) for RHIC 200 GeV *Au-Au* R_{AA}, from Abelev *et al.* (2013a) and Chatrchyan *et al.* (2012a) for LHC 2.76 TeV *Pb-Pb* v_2, and from Adare *et al.* (2010) for RHIC 200 GeV *Au-Au* v_2.

All of the plots are laid out as follows: The first column, comprising subplots (a) and (b), is for LHC 2.76 TeV *Pb-Pb* collisions, and the second column, with subplots (c) and (d), is for RHIC 200 GeV *Au-Au*. The first row is for the monopole density measured on the lattice, and the second row is for monopole density derived from the equation of state.

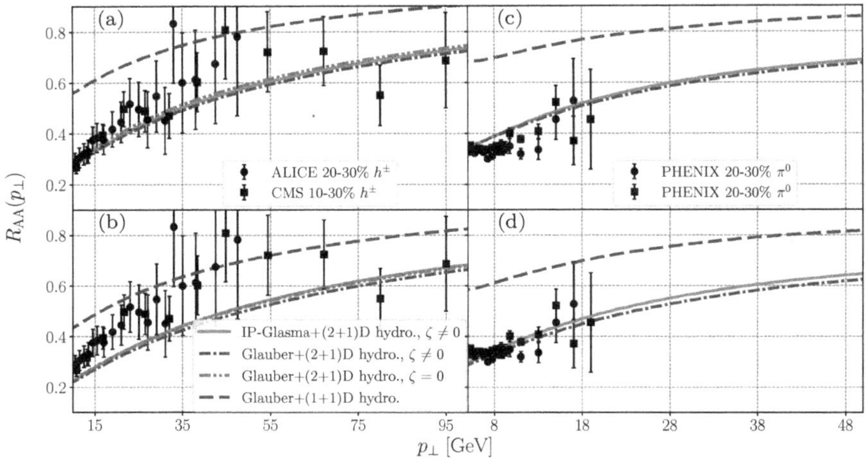

Fig. 19.8. Nuclear modification factor of charged hadrons in 2.76 *Pb-Pb* collisions (a), (b) and neutral pions in 200 GeV *Au-Au* collisions (c), (d). The first row (a), (c) is the result for monopole density from the lattice, while the second row (b), (d) is the result for the monopole density from the equation of state. The (red) solid curve is for IP-Glasma initial conditions and (2+1)D hydrodynamics with bulk viscosity ($\zeta \neq 0$), the (blue) dash-dot curve is for Glauber initial conditions and (2+1)D hydrodynamics with bulk viscosity ($\zeta \neq 0$), the (green) dash-dot-dot curve is for Glauber initial conditions and (2+1)D hydrodynamics without bulk viscosity ($\zeta = 0$), and the (purple) dashed curve is for the smooth Glauber initial condition with (1+1)D Bjorken evolution.

The Bjorken-evolving background (purple, dashed curve) causes far less energy loss than the scenarios with realistic hydrodynamic backgrounds. This is due to the fact that this is a one-dimensional expansion, and the matter does not expand in the transverse plane; the ellipse of above-T_c medium shrinks inward to the center of the fireball with time. This is unlike the (2+1)D hydrodynamic case, where the size of the above-T_c medium remains approximately constant with time, and all of the medium cools to approximately T_c by $\tau \sim 6$ fm/c.

At the end of the section, one may speculate, following Xu *et al.* (2016), that two kinetic coefficients may have similar T dependences, namely

$$\frac{\hat{q}(T)}{T^3} \sim \frac{s(T)}{\eta(T)} \tag{19.29}$$

with a peak near T_c since they are both related to the number of available scatterers. At the moment, we do not have sufficient accuracy to see whether this is indeed the case.

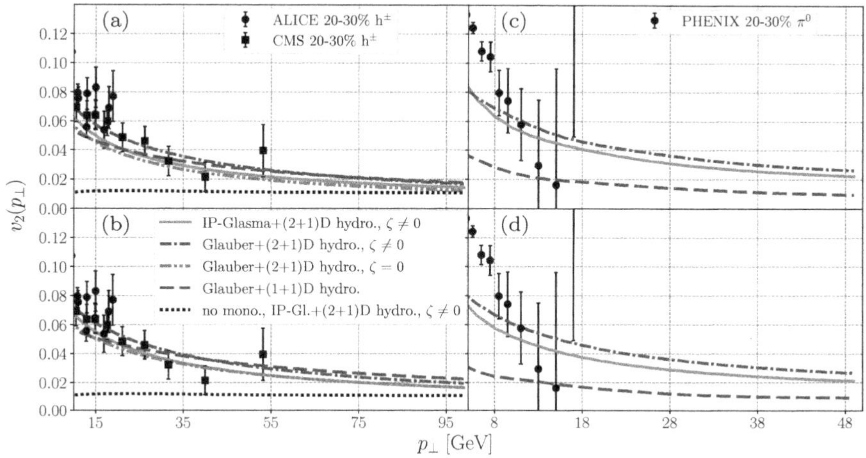

Fig. 19.9. (Color online) Azimuthal anisotropy of charged hadrons in 2.76 *Pb-Pb* collisions (a), (b), and neutral pions in 200 GeV *Au-Au* collisions (c), (d). The first row (a), (c) is the result for monopole density from the lattice, while the second row (b), (d) is the result for monopole density from the equation of state. The (red) solid curve is for the IP-Glasma initial conditions and (2+1)D hydrodynamics with bulk viscosity ($\zeta \neq 0$), the (blue) dash-dot curve is for Glauber initial conditions and (2+1)D hydrodynamics with bulk viscosity ($\zeta \neq 0$), the (green) dash-dot-dot curve is for Glauber initial conditions and (2+1)D hydrodynamics without bulk viscosity ($\zeta = 0$), and the (purple) dashed curve is for the smooth Glauber initial condition with (1+1)D Bjorken evolution. The (black) dotted curve is for for IP-Glasma initial conditions and (2+1)D hydrodynamics with bulk viscosity ($\zeta \neq 0$) with no monopoles.

Summary and connections: The radiative mechanism of jet quenching is an example of the Landau–Pomeranchuck–Migdal effect, in which the scattered quark collects transverse momentum from multiple scatterings. The data, especially those related to the elliptic dependence of jet quenching, indicate that the most effective scatterings occur on matter close to the phase transition conditions, with the corresponding parameter peaking there. A significant role for scattering on monopoles has been developed (see the discussion of magnetic quasiparticles in Chapter 24). We discuss jet quenching in strong coupling in holography in Chapter 21, which demonstrates the Mach shock cone, well known from hydrodynamical studies. Yet, direct observations of Mach cones near jets are still lacking.

Chapter 20

THE SMALLEST DROPS OF THE SQGP

20.1 The CMS Discovery of the "Ridge"

We previously discussed heavy ion AA collisions, with $A = Au, Pb$, and concluded that those result in a fireball, with a hydrodynamical explosion resulting in very well equilibrated matter at chemical and kinetic freeze-outs. We also discussed the usual (min-biased) pp collisions as a calibration case in which these phenomena are *not* observed. In some cases, such as jet quenching, the main observable was just the ratio of what is seen in those two extreme cases.

Somewhere in between those two extremes, there should be a transition between two regimes, which one may call "micro" (or transparent, a system that is small compared to the mean free path $L \ll l$) and "macro" (thermo- and hydrodynamics) regimes, in which the mean free path is small $L \gg l$. The expected transition may be gradual or sharp.

One way to explore this issue is by changing the centrality of AA from central to peripheral collisions. While this reduces the size of the fireball, it unfortunately also changes its density and geometry. Another approach can be to focus on central AA collisions, but with variable A, preserving the overall geometry. However, historically, the development was instead triggered by the unexpected discovery of the so-called "*ridge*" at the LHC in high-multiplicity pp collisions.

The original data from Khachatryan *et al.* (2010) are shown in Fig. 20.1. Plot (a) shows the two-particle correlation function for typical events with no restrictions (this is what is technically called *minimal-biased* sample) without a cut on the momentum.[1] One can see that the two particles

[1]Yes, the caption says $p_T > 0.1$ GeV/c, but it is only because the detector cannot see particles with smaller p_T. This cut is, at the momenta, significantly smaller than the average and is thus unimportant.

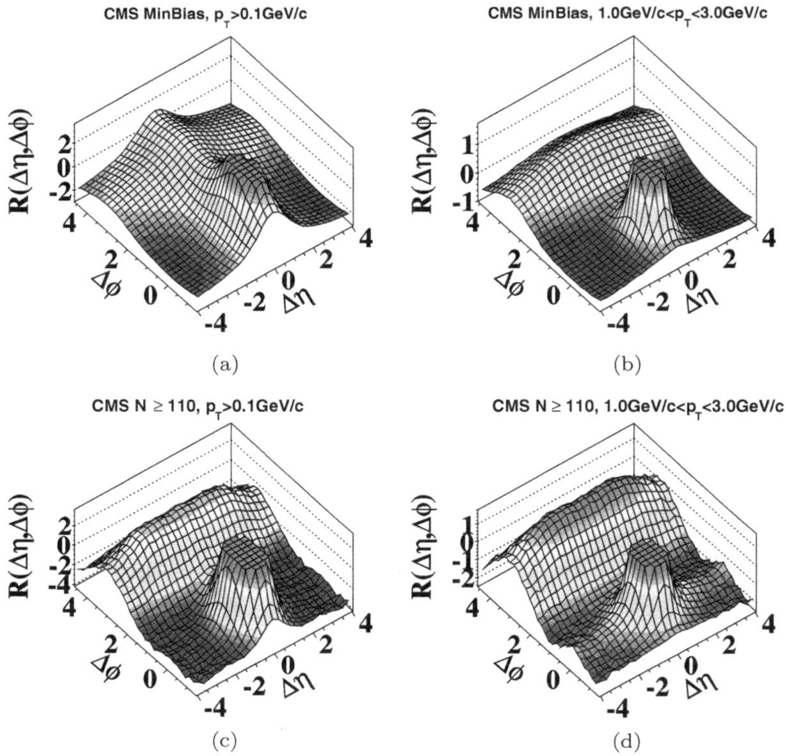

Fig. 20.1. 2D two-particle correlation functions for 7 TeV pp collisions versus pseudorapidity and azimuthal angle differences $\Delta\phi, \Delta\eta$. Plot (a) shows the minimum bias events with $p_T > 0.1$ GeV/c, plot (b) is for $1 < p_T < 3$ GeV/c, plot (c) for high-multiplicity events and $p_T > 0.1$ GeV/c, plot (d) for high-multiplicity events and $1 < p_T < 3$ GeV/c.

are strongly correlated if they are close in pseudorapidity $\Delta\eta < 1$, while they have a rather wide distribution in relative azimuthal angle $\Delta\phi$. If the transverse momentum is increased (plot (b)), one finds what is called a "jet peak" and an "antijet" one in the opposite direction in the transverse plane, $\Delta\phi = \pi$. Note that the rapidity of the antijet is not opposite that of the trigger particle because the rapidity of the two-parton center-of-mass frame is not fixed. Therefore, the antijet has a very wide (flat) distribution. So far, no surprises. These correlations are well described by pQCD processes and string fragmentation: In particular, these correlations are reproduced by event generators, such as those based on the Lund model.

Two other plots are taken with an *online high-multiplicity trigger*, selecting events with more than 110 charged tracks in the CMS detector. Plot (d) selecting a certain slice of transverse momenta shows a new "ridge," extending along the rapidity direction with a maximum at $\Delta\phi = 0$.

Before we discuss its physical meaning, let me provide some generic facts elucidating how extreme and admirable modern experiments have become. This special online trigger was required because, unfortunately, the effect was only observed in events with a tiny probability of $P \sim 10^{-6}$. Let me explain what it means in practical terms. Dividing the cost of LHC, $\sim 10^{10}$\$ by the number of recorded pp events $\sim 10^{10}$, one finds that an event costs about one dollar each! Taking the complexity of the events into account, with all secondaries identified and momenta determined, it is surprisingly cheap.

Yet, it also shows that the high-multiplicity pp events under consideration — appearing with a probability of 10^{-6} — cost about a million dollars each! Furthermore, in order to arrive at a meaningful two-dimensional correlation function with certain p_t cuts, one needs many such events.

Soon after the ridge discovery, it was also found that high-multiplicity pPb collisions also led to a similar "ridge." However, the probability of those was not $\sim 10^{-6}$ but several percent, which implies that they are simply the first bin in the centrality distribution. It made studying the ridge much easier.

Before leaving the subject and the historic plot reproduced in Fig. 20.1, let me point out one more observation, made by the ALICE collaboration and then confirmed by others. Let us compare plots (b) and (d): Both have a backward ($\Delta\phi \approx \pi$) structure, which was explained by a recoil (antijets, etc.). What happens if we want to remove such effects, by subtracting (b) from (d)? The answer is rather intriguing: The difference reveals an "anti-ridge," which is nearly exactly the same as a ridge. Both can then be described by a single $\cos(2\phi)$ correlation, revealing that we actually rediscovered the elliptic flow.

20.2 Possible Mechanisms of Ridge Formation

So, why is the ridge so exciting?

Note first that it extends by many units of rapidity. A simple causality argument — different rapidity slices are not in causal contact except at the very moment of collisions — tells us that it must come from the very early stages of the collision. Recall that elliptic and other azimuthal flows also have this property.

Ridge formation then implies two necessary conditions to be in place: (i) a nontrivial shape of the initial state, expressed in terms of ϵ_m parameters; and (ii) the collectivity of flows, which convert spatial anisotropies into momentum anisotropies v_n. Can they both be present in (high-multiplicity) pp, pA collisions, such as in heavy ion AA ones, or some other mechanism leads to them?

Let me start discussing the first ingredient first: the *shape* of the initial state, starting with pp. While the nucleon, like a nucleus, is on average azimuthally symmetric by itself, in the pp interaction dynamics, the direction of impact parameter (x or "in-plane") is obviously different from y or "out-of-plane."

Various proposed ideas about the shape of the initial state find it convenient to start with the stringy picture of a nucleon,[2] as a quark–diquark dipole connected by the QCD string (flux tube). As suggested by Bjorken *et al.* (2013), the chances of interaction, and especially the production of larger than average multiplicity, are increased if the orientation of these strings in both nucleons *happens to be the same.* The semiclassical theory of "stringy Pomeron" (Stoffers and Zahed, 2013; Basar *et al.*, 2012) starts with small-size dipoles but derives the Pomeron amplitude from a tunneling event, in which *two* flux tubes appear, both extended between the two dipoles in the in-plane direction. The distance between them, by necessity, is in the out-of-plane direction. Both of those ideas lead to a very large deformation of the initial state.

Furthermore, the LHC data on elastic pp profiles indicate a "black disk" regime for impact parameter $b < 1$ fm, interpreted as multi-Pomeron exchange. Another version of the stringy picture (Shuryak and Zahed, 2014) points at the Hagedorn phenomenon, leading to high excitation of the exchanged string. Both suggest a certain multi-string initial stage. The first attempt to understand its structure, in the so-called *"string ball"* picture, has been made by Kalaydzhyan and Shuryak (2014b). In contrast to the previous models, those seem to produce small deformations, $\epsilon_n \sim 1/\sqrt{N_{\text{strings}}}$, resembling those in AA collisions.

[2]See more in Chapter 24.

The alternative weak-coupling descriptions of the proton are provided by partonic models, which view it as a cloud of gluons and quarks. A dense cloud of them, known as color glass condensate (CGC), then in collisions turns into GLASMA, chaotic color fields.

One notable idea suggested in this framework by Altinoluk *et al.* (2015) was that of *Bose enhancement.* It is based on the well-known phenomenon of stimulated emission: In an ensemble of bosons, the emission of one more of them is proportional to $(1 + n_k)$, with n_k being the occupation number of the same momentum state. It is similar to the HBT and Bose correlations of the pions, except, of course, now we are discussing not the final state but the initial one. According to these authors, CGC must contain "domains" of strong coherent color fields. While their directions are random, in the *pp* scattering, those domains in one nucleon would lead to anisotropic scattering of partons of the other nucleons. The signature of this phenomenon should be its presence for the gluon momenta below the saturation scale $p_\perp < Q_s(N)$, where we emphasize the dependence of Q_s on the event multiplicity.

In summary, there is no shortage of models which can ensure rather anisotropic initial conditions. Let us now proceed to the second necessary part of the problem: the conversion of the initial shape into the final anisotropy of the particle momenta distribution.

The "mainstream" by now is to do what was already done for AA collision, which is to assume the QGP formation and use hydrodynamics. We will discuss it later in the book in substantial detail. Yet, there remain very interesting questions: Up to what critical multiplicity can it be used? And what anisotropies are there for even smaller N, all the way to minimally biased *pp* collisions? This part of the story is not included, as it is still "under construction."

20.3 Femtoscopy of "Small Systems," High-Multiplicity pA, pp

Before we go into details, let us see how small these systems really are. As we discussed in Chapter 14, the size of the system at freeze-out can be directly measured using *femtoscopy*. The corresponding data are shown in Fig. 20.2, where the left side combines the traditional two-pion correlation functions with the more novel three-pion

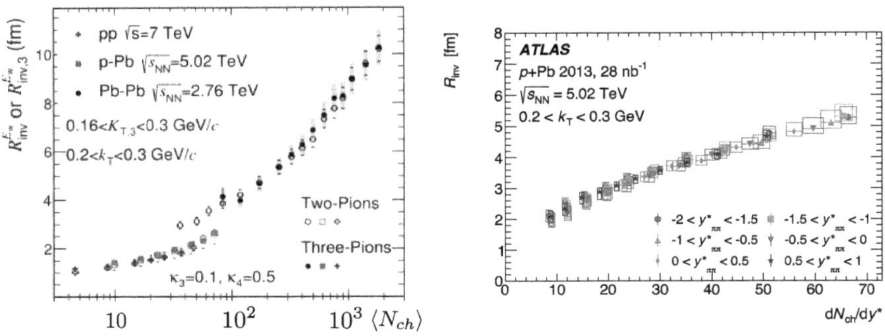

Fig. 20.2. (left) ALICE data on the femtoscopy invariant radius as a function of multiplicity for $pp, pPb, PbPb$ collisions. (right) ATLAS HBT data.

ones of identical pions. The right side shows the ATLAS data from Aaboud *et al.* (2017).

An overall growth of the freeze-out size with multiplicity is expected to be $\sim\langle N_{\rm ch}\rangle^{1/3}$, as expected if the freeze-out density is a constant. While for AA collisions, this simple idea indeed works, those for small systems (the pp, pA collisions) apparently fall into a different line, with a significantly smaller size. So, one can see that those two "small systems" get frozen at higher densities, compared to those in the AA collisions.

To understand that, recall the *freeze-out condition*: "The collision rate becomes comparable to the expansion rate,"

$$\langle n\sigma v\rangle = \tau^{-1}_{\rm coll}(n) \sim \tau^{-1}_{\rm expansion} = \frac{dn(\tau)/d\tau}{n(\tau)}. \tag{20.1}$$

Higher density means a larger left-hand side and thus a larger right-hand side. So, pp, pA high-multiplicity systems are more "explosive" with larger expansion rates! We indeed argue in the following that this conclusion is also confirmed by radial flow measured by particle spectra as well as the HBT radii.

But why are those systems "more explosive"? Where is the room for that, given that even the measured final size of these objects is *smaller* than in peripheral AA (which shows only a rather modest radial flow)? The only space left is at the beginning: Those systems must obviously *start accelerating earlier*, from an *even smaller* size than seen by femtoscopy, to produce strong collective flow at freeze-out.

Another puzzle is why central pA — a collision of a proton with about 16 others — appears to be so similar in size and collective flows to (same multiplicity) pp, a collision of only two protons. We turn to its discussion in the following section on "small systems."

20.4 Collectivity in Small Systems

Let us briefly recall the time sequence of the main events. The first discovery along this direction — in the very first LHC run — was due to the CMS collaboration (Khachatryan *et al.*, 2010), which found a "ridge" correlation in high-multiplicity pp events. Switching to central pA collisions, the CMS observed a similar ridge there, now with much higher probability, i.e. few percent instead of $P \sim 10^{-6}$ (Chatrchyan *et al.*, 2013). By subtracting central minus peripheral correlations, the ATLAS CMS and ALICE groups soon found that the observed "ridge" is accompanied by an "anti-ridge" in the other hemisphere, concluding that the phenomenon is a familiar elliptic flow, $v_2 \cos(2\phi)$ deformation.

The PHENIX collaboration also found a ridge-like correction in central dAu collisions at the RHIC, with the v_2 value about twice *larger* than in pPb at the LHC. This difference was soon attributed to different initial conditions: Projectile deuteron beams initiate a "double explosion" by their two nucleons. Hydrodynamical studies, such as those by Bozek (2012), quantified this simple idea. It was later extended to three-nucleon collisions in He^3Au as well.

However, are we sure that we are dealing with collective flows and not just some few-particle correlations? The set of detailed analysis which established the *collectivity* of the flow in pA, beyond any reasonable doubt, came from CMS (de Cassagnac, 2014). Their v_2 value was extracted using correlators of not only two but also four, six and even eight particles.

The results are shown in Fig. 20.3. Let me give here the definitions of these quantities following Bilandzic *et al.* (2011). They are averages of the following combinations of two-, four- and six-particle azimuthal angle phases, averaged over all combinations of these particles and also over all events (denoted therefore by the double-average brackets):

$$\langle\langle 2\rangle\rangle = \langle\langle e^{in(\phi_1-\phi_2)}\rangle\rangle, \quad \langle\langle 4\rangle\rangle = \langle\langle e^{in(\phi_1+\phi_2-\phi_3-\phi_4)}\rangle\rangle$$

$$\langle\langle 6\rangle\rangle = \langle\langle e^{in(\phi_1+\phi_2+\phi_3-\phi_4-\phi_5-\phi_6)}\rangle\rangle$$

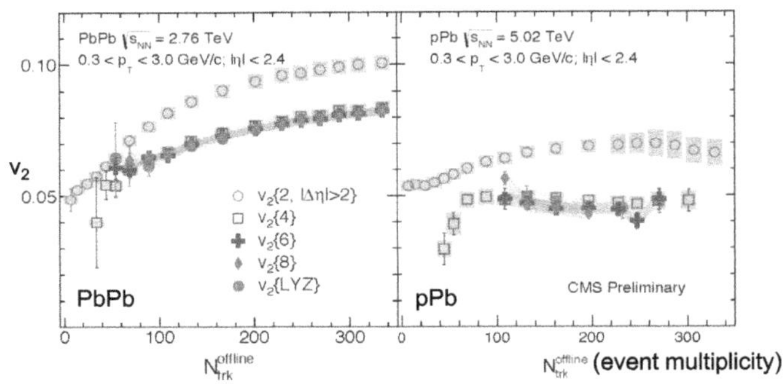

Fig. 20.3. CMS data for v_2 calculated using two-, four-, six- and eight-particle correlations, as well as Lee–Yang zeroes (basically all particles). A good agreement between those manifests the collectivity of the phenomenon.

etc., from which the cumulants are calculated via

$$c_n\{2\} = \langle\langle 2\rangle\rangle, \quad c_n\{4\} = \langle\langle 4\rangle\rangle - 2\langle\langle 2\rangle\rangle^2 \tag{20.2}$$

$$c_n\{6\} = \langle\langle 6\rangle\rangle - 9\langle\langle 4\rangle\rangle\langle\langle 2\rangle\rangle + 12\langle\langle 2\rangle\rangle^3 \tag{20.3}$$

$$\begin{aligned} c_n\{8\} &= \langle\langle 8\rangle\rangle - 16\langle\langle 6\rangle\rangle\langle\langle 2\rangle\rangle - 18\langle\langle 4\rangle\rangle^2 \\ &\quad + 144\langle\langle 4\rangle\rangle\langle\langle 2\rangle\rangle^2 - 144\langle\langle 2\rangle\rangle^4 \end{aligned} \tag{20.4}$$

Collective v_n coefficients are finally defined in terms of these cumulants by

$$\begin{aligned} v_n\{2\} &= (c_n\{2\})^{1/2}, \quad v_n\{4\} = (-c_n\{4\})^{1/4}, \\ v_n\{6\} &= \left(\frac{c_n\{6\}}{4}\right)^{1/6}, \quad v_n\{8\} = \left(-\frac{c_n\{8\}}{33}\right)^{1/8} \end{aligned} \tag{20.5}$$

While the two-particle correlation includes some non-flow correction (perhaps from resonance decays), those for four, six and eight particles give the same values. What it means is that such values are indeed shared by *all* secondaries: The case for the collectivity of elliptic flow is therefore proven.

Taking collectivity for granted, one can further ask if the v_2 observed is caused by the pre-collision correlations or the after-collision collective flows. A very nice control experiment testing this is provided by dA and He^3A collisions. Two nucleons in d are, on average, far from each other, and 2 MeV binding is so small that one surely can ignore their initial state correlations.

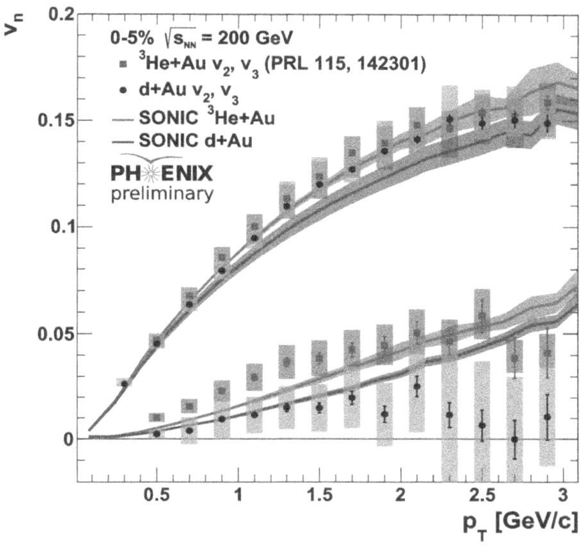

Fig. 20.4. The second and third harmonics of the flow as a function of p_T for dAu, He^3Au collisions from the PHENIX collaboration are compared with hydrodynamical calculations.

So, whatever the "initial shape" effect of pp in dA, it should be *reduced* by $1/\sqrt{2}$ because two shapes cannot be correlated. It should be reduced further in He^3A by $1/\sqrt{3}$, if the same logic holds.

Hydrodynamical predictions are the opposite: Double (or triple) initial explosions still lead to one common fireball, with the initial anisotropies *larger* than in pA. Data from the RHIC by PHENIX and STAR on dAu, He^3Au do indeed show such an *increase* in the v_2, v_3, relative to pAu, again in quantitative agreement with hydrodynamics (Bozek and Broniowski, 2014; Nagle *et al.*, 2014). In Fig. 20.4 from Velkovska (2017), those harmonics from the dAu, He^3Au PHENIX data are compared with hydrodynamical calculations. It looks like this issue is now settled.

ATLAS was able to perform the first measurements of the higher harmonics $v_n, n = 4, 5$ in central pPb, see Fig. 20.5. Except at very high $p_\perp$, those two harmonics seem to be comparable in magnitude: It is the first contradiction to "viscous damping" systematics.

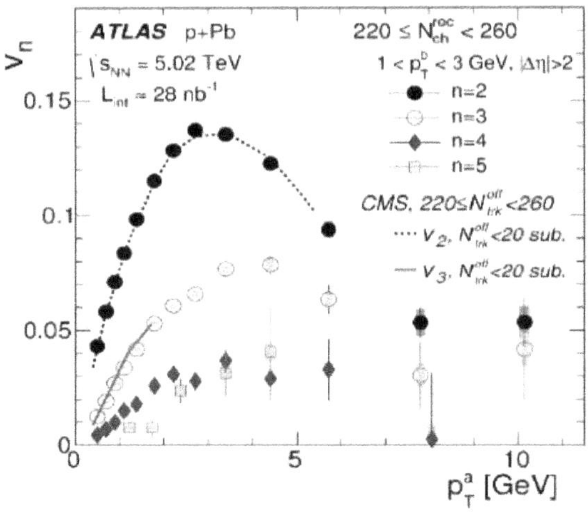

Fig. 20.5. v_n for $n = 2, 3, 4, 5$ versus $p_\perp$ in GeV for high-multiplicity bins indicated on the figure. The points are from ATLAS, and the lines are from CMS (presentation at QM2015).

20.5 Pedagogical Digression: Scale Invariance of sQGP and Flow Predictions for Small Systems

The acceptance of the hydrodynamical treatment of "small system explosions" was psychologically hard for many people in the field. Doubts were raised over whether it may be possible to treat a fireball, of 1 fm size or less, as a macroscopic system,[3] which can be subdivided further into "hydro cells."

So, let us step back from the data and consider the issue of scales on theoretical grounds. Taking smaller and smaller cells of ordinary fluid, such as water or air, one eventually reaches an atomic scale, beyond which water or air, as such, do not exist as individual molecules get resolved. However, a QGP is *not* like that: Being made of essentially massless quarks and gluons, which have no scale of their own, it has only one relevant scale given by its temperature T. Thus, a QGP is approximately scale-invariant. (The second scale Λ_{QCD} only enters via the logarithmic running of the coupling, which is relatively slow and can, in some approximation, be ignored.)

[3]Note that just 15 years ago, the same question was asked about systems of 6 fm size.

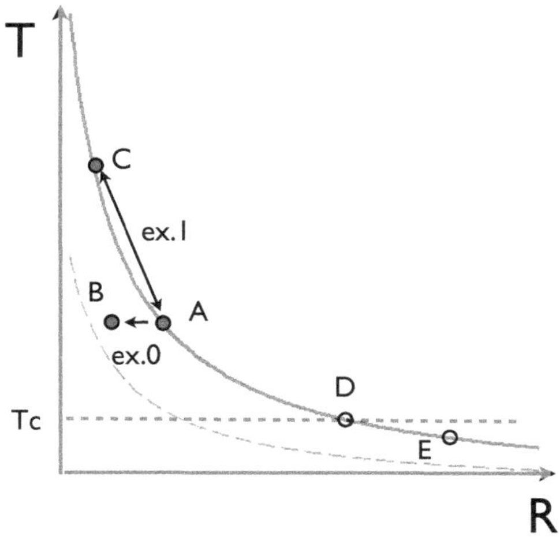

Fig. 20.6. Temperature T versus the fireball size R plane. The solid blue line is the adiabatic S = const., approximately TR = const. for sQGP. Example 1 in the text corresponds to reducing R, moving left, $A \to B$. Example 2 is moving up the adiabatic, $A \to C$. Example 3 corresponds to adiabatic expansion, such as $A \to E$, $C \to E$. If, in reality, C corresponds to pA, the freeze-out occurs at the earlier point D.

Indeed, as lattice numerical calculations show, at $T > 200$ MeV the QGP thermodynamics is approximately scale-invariant,[4] dimensionless ratios $\epsilon/T^4, p/T^4$ are approximately T-independent. A comparison of the data from the LHC and the RHIC further suggests that a similar scaling, $\eta/T^3 \sim$ const., holds for viscosity as well (although with less accuracy so far). Thus, as a first approximation, one may assume that the QGP does not have a scale of its own.

This means that it would show *exactly the same behavior* if conditions related by the scale transformation

$$R_A/R_C = \xi, T_A/T_C = \xi^{-1} \tag{20.6}$$

are compared. Consider *thought experiment 1*, in which we compare two systems on the same adiabatic A and C. For a scale-invariant sQGP, the

[4]At $T < 200$ MeV, there is no scale symmetry. There is a significant change instead. It is of course taken into account in hydrodynamics and will be discussed later, e.g. in the section on HBT radii.

points A, C are related by the scale transformation mentioned above and have the same entropy.[5]

Assuming the scale transformation is an approximate symmetry, one expects the same dynamical evolution. A smaller-but-hotter plasma ball C will explode in exactly the same way as its larger-but-cooler version A.

Let us now proceed to *thought experiment (2)*, which is the same as above but in QCD with a running coupling. In the sQGP regime, it leads to (very small, as lattice tells us) running of s/T^3, some (unknown) running of η/T^3, etc. The most dramatic effect is, however, not the running coupling *per se* but the lack of supersymmetry, which allows for the chiral/deconfinement phase transition from the sQGP phase at $T > T_c$ to the hadronic phase at $T < T_c$. The end of the sQGP explosion D thus has an *absolute scale* and is not subject to a scale transformation!

So, let us consider two systems A, C of the same total entropy/multiplicity, both initiated in the sQGP phase, with initial conditions related by the scale transformation. Their evolution would be the same until the larger/cooler one reaches $T \approx T_c$, where the scale invariance of their evolution ends! So, the final result of the explosions is not the same. In fact, the smaller/hotter system has an advantage over the larger/cooler one since the *larger* ratio of the initial and final scales T_i/T_f allows it to be accelerated more.

The hydrodynamic expansion does not need to stop at the phase boundary D. In fact, large systems, as obtained in central AA collisions are known to freeze-out at $T_f < T_c$, down to a 100 MeV order (and indicated in the sketch by the point E). However, small systems, obtained in peripheral AA or central pA collisions seem to freeze out at D.

Another meaning of the term "small systems" is applied when not the actual size but entropy/multiplicity is reduced. Deviations from hydrodynamics in such cases are seen as higher viscous corrections. For a discussion of those issues, see Spaliński (2016) and Romatschke (2017).

A brief summary: We expect hydrodynamics to work as well in smaller systems as it does in large fireballs of AA collisions due to the approximate scale invariance of a sQGP. In fact, observations and theory both agree that small systems under investigation explode even more violently since they spend a larger fraction of time in the sQGP phase.

[5]Note that the holographic models interpret the "RG scale" as an extra fifth coordinate. The evolution in scale is thus depicted as "gravitational falling" of particles, strings, fireballs, etc., along this coordinate. In this language, our two systems fall similarly in the same gravity; however, since the smaller system starts "higher," it gets a larger velocity at the same "ground level," given by a fixed scale T_c.

20.6 Comparison of the Peripheral AA, Central *pA* and High-Multiplicity *pp* Collisions

After the thought experiments with some ideal systems, let us return to reality. We will do it in two steps, beginning in this section with "naive" estimates for three cases at hand based on standard assumptions about the collision dynamics and then returning to a more model-dependent discussion in the following sections.

We want to evaluate the *initial* transverse radii and parton densities, not those of the fireball at freeze-out after a hydrodynamic expansion. The multiplicity is, however, the *final* one, but — due to (approximate) entropy conservation during hydrodynamics — we think of it as a proxy for the entropy at all times. The entropy generated by viscosity *during* expansion is relatively small and can be corrected for if needed.

(i) Our most studied case, the central *AuAu* or *PbPb* collisions, is the obvious benchmark. With the total multiplicity about $N_{AA} \approx 10^4$ and transverse area of nuclei $\pi R_A^2 \approx 100$ fm^2 one gets the density per area as

$$n_{AA} = \frac{N}{\pi R_A^2} \sim 100\,\text{fm}^{-2} \tag{20.7}$$

This can be transformed into entropy if needed in a standard way.

(ii) Central *pA* collisions (up to a few percent of the total cross-section) have a (CMS-defined) track multiplicity of about 100. Accounting for the unobserved range of $p_\perp, y$ and neutrals increases it by about a factor of 3, so $N_{pA}^{\text{central}} \sim 300$. The area now corresponds to the typical impact parameter b in *pp* collisions, or $\pi\langle b^2\rangle = \sigma_{pp} \approx 10\,\text{fm}^2$. The density per area is then

$$n_{pA}^{\text{central}} = \frac{N_{pA}^{\text{central}}}{\sigma_{pp}} \sim 30\,\text{fm}^{-2} \tag{20.8}$$

or one-third of that in central *AA* collisions. Using the power of LHC luminosity, CMS can reach — as a fluctuation with the probability 10^{-6} — another increase in the multiplicity by a factor of 2.5 or so, up to a density $N_{pA}^{\max}/\sigma_{pp}$ in *AA* collisions. Another approach used is a comparison of central *pA* with peripheral *AA* collisions of the same multiplicity, or more or less the same number of participants, or similar matter density.

(iii) Now, we move on to the last (and most controversial) case, of high-multiplicity *pp* collisions. (Needless to say, the density is very low for minimum-bias events.) "High multiplicity" at which CMS famously discovered the "ridge" starts from about $N_{pp}^{\max} > 100 * 3$ (again, 100 is the number of CMS recorded tracks and 3 denotes extrapolation outside the detector covered).

The big question now is: *What is the area?* Unlike in the case of central pA, we don't utilize the standard Glauber and full cross-sections (maximal impact parameters). We instead address a fluctuation which has a small probability. In fact, nobody knows the answer to this question. Based on the profile of pp elastic scattering (to be discussed in Section 26.2), it should correspond to the impact parameter b in the black disk regime. If so, $\pi b_{\mathrm{b.d.}}^2 \sim 1/2\,\mathrm{fm}^2$, which leads to a density per area of

$$n_{pp}^{\max} \approx \frac{N_{pp}^{\max}}{\pi b_{\mathrm{b.d.}}^2} \sim 600\,\mathrm{fm}^{-2} \tag{20.9}$$

Other evidence about glue distribution in a proton comes from the HERA diffractive production, especially of $\gamma \to J/\psi$. They also suggest an r.m.s. radius of only 0.3 fm, less than half of the electromagnetic radius.

A brief summary: In terms of the initial entropy density, one expects the following order of the densities-per-area involved:

$$\frac{dN^{pA}_{\mathrm{maximal}}}{dA_\perp} \sim \frac{dN^{AA}_{\mathrm{peripheral}}}{dA_\perp} \ll \frac{dN^{AA}_{\mathrm{central}}}{dA_\perp} \ll \frac{dN^{pp}_{\mathrm{maximal}}}{dA_\perp} \tag{20.10}$$

One may expect that the radial flow follows the same pattern; yet, the data show that this is *not* the case.

20.7 Size versus Radial Flow for Central pA Collisions

The simplest consequence of the radial flow is an increase in mean transverse momentum. CMS data as a function of multiplicity are shown in Fig. 20.7(a). While pp and pA data are denoted by points, the AA ones (from ALICE) are shown by shaded areas. (The most central ones correspond to the upper edge of this shaded region.)

The next experimental signature of the radial flow is that the blue-shifts it induces modify the spectra of secondaries of different masses differently. While light pions retain their exponential shape of the p_t spectra, only with a blue-shifted slope,

$$T' = T_f e^{\kappa} \tag{20.11}$$

the spectra of massive particles change their shape. Eventually, for very heavy particles (e.g. d or other nuclei), their thermal motion is negligible, and their spectra depend completely on collective velocities. Their distribution has a characteristic peak at the fireball's edge, with $v = v_{\max}$, and thus the $p_\perp$ spectra develop a peak at $p_\perp = m v_{\max}$.

More specifically, flow creates a "violation of the $m_\perp$ scaling" (Shuryak and Zhirov, 1979). The $m_\perp$ slopes T', defined by the exponential form

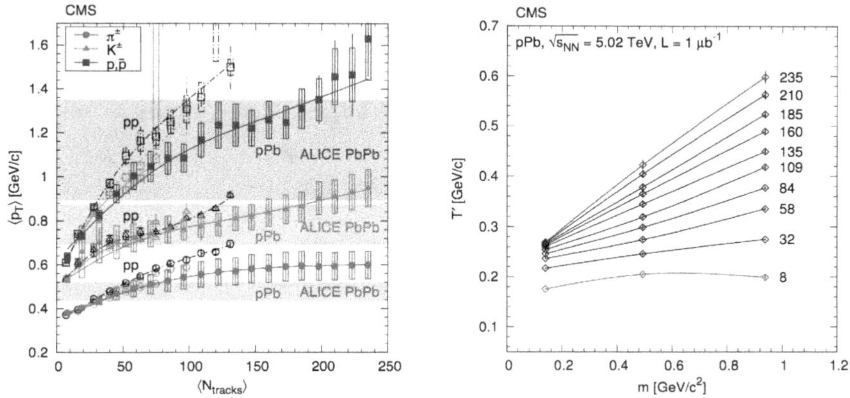

(a) (b)

Fig. 20.7. (a) Average transverse momentum of identified charged hadrons (pions, kaons, and protons; left panel) in the range $|y| < 1$ as a function of the corrected track multiplicity for $|\eta| < 2.4$ for pp collisions (open symbols) at several energies and for pPb collisions (filled symbols) at $\sqrt{s_{NN}} = 5.02$ TeV. (b) The slopes of the $m_\perp$ distribution T' (in GeV) as a function of the particle mass. The numbers on the right side of the lines give the track multiplicity.

(above certain p_t)

$$\frac{dN}{dydp_\perp^2} = \frac{dN}{dydm_\perp^2} \sim \exp\left(-\frac{m_\perp}{T'}\right) \tag{20.12}$$

are very sensitive indicators of the radial flow. A sample of such slopes for pA collisions, from CMS, is shown in Fig. 20.7. Note that for low-multiplicity bins (marked by 8 and 32 at the bottom right), one sees the same T' for all secondaries. This is the $m_\perp$ scaling: implying that flow is absent. This behavior is natural for independent string fragmentation, rescattering or the GLASMA models.

Flow manifests itself at higher-multiplicity bins, in which the slopes T' are mass-dependent. As seen from Fig. 20.7(b), they are growing approximately linearly with the particle mass. The effect becomes more pronounced with multiplicity: This is a sign of a stronger collective flow. Note that for the most central pA bin, this slope *exceeds*[6] those in central *PbPb* collisions at the LHC, the previous record-violent explosion!

This gives rise to what we call *the radial flow puzzle*. Indeed, naive estimates of densities in the previous section may suggest that the explosion

[6]Predicted to happen before the data; check version v1 of Shuryak and Zahed (2013).

in the highest-multiplicity pA case should still be *weaker* than in AA collisions. Indeed, both the system and the initial entropy density seem to be smaller. Yet, the data show the opposite: The observed radial flow strength (expressed via the magnitude of the transverse rapidity) follows a different pattern:

$$y_\perp^{AA,\text{central}} < y_\perp^{pA,\text{central}} < y_\perp^{pp,\text{highest}} \tag{20.13}$$

Hydrodynamics is basically a bridge between the initial and final properties of a system. For the radial flow dependence on the size of the system, it is convenient to follow Shuryak and Zahed (2013) based on Gubser's flow; see Section 10.7. One single analytic solution describes all cases considered: We proceed from the dimensional variables $\bar{\tau}, \bar{r}$ with the bar to the dimensionless variables

$$t = q\bar{\tau}, \quad r = q\bar{r} \tag{20.14}$$

rescaled by a factor q, with the dimension of an inverse length. In such variables, there is a single Gubser solution of ideal relativistic hydrodynamics for the transverse velocity (10.54) and energy density (10.55):

$$\begin{aligned} v_\perp(t,r) &= \tanh(y_\perp) = \frac{2tr}{1+t^2+r^2} \\ \frac{\epsilon}{q^4} &= \frac{\hat{\epsilon}_0 2^{8/3}}{t^{4/3}\left[1+2(t^2+r^2)+(t^2-r^2)^2\right]^{4/3}} \end{aligned} \tag{20.15}$$

The first equation — the radial velocity — has no parameters at all, but the second has the second dimensionless parameter of $\hat{\epsilon}_0$ related to the entropy and thus the multiplicity:

$$\hat{\epsilon}_0 = f_*^{-1/3}\left(\frac{3}{16\pi}\frac{dS}{d\eta}\right)^{4/3} \tag{20.16}$$

where the second equation fixes the shape of the freeze-out surface, usually an isotherm $T = T_f$. So, if for a particular case the parameters $q, \hat{\epsilon}_0$ are known, they can be found from the second equation.

Before turning to the actual plot of such surfaces, let us think about how two different collisions may be "conformal copies" of each other, with the size of the system and the inverse temperatures adjusted by the same factor. In the dimensionless variables, those merge into a single one. In fact, the blue line[7] marked AA in Fig. 20.8 corresponds not

[7] For the record, its parameters are $q = 1/4.3$ fm, $\hat{\epsilon}_0 = 2531, T_f = 120$ MeV.

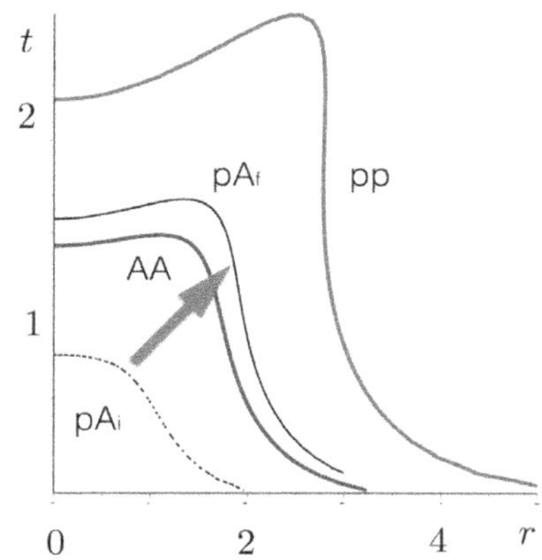

Fig. 20.8. The freeze-out surface in universal dimensionless time t and radial distance r coordinates. The (blue) thick solid line in the middle corresponds to central AA ($PbPb$) collisions, and the (red) thick solid line on the top corresponds to the highest multiplicity pp. Two (black) thin ones correspond to the central pPb case before and after collapse compression, marked pA_i, pA_f, respectively. The arrow connecting them indicates the effect of multi-string collapse.

only to central $PbPb$ collisions but actually to any other AA collisions at the LHC. The two black lines are for the pPb case: They both have $T_f = 170$ MeV and the same multiplicity but different scale parameters: $q = 1/1.6\,\text{fm}$ for the lower dotted line, but twice as small as the initial size $q = 1/0.8$ fm for the upper (thin solid black) line. As an arrow indicates, in order to explain the observed higher explosion velocity, one has to move the curve above the blue line. This implies that hydrodynamics must be initiated from a smaller "compressed" size, according to the "spaghetti collapse" scenario we will discuss in Chapter 26. If this is done, the freeze-out surface "jumps over" our AA benchmark blue line, and its radial flow gets stronger. The maximal transverse velocities on these curves (located near the turn of the freeze-out surface downward) are

$$v_\perp^{pAu} = 0.56 < v_\perp^{AA} = 0.81 < v_\perp^{pAu,f} = 0.84 \tag{20.17}$$

The upper red line is our guess for the maximal-multiplicity pp collisions, assuming its $q = 1/0.5$ fm. It has even stronger radial flow, with maximal $v_\perp^{pp} \approx 0.93$. So, paradoxically, small systems are in fact larger than AA in the appropriate dimensionless variables, and that is why their radial flow is better developed.

In summary, the observed pattern of radial flow can be explained, provided the initial size of the pA system is reduced as compared to the naive estimates of the preceding section.

20.8 Femtoscopy of Small Systems versus Radial Flow

Before we turn to femtoscopy, let us again formulate the logistics of what people try to do for small systems. Since we have checked the hydrodynamical description of flows in AA collisions in detail, we are confident that we understand both the initial and final stages of these collisions fairly well. For more uncertain cases of "small systems," pA, pp, one can then reverse the argument. Femtoscopy tells us about both the size and flow seen at the *final* stage of the fireball. Then, using hydrodynamics, one can find out the size and shape of the initial stage.

Let us then recall how the flow reveals itself in femtoscopy. As we already discussed previously, the observed radii show a decrease with an increase in the (total) transverse momentum $\vec{k_1} + \vec{k_2} = \vec{k_\perp}$ of the pair (Makhlin and Sinyukov, 1988). A modification of the argument is explained in a sketch shown in Fig. 14.4. At small $k_\perp$, the detector sees hadrons emitted from the whole fireball, but the larger the $k_\perp$, the brighter the small (shaded) part in which the radial flow is maximal in the direction of $\vec{k_t}$. This follows from maximization of the Doppler-shifted thermal spectrum $\sim \exp(p^\mu u_\mu / T_{\text{freezeout}})$. The other half of the fireball, moving in the opposite direction, is red-shifted and basically invisible.

This effect has been calculated (Hirono and Shuryak, 2015) and compared with the ALICE HBT data (Aamodt *et al.*, 2011a) in Fig. 20.9. The strong flow in high-multiplicity pp collisions is directly visible in the data. The effect is best seen in the "out"-directed radius R_{out} (the top plot). While the low-multiplicity data (e.g. the lowest-multiplicity ones, connected for visibility by the blue dashed line) are basically independent of the pair momentum, the data at high multiplicity (shown by the open red stars and red dashed line) show a significant *decrease* in the visible size of the system by a rather large factor. Another consequence of the flow is the anisotropy of the radii. In the bottom plot, the ratio of two radii is shown. At small multiplicity, it is always close to one, which means the source emits pions isotropically; however, at high multiplicity, the source becomes very anisotropic, and the radii in two directions, out and side or x and y, are quite different, with their ratio dropping to about $1/3$ at the largest k_t. Thus, a direct consequence of the flow, due to which the *visible*

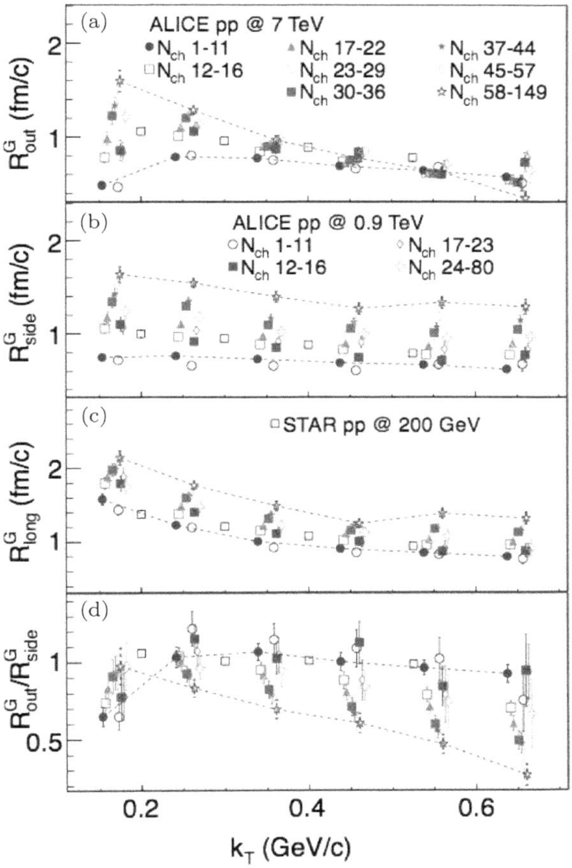

Fig. 20.9. HBT radii versus the pair transverse momentum k_T for various multiplicities of the pp collisions from ALICE.

size of the fireball, as seen by pairs of variable total momentum, can be different from the actual spatial size of it.

In Fig. 20.10, we show a series of calculations in which the initial QGP stage of the collision is modeled by a numerical hydrodynamics solution close to Gubser's analytic solution with the variable parameter q. (The late stages need to deviate from Gubser since, near T_c, the EOS is very different from conformal $\epsilon = 3p$ assumed in Gubser's derivation). Comparisons of these theoretical calculations with ALICE data suggest a very strong flow for the high-multiplicity pp.

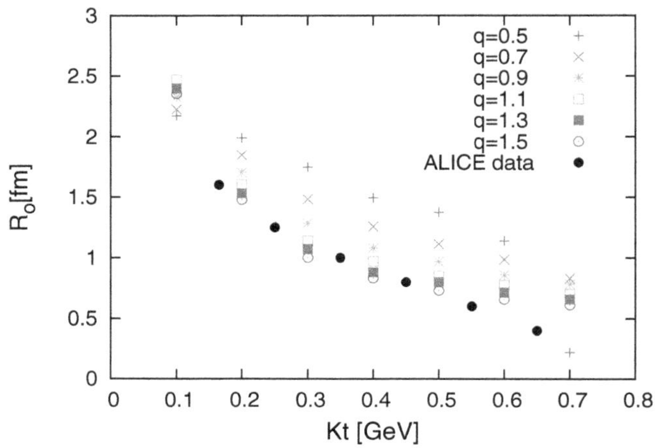

Fig. 20.10. HBT radii compared to the ALICE data (closed circles) for solutions starting with different initial sizes of the fireball, indicated by the Gubser scale parameter q (which is inversely proportional to the size).

In general, we therefore arrive at very different overall pictures of AA and pp collisions.

In the AA case, collisions of different centralities create a very similar initial matter density, and the flow is strongest for the *largest* systems, corresponding to central collisions. The dependence on centrality is related to different freeze-out conditions.

In the pp case, on the contrary, the flow dependence on multiplicity is due to variable initial matter density, while at the final state, the matter is about the same for any multiplicity.

The pA case remains somewhat uncertain. Only recently, rather complete femtoscopy data were obtained, see Fig. 20.11 from ATLAS (Aaboud *et al.*, 2017). Data points of different colors are for different centrality bins, and the radii are plotted versus total pair momentum k_T. First of all, the central collisions (as already mentioned above, including about 16 participant nucleons) lead to rather large fireballs of size $R \sim 4.5$ fm, thus challenging the denomination of such collisions as "small systems." Second, unlike the ALICE pp data, all centralities show a marked decrease in radii with k_T, indicating significant flow for all centralities. While quantitative theoretical analyses of these data have not been performed yet, it suggests that, in fact, pA collisions are more similar to AA than to pp.

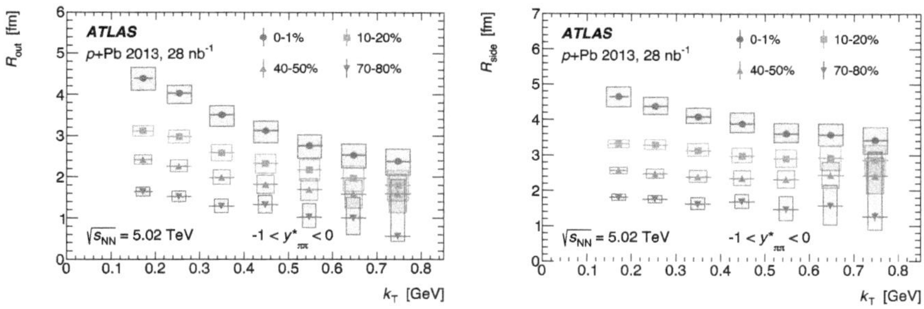

Fig. 20.11. The dependence of the R_{out} and R_{side} on k_T, ATLAS data.

Unlike central pA collisions, the highest-multiplicity pp events are significantly denser/hotter than central AA collisions. A very strong radial flow, seen in the spectra of identified particles and HBT radii, require a very small — sub-femtometer — initial size of the system. Despite the high cost associated with those events, their studies are justified because here we produce the most extreme state of matter ever created in laboratory conditions.

Summary and connections: In this chapter, we complemented our discussion of heavy ion collisions by considering systems of size ~1 fm. It has been shown that central pA and even high-multiplicity pp collisions show evidence of robust collective flows, consistent with hydrodynamical explosions. Femtoscopy confirmed this conclusion. From the perspective of the sQGP theory, it is a natural consequence of its quasi-conformal properties, due to which only the size times the temperature matters. The RHIC data with deuteron and He^3 collisions have given us additional confidence in the hydrodynamical explanation of the flows.

Chapter 21

HOLOGRAPHIC MODELS OF STRONGLY COUPLED QGP

21.1 Introduction to Holography

This chapter is devoted to the gauge–gravity duality, which allows us to study the *strongly coupled* limit of gauge theories in vacuum and/or at high temperatures/densities. It is a large field, requiring significant background in general relativity (GR), and it is not possible to present it here in full. Our aim would be to describe the logic of the construction, present the settings of the problem involved and discuss the main applications to strongly coupled QGP. Holography applications to QCD vacuum and hadrons are presented in Chapter 29.

The timeline of key discoveries is as follows:

- The discovery of AdS/CFT correspondence (Maldacena, 1998a), acting as starting point of this development, which, a decade later, became an important tool of nuclear theorists and high-energy phenomenologists, as well as the prime example of the "applications" in string theory.
- Black hole metric and the thermodynamics of the strongly coupled gluodynamics with extended[1] $N = 4$ supersymmetry (Gubser *et al.*, 1998).
- Shear viscosity and other transport properties derived from linearized hydrodynamics (Policastro *et al.*, 2002).
- General derivation of the full nonlinear hydrodynamics, from the gradient expansion of the Einstein equations (Bhattacharyya *et al.*, 2008a, 2008b; Natsuume and Okamura, 2008).

[1]This N, here and elsewhere in this chapter, indicates the number of supersymmetries. It should not be confused with the number of colors, which is here and elsewhere denoted by N_c.

- Specific studies of higher gradient effects and proposed resummation in "all order hydrodynamics" (Lublinsky and Shuryak, 2009).
- Formulation of the "top-down" equilibration mechanism (Lin and Shuryak, 2008b).
- Numerical solution to a number of out-of-equilibrium problems, such as high-energy collisions.
- The issue of entropy production and its evaluation via "trapped surfaces."
- A charge propagating in QGP and the jet quenching problem.

The subject is divided into this chapter, devoted to the properties of sQGP in thermal equilibrium, and another one, dealing with issues of equilibration.

21.2 The Building Blocks and the Setting

21.2.1 *The strings*

The historical origins of string theory can be traced to pre-QCD studies of hadronic scattering, which reveal Regge trajectories and Veneziano scattering amplitudes.[2] With the advent of QCD and the electric confinement phenomenon, physicists learned about the QCD flux tubes, also called "QCD strings."

The idea that strings can be fundamental objects, laying the foundation for a separate *string theory*, came next. String quantization has, however, shown that their consistent description can only be made in certain special space-time dimensions: 26 for bosonic strings, reduced to 10 or 11 for *superstrings*. String theory is a vast subject which we will not be able to discuss in full.[3]

Quantized closed strings form certain objects, some of which are *massless*. Those include the spin-2 *gravitons* as well as spin-1 vectors and spin-0 scalars. Such a set of massless states is the basis for the string theory's main aspiration: to be "a theory of everything" (TOE), combining gauge and gravity fields known in nature together. This aspect, however, is not discussed here.

---

[2]Which we briefly touched on in the Introduction chapter and then returned to in Chapters 26 and 24.

[3]Neither because of the limited space in this book nor due to my own qualifications.

Instead, we are interested in string theory as a *mathematical tool*,[4] which, via the gauge–string duality to be discussed, helps us solve certain problems related to strongly coupled four-dimensional gauge theories.

For now, it is enough to state that string histories in space-time are described by certain surfaces with a co-dimension of two, moving in the external space of dimension D and described by coordinates $X^\mu(\tau, \sigma)$. Here, τ, σ are internal time-like and space-like coordinates, and $\mu = 1, \ldots, D$, say $D = 10$.

21.2.2 *D-branes*

One of several "string theory revolutions" led to the discovery of solitonic objects of (nearly) any[5] dimensions p (ranging $p = -1, \ldots, 9$) called D_p *branes*. One should think of them as some p-dimensional membranes embedded in a higher dimension.

For our purposes, the most important example is the D_3 brane in $9 + 1$-dimensional space + time. It has infinite extension along three coordinates, x_1, x_2, x_3 (plus, of course, time denoted by x^0), while it has no extension (and thus is just a point object) in the remaining six coordinates, $x_4, \ldots, x_9$. For example, if one wants to plot it in the plane (x^1, x^9) (as shown in Fig. 21.1), it will correspond to a line parallel to the former

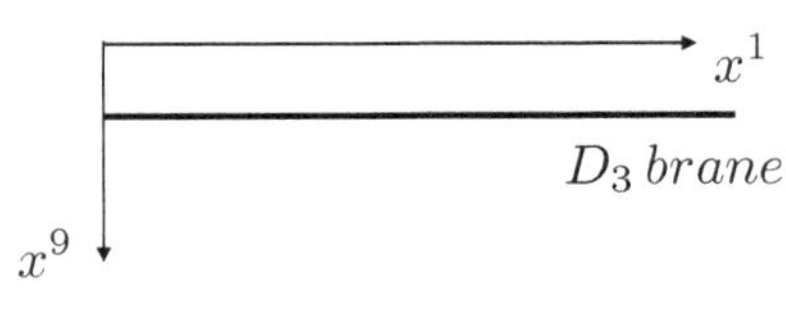

Fig. 21.1. A sketch of the D_3 brane (thick solid line) as it appears in the $x^1 - x^9$ plane.

[4]As an analogy, consider the extension of real numbers to a complex plane. As Cauchy had explained, some integrals, complicated on the real axes, become much simplified in the complex plane. Some can even be reduced to the sum of the point-like objects, the *poles*.

[5]The $p = 1$ object has time and its own space coordinate, so it is a string. The $p = 0$ object has only time, so it is a particle. What can then the $p = -1$ object possibly be? It is point-like in both *space and time*, the "instanton."

axes since it has no extension in x^9. In the plane (x^8, x^9), it would appear as a point since, for both these coordinates, the D_3 brane has a certain location but no extension. In this case, the first three coordinates are called *internal*, and the rest are *external* ones.

The branes are solitons, material objects with certain mass densities, tensions and also some charges, coupled to vector and gravity fields, as prescribed by their construction. Therefore, a static brane sources gravity/Coulomb/scalar fields around them. Finding those is similar to calculating the electrostatic field of a homogeneously charged slab with a certain number of internal and external dimensions. Because the bulk fields are weakly[6] interacting to the lowest order, it is just given by their propagator, inverting the Laplacian in the corresponding external dimensions.

Since the D_3 brane has no extensions in the six external coordinates $x_4 \ldots x_9$, this Laplacian is six-dimensional. Therefore, its analog of the familiar Coulomb law is $\sim 1/r^4$, where r is the distance from the brane in external coordinates. Such is the asymptotic behavior of their massless fields at large r. At finite r, the field becomes strong and self-interacting. Gravity, governed by GR, implies that they are static black hole Schwarzschild-like solutions in the six external coordinates. As there is no dependence on the $p = 3$ internal coordinates, the proper name of such objects is *black branes*.

To determine their fields in classical approximation (no loops or quantum fluctuations), one has to solve a system of coupled Einstein–Maxwell scalar equations in a static, spherically symmetric (in external coordinates) case. This is no more difficult than finding the Schwarzschild solution for a typical black hole, which, as we recall, has a metric tensor in spherical coordinates:

$$ds^2 = g_{\mu\nu} dx^\mu dx^\nu = -(1 - r_h/r)dt^2 + \frac{dr^2}{(1 - r_h/r)} + r^2 d\Omega^2 \tag{21.1}$$

with the horizon radius $r_h = 2G_N M/c^2$, depending on mass M and the Newton constant G_N. The "event horizon" is zero of g_{00}, it is limiting the part of space visible to a distant observer. Similar expressions, with powers of the distance appropriate for the number of external coordinates, describe gravity around p-branes.

There are, however, some important aspects in which those black branes are very different from the Schwarzschild black holes. One is related to

[6]This will be explained in the following.

their mutual interactions. Naturally, gravity and scalar exchanges lead to attraction, while exchanged vector fields between identical branes lead to repulsion. Surprisingly,[7] the charges/masses of D_p branes are such that all these forces mutually *cancel*, so there are no classical interactions between two D_p branes!

As a result, "classical brane engineering" is greatly simplified: One may simply stick a number of such branes to some random points, and they will stay there. It does not work with loop effects, but one may use the following trick. Let a large ($N_c \to \infty$) number of branes (a stack) be put into the *same* point. Since they combine their masses/charges and create strong gravity/fields, in this limit one justifies the use of *classical* Einstein–Maxwell equations for which loop corrections are not needed.

The next surprising feature is that the so-called *open strings* may keep their ends on the branes. If those two have numbers i and j in the stack, these strings generate effective gauge theory on the brane, with the non-Abelian $U(N_c)$ gauge group.

When the black hole has a nonzero vector charge, Gauss' law insists on constant flux through a sphere of any radius; thus, there are nonzero fields at large r and the Schwarzschild solution morphs into a "charged black hole." For asymptotically flat spaces, there is the famous statement, much emphasized by Wheeler: There are "no scalar hairs" of black holes, as there is no Gauss theorem to protect them. This happens to not be true in general, and for spaces which are AdS-like, the scalars should be surrounding the black hole as well.[8]

The other important difference from the familiar black holes is that the solution for D_3 branes happens to be the so-called *extremal* (six-dimensional, or 6D) charged black hole. It is called extremal because it has the lowest possible mass for a given charge. When the mass reaches its extreme value for a given charge, the black hole horizon shrinks to *zero*. Thus, branes are black holes *without* a horizon.

Spherical symmetry in 6D allows one to separate the five angles (making the five-dimensional (5D) sphere S_5) from the radius r (in 6D orthogonal to "our world" space $x_4, \ldots, x_9$) and write the resulting ten-dimensional

[7]To anyone not involved in monopoles/supersymmetry research, where this so-called BPS protection is usual. We will discuss it when we study monopoles.

[8]In fact, there is a whole recent direction based on a solution with a "scalar atmosphere" around black branes, providing a gravity dual to some "strongly coupled superconductors" on the boundary.

metrics as follows:

$$ds^2 = \frac{-dt^2 + d\vec{x}^2}{\sqrt{1 + L^4/r^4}} + \sqrt{1 + L^4/r^4}(dr^2 + r^2 d\Omega_5^2) \tag{21.2}$$

which here and below we would indicate by vector indices 1, 2, and 3. As usual, at large r, all corrections are small, and the solution tends to an asymptotically flat space.

At this point, Maldecena suggested further simplification of this metric. In the near-field region," at $r \ll L$, the 1 in both roots can be ignored. If so, in the last term, the two r^2 cancel out and the 5D sphere element gets a constant coefficient. This means that $4 + 1 = 5$ coordinates $t, \vec{x}, r$ get decoupled from five other coordinates, with the angles in 6D. Thus, quantum numbers or motion in S^5 become kinds of internal quantum numbers, like flavor in QCD: From now on, those will be mostly ignored. What is left for us to work with is a very simple 5D metric known as the anti-de-Sitter metric. Using a new coordinate $z = L^2/r$ (and thus $dz = -drL^2/r^2$), one can put it into the "standard AdS_5 form":

$$ds^2 = g_{\mu\nu}dx^\mu dx^\nu = \frac{1}{z^2}\left(-dt^2 + d\vec{x}^2 + dz^2\right) \tag{21.3}$$

(The reader is encouraged to check that this metric satisfies the Einstein equation of motion

$$R_{MN} - \frac{1}{2}g_{MN}R + \Lambda g_{MN} = 0 \tag{21.4}$$

with a certain nonzero cosmological constant Λ and find its magnitude. A recommended way to do it is to write in Maple or Mathematica the standard definitions of the Christoffel symbol and Riemann curvature tensor and then substitute this metric into those.)

Note that z counts the distance from "the AdS boundary" $z = 0$. This metric has no scale and is *not* asymptotically flat, even at the boundary. Performing dilatation on these five coordinates, we find that the metric remains invariant. In fact, it remains invariant for any conformal transformation.

21.2.3 *The AdS/CFT correspondence*

It is this metric which led Maldacena and others to the discovery of the "holographic gauge–gravity duality."

Holography, in general, is a d-dimensional representation of a D-dimensional object, with $D > d$. For example, holograms that the reader may have seen are $d = 2$ representation of a $D = 3$-dimensional object, e.g. a portrait of a person. In black holes physics, the idea of "holography" stemmed from the Bekenstein expression for its entropy: It is proportional to the horizon *area*, not the black hole volume, and yet it presumably contains all the information about it, even its formation.

In the problem at hand, the setting is the AdS_5 5d space-time of the extreme black hole described above (given above). It has neither a horizon area nor entropy, but it still has a boundary at $z = 0$, which is flat 4d space-time. On it, some gauge theory can be defined, and it is the vacuum of this theory, which turned out to be a *hologram* to the AdS_5 5d space-time. What could this this theory possibly be?

The initial argument by Maldacena was based on relating the symmetries. Since the AdS_5 space-time has conformal symmetry, its *hologram* must also possess the same (conformal) symmetry. One 4d gauge theory which is conformally invariant (to all loops, in full quantum formulation, or CFT) is that with $N = 4$ supersymmetries. Maldacena conjectured that string theory in AdS_5 and 4d $N = 4$ gauge theory are "holographically equivalent."

For readers not familiar with supersymmetric theories, it is perhaps beneficial to add a bit of information about the $N = 4$ super-Yang–Mills (SYM) theory. This is a cousin of QCD with non-Abelian gauge fields possessing the $SU(N_c)$ color symmetry. Instead of quarks, it has four species of fermions (numerated by an extra index $i = 1, \ldots, N$), which are supersymmetric partners of gluons, called gluinos. Unlike quarks, their color representation is adjoint with $N_c^2 - 1$ components, similar to gluons. There are also six scalar fields, with the adjoint color as well. Together with two polarizations of gluons, they make $2 + 6 = 8$ bosonic modes, the same number as as 2×4 fermionic ones. There are thus four supersymmetries (mixing gluons with corresponding gluinos), leading to cancellations of power and even logarithmic divergences. It has conformal symmetry, which includes dilatation or scale change, as a consequence of which the coupling *g does not run* at all to all loops!

How do we know this? Of course, one has to calculate first a few perturbative coefficients of the beta function and indeed see that the negative gauge contribution is nicely canceled by that of fermions and scalars, order by order. However, there are infinitely many coefficients, and one has to check them all! An elegant way to prove the case is based on another outstanding feature of the $N = 4$ SYM: This theory is *self-dual* under the electric–magnetic duality. As we discussed above, the

Dirac condition requires the product of electric and magnetic couplings to be constant; therefore, in QCD and other gauge theories, they indeed run in the opposite directions, with electric becoming weak in ultraviolet and magnetic becoming weak in infrared. But the multiplet of (lowest) magnetic objects of the $N = 4$ SYM theory includes six scalars (the monopoles), plus four fermions (monopoles plus one gluino zero mode occupied), plus two spin-1 (monopoles with two gluinos): This turns out to be exactly the same set of states as the original electric degrees of freedom (gluons–gluinos–Higgses). That means an effective magnetic theory has the *same* Lagrangian as the original electric formulation. Thus, it must have the same beta function. Since two couplings cannot run in the opposite direction following the same beta function, they cannot run at all!

Recall that in QCD-like theories, the fundamental scale λ_{QCD} comes into play because of its coupling "running." However, because the $N = 4$ SYM theory is conformal (its coupling constant does not run!), there is no analog of λ_{QCD} in this theory: All the fields are massless and all the coupling constant are dimensionless and constant. Thus, this $N = 4$ SYM theory has no scales at all. It is hence a conformal field theory — the CFT in the AdS/CFT correspondence.

(One consequence is that the finite-T version of this theory, which we will be mostly interested in, is the same whether T is large or small since there is no other scale to compare with! This is similar to QCD plasma in the so-called "quasi-conformal regime": At high enough $T > 2T_c$, all dimensionless ratios, such as (energy density)$/T^4$, are practically T-independent.)

Why are we interested in this rather complicated theoretical construction? Because AdS/CFT correspondence allows us to study gauge theory in the *strong coupling* regime via its holographic dual 5d gravity in weak coupling.

More precisely, we are using a large value of the so-called 't Hooft coupling, a combination of small gauge coupling g and a large number of colors which go together:

$$\lambda = g^2 N_c \sim \text{const} \gg 1 \tag{21.5}$$

Maldacena (1998a) had not just conjectured the correspondence but had shown how to use its consequences. For example, we need to find correlators of some gauge-invariant (colorless) operators in the 4d boundary theory, e.g. two stress tensors $T_{\mu\nu}$ at some points. These operators are coupled to certain fields in the 5d AdS "bulk," e.g. stress tensors coupled to the gravity

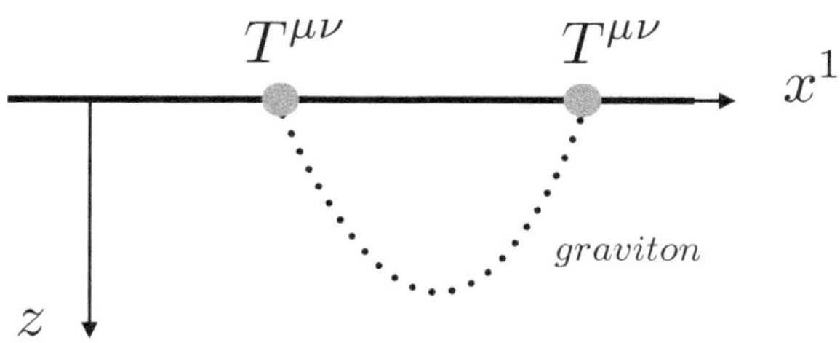

Fig. 21.2. A sketch indicating gravity-based calculation of the correlator of the two stress tensors on the AdS_5 boundary.

metric $g_{\mu\nu}$, see Fig. 21.2. Calculating correlators in both ways allows us to test the conjecture,[9] and it was shown to be true for certain propagators and even multi-point Green functions.

All tests performed were true, so the Maldacena conjecture was declared true in general. This created a large industry. The applications of holography can be divided into two categories. In the "top down" case, one looks for other cases in which an exact correspondence can be found, involving theories other than the $N = 4$ SYM theory; yet, it seems impossible to find such a correspondence for non-supersymmetric QCD-like theories. The opposite idea, known as the " down up" approach (also called AdS/QCD) builds holographic models imitating any theory without a string theory derivation. The $4 + 1$-dimensional background is, in this case, a solution to some phenomenologically selected action. For a review of this approach, see, for example, Gursoy and Kiritsis (2008). One bulk field in 5d — dilatons or gravity — generates many "radial excitations" of hadrons and their Regge trajectories. Such models were developed so that they include confinement and chiral symmetry breaking and even reproduce the key elements of QCD spectroscopy of mesons and glueballs.

At this point, many readers are probably very confused by a new (and curved) fifth-dimension AdS space. One should think of it as just a mathematical trick, somewhat analogous to a more familiar introduction to complex variables.[10]

[9]Which was a popular occupation from the late 1990s to early 2000. Note that the bulk diagram contains just a classical (no loops) propagator, while the corresponding correlator in the gauge theory starts with one loop since the stress tensor operator emits two gluons.

[10]Let me explain the analogy by a hypothetical situation. Suppose an Experimentalist measured some complicated cross-section which is approximately a sum of Breit–Wigner

However, the physical meaning of the fifth coordinate is, in fact, perfectly physical. One hint is provided by the fact that distance along it, $\int_a^b dl = \int_a^b dz/z = \log(b/a)$, is the logarithm of the ratio. Thus, its meaning is the "logarithmic scale," as we use in the renormalization group. If one takes a bulk object and moves it into a larger z, its hologram at the boundary ($z = 0$) grows in size: This direction can be said to correspond to RG flow into the infrared (IR) direction.

The running coupling constant would thus be a z-dependent field called "dilaton." Indeed, there are theories with a gravity dual in which this field (related to the coupling) does "run" in z. Unfortunately, known examples do not (yet?) include QCD! Despite that, there are efforts to build its gravity dual "bottom-up," introducing weak coupling at ultraviolet (small z) (Shuryak, 2007) and confinement in IR (large z) (Karch *et al.*, 2006b; Gursoy and Kiritsis, 2008) by certain modifications of the rules. These approaches — known as AdS/QCD — we will discuss later.

21.3 Falling Objects in AdS_5 and Their Holograms

The very first problem of physics was that taken up by Galileo, who studied how various objects fall (from the famous Leaning Tower of Pisa) under gravity. Recall that he famously found that objects made of any material fall with *the same acceleration.* Newtonian mechanics explained it by the fact that the gravity force is proportional to particle mass, so that it cancels from the second-order equation of motion (EOM). (We now call it the identity of gravitational and inertial masses.)

In Fig. 21.3, we show the setting of the same problem, but with a "stone" (or a piece of string, or something else) falling under gravity into the AdS_5 space.

In GR, there is no mass dependence because (according to Einstein) gravity is just geometry. All objects simply move by the geodesics, with zero covariant acceleration. The equation of geodesic, replacing Newton's EOM, has the form

$$\frac{d^2x^\alpha}{ds^2} + \Gamma^\alpha_{\beta\gamma}\frac{dx^\beta}{ds}\frac{dx^\gamma}{ds} = 0 \tag{21.6}$$

resonances. His friend Phenomenologist may be able to write the answer as an analytic function with certain pole singularities in the complex energy plane, which will help in fitting and evaluating integrals. Even better, their other friend Theorist cleverly developed a "bulk theory," deriving the pole positions from some interaction laws on the complex plane.

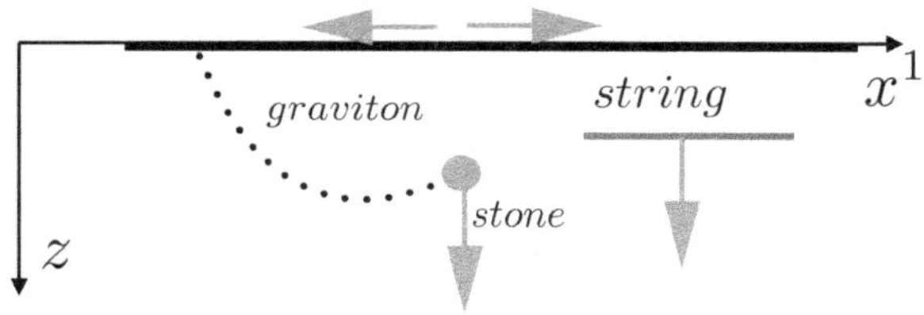

Fig. 21.3. A sketch of a "stone" (and a string) falling away from the AdS_5 boundary. The dotted line indicates gravity's influence on the boundary. The hologram of the falling stone is a certain exploding distribution of the energy/momentum density, indicated by blue arrows.

The role of the force is now given by the Christoffel symbol, related in the standard way to the metric

$$\Gamma^{\alpha}_{\beta\gamma} = \left(\frac{1}{2}\right) g^{\alpha\sigma} \left(g_{\sigma\beta,\gamma} + g_{\sigma\gamma,\beta} - g_{\beta\gamma,\sigma}\right) \tag{21.7}$$

Solving the problem this way, using this equation either for its $\alpha = 0$ (i.e. the time component) or 5 (i.e. z-component) is left to the reader as an exercise.

When we started thinking about this problem (Lin and Shuryak, 2008c) in AdS_5, we started with the action for a particle falling in the $z(t)$ direction:

$$S \sim \int dt \frac{\sqrt{1-(dz/dt)^2}}{z} \tag{21.8}$$

which is different from the usual action of a relativistic particle by $1/z$ coming from $\sqrt{-g_{00}}$ for the time line element. Euler's equations of motion for this action are then

$$\frac{d^2 z}{dt^2} = \frac{1-(dz/dt)^2}{z} \tag{21.9}$$

However, there is also a third and simplest way to address the problem: using *energy conservation*. In fact, since the metric $g_{\mu\nu}$ does not depend on x^0 (and also x^1, x^2, x^3), energy and the 1, 2 and 3 components of momentum are conserved. In GR, the conserved energy is mass times the *lower* component of the *zeroth* component of the four-velocity, so

$$u_0 = g_{00}\frac{dt}{ds} = \frac{1}{z^2}\frac{z dt}{\sqrt{dt^2 - dz^2}} = \text{const} \tag{21.10}$$

All three lead to the same solution:

$$z(t) = \sqrt{t^2 + z_0^2} \tag{21.11}$$

demonstrating that a particle velocity is approaching the speed of light, which (as is clear from $ds = 0$) in the z direction is also 1, as in others.

The falling happens for $z \to \infty$, which corresponds to $r \to 0$ or the location of the center of the extreme black hole we discussed above.

Before moving forward, let us discuss a slightly more complicated problem, also discussed by Lin and Shuryak (2008c). Suppose we drop into AdS_5 a piece of horizontal string, instead of a particle. As it is described by its location in the fifth coordinate $Z(t,x)$, the action changes to a 2d integral:

$$S \sim \int dt \int dx \frac{1}{Z^2}\sqrt{1 + \left(\frac{\partial Z}{\partial x}\right)^2 - \left(\frac{\partial Z}{\partial t}\right)^2} \tag{21.12}$$

now with a 2d volume element $\sqrt{-g_{00}g_{11}} \sim 1/z^2$. The EOM is therefore different: How can it be? The answer is that even a straight horizontal piece of string, falling in z, changes its length (and thus mass) on the way. Energy conservation applies differently to objects with variable mass (or changing internal energy).

Now, suppose the gauge theory (and us[11]) live on the $z = 0$ boundary. How would the event of a stone falling in the bulk be observed? The stone has the stress tensor $mu^\mu u^\nu$, which couples to gravitons, which can travel to the boundary and be sourced by another stress tensor $T^{\mu\nu}(x^0, x^1, x^2, x^3)$ there. This would generate *a hologram of the falling stone.*

Now, since the stone (or any other falling object) goes toward a larger z away from the boundary, its hologram expands in size, roughly as $r \sim t$ at a later time. This implies that the boundary observers (i.e. us) would see a spherically symmetric *relativistic explosion.* That looks similar to what we study in the course: heavy ion collisions. This hints that the complicated relativistic explosions we study in this book may have some *dual description in bulk theory,* perhaps as simple as a falling stone!

Is this explosion *hydro-like*? In particular, can the resulting stress tensor be parameterized by flow velocity and a single parameter of temperature? We did find the explicit form of $T^{\mu\nu}(x^0, x^1, x^2, x^3)$ and verified that, while the energy–momentum conservation $\partial_\mu T^{\mu\nu} = 0$ holds, the diagonalized $T^{\mu\nu}$ is *not* spatially isotropic. So, it is *not* an explosion describable by hydrodynamics, at least not in the *ideal* (local) approximation.

Here is another interesting aspect of the falling stone problem: One can think of a stone moving from infinity "upward," flying a certain distance

[11]To make it more interesting, think of the "brane world" cosmologies, in which we actually do live on D_3 embedded in some larger space. One can think of the Big Bang as a consequence of some large mass (another brane?) moving away from us.

toward the boundary and then falling back to the black hole centrum. This setting was discussed by Gubser *et al.* (2008a), who found a stress tensor distribution different from that in our paper (where only $t > 0$ part was calculated). All bulk EOMs above are obviously time-reversal ($t \to -t$) invariant and thus should be the boundary $T^{\mu\nu}(t)$. The explosion at $t > 0$ must be complemented by a symmetric implosion at $t < 0$. This consideration implies that *no amount of dissipation* (and thus entropy production) can, in principle, occur in such an explosion! How entropy growth can be explained by holography we will discuss later.

21.4 Using the Correspondence via the Bulk Strings

The colored degrees of freedom, together with their gauge invariance, are properties of the $d = 4$ boundary gauge theory. In the $D = 5$ dimensional bulk, there are no colored fields at all since the bulk fields can be sourced only by *colorless* operators.

Let me begin with our first example of the "AdS/CFT at work," related to the strong-coupling version of the Coulomb law calculated by Maldacena (1998b). The setting — to be called "the Maldecena dipole" — is shown in Fig. 21.4(a), which includes two static charges (heavy fundamental quarks) separated by the distance R.

At weak coupling, i.e. the usual QED, we think of one charge creating the electric potential in which the other is placed, leading to the usual

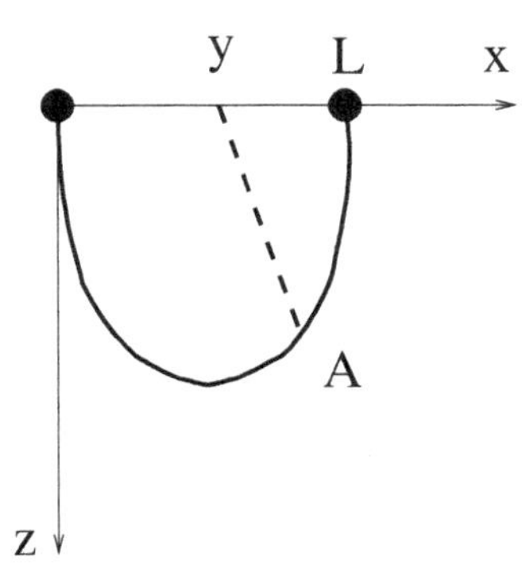

Fig. 21.4. Setting of the Maldacena dipole: Two charges at the boundary (black dots) are connected by the string (shown by a solid curve) pending under gravity toward the AdS center $z \to \infty$. Classical graviton propagator (the dashed line) should be used to calculate the "hologram": the stress tensor at the observation point y. The string is the gravity source; point A has to be integrated over.

Coulomb law, which in our notation is $V(L) = -(g^2/4\pi L)$ which is a sum of two Coulomb fields:

$$E_m(y)\left(\frac{g}{4\pi}\right)\left(\frac{y_m-(L/2)e_m}{|y_m-(L/2)e_m|^3}-\frac{y_m+(L/2)e_m}{|y_m+(L/2)e_m|^3}\right) \tag{21.13}$$

Because of a cancellation between two terms, at large distances from the dipole, the field decays as $E \sim L/y^3$. The corresponding energy density (and other components of the stress tensor) are thus of the order $T_{00} \sim g^2L^2/y^6$, with a certain "dipole" angular distribution.

In the $N = 4$ theory at *weak* coupling, the only difference is that one can exchange massless scalars on top of gluons. It is always attractive, and for two heavy quarks, it leads to cancellation of the force, with doubling for quark–antiquark (which we discuss now). The QED coupling g^2 changes to the 't Hooft coupling $\lambda = g^2N_c$ proportional to the number of colors N_c.

Now, we turn to the AdS/CFT for the N=4 theory at *strong* coupling (Maldacena, 1998b). The electric flux in the bulk forms a singular object — the string (shown by the solid curve in Fig. 21.4(a)) — which pends from the boundary $z = 0$ due to gravity force into the fifth dimension, as in the famous catenary (chain) problem.[12] The calculation thus follows from the Nambu–Goto action for the string containing the world volume area:

$$S = \frac{1}{2\pi\alpha'}\int d\sigma d\tau\sqrt{\det G_{MN}\partial_\alpha X^M\partial_\beta X^N} \tag{21.14}$$

where the two coordinates σ, τ parameterize the string world line $X^M(\tau,\sigma)$, with M, N being the space-time indices in the whole space (10d reduced to 5d in AdS/CFT). G_{MN} is the space metric, and det stands for 2×2 matrix with all α, β. In the AdS_5 metric, we need the components $-G_{00} = G_{11} = G_{55} = 1/z^2$, and we can think of σ, τ as our coordinates x, t: The string is then described by only one function $z(t, x)$, and its action is reduced to

$$S \sim \int dtdx\frac{1}{z^2}\sqrt{1+(\partial z/\partial x)^2-(\partial z/\partial t)^2} \tag{21.15}$$

We use this action for "falling strings" in the following; now, we proceed to further simplifications for a static string, for which there is no time derivative and the function is $z(x)$. Maldacena uses $u(x) = 1/z(x)$, and

---

[12]Another — more Einsteinian — way to explain it is to note that this is simply the shortest string possible: It is not straight because the space is curved, and going to a larger z corresponds to a shorter distance, as the AdS_5 metric tells us. It is the same reason why the shortest path from New York to London does not look straight on the map.

thus the Lagrangian becomes $L = \sqrt{(u_{,x})^2 + u^4}$ with the comma denoting the x-derivative. One more simplification comes from the fact that x does not appear in it; thus, an "energy" is conserved:

$$H = p\dot{q} - L = \frac{\partial L}{u_{,x}} u_{,x} - L = E = \text{const} \tag{21.16}$$

which reduces the EOM from the second-order equation to just

$$(u_{,x})^2 = u^4(u^4/E^2 - 1) \tag{21.17}$$

which can finally be directly integrated to

$$x(u) = \int_{u_m}^{u} \frac{du'}{(u')^2\sqrt{(u')^4/E^2 - 1)}} \tag{21.18}$$

The minimum position of the string u_m is related to E by the relation following from this formula at $x = L/2$. Plugging the solution back into action and removing divergence (which is independent of L) gives finally the total string energy, which is the celebrated *new Coulomb law at strong coupling*:

$$V(L) = -\frac{4\pi^2}{\Gamma(1/4)^4} \frac{\sqrt{\lambda}}{L} \tag{21.19}$$

The power of distance $1/L$ is, in fact, the only one possible by dimension, as the theory is conformal and has no scales of its own. What is remarkable is the (now famous) $\sqrt{\lambda}$ appearing instead of λ in the weak coupling. (The numerical coefficient in the first bracket is 0.228, to be compared to the result from a diagrammatic resummation in the following.)

What is the reason for this modification? For pedagogical reasons, let me begin with two "naive but reasonable guesses," both to be shown wrong later in this section:

(i) One idea is that a strongly coupled vacuum acts like some kind of a space-independent dielectric constant, $\epsilon \sim 1/\sqrt{\lambda}$, which reduces the effect of the Coulomb field similarly at all points.
(ii) Perhaps, such a dielectric constant has nonlinear effects and is thus not the same at different points. However, the fields created by the static dipole are still just the electric field $\vec{E}$.

A good feature of the AdS/CFT is that one can get many further details about the problem and test ideas like the one just expressed, but we cannot get *all* features. In order to understand the difference between dipole fields in weak and strong regimes, it would be nice to calculate the electric and scalar field distributions in both limits. Unfortunately, we cannot do that: Those fields do not belong to the limited set of quantities one can calculate via the AdS/CFT correspondence. In general, the AdS/CFT rules let us find *holograms* on the boundary of whatever happens in the bulk. In particular, this can be done after solving bulk wave equations for the bulk fields, scalars, vectors or gravitons. For example, consider our results for the boundary stress tensor obtained from the (linearized) Einstein equation for the gravity perturbations, shown by the dashed line in Fig. 21.4(a).

For example, the *stress tensor* of matter $\langle T_{\mu\nu}(y)\rangle$ at any point y on the boundary, induced by the Maldacena dipole, was found by Lin and Shuryak (2007). The solution is too technical to be presented here, and even the resulting stress tensor expression is too long[13]: Let me just show only the leading terms far from the dipole $y \gg L$.

$$T_{00} = \sqrt{\lambda}L^3 \left(\frac{C_1 y_1^2 + C_2 y^2}{|y|^9} \right) f(\theta) \tag{21.20}$$

where C_1, C_2 are numerical constants whose values can be looked up in the paper and $f(\theta)$ is the angular distribution shown by the solid line in Fig. 21.5, quite different from that in weak coupling (the dashed line). Note that in weak coupling, the energy density from the dipole is just the square of the electric field (21.13), leading to a different power: $T_{00} \sim \lambda L^2/y^6$.

Unlike lattice QCD, which can only calculate QFT properties in Euclidean space-time, the AdS/CFT correspondence allows us to study time-dependent processes. We already studied the falling stone and now proceed to moving color charges on the boundary studied by Lin and Shuryak (2008a, 2008c).

Since one can always go to the center-of-mass frame, one can consider two charges to move with velocities $\pm v$ away from each other. In those papers, the charges are held by "invisible hands" and do not slow down, as would happen if one included the backreaction of the radiation losses: This was done in the later papers. So, the string ends move in a predescribed manner. The former paper solved for the moving string, and the latter one for the stress tensor hologram on the boundary.

[13] By the way, the stress tensor should always be traceless in conformal theory and have zero divergence: Those conditions were used to explicitly verify that no mistake in the calculation was made.

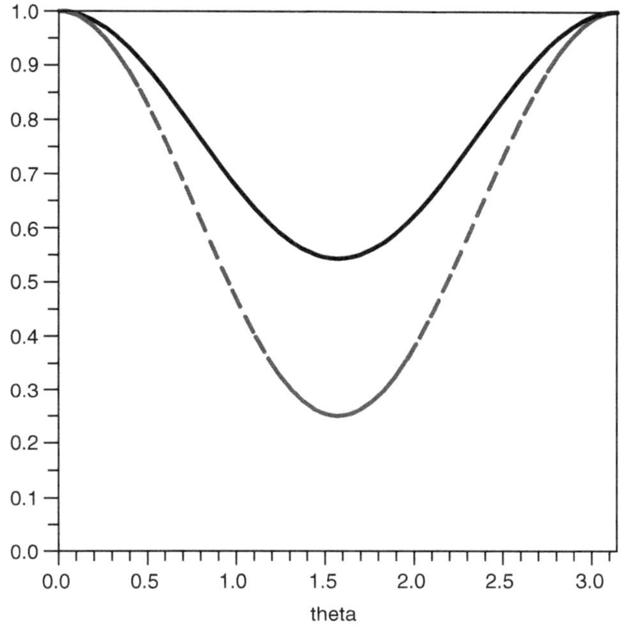

Fig. 21.5. Angular distribution of the far-field energy versus the polar angle ($\cos(\theta) = y_1/|y|$). The solid black line is the AdS/CFT result, compared to the perturbative dipole energy $(3\cos^2\theta + 1)/4$ (the dashed blue line), both normalized at zero angle.

If the charges move slowly, one can think of a small modification of the string shape. Indeed, we found a scaling solution which gives the strong-coupling generalization of the *Ampere law*, the magnetic interaction of two currents. Like the modified Coulomb, it turned out to be proportional to the *square root* of the product of two currents. For dimensional reasons, its dependence on distance is still $\sim 1/r$, as in weak coupling.

If the charges move ultrarelativistically (close to the speed of light), the problem resembles the process

$$e^+e^- \to \bar{q}q \to 2\,\text{jets}$$

which is well studied in QCD. The question was, what happens at strong coupling. Some of the resulting holograms, calculated in the second paper, are shown in Fig. 21.6 from Lin and Shuryak (2008a). As one can see, there are *no jets at strong coupling!* Later, this conclusion was also confirmed through the analysis of diagrams in which multiple gluons are emitted in strong coupling.

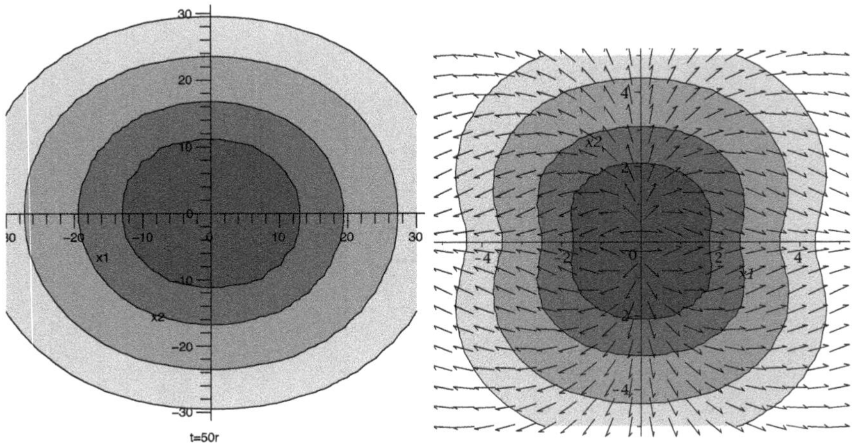

Fig. 21.6. Holograms of the falling string with two ends moving away from each other. The left figure shows the energy density distribution T^{00}, and the right figure shows the momentum density T^{0i}, with the direction shown by arrows. Both are for time $t = 10r_0$.

21.5 The AdS/CFT Correspondence at Finite T via *Non-extreme* Black Brane Solution

As we have seen, despite strong coupling, the vacuum of the $N = 4$ supersymmetric Yang–Mills theory is very different from that of QCD. In particular, in the former, the charges interact via the Coulomb law $V \sim 1/r$, and there is no confining potential $V \sim r$. Therefore, the AdS/CFT correspondence is of no use in the vacuum, unless it is strongly modified (as done in AdS/QCD models).

The situation is different at high T: In this case, both theories have QGP phases with near-conformal properties. So, in this case, the AdS/CFT correspondence is expected to be much more useful. As explained by Witten, the thermal states of the brane construction are described by non-extreme black holes with a *nonzero horizon*. Since the metric still depends on the fifth coordinate only, the horizon is located at some value z_h of it, and the physically accessible region ranges from $z = 0$ to z_h.

(This idea may look controversial because Schwarzschild-like black holes with horizons are known to emit Hawking radiation. In a locally flat 4d universe, they are therefore nonstatic. Eventually, they would emit all their energy and disappear. Yet, "black branes" in AdS_5 have a different fate: Their Hawking radiation heats up their curved universe to a certain equilibrium temperature, reaching a static thermal solution.)

Projecting this solution to the boundary, one can access the thermodynamics of the strongly coupled $N = 4$ plasma at a nonzero temperature T. Unlike QCD-like theories (which develop a scale $\lambda_{\rm QCD}$ via running coupling), the $N = 4$ SYM has a zero beta function and thus has no scale of its own. So, its properties must obey trivial scaling given by the dimension. In particular, its energy density can only be $\epsilon = C(g)T^4$, with some dimensionless coefficient depending on the coupling. At zero coupling, one has noninteracting gas or the Stefan–Boltzmann generalization of thermal radiation, famously explained by Planck. It was famously found by Gubser *et al.* (1996) that

$$C(g^2 N_c \to \infty) = \frac{3}{4} C(0) \tag{21.21}$$

The unexpected numerical coefficient 3/4 is, in fact, in good agreement with the lattice equation of state at relevant T, better than 1 corresponding to the noninteracting QGP.

The metric for a black hole AdS_5 solutions can be written in different coordinates. The most widely used ones, similar to the Schwartzschild one, have the form

$$ds^2 = L^2 \left(-h(u)dt^2 + u^2 \vec{dx}\vec{dx} + \frac{du^2}{h(u)} \right), \quad h(u) = u^2(1 - u_h^4/u^4) \tag{21.22}$$

with the horizon location related to the temperature at the boundary $T = u_h/\pi$.

As for the 4d Schwarzschild black hole, one can switch to the so-called Eddington–Finkelstein metric

$$ds^2 = -2dtdr - r^2 f(br)dt^2 + r^2 \vec{dx}\vec{dx} \tag{21.23}$$

eliminating the singular clock red-shift at the horizon and allowing us to follow geometry inside the black hole as well.

Let me briefly recall a few more facts from the Black Hole toolbox. Studies of how quantum field theory can sit in the background of classical black hole metrics have resulted in two major discoveries: the *Hawking radiation* and the *Bekenstein entropy*, related to horizon radius and area, respectively. Hawking radiation renders black holes in asymptotically flat spaces unstable: It heats the Universe until it disappears. Putting black holes into a finite box also does not help: They are generically thermodynamically unstable and get smaller and hotter until they finally burn out. Only in appropriate curved spaces can black branes be in thermal equilibrium with their "universe," filled with radiation at some finite temperature T.

Let me now provide an "intuitive picture" for nonexperts, explaining the finite-temperature Witten's settings, in which most pertinent calculations are done, as shown in Fig. 21.9. The upper rectangle is the three-dimensional (3d) space boundary $z = 0$ (only two dimensions shown), which is flat (Minkowskian) and corresponds to "our world," where the gauge theory lives. Lower black rectangles (reduced in area because of the curvature induced by $1/z^2$ in metric) are the corresponding (same in $x-1, x-2, x-3$) patches of the horizon (at $z = z_h$) of a black hole, whose center is located at $z = \infty$. Studies of finite-T conformal plasma by AdS/CFT famously began exactly with the evaluation of the Bekenstein entropy (Gubser *et al.*, 1998), $S = A/4$ via calculating the horizon *area* A.

Now comes the promised "intuitive picture": This setting can be seen as a *swimming pool*, with the gauge theory (and us, to be referred to in the following as "distant observers") living on its surface at zero depth $z = 0$, enjoying the desired temperature T. In order to achieve that, the pool's bottom looks *infinitely* hot for observers who are sitting at some fixed z close to its coordinate z_h, thus diving to such a depth is not recommended. Strong gravity takes care of and stabilizes this setting thermodynamically: Recall that time units, as well as those of energy and temperature, are subject to "warping" with the g_{00} component of the metric, which vanishes at z_h.

When astronomers found evidence for black holes and accretion into them, the physics of black holes became a regular part of physics since lots of problems had to be solved. Here, an important step forward was the so-called *membrane paradigm*, developed by many people and best formulated by Thorne and collaborators (Thorne *et al.*, 1986), known also under the name of "stretched horizons." Its main idea is to imagine that there is a physical membrane at some small distance ϵ away from the horizon and that it has properties exactly such that all the equations (Maxwell's, Einstein's, etc.) would have the same solutions outside it as *without* a membrane but with a continuation through the horizon. For example, a charged black hole would have a membrane with a nonzero charge density to terminate the electric field lines. The fact that a Poynting vector at the membrane must be pointed inward means there is some (time-odd!) relation between

$\vec{E}$ and $\vec{B}$: This is achieved by giving the membrane a finite *conductivity*, which in turn leads to ohmic losses, heat and entropy generation in it. Furthermore, as shown by Damour back in the 1980s, the displacements of the membrane and relations for the gravitational analog of the Poynting vector leads to a nonzero *viscosity* of the membrane, and its effective low-frequency theory take the form of Navier–Stokes hydrodynamics. As we will see later in this chapter, all of those ideas have resurfaced now in the AdS context. In particular, the hydrodynamics of the gauge theory plasma is simply a "hologram" of the more fundamental "hydrodynamics of the horizon membrane."

21.6 Deriving Hydrodynamics from Holography

The AdS/CFT correspondence gives one rare opportunity to derive the hydrodynamics from the first principles of the theory, using nothing else but the gradient expansion. In principle, it can be followed in any order, but we will (following the original paper by Bhattacharyya *et al.*, 2008b) only do it up to the first nontrivial order (they did it for two orders).

We already know that the finite-T QGP is dual to the black brane solution in the bulk. Let us now introduce the so-called "boosted black brane" via the (Eddington–Finkelstein) metric

$$ds^2 = -2u_\mu dx^\mu dr - r^2 f(br) u_\mu u_\nu dx^\mu dx^\nu + r^2 P_{\mu\nu} dx^\mu dx^\nu \qquad (21.24)$$

with

$$f(r) = 1 - \frac{1}{r^4}, \quad u^i = \frac{\beta_i}{\sqrt{1-\beta^2}}, \quad u^v = \frac{1}{\sqrt{1-\beta^2}}, \quad P^{\mu\nu} = u^\mu u^\nu + \eta_M^{\mu\nu}$$

Note that the projector to the rest frame of matter contains the flat Minkowski metric $\eta_M^{\mu\nu} = \mathrm{diag}(-1, 1, 1, 1)$. Note further that the zero of the function f gives the black hole horizon, and thus, $b = 1/\pi T$ is the only place indicating the matter temperature.

The hologram of the brain moving with a velocity u^μ is the QGP moving with the same velocity: It is the same solution as studied before, just boosted. Now, we make the crucial (and very typical) step of promoting T, u^μ to be a nonconstant:

$$b \to b(x), \quad u^\mu \to u^\mu(x)$$

Of course, new derivatives of them would upset the Einstein equations; therefore, after such a substitution, the metric is no longer a solution.

Imagine, however, that their dependence on coordinates is very slow, so they are *nearly* constant. This means one has a *nearly* good solution. The point is that one can correct for such derivatives and get an *approximate* solution.

Let me begin with the answer. We know that on the boundary, gravity is coupled to $T_{\mu\nu}$, which must be conserved, $\partial_\mu T^{\mu\nu} = 0$. The local terms are known:

$$T^{\mu\nu}_{\text{local}} = (\pi T)^4 (\eta^{\mu\nu}_M + 4u_\mu u_\nu) \tag{21.25}$$

and the question is to find terms in the stress tensor of the first, second, etc., order in small gradients. The first nontrivial term is

$$T^{\mu\nu}_{\text{first}} = -2(\pi T)^3 \left(P^{\mu\alpha} P^{\nu\beta} (\partial_\alpha u_\beta + \partial_\beta u_\alpha)/2 - P^{\mu\nu} \partial_\alpha u^\alpha /3\right) \tag{21.26}$$

Note that projectors ensure that we have spatial gradients of the flow velocity in the matter rest frame, so it is basically the Navier–Stokes term with the shear viscosity $\eta/s = 1/4\pi$ written in front. (The bulk viscosity is zero because the matter is conformal.)

The method of solution is as follows. The metric is written as the one above, $g^0_{\mu\nu}$, plus the correction of the first order in derivatives $g^1_{\mu\nu}$, which contains small derivatives of T, u^μ from $g^0_{\mu\nu}$ and new corrections which cancel them in the Einstein equations. The procedure is logically consistent but technically complicated because there are more Einstein equations in 5D than functions, but one needs to be sure that *all* of them are satisfied.

21.7 Waves in Plasma as Black Hole Quasinormal Modes

Gravity waves propagating in the AdS background metric have certain dispersion relations $\omega(\vec{k})$ with calculable real and imaginary parts; an example is shown in Fig. 21.7 from Kovtun and Starinets (2005). Such quasinormal[14] modes are known for various examples of black holes. In fact, they were even experimentally observed by LIGO as "late time afterglow" following black hole merger events.

Returning to the AdS black hole and Fig. 21.7, note that the lowest eigenvalues shown by the larger dots are close to the origin. They are nothing else but the *sound modes* of strongly coupled plasma. For reference,

[14]Like the wave functions of α-decaying nuclei, when energy is complex, the wave function grows in space and is not normalizable, thus the name. In nuclear physics, these are called quasistationary states.

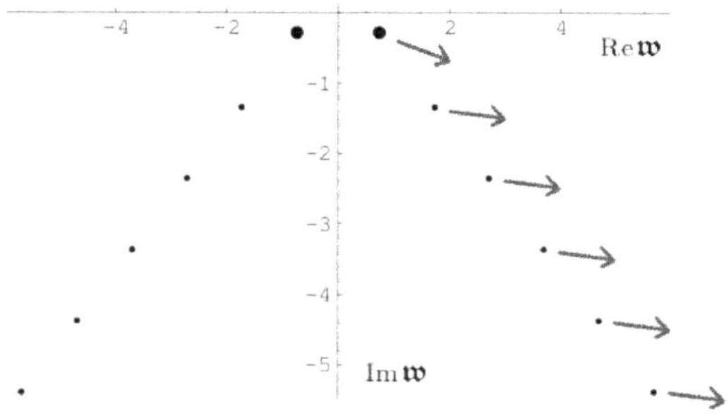

Fig. 21.7. A set of frequency modes on the complex ω plane. Dots denote a particular value of the wave vector k, and the arrows indicate the direction of motion as k increases.

let me mention several more known terms at small k (following Lublinsky and Shuryak, 2009):

$$\frac{\omega}{2\pi T} = \pm\frac{\tilde{k}}{\sqrt{3}}\left[1+\left(\frac{1}{2}-\frac{\ln 2}{3}\right)\tilde{k}^2 - 0.088\,\tilde{k}^4\right] - \frac{i\tilde{k}^2}{3}\left[1-\frac{4-8\ln 2+\ln^2 2}{12}\,\tilde{k}^2 - 0.15\,\tilde{k}^4\right], \tag{21.27}$$

where momentum is made dimensionless via $\tilde{k} \equiv (k/2\pi T)$. First, note that at small k, the imaginary viscous term is very small[15]: $\mathrm{Im}\,\omega \sim k^2$. Second, the dispersive correction to the speed of sound — the k^2 term in the first square brackets — has a positive coefficient. Thus, AdS/CFT predicts that one sound wave can decay into two. Third, note that both higher-order corrections to viscosity are negative. This is in contrast to some popular second-order *ad hoc* schemes, such as Israel–Stuart.

All other "non-hydrodynamical" modes have large $\mathrm{Im}(\omega)/(2\pi T) = O(1)$. During their relaxation times, of the order of $z_h \sim 1/(2\pi T)$, they all disappear since they "fall into the black hole."[16]

The essence of the AdS explanation for the rapid equilibration is thus simple: Any objects (non-hydro modes) become invisible as they are absorbed by the black hole. The only[17] remaining memory is their total

[15]We used that fact above in the section on the "acoustic damping" phenomenology.

[16]It is amusing to note that a puzzling process of QGP equilibration is, in the AdS/CFT setting, simply "problem number one" in physics, the Galilean stones falling in a gravity field.

[17]Plus all conserved charges if they are there.

mass, which the black hole transforms into the appropriate amount of Bekenstein entropy.

21.8 A Trailing String and Jet Quenching

The finite-T setting for AdS/CFT is different from the zero-T one discussed so far owing to the existence of a horizon. Relaxation corresponds to all kinds of objects falling into the horizon, with their energy and information lost. One of the simplest is a heavy quark moving with a fixed velocity v, see Fig. 21.9(a). It is used in the calculation of heavy quark quenching (Sin and Zahed, 2005; Herzog *et al.*, 2006; Gubser, 2006; Buchel, 2006; Sin and Zahed, 2007): The electric flux goes into the so-called "trailing string" falling toward the horizon.[18] The drag force is

$$F_{\text{drag}} = -\pi T^2 \sqrt{g^2 N_c} \frac{v/2}{\sqrt{1-v^2}}$$

and it can be related to the diffusion constant via the Einstein relation, a nontrivial successful check on two very different calculations.

Let me jump to the next set of papers, showing the hologram of the trailing string in the nonzero T background. In Fig. 21.8 from Chesler and Yaffe (2007), one can see a Mach cone of sound, with an expected opening half-angle of about 50°.

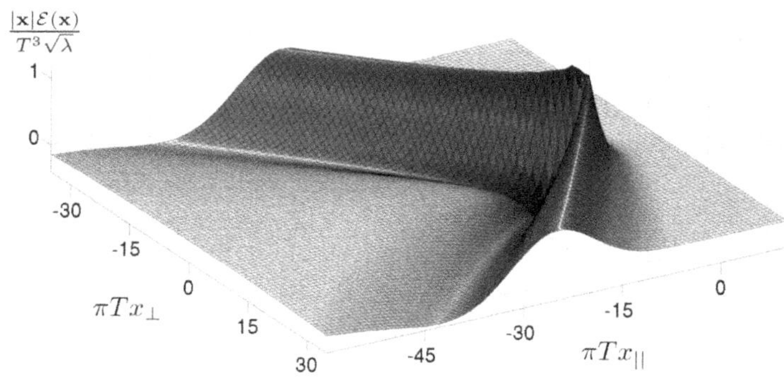

Fig. 21.8. Plot of $|x|E(x)/(T^3\sqrt{\lambda})$ for $v = 3/4$, with $T = 0$ and near-zone contributions removed.

[18]Think of the heavy quark as a ship diligently laying underwater cable at the sea's bottom.

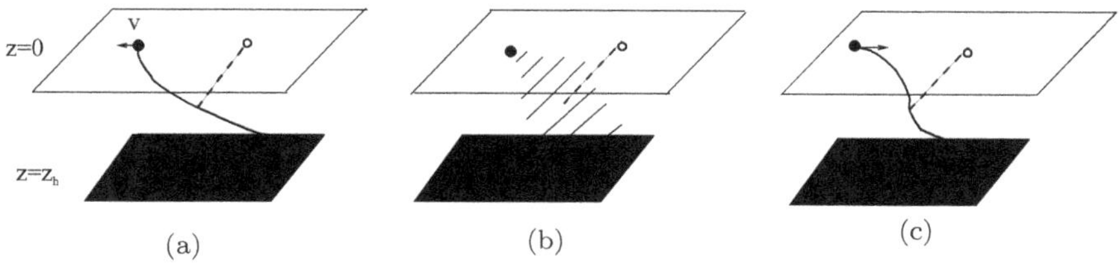

Fig. 21.9. Schematic view of the relaxation settings: A string (a), a wave (b) or a particle (c) falling into the fifth dimension toward the black hole.

The dashed lines in all plots of Fig. 21.9 correspond to the next-order diagram, describing the backreaction of the falling bulk objects onto the boundary, the observation point denoted by a small open circle. This field may also have spins of 0, 1 or 2, providing three pictures — the (4d) "holograms" — of the boundary. Contrary to our intuition (developed from our limited flat-world experience), the hologram is *not* a reduced reflection of more complete 5d dynamics in the bulk, but in fact represents it fully. This phenomenon — the AdS/CFT duality — is a miracle occurring due to the near-black-hole setting.

These holographic images are what the surface observer will see. An image of the trailing string was calculated by Friess *et al.* (2006) and Chesler and Yaffe (2008): it accurately displays a hydro conical flow. For a hologram of the stone (Fig. 21.9(c)) see Gubser and Yarom (2009); however, to our knowledge, the holographic "backreaction" of the falling waves remains to be demonstrated.

How are these predictions related to experiments? One important test should be to see whether the drag force does indeed depend only on the *velocity* (rather than momentum): Can it be done via single electrons from c and b decays. Another challenge is to test if the effective viscosity is indeed *decreasing* with increasing gradients, as AdS/CFT nontrivially indicates (Lublinsky and Shuryak, 2007): It can be inferred from elliptic flows at more peripheral collisions (a thinner "almond".)

Last but not least, the AdS/CFT correspondence predicts a particular ratio of the excitation amplitudes of the sound (conical flow) and "diffuson" mode. So far, we seem to have observed experimentally only the conical flow but not the second mode (which would result in a peak in the original direction of the jet). Much more detailed experimental and theoretical studies are required to see if there is indeed a potential discrepancy here or not.

21.9 Holography Out of Equilibrium 1: The Shocks

Shocks are classic examples of out-of-equilibrium phenomena. They are traditionally divided into two categories: weak and strong. In the former case, the difference between matter before and after the shock is small. Weak shock can be treated hydrodynamically, e.g. using the Navier–Stokes approximation. Strong shocks have finite jumps in matter properties. Their profiles have large gradients, so one needs some more powerful means to solve the problem without relying on hydrodynamics, which is just an expansion of gradients.

The reason we put this example as number one is because it is the only one which can be considered in a *stationary approximation*. Indeed, in the frame which moves with the velocity of the shock, its profile is *time-independent*.

Strong shocks in an AdS/CFT setting were discussed by Shuryak (2012b). An example from this work, the Navier–Stokes shock profile, is shown in Fig. 21.10. The fluxes of total energy and momentum are tuned to be the same, from the left to the right side of the picture. One may think of it as a low-density QGP entering on the right with higher rapidity and getting suddenly excited into a higher-density QGP, floating out more slowly. It resembles a picture seen from the cockpit of a supersonic jet.

In the AdS/CFT setting, one can solve the problem from first principles by solving the Einstein equations. Since the setting has an extra holographic dimension z, even the static solution depends on not one but two variables: the longitudinal coordinate x and z. Not going into details, the surprising conclusion was that all corrections to the Navier–Stokes profile of the shock happen to be rather small, at a scale of a few percent, even without any apparent small parameter in the problem.

(Another tool used to correct the Navier–Stokes solution was the so-called "resummed hydrodynamics" (Lublinsky and Shuryak, 2009): It also led to corrections at the percent level. Unfortunately, the accuracy of the particular AdS/CFT solution was not sufficient to tell whether both agreed or not.)

The lesson is that all higher-order gradient corrections to the Navier–Stokes solution have a strong tendency to cancel each other.

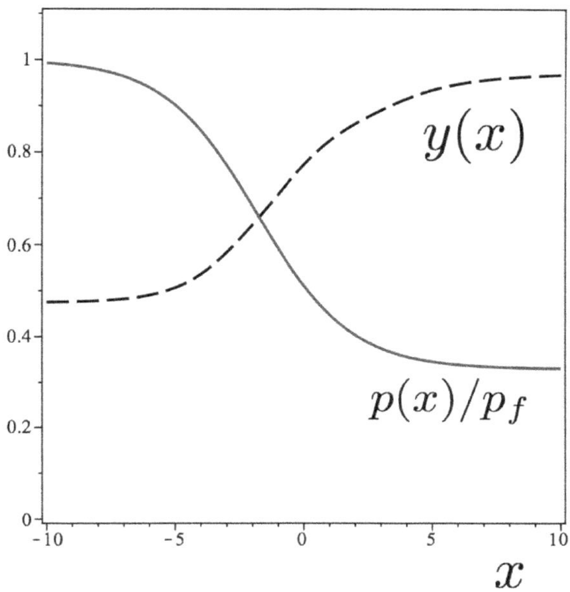

Fig. 21.10. Profile of a strong shock in a QGP in its rest frame, according to Navier–Stokes hydrodynamics, as a function of the coordinate normal to the shock front. The time goes from right to left, so the left-hand side shows the final values of the observables, while the right-hand side shows the initial ones. The pressure, shown by the (blue) solid line, is in units of its asymptotic value: Thus, the curve jumps to 1 on the left side. The flow rapidity, shown by the (black) dashed line, is reduced. The process is thus the rapid formation of a hotter/denser QGP by the influx of a cooler and more dilute one.

21.10 Holography Out of Equilibrium 2: The Falling Shell

This setting of a falling matter shell was proposed by Lin and Shuryak (2008b). It is, in a way, complementary to the previous one: There is dependence on time t but not on space x because the shell's motion occurs along the holographic fifth direction z.

The physical meaning of this motion is as follows. First of all, recall that the fifth coordinate $z = 1/r$ corresponds to a momentum "scale." Small values near the boundary (large r) correspond to the UV end of the scales, while large z and small r correspond to the IR or small momenta. Since

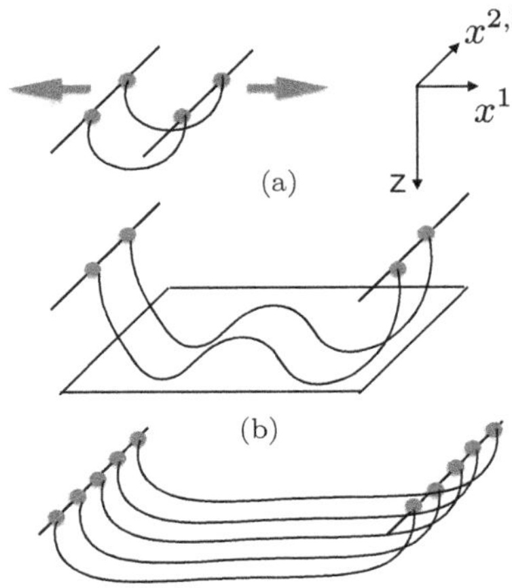

Fig. 21.11. (a) An early-time snapshot of a pair of strings created after one color exchange. The coordinates are explained on the right: The colliding objects move with the speed of light away from each other, and the strings are stretched. (b) A later-time snapshot in the AdS/QCD background. After strings reach the "levitation surface," shown by a rectangular shape, they begin oscillating around it. (c) In the case of a high density of many strings, they can be approximated by a continuous membrane.

everything happens much quicker in the UV as compared to the IR, the equilibration process naturally proceeds from UV to IR, also known as "top-down" equilibration. The gravity force in the AdS is directed accordingly.

One can imagine that this process can, in some sense, be reduced to a thin "equilibration shock wave," propagating in the z direction. The key idea of the paper by Lin and Shuryak (2008b) was that this shock can be considered a certain external object — a shell or an elastic membrane — falling *under its own weight* (see the sketch in Fig. 21.12(a)). If this is the case, the total energy of the membrane is *conserved* (potential energy goes into kinetic energy). The consequences of this are very important: While the shell is falling toward the AdS center, the metric — both above and below the membrane — is actually time-independent, as it depends only on its total mass. So, there is no need to solve the Einstein equations.[19] In the case of an extreme black hole at the AdS center (the blue

[19]It is instructive to recall Newton's proof that a massive sphere has an outside field that is the same as a point mass and that there is no gravity inside the sphere. This is also true if the sphere is falling. It also remains true for general relativity.

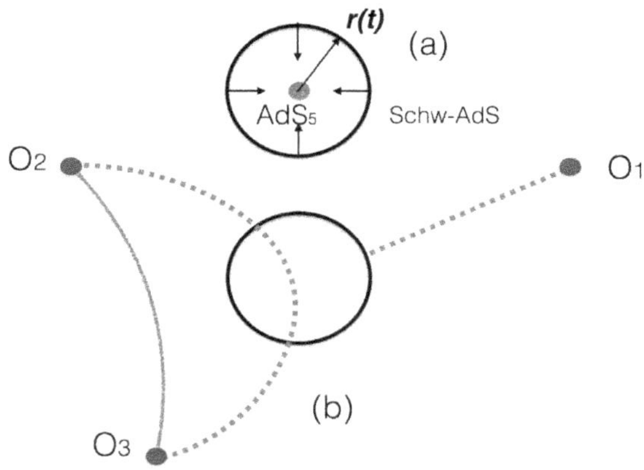

Fig. 21.12. (a) A sketch of falling shell geometry. Its radius $r(t) = 1/z(t)$, which is used in the text. (b) Single-point observer O_1 and the two-point observers O_2, O_3.

dot in Fig. 21.12(a)), the solution consists of the following: (i) a thermal Schwarzschild-AdS metric above the shell; and (ii) an "empty vacuum" or the AdS_5 solution below it.

The only equation that needs to be solved is the one describing the motion of the shell itself, $r(t) = 1/z(t)$. It is not so trivial to derive since the coordinates used below and above the shell are discontinuous. Fortunately, a thin-shell collapse has already been solved in general relativity. To derive it, one needs to connect two metrics via the so-called "Israel junction condition." The shell equation of motion in time for the distant observer t is given by

$$\frac{dz}{dt} = \frac{\dot{z}}{\dot{t}} = \frac{f\sqrt{\left(\frac{\kappa_5^2 p}{6}\right)^2 + \left(\frac{3}{2\kappa_5^2 p}\right)^2 (1-f)^2 - \frac{1+f}{2}}}{\frac{\kappa_5^2 p}{6} + \frac{3}{2\kappa_5^2 p}(1-f)} \tag{21.28}$$

where $\kappa_5^2 p$ is the product of the 5d gravity constant and the shell elastic constant, and $f = 1 - z^4/z_h^4$ is the standard black hole function of the thermal AdS background.

The shell starts falling with zero velocity from a certain height and is then accelerated to nearly the speed of light. Finally, near the horizon position $z \to z_h$, there appears a "braking phenomenon": The shell slows down to a velocity of zero. This breaking is a standard feature stemming from the use of distant observer time, familiar from the Schwarzschild metric.

After a solution is found, one can calculate what different observers at the boundary, i.e. in gauge theory, will see. In particular, one may ask if/how such an observer can tell whether it is a static black hole (the thermal state with a stationary horizon) or a falling shell.

A *"one-point observer"* O_1 in Fig. 21.12(b) would simply see that the stress tensor perturbation induced a gravitational propagator, as indicated by the red dashed line. Since the metric above the shell is thermal-AdS, such an observer will see the *time-independent* temperature, pressure and energy density, corresponding to the static final equilibrium. Yet, more sophisticated *"two-point observers"* O_2, O_3 can measure certain correlation functions of the stress tensors. They will see contributions both from gravity waves propagating along the line shown by the solid line above the shell, i.e. in the thermal metrics, as well as from waves propagating along the path shown by the dashed line which penetrate *below* the shell: They would notice deviations from equilibrium. Solving for various two-point functions in the background with falling shell/membrane we found these deviations. They happen to be oscillating as a function of the wave frequency. This observation, at first puzzling, is explained (Lin and Shuryak, 2008b) by finite "echo" times due to a signal reflected from the shell. So, one can experimentally observe an echo coming from the fifth (nonexisting) dimension!

For further discussion of the scenarios of top-down equilibration within falling scalar fields, etc., the reader is referred to subsequent literature, e.g. Balasubramanian *et al.* (2011).

21.11 Holography Out of Equilibrium 3: Anisotropic Plasma

Our next example is a setting in which one starts with some *anisotropic* but homogeneous metric and follows its relaxation to equilibrium; for a summary, see Chesler and Yaffe (2014). The metric is diagonal, with time-dependent but space-independent components, the resulting Einstein equations are solved numerically.

Rapid relaxation of the equilibrium thermal-AdS solution is observed. A number of initial states can be compared, selected with the same equilibrium energy density (or horizon or T) at late times. While initially, the momentum asymmetry can be very large, say, an order of magnitude, it becomes exponentially smaller with time. Any deviations from equilibrium are strongly red-shifted as they approach the horizon.

There are no hydrodynamical modes since the setting is homogeneous ($\vec{k} = 0$). The lowest mode has a frequency of $\omega = (2.74 + i3.12)\pi T$. So, the strongly coupled QGP has an "isotropization time" as short as

$$\tau_{\text{isotropization}} \sim \frac{0.1}{T}. \tag{21.29}$$

21.12 Holography Out of Equilibrium 4: Rapidity-Independent Collisions

The holographic AdS_5 setting allows us to study ultrarelativistic collisions and to see, by solving the Einstein equations, when and how the hydrodynamical description becomes valid in a strong coupling limit.

The picture of "debris" created in the bulk after a high-energy collision, forming a small black hole, falling toward the AdS center (Shuryak *et al.*, 2007) relates the exploding/cooling fireball in the real world to a hologram of the black hole horizon moving away from the boundary. However, the specific solution discussed in that paper was spherically symmetric, and thus, it is more suited for cosmology than for heavy ion applications.

Another setting with "falling" horizons, corresponding to the Bjorken rapidity-independent flow (see Appendix), was developed by Janik and Peschanski (2006). In this case, the horizon is rapidity-independent and has a time-dependent location $z_h(\tau)$. At later times, the solution to the Einstein equation reproduces Bjorken hydrodynamical flow because at $\tau \to \infty$,

$$z_h(\tau) = 1/\pi T(\tau) \sim \tau^{1/3}. \tag{21.30}$$

Note that the variable $w = T\tau$ has the meaning of the macro-to-micro scale ratio. At late time $\tau \to \infty$, it grows, indicating that the system becomes more macroscopic and hydrodynamics becomes more accurate. The question is when exactly the hydrodynamical description becomes valid and *with what accuracy.*

Figure 21.13 from Heller *et al.* (2012) displays the time evolutions of many initial states, all approaching the same hydrodynamical solution. In Fig. 21.13 (left), it is shown that this happens via convergence to a certain *universal function* of the variable $w = \tau T$, defined by

$$\frac{dw}{d\ln\tau} = F(w), \tag{21.31}$$

Note that the existence of such universality is the essence of "resummed hydro" (Lublinsky and Shuryak, 2009).

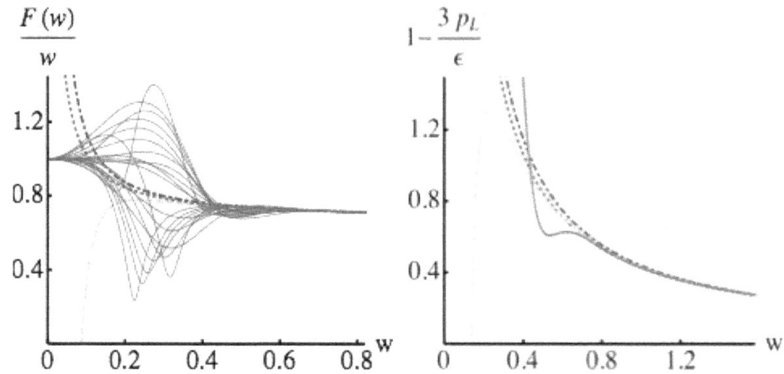

Fig. 21.13. (left) The temperature evolution combination $d\log(w)/d\log\tau$ for different initial conditions (black thin curves) converging into a universal function of $w = T\tau$ compared to hydro. (right) The pressure anisotropy for one of the evolutions compared to first- (Navier–Stokes), second- and third-order hydrodynamics.

Depending on the required degree of accuracy, one may assign a specific initial value of w at which the "hydrodynamics starts," in the range of $w_i = 0.4, \ldots, -0.6$. The plot on the right demonstrates that at such a time, the anisotropy is still large and viscosity is important.

The lesson from this work can be better explained by comparing its result to naive expectations that hydrodynamics starts when macro and micro times are the same, $w = \tau T > 1$, and that the accuracy of hydrodynamics becomes bad, $O(1)$. Calculations show instead that at *twice the smaller* time $w \sim 1/2$,- the accuracy of (the lowest-order Navier–Stokes) hydro suddenly becomes quite good, to a few percent!

Why is it so? While gradients are not yet small at that time, the combined effect of all of them is. The Lublinsky–Shuryak resummation provides an answer: The higher-gradient series has alternating signs, and it can be Pade resummed *a la* a geometric series to a decreasing function.

The issue has a practical aspect, related to one of the first observations made at the first LHC *PbPb* run. It was found that the (charged hadron) multiplicity in *PbPb* collisions grows with energy a bit more rapidly than in *pp*:

$$\frac{dN^{PbPb}}{dy}(y=0,s) \sim s^{0.15} \quad \frac{dN^{pp}}{dy}(y=0,s) \sim s^{0.11} \tag{21.32}$$

From the RHIC energy ($E = 0.2$ TeV) to the LHC, the double ratio

$$\frac{\frac{dN}{d\eta}|_{PbPb,\mathrm{LHC}}/\frac{dN}{d\eta}|_{pp,\mathrm{LHC}}}{\frac{dN}{d\eta}|_{AuAu,\mathrm{RHIC}}/\frac{dN}{d\eta}|_{pp,\mathrm{RHIC}}} = 1.23. \tag{21.33}$$

shows a noticeable change with energy, which calls for an explanation.

A simple form for the function $F(w)$ was proposed by Lublinsky and Shuryak (2011). If known, one can calculate the entropy produced, beginning from the time w_i: It turns out to be about 30%. Furthermore, we get the following expression for the contribution to this double ratio: $\approx 1 + \frac{3[\bar{\eta}(\mathrm{LHC}) - \bar{\eta}(\mathrm{RHIC})]}{2w_i + 3\bar{\eta}(\mathrm{RHIC})}$ and show that the observed growth can be naturally explained by the viscosity growth, from the RHIC to the LHC, predicted by a number of phenomenological models.

21.13 Holographic Collisions of Small Systems and the Accuracy of Hydrodynamics

We end the series of out-of-equilibrium problems with a direct numerical solution to holographic collisions worked out by Chesler (2016). The metric used has the form of the Fefferman–Graham type:

$$ds^2 = r^2 \left[-dt^2 + d\vec{x}^2 + \frac{dr^2}{r^4} + h_\pm(x, y, z_\mp, r) dz_\mp^2 \right] \tag{21.34}$$

where r is the AdS radial coordinate (inverse to z used earlier), in which the boundary is $r = \infty$, and $z\pm \equiv z \pm t$ are the light cone directions along which two objects move.

The numerical solution to the Einstein equation in the bulk leads to the distributions for several observables on the boundary for central (head-on) and off-center collisions are shown in Fig. 21.14.

What we are actually interested in is not only the exact solution of the Einstein equation itself but also its comparison with the predictions of the (various versions of) the hydrodynamics. The time and space of their normalized residual, defined by

$$\Delta(t, x) = \max_{t' > t} \left[\frac{||T^{\mu\nu}(t', \vec{x}) - T^{\mu\nu}_{\mathrm{hydro}}(t', \vec{x})||}{p(t', \vec{x})} \right]$$

is shown in Fig. 21.15 from Chesler (2016). The time dependence is very curious: The drop is rapid due to relaxation of non-hydrodynamical modes,

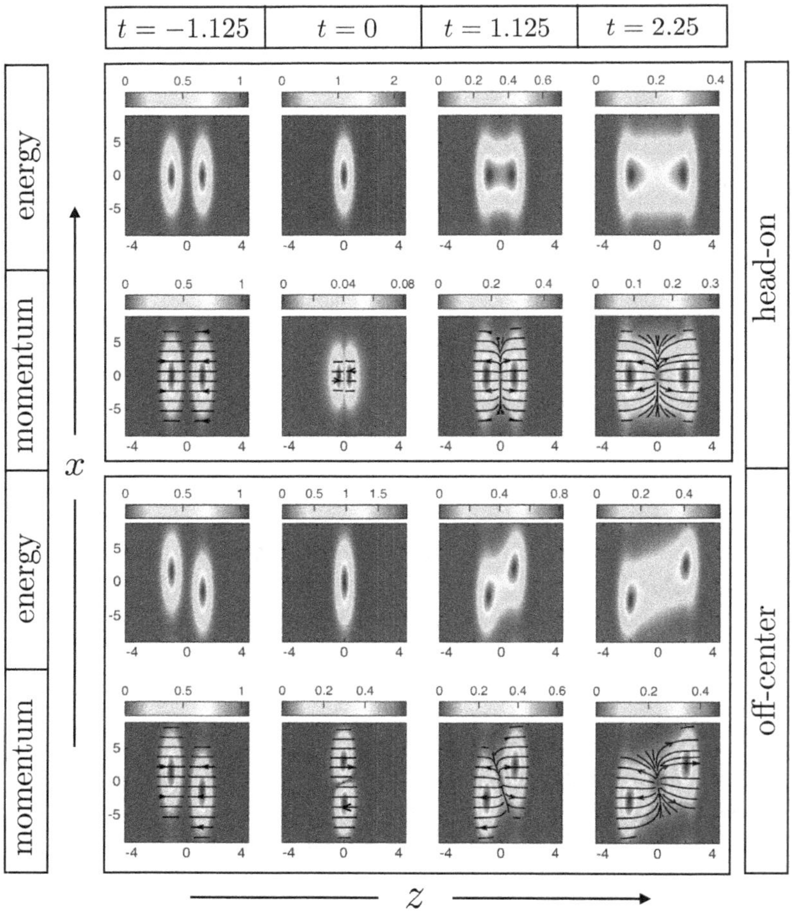

Fig. 21.14. Holographic collisions lead to the boundary distribution of the stress tensor displayed.

but it does not go all the way to zero but stays near $\Delta \sim 1/10$, decreasing very slowly. The spatial distribution shows that Δ is small, nearly in the whole fireball.

We follow Casalderrey-Solana *et al.* (2013), who studied both full stopping and "flying through" scenarios. The colliding shock waves have an initial profile of

$$T_{\pm\pm} = \frac{2\pi^2}{N_c^2}\rho^4 \exp\left((t \pm z)^2/2w^2\right) \tag{21.35}$$

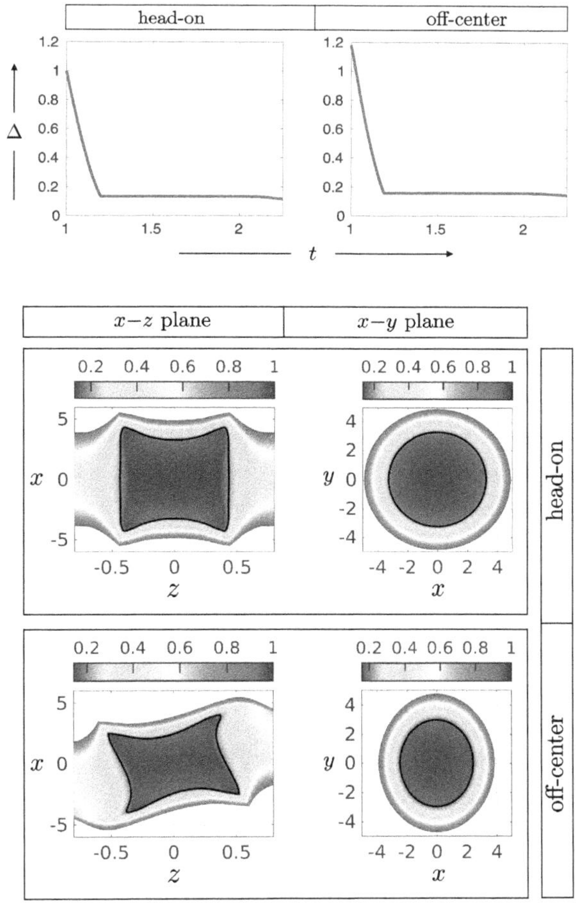

Fig. 21.15. The time and space dependence of the normalized residual Δ.

with two parameters. The conformal symmetry of the setting tells us that the results depend only on their dimensionless product ρw. Increasing collision energy means decreasing this product: "Thick shocks" would be when it is 2, and "thin shocks" when it is 1/8. Among the many nice plots in the paper, perhaps the most informative one is Fig. 21.16. While thick shocks (left) lead to nearly complete stopping of matter, with relatively small anisotropy of pressure and subsequent Landau-like hydrodynamical explosion (compare the solid and dotted lines), the thin case shows anisotropy

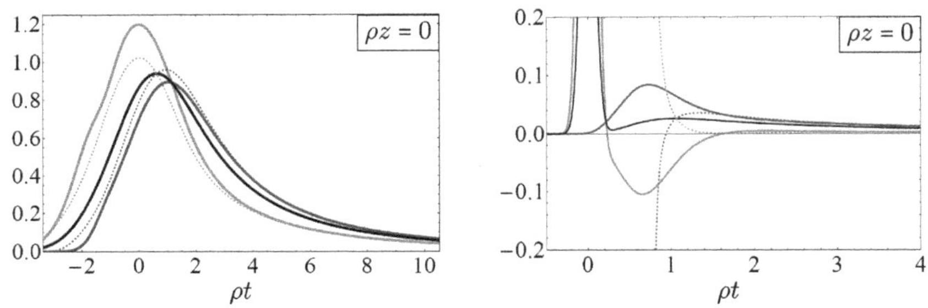

Fig. 21.16. Center values of $\epsilon/3\rho^4$ (black), P_L/ρ^4 (red) and P_T/ρ^4 (blue) as functions of time for a collision of thick (left) and thin (right) shocks. The dotted curves show the hydrodynamic approximation.

of pressure, with strong negative longitudinal pressure and then rapidly going to zero. Yet, even in this case, at $\rho t > 1$, hydrodynamics is established.

Symmetric collisions of relativistic shocks were further studied by Chesler *et al.* (2015) and extended to asymmetric collision by Waeber *et al.* (2019). The latter paper has an extensive explanation of technical details.

The basic conclusions of these studies are as follows:

(i) flow rapidity and space-time rapidity are very close to each other, as in the Bjorken flow;
(ii) the energy density at a proper fixed time τ has a Gaussian shape, as in the original Landau solution;
(iii) the time when the stress tensor takes a hydrodynamical form depends little on the asymmetry and is sensitive only to the local energy density.

21.14 "Trapped Surfaces" and Entropy Production

The simplest geometry to consider is wall-on-wall collisions, in which there is no dependence on the two transverse coordinates and only the remaining three — time, longitudinal (rapidity) and the holographic direction — remain in play. Needless to say, it is a very formidable problem, solved by Chasler and Yaffe via clever "nesting" of the Einstein equations. The reader can find explanations in Chesler and Yaffe (2014).

Collisions of finite-sized objects are even more difficult to solve, but those historically brought about a discussion of the very important issues of *trapped surface* formation and *entropy production.* It was pioneered by Gubser *et al.* (2008b), who considered head-on (zero impact

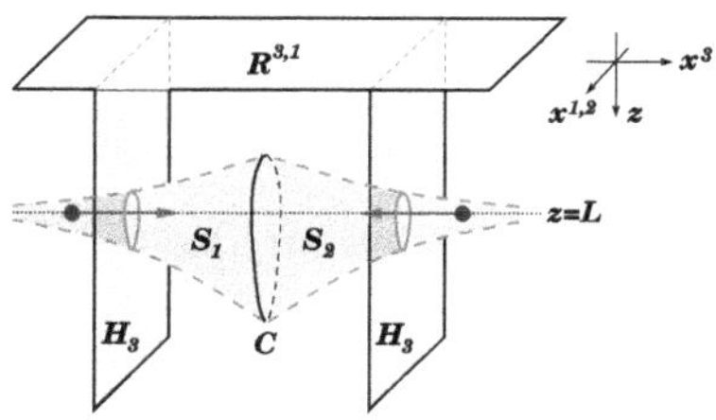

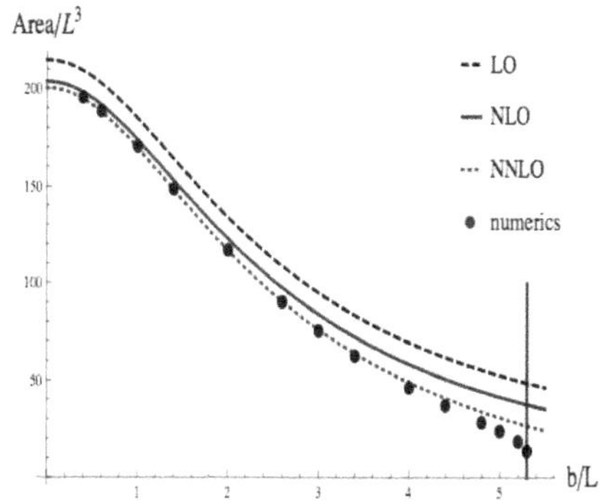

Fig. 21.17. (a) From Gubser *et al.* (2008b): A projection of the marginally trapped surface onto a fixed time slice of the AdS geometry. (b) The area of the trapped surface versus the impact parameter, with the comparison of the numerical studies (Lin and Shuryak, 2009) shown by points and analytic curves from Gubser *et al.* (2009). The vertical line shows the location of the critical impact parameter b_C, beyond which there is no trapped surface.

parameter) collisions of point black holes. The setting is shown in Fig. 21.17(a).

"Trapped surface" is a technical substitute for the horizon — for this review, it is not necessary to discuss the difference — and its appearance in the collision basically means that there exists a black hole inside it. Classically, all information trapped inside it cannot be observed from outside, and the lost information is *entropy*. For known static black hole solutions, this area does give the black hole Bekenstein entropy.

Locating this surface allows one to limit the produced entropy *from below* by simply calculating *its area*. The reason this entropy estimate is from below is because the trapped surface area is calculated at the *early* time $t = 0$ of the collision, not at its end. No particle can get out of a trapped surface, but some can get into it *during* the system's evolution, increasing the black hole mass and thus its entropy.

Gubser *et al.* denote the distance separating the colliding black hole from the boundary by L; we will discuss its physical meaning in the following. Naively, central collisions have only axial $O(2)$ symmetry in the transverse plane $x_\perp$; however, using global AdS coordinates, these authors found a higher $O(3)$ symmetry of the problem, which manifests if a coordinate

$$q = \frac{\vec{x}_\perp^2 + (z-L)^2}{4zL} \tag{21.36}$$

is used. It was shown that the 3d trapped surface C at the collision moment is in this coordinate a three-sphere, with some radius q_c. If q_c is determined, the relation between the CM collision energy and the Bekenstein entropy can be calculated.

For large q_c, these expressions are

$$E \approx \frac{4L^2 q_c^3}{G_5}, \quad S \approx \frac{4\pi L^3 q_c^2}{G_5}, \tag{21.37}$$

from which, by eliminating q_c, one finds the main conclusion: The entropy grows with the collision energy as

$$S \sim E^{2/3} L^{5/3} \tag{21.38}$$

Note that this power has, in general, a dimension of $(d-3)/(d-2)$, so it is directly related to $d = 5$, the dimension of the AdS space. Note also that it is different from the Fermi/Landau 1950s' prediction of $S \sim E^{1/2}$ as well as from the data, which according to (21.32) indicate the power of about 0.30.

Let us now return to the meaning of the parameter L. Gubser *et al.* relate the "depth" of the colliding objects with the nuclear size, $L \sim 1/R_A$, which cannot depend on energy. An alternative idea suggested by Lin and Shuryak (2009) ascribe L to the (inverse) "saturation scale," the typical parton's momenta in the wave function of the colliding objects. In this case, it is related to the collision energy by

$$L \approx \frac{1}{Q_s(E)} \sim E^{-\alpha} \tag{21.39}$$

where $\alpha \approx 1/4$ is an empirical index of the PDFs. It is especially clear if one would consider wall-on-wall collisions, in which the nuclear size R_A tends to infinity, while Q_s characterizes the material of the wall and remains fixed. L is not of the $O(10\,\text{fm})$ scale, but rather two orders of magnitude smaller $O(0.1\,\text{fm})$. Furthermore, it is expected to *decrease* with the energy $L \sim 1/Q_s(E)$. Including this modifies (21.38) to

$$S \sim E^{(2/3)-(5/3)\alpha} \sim E^{0.25} \tag{21.40}$$

which is in reasonable agreement with the observed multiplicity.

The generalization of this theory to non-central collisions (Lin and Shuryak, 2009) led to the results shown in Fig. 21.17(b). The figure is from Gubser *et al.* (2009); it shows the dependence of the trapped surface area on the impact parameter. Specifically, numerical results from Lin and Shuryak (2009) (points) are compared with the analytic series of curves (Gubser *et al.*, 2009), which are in excellent agreement.

From a gravity point of view, the qualitative trend shown is clear: Two colliding objects may merge into a common black hole provided that the impact parameter is less than some critical value $b_c(E)$, depending on the collision energy. Indeed, with b rising, the trapped energy decreases while the total angular momentum increases, so at some point, the Kerr parameter exceeds 1 and thus no black hole can be formed. Interestingly, the calculation shows that it happens with a finite jump — a first-order transition[20] as a function of the impact parameter.

A related CERN experiment, called SHINE, is devoted to systematic studies of nuclear AA collisions with the A variable ranging from small to large. It indeed observes rather rapid changes in multiplicity and other properties at $A \sim 20$, from the pp-like to heavy-ion-like regimes. This "onset of a fireball" (as they call it) can be holographically related to the formation of a common trapped surface.

Just a bit below the critical value of the impact parameter, the trapped surface and black hole exist, and nothing indicates that at a larger b none is formed. So, at $b < b_c$, a creation of a QGP fireball occurs, while for peripheral $b > b_c$ collisions, the system remains in the hadronic phase. A jump in entropy as a function of impact parameter is rather surprising, but in fact the experimental multiplicity-per-participant plots do indeed show a rapid change between non-QGP small systems and QGP-based not-too-peripheral AA collisions. In Fig. 21.18 from Lin and Shuryak (2009), we show the data of the final multiplicity per participant as a function of collision centrality for three energies. Although the very peripheral data are missing here, all of them suggest the same entropy production for central heavy ion collisions, but a different value for pp and perhaps very peripheral collisions.

It would be interesting to compare it with all the available information on small systems undergoing transition to the explosive regime.

In the same paper (Lin and Shuryak, 2009), we pointed out that the simplest geometry of the trapped surface would be that for a wall–wall collision, in which there is no dependence on transverse coordinates x^2, x^3. Thus, a sphere becomes just two points in z, above and below the colliding bulk objects. We elaborated on this, considering the collision of two infinite

[20]The first-order transition stems from the large N_c approximation: This conclusion may perhaps be modified at finite N_c. Furthermore, the problem of a trapped surface in quantum gravity is way too complicated, and to our knowledge, it has not been studied so far.

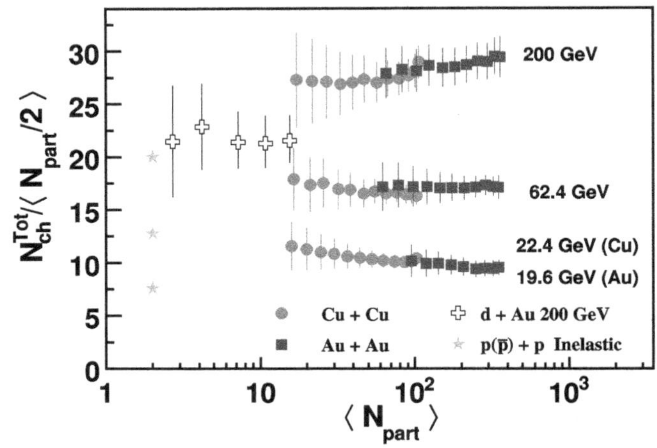

Fig. 21.18. PHOBOS data on the integrated number of charged particles, scaled by $N_{\rm part}/2$, in $p+p, d+Au, Cu+Cu$ and $Au+Au$ collisions as a function of the mean $\langle N_{\rm part} \rangle$).

walls made of materials with different "saturation scales" (Lin and Shuryak, 2011) and studied the conditions for trapped surface formation.

Summary and connections: In our brief introduction to the AdS/CFT correspondence approach, we discussed a number of particular examples, rather than following the traditional general theory path. It began with a "pending string," giving a strong coupling version of the Coulomb law, goes to "falling string" and then hydrodynamical modes. A number of out-of-equilibrium situations are discussed, all of them displaying rapid equilibration in strongly coupled plasmas. At the end, we discussed the connection between the geometry of the "trapped surface" and the amount of entropy produced. While this chapter is devoted to the applications of holography to strongly coupled matter (sQGP), its other applications are discussed in Chapter 29 in the third part of the book.

Chapter 22

WHAT HAVE WE LEARNED ABOUT HOT QCD MATTER?

The big picture is, of course, the discovery that there is a phase transition from "resonance gas" to quark-gluon plasma. We have learned a lot about its thermodynamics and, even more importantly, its unusual kinetics.

22.1 Thermo- and Hydrodynamics

On the theoretical side, the developments have progressed in two main directions. One of them is based on lattice gauge theory *at finite temperatures*, which we discuss in Chapter 8. In this case, the transition from weak coupling at the lattice scale a to a stronger one occurs naturally, following QCD renormalization group flow. This change in scale, however, makes the calculations numerically expensive.

The theory provided the thermal *equation of state*, which has been used quite successfully in hydrodynamics as well as for *higher-order susceptibilities* to describe event-by-event fluctuations, as we discussed in Chapter 6.

We previously discussed in detail that particle composition is defined at a stage called the *chemical* freeze-out, at which it shows a rather remarkable degree of thermal equilibrium. Simple exponential dependence of yield (per degree of freedom) on particle mass extends to light nuclei over about eight orders of magnitude. It also described states with charm quarks if proper fugacity of those is included.

We also discussed the remarkable successes of (lattice-based) thermodynamics coupled with hydrodynamics. Experimental discoveries at the RHIC and LHC left no doubts that a novel (but "normal") phase of QCD, the QGP, has been observed. The observations of *collective flows* and their theoretical modeling convinced us that the microscale of matter ("mean free path" or "correlation length") is indeed surprisingly small compared to the system sizes, so that these flows proceed in a *hydrodynamical regime.*

Furthermore, it was shown that such a regime does extend to systems smaller than fireballs created in heavy ion collisions of $AuAu, PbPb$. The hydro systematic is valid at least until the azimuthal harmonics of number 7 or 8. There are radial and elliptic flows in "small systems" produced in pAu, dAu, He^3Au collisions (at RHIC) as well.

It is also necessary to give some perspective here. At the time of this writing, the Beam Energy Scan program at RHIC is not yet complete, and while we discussed proposed signals in Chapter 6 (such as kurtosis of proton multiplicity distribution and light nuclei production), the data at hand only provide some hints on the possible location and existence of the QCD critical point. Better accuracy is needed to make statistically significant statements on this important issue.

22.2 Kinetic Properties of QGP/Hadronic Matter

As soon as QCD is applied to finite temperatures, its perturbative version is applied to quark and gluon kinetics. The lowest-order cross-sections have been calculated and applied to the estimates of the mean free paths already in the first review (Shuryak, 1980). While differential transport cross-sections were easily obtainable, their integral diverges logarithmically at large impact parameters, unless those are cut off by screening masses. Some appropriate resummations of the perturbative series were needed.

These have been carried out by Arnold *et al.* (2000). The viscosity is proportional to the mean free path and thus inversely proportional to the cross-sections; therefore, it is more appropriate to discuss its inverse, and the leading log can be written as

$$\frac{T^3}{\eta} = \frac{1}{\kappa} g^4 \log\left(\frac{1}{g}\right) \tag{22.1}$$

and the coefficient κ for pure gauge theory and $N_f = 2$ QCD are

$$\kappa_{N_f=0} = 27.126, \quad \kappa_{N_f=2} = 86.473 \tag{22.2}$$

The second coefficient is larger because there are more degrees of freedom and, respectively, processes. It is customary to use entropy s instead of T^3, which partially absorbs this effect. Anyway, the assumed small coupling should lead to *small* s/η values. In physical terms, one expects the mean free path to be *longer* than the inter-particle distance.

Remarkably, this is *not* what is experimentally observed!

The viscosity values used are $\eta/s = 0.12$ for the RHIC data and $\eta/s = 0.2$ for the LHC data (see, for example Gale *et al.*, 2013), which suggested some movement toward perturbative QGP. However, in more recent years, the heavy ion community has been attracted to various forms of "global analysis," which combine certain models of the initial stage, interpolating them into hydro and then into hadronic cascades together with a large set of data, extracting empirical matter parameters with their possible variations. As an example of such, let me mention the *Trajectum* program, developed by a group in Holland (Nijs *et al.*, 2021), and JETSCAPE (Everett *et al.*, 2021). The fine details can be found in these works; as an example, their posterior distribution of shear and bulk viscosities versus temperature is presented in Fig. 22.1.

In this plot, one can see that by now η/s is relatively well constrained to 0.12–0.15 but there is hardly any tendency for it to grow with T. The bulk viscosity is smaller by, say, a factor of two, and there seems to be a hint at a high T toward conformal QGP, with zero bulk viscosity.

The inverse ratio s/η, as shown in Fig. 22.2, combines the empirical value (the red point) with that obtained from lattice calculation

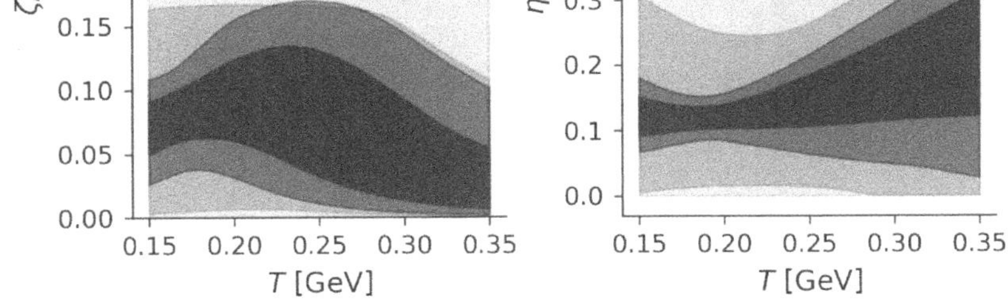

Fig. 22.1. The posterior for the dependency of specific bulk (left) and shear (right) viscosities (over entropy density) on the temperature, resulting from model parameter estimation using ALICE data for *Pb-Pb* collisions at $\sqrt{s}_{NN} = 2.76$ TeV.

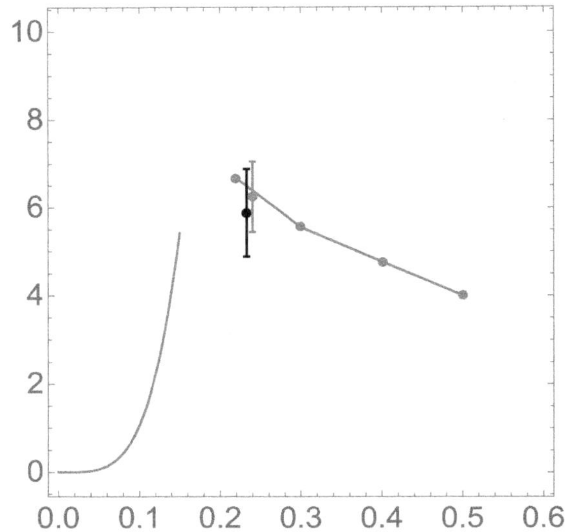

Fig. 22.2. The ratio s/η versus temperature T (GeV). The red and black points with error bars denote the empirical value and lattice determination, and the four points connected by a line are from monopole-gluon scattering. The line on the left corresponds to the estimates for hadronic resonance gas.

(Nakamura and Sakai, 2005) as well as from the monopole–monopole scattering model (Ratti and Shuryak, 2009), as indicated by four points without error bars and connected by a line. The curve on the left corresponds to estimates in hadronic resonance gas. It is clear that this ratio has a peak near T_c and that the value is well above one. It means that, contrary to perturbative ideas, the mean free path is *smaller* than the inter-particle distance. In order for this to be possible, one needs to use the *strong coupling regime* instead!

On the other hand, as we discussed in Section 21.7, strong coupling methods based on the AdS/CFT correspondence suggest the famous relation

$$\frac{s}{\eta} = 4\pi \tag{22.3}$$

(also claimed to be the maximal possible value for this ratio). It is larger than empirical values by a factor of about two, as observed from Figs. 22.1 and 22.2. Also, as we discussed in other sections on holography, the strong coupling limit describes well a rapid onset of hydrodynamical behavior, as is indeed observed.

Another example of a kinetic coefficient (which we discussed in Section 19.3 and below it) is the so-called $\hat{q}$, which describes the rate with which a quenching jet accumulates momentum normal to its direction. The empirical temperature behavior of the dimensionless ratio $\hat{q}/T^3$ on T shown in Fig. 19.5 is remarkably similar to that of s/η. Both suggest a peak near T_c and even of similar magnitude.

22.3 A Wider Look at Weak versus Strong Coupling Approaches

The discussion in the preceding section of two kinetic coefficients, s/η and $\hat{q}/T^3$, strongly suggests that they are *not* described by the perturbative expansion in the coupling. However, even while confronting these facts, one may perhaps think that those are some special cases and that the more basic properties of QGP are still in (at least qualitative) agreement with the perturbative approach. (Such statements are prolific in the literature.)

Unfortunately, even such more modest claims are not true. Consider, for example, the *screening masses*. The original pQCD prediction (Shuryak, 1978b) for the electric mass is

$$\frac{M_E}{T} = g\sqrt{1 + \frac{N_f}{6}} \tag{22.4}$$

and zero for the magnetic one, $M_M = 0$. So, if true, the electric screening length would directly give the pertinent coupling value.

Jumping over decades of work, let us discuss the values of the electric and magnetic screening masses extracted from various approaches currently adopted. Those are listed in Table 22.1, including predictions from various strong coupling approaches: The first line corresponds to a (large N_c) holographic model, the next two to a lattice (the last with small physical quark masses), and the last to the dimensionally reduced 3D effective theory for $N_f = 3$ light quarks. Looking at this table, one finds that the electric

Table 22.1. The electric and magnetic screening masses, normalized to the temperature. The last column lists the squares of their ratios.

References	M_E/T	M_M/T	M_E^2/M_M^2
Bak *et al.* (2007)	16.05	7.34	4.80
Maezawa *et al.* (2010)	13.0(11)	5.8(2)	5.29
Borsányi *et al.* (2015b)	7.31(25)	4.48(9)	2.66
Hart *et al.* (2000)	7.9(4)	4.5(2)	3.10

mass is *not* much smaller than the temperature: instead, $M_E/T > 1$. This means the coupling is not small and pQCD is not applicable. The second important observation is that, while the magnetic mass is still smaller than the electric one, it is smaller only by a factor of two or so. This means magnetic charges play a significant role, comparable to that of electrically charged quasiparticles, quarks and gluons. In the following, we discuss the role of magnetically charged quasiparticles, the *monopoles*, which are believed to play an important role in QGP dynamics.[1]

Furthermore, results from the lattice (Borsányi *et al.*, 2015b) on the temperature dependence of screening masses are shown in Fig. 22.3. This and similar plots indicate only a very gradual decrease toward high T, suggesting that the expected transition to a weak coupling regime is not seen in these calculations.[2]

We think that those observations are sufficient to conclude that QGP (as produced in experimental conditions) is in a very peculiar *strong coupling regime*, or sQGP.

The mainstream approach to its theory is *holography*, or AdS/CFT correspondence, discussed in Chapter 21. In most of these works, one used

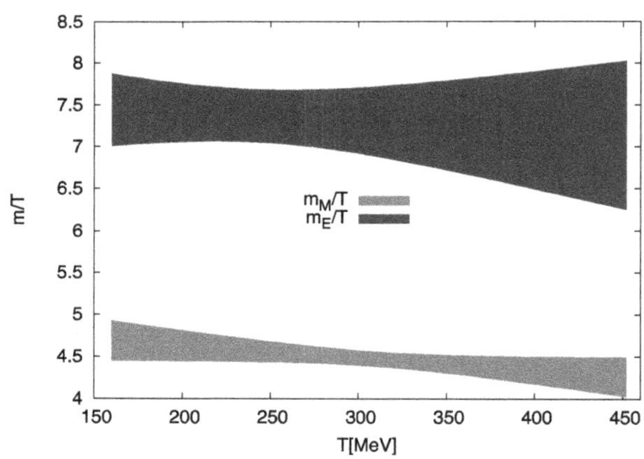

Fig. 22.3. Electric and magnetic screening masses in units of temperature from lattice calculations based on Polyakov line correlators.

[1]Some people even argue that the name itself should be upgraded to QGMP.

[2]Special dedicated lattice simulations with extremely large T have been performed, in which that indeed happens. However, there is no hope of reaching this regime in heavy ion collisions.

the $\mathcal{N} = 4$ SYM theory instead of QCD, in which the coupling constant does not run and can be fixed at any value one needs.

It is fair to say that the holographic approach has *opened a whole new world* for QGP theory, with general-relativity-driven multidimensional curved spaces. Theorists with a background in nuclear or high-energy physics needed to learn a lot to apply this approach to their problems of choice. And they did indeed apply it to a wide range of problems, from static thermal matter to the derivation of hydrodynamics, the evaluation of many kinetic coefficients and even spectacular out-of-equilibrium settings. Without pointing out specific applications here, let me just say that we now have a rather extensive theory describing QGP in a strongly coupled regime.

Finally, let me mention one important issue which remains open, as the strong coupling approach to it has not yet been developed. It is the issue of *initial chemical equilibration* (discussed in the chapter on photon/dilepton production).

The initial state is composed mostly of gluons, while the equilibrium QGP has two-thirds of the degrees of freedom provided by quarks. The pertinent perturbative QCD cross-sections are

$$\begin{aligned}\frac{d\sigma_{gg\to gg}}{dt} &= \frac{\pi\alpha_s^2}{s^2}\frac{9}{2}\left(3 - \frac{ut}{s^2} - \frac{us}{t^2} - \frac{st}{u^2}\right)\\ \frac{d\sigma_{gg\to\bar{q}q}}{dt} &= \frac{\pi\alpha_s^2}{s^2}\left(\frac{u^2+t^2}{6ut} - \frac{3u^2+3t^2}{8s^2}\right)\end{aligned} \tag{22.5}$$

and at large angles $|t| \sim s$, the ratio of the two is very large:

$$\frac{d\sigma_{gg\to gg}}{dt} / \frac{d\sigma_{gg\to\bar{q}q}}{dt} \sim 200 \gg 1 \tag{22.6}$$

This observation led us to suggest a "*hot glue scenario*" (Shuryak and Xiong, 1993), in which a suppressed quark fraction leads to suppressed photon/dilepton production.

While we have argued strongly in this section against weak coupling methods, it is also logical to suspect large values of the ratio (22.6) underlying the "hot glue scenario." Yet, we neither have strong phenomenology nor strong coupling theory to decide on this important issue. Can we (or maybe not) separate kinetic equilibration from the chemical one when *entering* the QGP, as we can do so when *exiting* resonance gas at the collision end?

Part 3

NONPERTURBATIVE QCD AND HADRONS

Chapter 23

INSTANTONS, SPHALERONS AND OTHER CHIRAL EFFECTS

In this part of the book, we discuss nonperturbative dynamics in QCD (and related theories). We start with the general topological features of the gauge fields. Then comes the presentation of certain topological solitons, followed by a description of the QCD vacuum in terms of soliton ensembles. The main application is related to nontrivial states of light fermions coupled with topology and chiral anomaly relations. We discuss QCD chiral symmetries and their breaking in vacuum and at sufficiently low temperatures. Of particular importance is the "gapping" of light quarks, turning them from near-massless fermions into *constituent quarks* with effective masses of $M_q \approx 330$–400 MeV. A brief presentation of its theory is the subject of this chapter. The phenomenon of *color confinement* and an extensive discussion of interquark forces are the subjects of Chapter 24.

23.1 Gauge Topology, Instantons and Sphalerons

Gauge topology is a vast subject, to which we cannot give full justice in this book.[1] Nonetheless, we need to present some main ideas and expressions which are necessary to understand this chapter.

The gluon fields have a nontrivial topological structure, which can be best explained using the following function of the gauge fields, known as the "topological current":

$$K_\mu = \frac{1}{16\pi^2}\epsilon^{\mu\alpha\beta\gamma}\left(A^a_\alpha \partial_\beta A^a_\gamma + \frac{1}{3}\epsilon^{abc}A^a_\alpha A^b_\beta A^c_\gamma\right) \tag{23.1}$$

[1]See my other book on the subject, *Nonperturbative topological phenomena in QCD and related theories*, Springer-Nature 2021.

Its space integrated fourth component is known as the Chern–Simons number, introduced by Chern and Simons (1974):

$$N_{\rm CS} = \int d^3x K_4 \tag{23.2}$$

This topological current is not conserved; its divergence is nonzero and equal to

$$\partial_\mu K_\mu = \frac{1}{32\pi^2} G^a_{\mu\nu}\tilde{G}^a_{\mu\nu} \tag{23.3}$$

The integral over the right hand side,

$$Q_{\rm top} = \frac{1}{32\pi^2}\int d^4x G^a_{\mu\nu}\tilde{G}^a_{\mu\nu} \tag{23.4}$$

is the so-called 4D *topological charge*. If the gauge fields vanish at all infinities, one can show that $Q_{\rm top}$ can only take integer values, positive or negative.

Let us begin by exploring gauge field structure with the simplest "pure gauge" case, $A_\mu(x) \neq 0$, but with $G_{\mu\nu} = 0$. It has the form[2]

$$A_\mu = U^+(\vec{x}) i\partial_\mu U(\vec{x}) \tag{23.5}$$

and we consider the time-independent case with $A_0(x) = 0$, so that, here, the indices are $\mu = 1,2,3$ only. Substituting it in N_{CS}, one finds the following expression:

$$N_{\rm CS} = \frac{1}{24\pi^2}\int d^3x \epsilon^{ijk}(U^+\partial_i U)(U^+\partial_j U)(U^+\partial_k U) \tag{23.6}$$

Now, the unitary $SU(N_c)$ matrix $U(\vec{x})$ is a map from the 3D space to a gauge group. If the group is $SU(2)$, with three generators and three Euler angles, it is a map from one 3D sphere to another. The topological integer *winding number* is the number of times one covers the other. Obviously, a topological number cannot be changed by smooth deformations, so all possible gauge maps are actually split into separate disconnected classes.

Summary so far: When N_{CS} is equal to any integer, the lowest energy and the field strength $G_{\mu\nu}$ are zero. When it is not an integer, there must be nonzero field strength, or energy barriers (or "mountains") separating such zeros (or "valleys"), see Fig. 23.1. The shape of it was obtained by *conditional* energy minimization, keeping both $N_{\rm CS}$ and the r.m.s. size ρ

[2]In QED, in which the gauge symmetry group is Abelian and $U = \exp(i\phi(x))$, the "pure gauge" would correspond to $A_\mu = \partial_\mu\phi$, which obviously produces zero field strength.

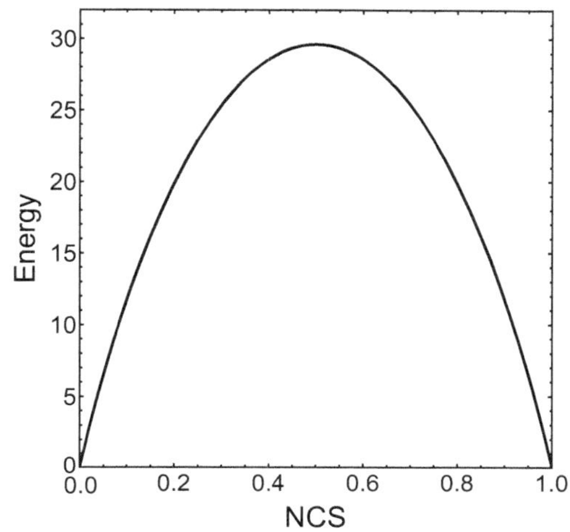

Fig. 23.1. The shape of the barrier separating two "valleys": the potential energy E (in units of $1/g^2\rho$) versus the Chern–Simons number $\tilde{N}_{\rm CS}$, for the "sphaleron path" solution to be derived in the chapter on sphalerons.

constant (Ostrovsky *et al.*, 2002). A particular gauge field configuration at the very "top of the barrier" is called a "sphaleron."[3] Being an extremum of the action, it satisfies the Yang–Mills field equations. However, being a maximum rather than an action minimum, it is unstable: With small perturbation, it would "roll down the hill."

Another solution to the classical Yang–Mills equations (Belavin *et al.*, 1975), now in Euclidean time, describes tunneling *through* the mountain; those are called "instantons." They will dominate the discussion in this chapter.

An infinitely repeating periodic mountain chain resembles crystals. Like them, the problem admits "propagating wave" solutions with nonzero quasi-momenta. The wave function in each period repeats itself except for an extra phase factor. In our case of gauge theories, this quasi-momentum is denoted by the parameter θ, with the phase being $\exp(i\theta)$. Observations[4] show that QCD does not violate T and CP parities, which means we live in a world in which $\theta = 0$.

[3]First discussed in an electroweak theory context by Klinkhamer and Manton (1984), the name means "ready to fall."

[4]In particular, the strongly constrained electric dipole moment of the neutron.

23.2 "Chiral Anomaly" and Its Consequences

This important phenomenon has several names: While we will call it $U_a(1)$ *chiral anomaly*,[5] it is also called in literature by the term "axial anomaly," "triangular anomaly" (in the Feynman diagrams in which it was first found), or the Adler–Bardeen–Jackiw anomaly (after the names of the people who discovered it).

The "chiral symmetries" appear in the case when quarks are massless, so that left- and right-handed components are not mixed in the Lagrangian. The flavor symmetry $U(N_f)$ gets doubled into $U(N_f)_L \times U(N_f)_R$. This doubled symmetry can be rewritten as vector and axial ones, and the latter can be further split into $U(N_f)_A \to SU(N_f)_A \times U(1)_A$. Both symmetries are broken in the QCD vacuum but in a different manner. $SU(N_f)_A$ is broken *spontaneously*, as will be explained later in this chapter.

$U(1)_A$ is broken explicitly by the phenomenon of chiral anomaly. Although a symmetry of the action, it is *not* a symmetry of quantum field theory (QCD) because it is not a symmetry of the path integral measure. We will not have space here to derive this relation and only state that the corresponding axial current is *not* conserved. Its divergence is nonzero:

$$j^5_\mu = (\bar{u}\gamma_\mu\gamma_5 u + \bar{d}\gamma_\mu\gamma_5 d + \bar{s}\gamma_\mu\gamma_5 s) \tag{23.7}$$

$$\partial_\mu j^5_\mu = \frac{1}{32\pi^2}\epsilon^{\alpha\beta\gamma\delta} G^a_{\alpha\beta} G^a_{\gamma\delta} \tag{23.8}$$

Note that the right-hand side of this relation is proportional to the 4D topological charge of the gauge field Q_{top} which we already came across in the previous section. Note further that the term on the right-hand side can also br rewritten as the product of $(\vec{E}\vec{B})$ fields, and it is not only P-odd but also CP-odd.

With two non-conserved currents, K_μ and j^5_μ, with the same divergence (up to a factor), one can construct their combination, which *is conserved*. It implies that any changes in the Chern–Simons number N_{CS} and the axial charge must strictly follow each other! Furthermore, the normalization of this expression is such that a change in the topological charge by

---

[5]The term "anomaly" does not sound very attractive, which indeed reflects the historical attitude toward this phenomenon. It was found unexpectedly during calculations of the triangular Feynman diagrams with a fermion loop, and despite significant efforts by many theorists to eliminate it through various tricks, it has survived. Now, its nature is well understood, and the term "anomaly" no longer creates any negative feelings.

one unit,

$$\Delta Q \equiv Q_{\text{final}} - Q_{\text{initial}} = 1$$

corresponds to a change in the axial charge by $2N_f = 6$ units.

In particular, if N_{CS} changes from 0 to 1, the minimal required number of produced quarks and antiquarks is

$$|0\rangle \to |(\bar{u}_R u_L)(\bar{d}_R d_L)(\bar{s}_R s_L)\rangle \tag{23.9}$$

with L and R indicating the left- and right-polarized light fermions.

As we already mentioned, one mechanism of such a transition of the gauge fields is tunneling, even between topologically different configurations with $\Delta N_{\text{CS}} = 1$. It can be described semiclassically by the so-called *instanton* solution found by Belavin *et al.* (1975). As a consequence of the chiral anomaly relation, each instanton *must* create six units of chiral charge in a quark state, as described by this equation. The detailed description of quark production via instantons was provided by 't Hooft (1976a). While the chirality of a particular massless quark cannot be flipped by gauge fields, it can be "dumped into the Dirac sea," with a quark having the opposite chirality pulled out of it.

At finite temperatures or in some collisions in which enough energy is available, there is also *another* mechanism to induce transitions with nonzero $\Delta N_{CS} \neq 0$: Instead of tunneling *through* the potential barrier, one can go over *its top*. This is known as the *sphaleron mechanism*. Again, in this case, the chiral anomaly requires certain fermion production.

23.3 Instantons in the Euclidean QCD Vacuum

The phenomenology of the "QCD vacuum structure" began with the QCD sum rules (Shifman *et al.*, 1979), where the short-distance description of *Euclidean point-to-point correlation functions* via the operator product expansion (OPE) was related to their long-distance description in terms of the lightest hadrons. The (Euclidean) correlators of QCD operators were calculated in terms of the leading *vacuum condensates*, $\langle G^2\rangle$, $\langle \bar{q}q\rangle$, etc. Yet, it was soon realized that the strongest nonperturbative effects (in the scalar and pseudoscalar channels) are *not* described by the "mean field" OPE predictions (Novikov *et al.*, 1981).

Extensive studies of point-to-point (Euclidean) correlation functions in multiple mesonic channels based on phenomenology (Shuryak, 1993) have indeed found striking differences between correlators of operators with different quantum numbers. While for vector currents ($\bar{q}\gamma_\mu q$) made of *light* quarks $q = u, d, s$, only small deviations from free quark propagation is observed, with the correlators of the scalar and pseudoscalar operators ($\gamma_\mu \to 1, \gamma_5$) showing strong splittings from them at surprisingly small distances. As shown in detail in the review by Shuryak (1993), all these splittings found their explanations in terms of topological fluctuations of gluonic fields, described semiclassically by instantons.

The resulting semiclassical model of the QCD vacuum structure based on instantons (Shuryak, 1982a) is known as the "instanton liquid model" (ILM). It introduced an important scale parameter of the nonperturbative vacuum — the typical instanton size:

$$\rho \sim \frac{1}{3}\,\text{fm} \tag{23.10}$$

Because it is rather small compared to the sizes of most hadrons (except for the lowest Υs and the pion, see the following), where possible, one can use a *quasi-local* approximation in which it tends to zero.

In the ILM, the vacuum fields are very inhomogeneous: Spots of rather strong gauge fields at the instanton centers are surrounded by plenty of "empty" space-time, free from such fields. Since the mid-1990s, such field distributions have been obtained from lattice configurations by means of various "cooling" methods. A picture is worth a thousand words, so we reproduce in Fig. 23.2 a visualization of the topological charge distribution from Leinweber (1999) and Biddle *et al.* (2020a, 2020b), which reveals a number of instantons and anti-instantons.

Note that the topological clusters are also threaded by thin center vortices or Z_3-fluxed strings, which span world-sheet surfaces in four dimensions. While the center vortices are important for enforcing confinement at long distances, Fig. 23.2 shows that they are, on average, decoupled from the inhomogeneous and strong topological fields. Moreover, their field strength in the vicinity of these topological fields is about $\sigma_T\bar{\rho} \approx 0.3\,\text{GeV}$, which is weaker than the typical chromo-electric or chromo-magnetic field in the instanton center $\sqrt{E} = \sqrt{B} \approx 2.5/\bar{\rho} \approx 1.5\,\text{GeV}$, but crucial for long-distance color correlations.

Two parameters of the original ILM (also to be referred to as the "dilute instanton model") are the size and density of instantons:

$$\rho = \frac{1}{3}\,\text{fm}, \quad n_{I+\bar{I}} = \frac{1}{R^4} = 1\,\text{fm}^{-4} \tag{23.11}$$

which, in turn, defines the so-called *diluteness parameter*:

$$\kappa \equiv (\pi^2\rho^4/2)n_{I+\bar{I}} \tag{23.12}$$

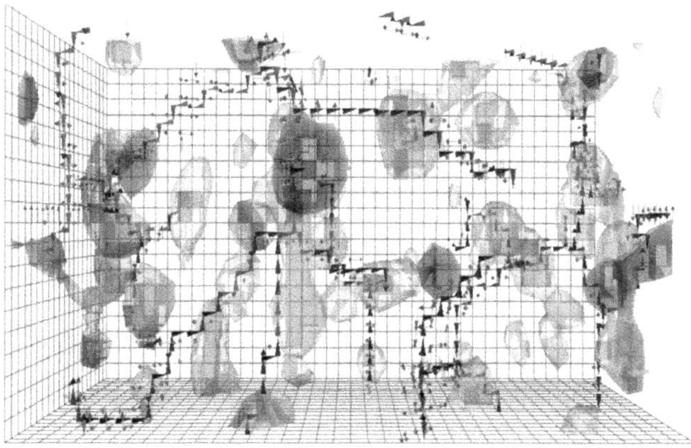

Fig. 23.2. Instantons (yellow) and anti-instantons (blue) configurations in the "deep-cooled" Yang–Mills vacuum. After center projection, they are threaded by center P-vortices (Biddle *et al.*, 2020a, 2020b). They constitute the primordial gluon epoxy (hard glue) at the origin of the light hadron masses (Zahed, 2021). The center P-vortices are responsible for confinement. See the text for details.

where the term within the brackets is the volume of the 4D sphere. ILM predicts it to be small, $\kappa_{\rm ILM} \ll 1$. These parameters have withstood the scrutiny of time and describe rather well the chiral dynamics related to pions, the Euclidean correlation functions in the few femtometers range, and much more, see Schäfer and Shuryak (1998) for a review. The main consequence of ILM is its role in explaining chiral symmetry breaking, to which we turn in the following section.

Well-isolated instantons seen on the plot above are not the only fluctuations of the gauge field along the Chern–Simons coordinate. There are also "tunneling incomplete," which are technically described as overlapping instanton–antiinstanton pairs or *molecules*.

Close instanton–antiinstanton pairs are, of course, known and, in particular, were observed on the lattice. They are, however, not seen in Fig. 23.2, obtained by the so-called "deep cooling" of configurations, during which close instanton–antiinstanton pairs are already annihilated. The application of the "molecular component" of the vacuum was made previously in connection to phase transitions in hot/dense matter. Indeed, (if quark masses are neglected), this component is the only one which survives at temperatures $T > T_c$, where chiral symmetry is restored. The accounting of "atomic" and "molecular" components together began with Ilgenfritz and Shuryak (1989). The "molecular component" was also shown to be important at high baryonic densities, where it contributes to quark pairing and color superconductivity (Rapp *et al.*, 1998). The $I\bar{I}$ interaction was first done for quantum-mechanical instantons by Shuryak (1988b); for gauge theory, the "streamline equation was derived by Balitsky and Yung (1986), which was solved approximately by Yung (1988) and numerically by Verbaarschot (1991). The surprising finding of the latter paper was that "Yung ansatz" was rather accurate not only at large distances $R \gg \rho$ between instanton and antiinstanton, where it was derived, but also for any distance up to zero.

The "molecular component" of the instanton ensemble does not generate near-zero Dirac eigenvalues and is therefore not important for spontaneous breaking of the $SU(N_f)_A$ symmetry. Yet, as they consist of fields, there are observables to which they contribute. The simplest of which is the so-called *gluon condensate* $\langle G^2_{\mu\nu} \rangle$ introduced in the context of the QCD sum rules framework (Shifman *et al.*, 1979). The accuracy of this number was later questioned, and it was revised to a larger value. Further discrepancies were shown in lattice studies, which attempted to extract local (or nonlocal) observables with powers of the gauge field strength $G_{\mu\nu}$.

On the lattice, the issue was revived when "cooling" by the gradient flow method was consistently related to the renormalization group (RG) flow (Luscher and Weisz, 2011). Using it, one can set the appropriate scale dependence of the "molecular component" on a firm basis. Skipping several decades, let us take as an example the work by Athenodorou *et al.* (2018), which studied the three- and four-point gluon field correlators and related their evolution to topology during cooling. Their original motivation was to extract the gluon coupling $\alpha_s(k)$, so the observable was chosen to be the ratio of the three-point to two-point Green function, leading to effective coupling:

$$\alpha_{\mathrm{MOM}}(k) = \frac{k^6}{4\pi} \frac{\langle G^{(3)}(k^2) \rangle^2}{\langle G^{(2)}(k^2) \rangle^3} \tag{23.13}$$

In the "uncooled" quantum vacuum (with gluons), the effective coupling starts running downward at large $k > 1\,\mathrm{GeV}$, as expected by asymptotic freedom. However, at low $k \to 0$, one finds another persisting positive power of k, with a slope that matches exactly the one following from an instanton ensemble (Boucaud *et al.*, 2003):

$$\alpha_{\mathrm{MOM}}(k) \to \frac{k^4}{18\pi n_{I+\bar{I}}} \tag{23.14}$$

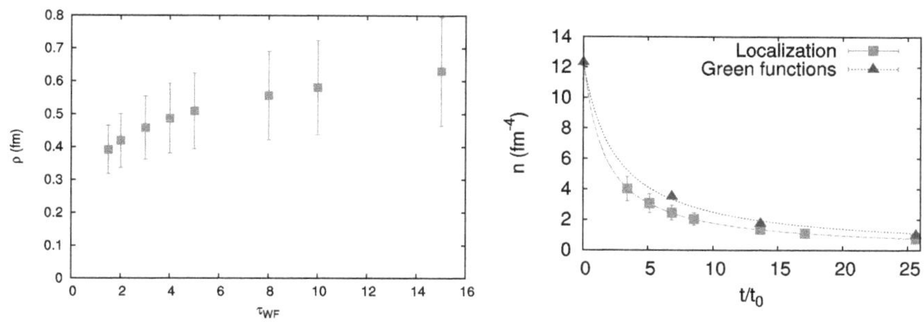

Fig. 23.3. The dependence of the mean instanton sizes (upper plot) and density (lower plot) on the gradient flow cooling time τ (arbitrary units). The quantum vacuum, of course, corresponds to an extrapolation to $\tau \to 0$.

Furthermore, it was observed that with increasing cooling time τ, the same power spreads to higher momenta, $k > 1\,\text{GeV}$. It was observed that "cooling" eliminates not only perturbative gluons but close instanton–antiinstanton pairs as well.

The dependence of the mean instanton size and density as a function of the (gradient flow) cooling time is shown in Fig. 23.3, adapted from Athenodorou *et al.* (2018). Their main conclusion from this analysis was that the size and density — extrapolated to *zero* cooling time ($\tau \to 0$) — are

$$\rho \to \frac{1}{3}\,\text{fm}, \qquad n \to O(10\,\text{fm}^{-4}) \tag{23.15}$$

So, the *total* instanton density is significantly larger than observed earlier by a "deep cooling" (large τ) of $1/\text{fm}^4$, as in the original ILM. (In fact, the density may even be larger than that because the definition of instantons used in this paper excludes pairs ("molecules") which are too close.)

The so-called *"dense instanton liquid model"* (dILM) was proposed, with such a density value. We will return to its application to interquark forces, mesonic form factors (Shuryak and Zahed, 2021), matching kernels (Liu and Zahed, 2021) in Chapter 28.

23.4 Spontaneous Breaking of $SU(N_f)_A$ Chiral Symmetry

Long before the development of QCD, Nambu and Jona-Lasinio (1961a) (NJL) discussed near-massless quarks, introduced chiral symmetry and suggested that it must be broken in the vacuum. Inspired by the Bardeen–Cooper–Schriffer (BCS) theory of superconductivity, they have demonstrated that an attraction between a quark and an antiquark, if strong enough, can produce a Bose condensate of quark-antiquark pairs (analogous to) Cooper pairs $\langle \bar{q}q \rangle \neq 0$. It then produces an effective quark mass, analogous to the BCS superconductor gap at the surface of the Fermi sphere.

Since this scalar condensate relates the left- and right-handed components of quark fields, this phenomenon spontaneously breaks the $SU(N_f)_A$ chiral symmetry. Among many other effects, NJL had shown that this phenomenon must create $N_f^2 - 1$ (near) *massless pions.*

Instantons are semiclassical solitons made of glue but not only. As discovered by 't Hooft (1976a), they generate a $2N_f$-fermion effective Lagrangian. We already discussed above that, as dictated by the axial anomaly, this Lagrangian violates the $U(1)_A$ chiral symmetry explicitly. This solves the famous "$U_A(1)$ problem" by making the remaining ninth pseudoscalar meson η' heavy, unlike the octet members π, K, η.

Here, we focus on another effect of instanton-induced Lagrangian, namely in the scalar σ channel. Like NJL attraction, it can create a scalar condensate and make π, K, η massless Nambu–Goldstone modes, provided it is strong enough. In the framework of the ILM (Shuryak, 1982b), one can show this to be the case and obtain a microscopic understanding of chiral symmetry breaking, chiral perturbation theory, pion properties, etc. So, the 't Hooft instanton-induced effective Lagrangian happens to be a QCD-based substitute for the hypothetical Nambu–Jona–Lasinio (Nambu and Jona-Lasinio, 1961a) four-quark interaction.

We will not go into corresponding formulae, providing instead an alternative and more intuitive way to explain this theory. An instanton field can bind one quark (per flavor) technically described by the "zero mode." Consider an ensemble of instantons/antiinstantons as a 4D gas of "atoms," each having one bound state. Depending on the density of those, the spectrum of quark states can take two basic forms. One is "metallic," with no gaps in the Dirac eigenvalues near zero. Another is an "insulator," with a gap and localized states. There should be a Mott-type transition between them. Indeed, the former corresponds to the QCD vacuum with broken chiral symmetry. The latter, occurring at higher temperatures of $T > T_{\rm ch}$, corresponds to a QGP phase in which chiral symmetry is unbroken.

A technical description of fermions is obtained via the calculation of the "fermionic determinant" in the QCD partition function. It can be represented as a sum over quark loop diagrams. As bound quarks are "hopping" from one instanton to another, the resulting loops in the quark determinant can become very long, even infinite in the thermodynamic limit of $V_4 \to \infty$. If so, there appear to be Dirac eigenvalues close to zero,

$\lambda = O(1/V_4)$. The chiral symmetry breaking is therefore ascribed to the *collectivization* of the instanton zero modes into the so-called *"zero mode zone"* (ZMZ). Due to nonzero matrix elements of the Dirac operator, near-zero Dirac eigenvalues reside within a strip of small width,

$$|\lambda| \sim \text{width}\,(\text{ZMZ}) \sim \frac{\rho^2}{R^3} \sim 20\,\text{MeV} \tag{23.16}$$

as predicted by the original ILM with the parameters given above (23.11).

Glozman *et al.* (2012) studied on the lattice what happens in meson and baryon spectroscopy if all Dirac eigenstates inside a certain strip of eigenvalues $|\lambda| < \Delta$ get eliminated. A strong restructuring of the light hadronic spectra was indeed observed, provided $\Delta >$ width (ZMZ). In particular, in this case, the Nambu–Goldstone modes (pions) totally disappear from the spectra, as expected. It would be interesting to extend the same analysis to the heavy-light sectors and to address more generally what happens with all forms of spin-dependent forces.

A statistical description of interacting instanton ensembles, including the dynamical quarks along such lines, was developed both by means of analytical methods and statistical numerical simulations. For a review, see Schäfer and Shuryak (1998). For related lattice studies, see Chu *et al.* (1994) and Faccioli and DeGrand (2003).

23.5 Sphaleron Production in Hadron–Hadron Collisions

We begin this section by explaining the connection between instantons and (3D) sphalerons, as elucidated by Ostrovsky *et al.* (2002).

Note that the former are 4D solitons, whereas the latter are static 3D solitons. How they can be related is explained via the setting shown in Fig. 23.4. Let us discuss gauge field configurations, including an instanton and an antiinstanton, placed at a certain (Euclidean time) positions $\tau = -T/2$ and $\tau = T/2$. Note that the instanton fields are self-dual, $\vec{E} = \vec{B}$, while the antiinstanton ones are anti-selfdual, $\vec{E} = -\vec{B}$. By symmetry, at the 3D plane of $\tau = 0$ in between, the electric field must vanish. Therefore, by symmetry, at this 3D plane $t = 0$, the fields should be purely magnetic. So, there exists a one-dimensional set, parameterized by the time distance T between the instanton and the anti-instanton.

Out of all $I\bar{I}$ configurations, one would like to select "the best ones," minimizing their action but keeping the I and $\bar{I}$ in place. Those are obtained

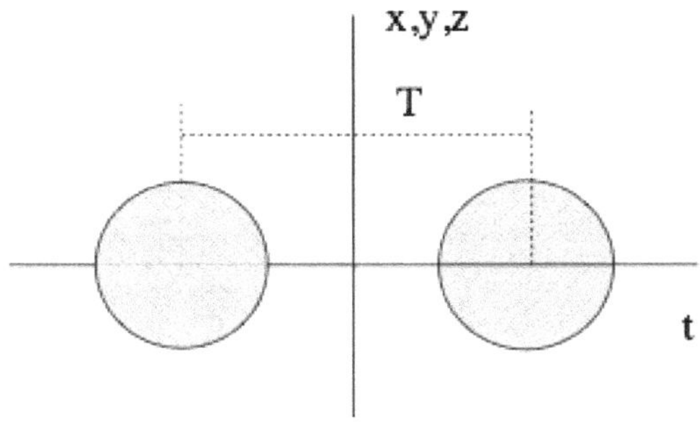

Fig. 23.4. The instanton and antiinstanton (shown by shaded circles) are placed at $t = -T/2$ and $t = T/2$ along the horizontal (Euclidean) time axes. The vertical line should substitute for three spatial directions. Purely magnetic (3D) configurations are located at the middle plane $t = 0$, which defines the "sphaleron path."

by a "streamlike" equation,[6] which was solved by Verbaarschot (1991). The solution was found to be very close to a guess, known as "Yung ansatz."

For each of those, one can define their energy normalized by size:

$$ER = \frac{1}{2}\left[\int d^3 r r^2 \mathcal{B}^2 \times \int d^3 r \mathcal{B}^2\right]^{1/2} \tag{23.17}$$

and the Chern–Simons number $N_{\rm CS}(T)$. It has been calculated by Ostrovsky *et al.* (2002) for the "streamline" Yung ansatz, which is shown in Fig. 23.5. Note that at $N_{\rm CS} = 0$ (corresponding to a large distance T), the energy is zero, and it also tends to be zero at $N_{CS} = 1$. The maximum is in the middle, at $N_{\rm CS} = 1/2$, when the instanton and anti-instanton "half-overlap." These configurations constitute the so-called *sphaleron path*, leading from one classical vacuum to the next. Recall that the shape of the barrier separating the "Chern–Simons valleys," we already discussed at the beginning of the chapter, see Fig. 23.1: It can indeed be obtained in several other ways.

We returned to it using the $I\bar{I}$ plane (Fig. 23.4) because we wanted to reinterpret this figure, putting it into a new light. Half of the plane $t < 0$ described the amplitude of a process creating out of an "empty vacuum" a magnetic ball, the sphaleron. The other halfplane $t > 0$ is a conjugated amplitude. The line $t = 0$ is the "unitarity cut," which, in

[6]These configurations in mathematics are called Lefschitz "timbles," the lines in functional space connecting the singular points along the gradient lines.

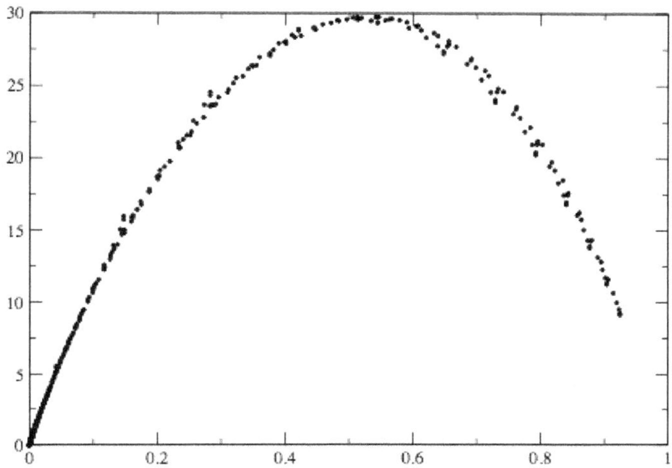

Fig. 23.5. The energy times the r.m.s, radius $E \cdot R$ versus the Chern–Simons number $\tilde{N}_{\mathrm{CS}}$ for the 3D configurations obtained from the instanton–antiinstanton Yung ansatz, as described in the text. The curve should be compared to the sphaleron path we soon obtain from constrained minimization.

quantum mechanics, is also known as the "turning point." Indeed, the meaning of an electric field in gauge theory is that of the field's momenta; its vanishing at $t = 0$ means that from that configuration on, the system is ready for propagation in the Minkowskian world. In our case, that would be a process of "sphaleron explosion." It does not appear in the production probability because its action is real and $|\exp(iS_M)| = 1$, meaning that the explosion happens with unit probability. A semiclassical description of sphaleron explosion and quark production has been solved by Shuryak and Zahed (2003a).

Adopting this interpretation of instanton–antiinstanton configurations, one is ready to recognize that the action over the whole plane in the exponent gives the probability of the production of magnetic balls on the "sphaleron path." When T separating the instantons is large, it is close to $\exp(-2S_{\mathrm{inst}})$: This is the "instanton tunneling" mechanism, in which when T is small, the action is also small, and the mass of the ball is small. The most interesting case is when, at the maximum of the potential, the sphaleron, is produced: In this case, the action is close to half of $\approx 2S_{\mathrm{inst}}/2$. This result has also been earlier argued by multi-gluon resummation for "anomalous" gluon production (Arnold and Mattis, 1991).

What if the process just described creates a sphaleron, which then decays into a Chern–Simons valley *different* from the one it originated? The anomaly relation would tell us that during this process $\Delta N_{\mathrm{CS}} = \pm 1$ and therefore ± 6 units of axial charge should be created (destroyed).

An experiment suggested by Shuryak and Zahed (2003b) is the double diffractive production of a single sphaleron in the middle. This reaction can also be described as Pomeron–Pomeron collisions with an invariant mass M. As we discuss in the following, such studies have been experimentally done, in particular at the CERN $Sp\bar{p}S$ collider.

The expected sphaleron mass, according to the theory presented above, should be

$$M_{\text{sphaleron}} = \frac{3\pi^2}{g^2(\rho)\rho} \tag{23.18}$$

where ρ is its r.m.s. radius. Assuming that it is the same as the radius of instantons in the QCD vacuum, it should then peak at $\rho \sim 1/3$ fm, which leads to $M_{\text{sphaleron}} \sim 3$ GeV.

Unfortunately, so far, there has been no decisive experimental evidence of sphaleron production in double-diffractive processes. The best set of data came from the CERN UA8 experiment (Brandt *et al.*, 2002), which was a combination of forward–backward "roman pot" detectors, triggering on diffractive protons, and a UA2 central calorimeter. In Fig. 23.6, we show some of their results. The left plot is the distribution over the cluster masses. A large yield at small masses ($M < 8$ GeV) was rather unexpected, while the higher mass region is fitted by conventional Regge phenomenology.

The other two plots show that the small and large mass clusters have completely different angular distributions. While the former have perfectly isotropic distribution (in their rest frame, of course), the latter are peaked along the beam directions.

These data fit well with our predictions for QCD sphaleron production. Note that the peak in mass is at the 2–4 GeV bin, which is heavier than normal hadronic resonances but agrees well with the sphaleron mass for $\rho \sim 1/3$ fm. Also, the isotropy of the decay fits well to the spherical sphaleron explosion. Yet, we cannot claim a discovery yet; for that, one needs to detect that the decay includes a specific spin-flavor operator $(\bar{u}_R u_L)(\bar{d}_R d_L)(\bar{s}_R s_L)$ (or its $L \leftrightarrow R$ conjugate) with six units of axial charge.

Fortunately, a part of this program has already been performed in a specific case of η_c decays. This pseudoscalar brother of the more famous J/ψ is expected to couple strongly to the $G_{\mu\nu}\tilde{G}_{\mu\nu}$ operator, the topological

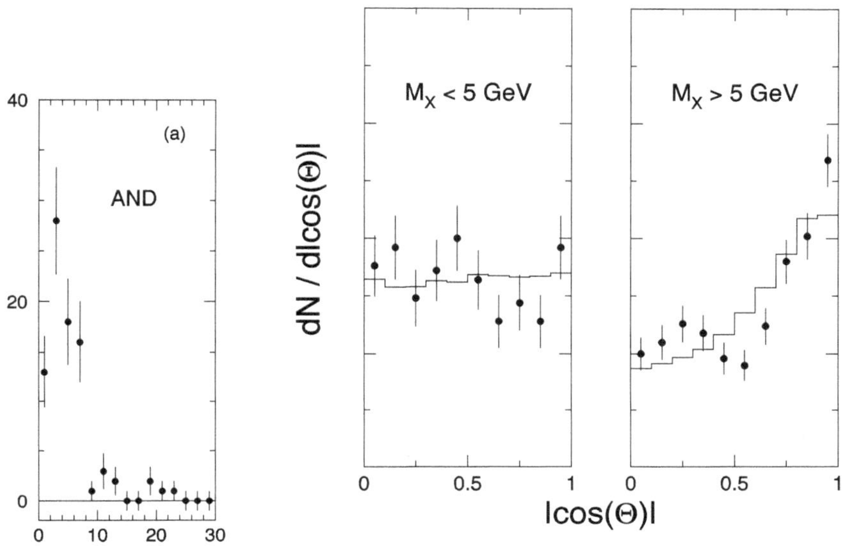

Fig. 23.6. The results of the UA8 experiment on double-Pomeron central cluster production. The left plot is the distribution of (acceptance corrected) cluster mass M (GeV). "AND" at the plot indicates that a subset of events in which *both* diffracted protons were detected (which provides better kinematical control) are used. The right plot shows the angular distribution of cluster particles in the cluster rest frame for small and large masses.

charge. It was noted by Bjorken (2000) that decays of η_c have three large three-body modes, each about 5% of the total width:

$$\eta_c \to KK\pi; \quad \pi\pi\eta; \quad \pi\pi\eta'$$

Note that there is no $\pi\pi\pi$ decay mode or any other decay mode one may think of. This is because the 't Hooft vertex *must* have all light quark flavors, including $\bar{s}s$. The vector particle J/ψ is our "control group," which has the same mass and wave function but quantum numbers which do not allow decays via $G_{\mu\nu}\tilde{G}_{\mu\nu}$. Such decay modes are, in this case, indeed absent. The actual calculations were done by Zetocha and Schäfer (2003), which included the following two- and three-meson decay channels of the lowest charmonium state:

$$\eta_c \to \pi\pi, KK, \eta\eta, KK\pi, \eta\pi\pi, \eta'\pi\pi \tag{23.19}$$

using the three-flavor instanton-induced 't Hooft effective Lagrangian, and all decay rates came out right.

Another lucky coincidence is that the mass of η_c is 3 GeV, emerging to be the same as the expected peak of the sphaleron mass distribution. Therefore, we do expect that the distribution in exclusive decay channels of QCD sphalerons should be very close to that of η_c. Unfortunately, the calorimeter of the UA2 detector used in the data we just discussed was not sophisticated enough to separate the exclusive decay channels. It would be a pity to miss opportunities to study double-differactive sphaleron production at existing colliders. Both RHIC and LHC have forward (roman pots) groups and sophisticated central detectors (STAR and ALICE) capable of doing it. A combination of those would solve the problem; unfortunately, it has not happened so far.

23.6 Looking for Chiral Magnetic Effect in Quark-Gluon Plasma

Another approach to topologically nontrivial ("anomalous") transitions in gauge fields in experiments was suggested by Kharzeev *et al.* (2008). Instead of looking for individual sphaleron transitions, they proposed to observe the *global chiral L–R disbalance* of the fireball produced in heavy ion collisions by fluctuating multiple sphaleron transitions. Note that the chiral charge is a CP-odd quantity, so its average value is, of course, zero in strong interactions. Therefore, we have to detect event-by-event fluctuations.

We already discussed sphaleron production in *pp* collisions and hence will not consider their production in heavy ion collisions in any detail. In the so-called GLASMA model of the initial stage, one follows the classical Yang–Mills equation for gauge fields. For a simulation of sphaleron processes in this framework, see, for example Mace *et al.* (2016). Let me just say that they observed that the axial charge of the fireball has Gaussian-like distribution with a mean of approximately $Q_5^{\text{r.m.s.}} \sim 20$.

To explain how it was proposed to be observed, we need to explain the so-called chiral magnetic effect (CME) (Fukushima *et al.*, 2008), according to which the CP-odd chiral disbalance leads to an *electric* current along the applied *magnetic* field.[7] A very strong pulse of magnetic field is naturally there due to the through-going current of the colliding protons. Its direction is normal to the collision plane, and CME should lead to electric dipole and

[7]Note that the coefficient between the current and the field is, in this case, T-even. So, unlike the usual Ohmic current, this one is a *non-dissipative* one. This observation will lead to multiple applications of the CME in condensed matter physics and perhaps even electronics.

quadrupole components of charge distribution, which can in principle be observed.

For this idea to work, it is important that ambient matter should not be ordinary hadronic matter, but quark-gluon plasma (QGP) at $T > T_c$. Only in this case is the chiral symmetry unbroken, and thus the chiral disbalance remains *conserved*[8] for a long time. This suggestion has led to significant experimental activity. We, of course, cannot describe it here in detail: The effect is clearly seen, but possible backgrounds are not yet completely understood.

The experimental program continues at both the RHIC and the LHC. RHIC has made a dedicated run with two beams of nuclei,

$$Ru^{96}_{44} + Ru^{96}_{44} \quad \text{and} \quad Zr^{96}_{40} + Zr^{96}_{40}$$

having the same number of nucleons ($A = 96$) but a rather different number of protons ($Z = 44$ and 40, respectively). It was assumed that the same A would reduce the differences in background effects (unrelated to electric charges) to $\sim$1%. Unfortunately, this assumption proved to be wrong because these two nuclei happened to have rather different shapes for reasons we already discussed in Chapter 17, the backgrounds are different and, inside uncertainties, the CME has not been observed so far.

Hopefully, eventually, sphaleron production rates will be studied in *pp* and nuclear collisions, and the magnitude of the CME will be measured. Experimental testing of the sphaleron theory in QCD will, no doubt, strengthen our understanding of electroweak sphalerons as well, which may play an important role in cosmological baryon asymmetry.

Summary and connections: Gauge topology ascribes to gauge field configurations the Chern–Simons number. Its integer values correspond to pure gauge configurations with zero field and energy. Those are separated by barriers. Tunneling through those is described by instantons, excitation over the barriers by sphalerons. In both cases, changes in the Chern–Simons number must be followed by changes in fermionic chiral charges. As fluctuations, heavy ion collisions can produce *chiral disbalance*, and a search for the CME, induced by them, is going on at RHIC. Discussions of single sphaleron observations in double-diffractive scattering are under consideration.

[8]Except for quark mass terms, leading to chirality flips. Those masses are quite small to be important at the time scale considered.

Chapter 24

CONFINEMENT, MAGNETIC MONOPOLES AND THE ELECTRIC–MAGNETIC DUALITY

24.1 Confinement and Flux Tubes

The permanent confinement of color-electric charges (or "confinement," for short) is the most famous nonperturbative feature of the gauge theories. The Lagrangian of QCD-like gauge theories is similar to that of QED, with massless photons substituted by massless gluons and massive electrons by (very light, or even massless) quarks. There are multiple definitions of the term itself, such as the statement that no object with a color charge can appear in the physical spectrum. There is a pending million-dollar prize offered for a mathematical proof that pure gauge theory has a finite mass gap.

Physicists are already sure that it is the case beyond any reasonable doubt. Billions of high-energy collisions of hadrons and nuclei (which we already briefly discussed above) were observed produced large number of secondaries, and none of them ever was a "colored object," such as a quark or a gluon. The formal limits on their production are so restrictive that it makes no sense to even mention them.

We already discussed above the theoretical (although numerical) "proof," the first demonstration[1] on the lattice and the pioneering work by Creutz (1980) indicating that the confining string tension scales correctly, as the weak coupling renormalization group scaling requires.

[1]Needless to say, no mathematician would take some numerical evidence as a proof. In fact, the very existence of any nontrivial QFT, such as QCD itself, is based on the path integral formulations which include taking a limit of the vanishing lattice spacing of $a \to 0$. Since the very existence of this limit was never rigorously proven, all that was done numerically was to use a few values of decreasing a and see if the results looked convergent to a certain limit. Most physicists trust this procedure, whereas mathematicians don't.

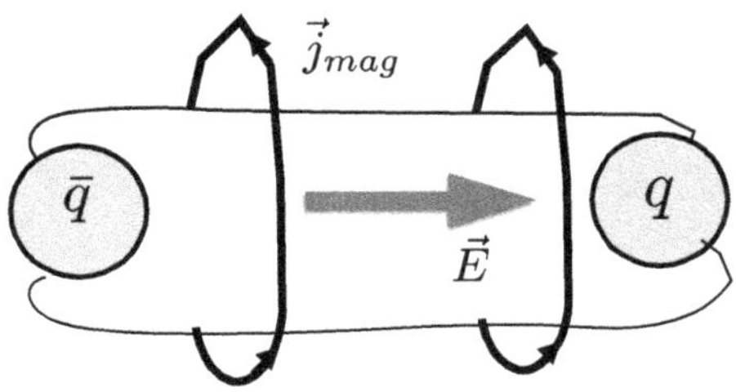

Fig. 24.1. Schematic diagram of the confining electric flux tube between quark and antiquark. The black circular lines around it indicate the presence of a magnetic current $\vec{\jmath}_{\rm mag}$ needed to stabilize it.

What was further observed on the lattice was that the electric flux from a color charge is not distributed radially outward, as in electrodynamics, but instead, being expelled from the QCD vacuum, is confined into a *flux tube* between the charges. The energy per length of the tube is known as its tension:

$$\sigma_T \approx (420\,\mathrm{MeV})^2 \approx 1\,\mathrm{GeV/fm} \tag{24.1}$$

So, a Coulomb potential $V(r) \sim 1/r$ at small r changes to a linear potential between charges $V(r) = \sigma_T r$ at large r.[2]

Its most intuitive explanation was provided by the so-called *dual superconductor model* (Nambu, 1974; Mandelstam, 1976; 't Hooft, 1976b). The "dual superconductor" is a Bose–Einstein condensate of monopoles. (So, it is *magnetically* charged!) The word "dual" means that, while it is the *magnetic* field confined into flux tubes in a superconductor, containing *electrically* charged Cooper pairs, in the QCD vacuum the magnetic condensate which expels the *gluo-electric* fields into the flux tubes. So, the QCD electric flux tubes can simply be a *dual copy* of the electric flux tubes in superconductors.

If so, the "coil" of the tube — the solenoidal-shaped current around it, which keeps the field inside and counterbalances its pressure — should also be dual, providing a "magnetic current" is created by the motion of *magnetic charges*. We however postpone the discussion of what those charges are until the following section, addressing first the subject more phenomenologically using the results of the lattice gauge theory simulations.

[2]It is amusing to note that, converting the tension σ into ordinary mechanical units, one finds that it can hold about 10 ton of weight!

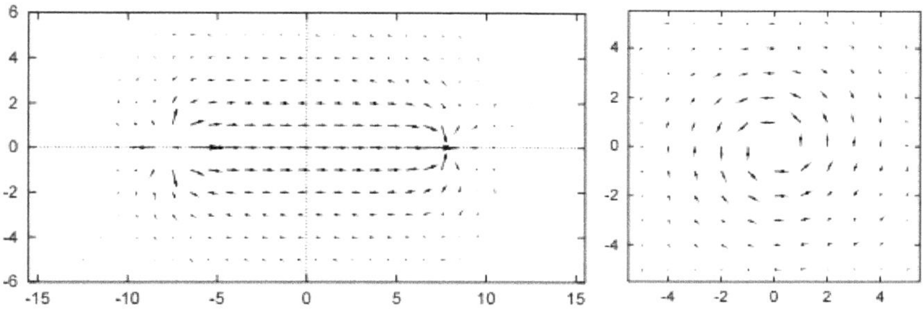

Fig. 24.2. Lattice data of the distribution of the electric field strength (left) and the magnetic current (right) for two static quark–antiquark external sources.

Figure 24.2 (displaying the result of lattice simulations summarized by Bali, 1998) shows the distribution of the electric field (left) along the flux tube and the magnetic current (right) in a transverse plane. So, at least in numerical simulations, the flux tube is indeed clearly visible. At this point, one would like to see whether its shape and other properties can be described by some generic effective model, such as the Ginzburg–Landau theory.[3]

The profile of the electric field is shown by squares in Fig. 24.3 (left), where the lines are just fits.

Its curl shown in Fig. 24.3 (right) coincides well with the separately measured magnetic current k. So, at least the second equation — basically the dual Maxwell equation — is indeed satisfied. With some (not too good) accuracy, the first equation is also satisfied, and $f(x)$ shown in Fig. 24.3 (left) is the resulting profile of the condensate obtained. Note that, as in the Abrikosov vortex, it vanishes at the vortex center. The resulting parameters for two basic lengths, in "physical units" obtained by fixing the lattice scale to physical σ, are

$$\lambda = 0.15 \pm 0.02\,\text{fm}, \quad \xi = 0.251 \pm 0.032\,\text{fm}, \quad \kappa = \frac{\lambda}{\xi} = 0.59 \pm 0.14 < \frac{1}{\sqrt{2}} \tag{24.2}$$

Their ratio is less than the critical Ginzburg–Landau value (right). So, one can conclude that we all live inside a *dual superconductor of type I.*

[3]Let me recall that when the Ginzburg–Landau paper was written, the physical nature of an electric object which makes the condensate was also unknown: They argued for the form of effective action on general grounds, as we do now.

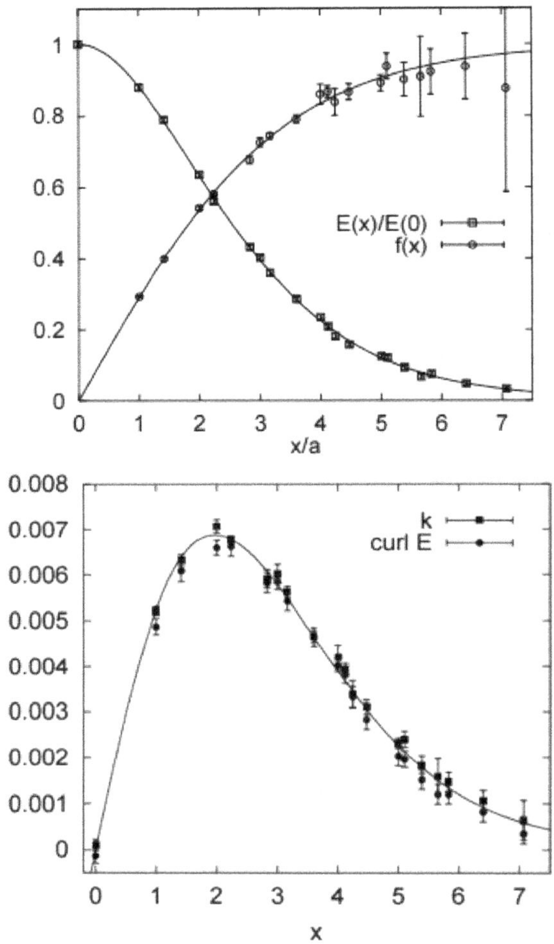

Fig. 24.3. (upper) Transverse profile of the electric field and the condensate. (lower) The transverse distribution of the magnetic current. The points in both are lattice data, and the lines are fits using Abrikosov's vortex solution.

Those who are not convinced by the not-too-impressive accuracy of this important statement may wonder if there are more direct manifestations of it. Indeed, there are; however, they remain controversial. Multiple flux tubes are routinely produced in high-energy collisions; for statistical reasons, the best studied cases are the LHC p-Pb collisions. When the impact parameter changes from peripheral to central collisions, the number of flux tubes is believed to change from about a dozen to 40 or so. Kalaydzhyan and Shuryak (2014a) suggested that not only is the log-distance interaction of strings attractive, as it should be in type-I cases, but at a large enough density of the flux tubes, it produces a collective implosion. This leads to the formation of QGP and its subsequent explosion, of which indeed there are observed consequences, namely collective flows.

24.2 Dirac Monopoles

Let us start discussing the concept of electric–magnetic duality for pure Maxwellian electromagnetic fields. In this case, it is almost trivial: Introducing the usual vector potential A_μ, it has been decided to define the electric field as its *gradient*, while the magnetic field is defined as its *curl*. Introducing the so-called *dual potential* C_μ, one can define the fields differently: $\vec{E}$ to be a curl and $\vec{B}$ to be a gradient. As long as those dual descriptions reproduce the same fields, one definition is as good as the other.

The duality seems to be lost when one introduces sources for those fields. QED has electrically charged electrons, muons and many other electrically charged particles, yet one has not observed any magnetic charges. As explained in elementary physics textbooks, when trying to cut magnetic dipoles (e.g. a piece of magnetized iron) into two parts, one always gets two dipoles, never succeeding to separate the north and south poles!

Yet, as famously shown by Dirac (1931), those can exist in principle, provided that the electric and magnetic charges, e and g, satisfy the *Dirac condition*

$$\frac{g_{\text{electric}} g_{\text{magnetic}}}{2\pi} = \hbar n \tag{24.3}$$

where n is an integer.

The existence of the magnetic monopoles would not contradict Maxwell's equation $\text{div}\vec{B} = 0$ if there were a singular magnetic flux, known as the Dirac string, which could carry the needed flux to the monopole location. See a sketch of it in Fig. 24.4.

An example of a potential which includes a monopole and a Dirac string is

$$\vec{A} = g\left(\sin(\phi)\frac{1+\cos(\theta)}{r\sin(\theta)}, -\cos(\phi)\frac{1+\cos(\theta)}{r\sin(\theta)}, 0\right) \tag{24.4}$$

where $\vec{A}$ is in Cartesian coordinates, but r, ϕ, θ are in spherical coordinates. Using curl in such coordinates, calculate the corresponding magnetic field.

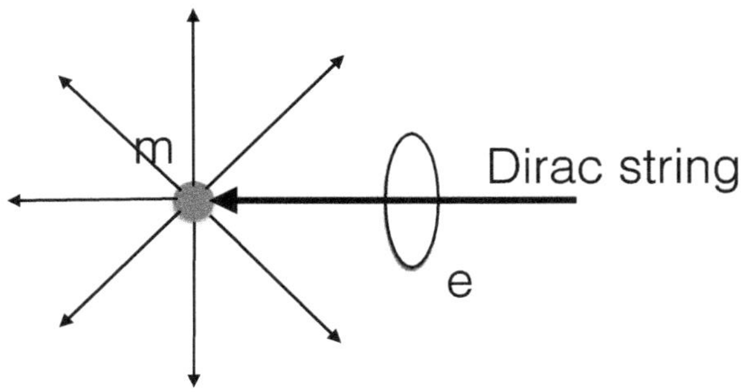

Fig. 24.4. The magnetic monopole with the Dirac string (thick line) carried the ingoing magnetic flux, equal to the total flux carried out by the Coulomb-like field (thin lines with arrows). The circle around the Dirac string is the path of an electron e moving around it.

Check that the singularity of the vector potential does carry the needed magnetic flux.

The wave function of an electron going *around the Dirac strings* (a circle in Fig. 24.4) would get a certain phase ϕ. If the Dirac strings were "visible" and had finite energy, they would produce a confining potential. However, Dirac argued that if the phase $\phi = 2\pi n$, which happens when his condition written above is satisfied, the Dirac strings are *invisible* and thus unphysical. Note that this argument has a dual: If one would prefer to use the other potential C_μ, magnetic objects would not have Dirac strings, but electric ones would. All such singularities, either in the vector potential A_μ or in C_μ, must be unphysical pure gauge artifacts. Dirac emphasized that his relation is the *only known reason why electric charge should be quantized* and expressed hope that nature would make use of it.

Needless to say, magnetic monopoles were looked for in all kinds of substances, taken from the bottom of the oceans, the surface of the Moon and secondaries produced at all accelerators, but in vain; QED magnetic monopoles were never found.

24.3 't Hooft–Polyakov Monopoles

The surprising finding was made not by experimentalists but by two distinguished theorists, 't Hooft (1974) and Polyakov (1974), so to speak "by the tip of their pen." The monopole solution has been found for the non-Abelian gauge theories interacting with scalar fields having color in the adjoint color representation.[4] As a good pedagogical introduction to monopole solutions, the reader may consult the book by Shnir (2005).

Such fields and monopoles need to appear in theories with extended supersymmetry, such as the so-called Seiberg–Witten theory (Seiberg and Witten, 1994). Another set of theories with monopoles are models aimed at the grand unification of the electroweak sector with QCD. All of that led to a vast literature on the physics of monopoles in non-Abelian theories.[5]

Unfortunately, there are no scalar fields in QCD, which is the main focus of this and subsequent chapters. Thus, it is unclear how to use the 't Hooft–Polyakov monopole solution in its framework. At finite temperatures, the Lorentz invariance is anyway broken by the matter rest frame, and the fourth component of the gauge field A_4, which is of course in adjoint color representation, can be used as a substitution for the adjoint scalar. We discuss this option in Chapter 27.2, where we demonstrate an explicit solution and its properties.

The Lagrangian of the Georgi–Glashow model:

$$L = -\frac{1}{4}(G^a_{\mu\nu})^2 + \frac{1}{2}(D_\mu \phi)^2 - \frac{\lambda}{4}(\phi^2 - v^2)^2 \tag{24.5}$$

which is a direct descendant of the Ginzburg–Landau free energy of a superconductor. It differs from the electroweak sector of the standard model, the Weinberg–Salam model, exactly by the fact that ϕ has an adjoint color representation. Nonzero VEV, shown as v, provides different "Higgsing" of the gauge fields: Their mass is proportional to the commutator of their color generator with that of the VEV. In this model, the $SU(N_c)$ color group is divided into its $N_c - 1$ diagonal subgroups.

The simplest case is $N_c = 2$, in which there are three generators τ^a — three Pauli matrices — with $a = 1, 2, 3$. Let us take the nonzero VEV to be along the diagonal, so $\langle \phi^3 \rangle = v$. Then, two gauge bosons ($W^{+,-}$) get nonzero masses, while boson number 3 (neutral "photon") remains massless.

[4] A model of this kind was previously suggested as a candidate for the theory of electroweak interactions by Georgi and Glashow. Subsequent discoveries, however, pointed to another option: scalar fields with the *fundamental* color representation, as suggested by Weinberg and Salam.

[5] If there are monopoles at the grand unification scale of energies, they must be stable and thus generated in the Big Bang. Since we do not find them around us today, it was a problem. One possible solution to it was the invention of the "cosmic inflation stage" with a huge exponential expansion of the Universe. Today, there is other cosmological evidence for inflation. So, it can now stand on its own legs. The fact that it solves the "monopole problem" is now a footnote.

(The Georgi–Glashow model was designed to avoid the existence of the Z boson, which was then unknown.)

Since there is such a drastic difference between those components, one may like to introduce special notations for the Abelian-projected fields (without color indices):

$$A_\mu = A^a_\mu \hat{\phi}^a, \quad \hat{\phi}^a = \frac{\phi^a}{|\phi^a|} \tag{24.6}$$

where we introduced the unit vector indicated by a hat. In order to also define the Abelian field strength, the field definition should not only be the usual Abelian expression based on A_μ because it should be supplemented by a term canceling possible derivatives of the Higgs color direction. The definition is

$$F_{\mu\nu} = \partial_\mu A_\nu - \partial_\nu A_\mu - \frac{1}{e}\epsilon_{abc}\hat{\phi}^a \partial_\mu \hat{\phi}^b \partial_\nu \hat{\phi}^c. \tag{24.7}$$

This last term is, of course, zero for a constant Higgs field. Furthermore, it in fact vanishes for all topologically trivial configurations of $\phi^a(x)$: It will be nonzero for topologically nontrivial ones, and the possibility of having a gauge in which there is no Dirac string is based on this observation.

The magnetic current can now be defined from the definitions given above:

$$k_\mu = \partial_\nu \tilde{F_{\mu\nu}} = \epsilon^{\mu\nu\rho\sigma}\epsilon_{abc}\partial_\nu \phi^a \partial_\rho \phi^b \partial_\sigma \phi^c \left(\frac{1}{2v^3 e}\right) \tag{24.8}$$

Unlike the usual nether currents, this magnetic current k^μ is conserved by definition, without any underlying symmetry. The integral of its density is known in mathematics as the *Brouwer degree*. As with any other topological quantity, it gives, in appropriate normalization, an integer, which defines a topologically distinct Higgs field.

How it may happen is clear from an example of a "hedgehog"-like field,[6]

$$\phi^a(r \to \infty) \to v\frac{r^a}{r} \tag{24.9}$$

in which the "needles" go radially. Its magnetic charge is

$$g = \int d^3x k_0 = \frac{4\pi}{e} \tag{24.10}$$

[6]The relation between such a field configuration and this cute animal has appeared for the first time in Polyakov's paper.

Let us now look for a solution consistent with that asymptotical trend in terms of two spherically symmetric functions:

$$\phi^a = \frac{r^a}{er^2}H(ver); \quad A_n^a = \epsilon_{amn}\frac{r^a}{er^2}[1 - K(ver)]; \quad A_0^a = 0 \tag{24.11}$$

When those are plugged back into the expression for the Hamiltonian, one finds the following expression for the monopole mass:

$$E = \frac{4\pi v}{e}\int_0^\infty \frac{d\xi}{\xi^2}\left[\xi^2\dot{K}^2 + (1/2)(\xi\dot{H} - H)^2 \right.$$
$$\left. + (1/2)(K^2 - 1)^2 + K^2H^2 + \frac{\lambda}{4e^2}(H^2 - \xi^2)^2\right]$$

Here, we rescaled the radial coordinate $\xi = evr$, and the derivative over it is denoted by a dot. This expression can be used as an effective action corresponding to classical motion in the K, H plane of a particle, with ξ being the time. The standard equations of motion are obtained from their variations.

Since a monopole is a static classical solution, its energy is minimal for all functions K, H that satisfy the boundary conditions. Using a variation of the Hamiltonian, derive the equation of motion for K, H. At small r, these conditions are $H \to 0$ and $K \to 1$. Then, the Higgs field is smooth despite the "hedgehog" direction. At large distances, $H \to \xi$, and the Higgs attains a magnitude of v, while $K \to 0$.

At a large distance, it is a Dirac monopole with a magnetic charge:

$$g = \frac{4\pi}{e} \tag{24.12}$$

which is consistent with the Dirac quantization condition. In fact, we prefer to write more symmetrically using electric and magnetic "fine structure constants":

$$\frac{g^2}{4\pi}\frac{e^2}{4\pi} = \alpha_e\alpha_m = 1 \tag{24.13}$$

which are then inverse to each other.

The solution for the EOM can generally be obtained numerically. Two of them are shown in Fig. 24.5 taken from Shnir (2005). On general grounds, the monopole mass can be written as

$$M = \frac{4\pi v}{e}f\left(\frac{\lambda}{e^2}\right) \tag{24.14}$$

with f being smooth function. Some of its values are $f(0) = 1$ and $f(1) = 1.787$.

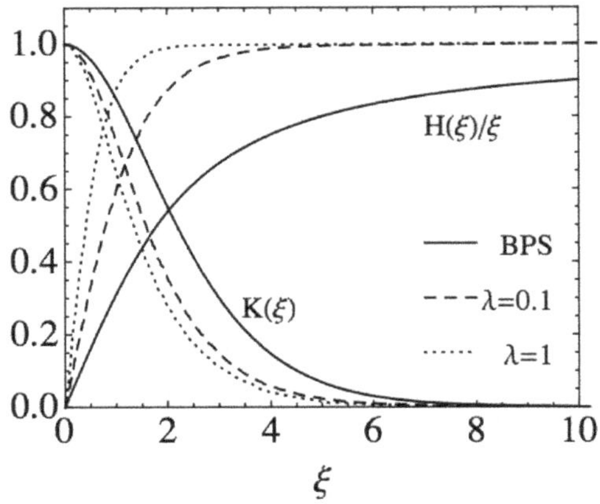

Fig. 24.5. Classical solutions $K(\xi)$ and $H(\xi)$ for the 't Hooft–Polyakov monopole at $\lambda = 0$ (the BPS limit) and $\lambda = 1$.

The former case of $\lambda = 0$ is the so-called Bogomolny–Prasad–Sommerfeld (BPS) limit, in which there is no scalar potential at all. In this limit, both functions are known analytically:

$$K = \frac{\xi}{\sinh(\xi)}, \quad H = \frac{\xi}{\tanh(\xi)} - 1 \tag{24.15}$$

Using these values for K, H, it is straightforward to calculate the Abelian magnetic field for the BPS monopole defined as $\tilde{F}_{\alpha\beta} = \frac{1}{2}\epsilon_{\alpha\beta\gamma\delta}F_{\gamma\delta}$, with the Abelian F defined in (24.7). Substituting it into the Hamiltonian, one gets the energy profile of the monopole.

24.4 Electric–Magnetic Duality and RG Flow

Let us follow qualitatively the dynamics along the RG trajectory, starting with large momenta (UV) and ending at low ones (IR). In particular, this can be seen as the temperature dependence, from high to low T, or alternatively, as the chemical potential dependence, from high to low μ.

In UV, we begin with a weak coupling regime, $g_{\text{electric}} \ll 1$, because of asymptotic freedom. Monopoles must have their field to be strong $A \sim 1/g_{\text{electric}}$, in order for the nonlinear commutator term in the field to be comparable to the derivatives. Therefore, the mass

$$M \sim \frac{1}{g^2_{\text{electric}}} \tag{24.16}$$

is large compared to the temperature, whichever is defining the overall scale, and the monopole density

$$n \sim \exp(-M/T) \sim \exp\left(-\frac{\text{const}}{g^2_{\text{electric}}}\right) \tag{24.17}$$

is exponentially small. Thus, at high T, monopoles (and other similar nonperturbative phenomena) are unimportant. Yet, from the Dirac condition, we see that their magnetic coupling is large, so they interact strongly. Now, by decreasing T, we try to move toward the IR. The electric coupling grows, the soliton mass M/T decreases, and their density grows. At some point, the density of the magnetic monopoles and electric charges (quarks and gluons) becomes comparable: This state was called the electric–magnetic equelibrium. In QCD-like theories, at this point, a number of phase transitions occur: Due to confinement and chiral symmetry breaking, the low-T phase of QCD becomes a hadronic phase (in which we live). We discuss how the transitions happen in the following section.

However, before that, let me describe what happens in some extended supersymmetric theories. Supersymmetry is a powerful Bose–Fermi symmetry, which does not seem to exist in nature, but which considerably simplifies problems, occasionally to the degree that those can be solved. The word "extended" means that it has more than one supersymmetry. The simplest of those, called $N = 2$, has two fermionic partners of a gluon; let us call them gluinos of the types ψ and χ. Yet, this is still an incomplete set (multiplet) of extended supersymmetry groups since gluons exhibit two polarizations, as does each Majorana fermion, with $2 \times 2 = 4$ fermionic degrees of freedom. To balance the number of fermions and bosons, one needs one more *complex* scalar, ϕ, also with two degrees of freedom. (Times for all of them $N_c^2 - 1$ color components of the adjoint representation of $SU(N_c)$.)

The existence of scalars is crucial because, by Lorentz symmetry of the vacuum state, only scalars are allowed to have VEV. So, this theory has a complex parameter $v = \langle\phi\rangle$ characterizing a whole plane of possible vacua. None of them is better than the other since supersymmetry requires the vacuum energy of any of them to be zero. They are nevertheless all different since v defines a scale, and a running coupling $g(v)$ depends on it. In particular, far from the center of the v-plane, where VEV is large, the coupling is small (due again to the asymptotic freedom).

Naively, one may think that the opposite strong-coupling regime would be at the center, where $v \to 0$. As shown by Seiberg and Witten (1994), this is wrong, and in fact, there are *two singular points* on the v plane. In one of them, the monopole ($Q_{\text{electric}} = 0, Q_{\text{magnetic}} = 1$) becomes massless, while in another, the dyon ($Q_{\text{electric}} = 1, Q_{\text{magnetic}} = 1$). Both are perfect examples of what we are looking for: the ultimate IR end of the RG flow. Indeed, in those vacua,

$$g_{\text{electric}} \to \infty, \quad g_{\text{magnetic}} \to 0 \tag{24.18}$$

which is the opposite of the electroweak coupling in UV! In the vicinity of those points, the low-energy effective theory is the *magnetic theory* of light and weakly coupled monopoles (or dyons). Seiberg and Witten were able to show that the beta function of this theory and the effective theory just mentioned indeed match exactly.

What was observed in $N = 2$ SYM is quite remarkable: A non-Abelian theory with asymptotic freedom in UV turned out to be continuously connected by the RG flow with a completely different *magnetic* theory, which is Abelian and has a beta function of the opposite sign! It turned out that they are, in fact, one and the same theory on the whole v plane, and whether we use the electric or magnetic *formulation* of the theory as input is just a matter of convenience: *both* are true!

What is remarkably different for $N = 2$ SYM from QCD is that in this theory, the weak and strong coupling regimes are *not* separated by any phase transitions. (This happens due to supersymmetry, and it would be hard to explain why here.)

Returning from the hypothetical supersymmetric theory back to our real world, where we seem to lack this powerful tool, we are back to QCD-like theories of quarks and gluons. *There are no scalar fields* in these theories at all, and thus, one should either (i) give up on monopoles entirely; (ii) create some composite scalar, perhaps a gauge-dependent one; or (iii) shift to the finite T theory, which lacks Lorentz invariance and allows for a nonzero VEV of A_4 as a substitute scalar.

Option (ii) has been intensely studied on the lattice. Furthermore, these monopoles (still defined in certain gauges) do indeed form Bose–Einstein condensates, precisely at $T < T_c$ in the confined phase. This is observed by following their Bose clusters, similar to our discussion of liquid He; see, for example, D'Alessandro *et al.* (2010). Spatial correlations between lattice monopoles have been measured. They correspond well to those in Coulomb plasmas.

More recent developments in the deconfinement phase transition are related to option (iii). Using A_4 instead of a scalar field, one finds well-defined topological solitons called *instanton-dyons*, to which we will turn in Chapter 27.

Instead of going into details, let us outline a wider picture of the electric–magnetic duality. An important motive of this book has been the "running coupling constants" described by the renormalization group. One may ask whether the "naive" Dirac condition just given above is in fact in agreement with such "running" (scale dependence). Remarkably, it does!

The usual gauge coupling $g_{\text{electric}}(k)$ becomes weak at large momenta k but grows as momentum scale is reduced. The density of topological objects, such as monopoles, at a temperature of T,

$$n(T) \sim \exp\left(-\frac{\text{const}}{g^2_{\text{electric}}(T)}\right) \tag{24.19}$$

grows as T decreases. At $T \sim T_c$, their density tends to become comparable to that of "electric charges," such as quarks. In some theories, such as the $N = 2$ supersymmetric gauge theory, one can go all the way to the dual phase, in which magnetic objects are weakly coupled and electric ones are strongly suppressed. In non-supersymmetric theories, we can only rely on the lattice data, but one can still see evidence (Liao and Shuryak, 2008b) of the fact that magnetic charges flow in the opposite direction, in perfect agreement with the Dirac condition.

I would like to end with a general point, perhaps surpassing the whole subject of this chapter — the QCD vacuum problem — in importance. Quite different theories, such as the electric gauge theory, weakly coupled in the UV, and the magnetic theory, weakly coupled in the IR, may in fact simply be the limits of a much larger common theory. By following continuously the RG flow, it turned out to be possible to connect QFTs, which were studied separately and considered completely different in their properties. In particular, it is possible to connect a QCD-like theory with a QED-like one. Note that the Dirac condition retains the product of the two couplings and thus *forbids both to be weakly coupled.* Therefore, it is very nontrivial to see the RG relation, as one needs to pass through the region where both couplings are strong. Yet, such a connection can be traced nonperturbatively, or at least be indicated by the dualities!

24.5 Strongly Coupled QGP as a "Dual" Plasma

The discovery that the QGP is a "perfect liquid" with extremely small *viscosity* (to be covered in the following chapter) had led to significant interest in strongly coupled systems in general. (It is even now referred to as strong QGP, or sQGP for short, in the literature.)

Two research directions developed as a result of curiously following the ideas of two "dualities":

(i) the duality of conformal strongly coupled gauge theory (CFT) to weakly coupled string (gravity) theory in a special five-dimensional geometry known as anti-de-Sitter (AdS_5) space, AdS-CFT correspondence.
(ii) the electric–magnetic duality concept, emphasizing the RG flow and transition from weakly coupled electric theory in UV to mixed strongly coupled electric–magnetic sectors around T_c.

The first one now can be characterized as mainstream. The second one, on the contrary, is rather accessible, so we included it here.

In very general terms, we are going to discuss systems containing both electrically and magnetically charged particles. For brevity, we will call them "dual plasmas." We begin with classical behavior first, considering only a few particles and then proceed to a many-body theory. Next, we discuss quantum theory; so far, it is developed at the level of binary scattering only, not yet as a many-body theory.

Unexpected things appear along the way, starting with the very first step. As early as 1904, J. J. Thomson[7] observed that even *two static charges*, electric e and magnetic g, have a nonzero angular momentum J carried by a rotating electromagnetic field. Indeed, two Coulomb fields from these charges at a generic point have a nonzero Pointing vector $[\vec{E} \times \vec{B}] \neq 0$, rotating around the line connecting two charges. The field is rotating[8] even when both charges are not themselves moving!

This has implications for cases when those charges start moving. Let us start with a magnetic charge moving in an electric Coulomb field of a static charge.[9] In fact, even earlier than the time of Thomson, in 1896, Poincare had already thought about this problem. The only force acting is the (dual)

[7]Recipient of the 1906 Nobel Prize in Physics, famous for first measuring the charge-to-mass ratio of the electron and thus discovering the first elementary particle. His other high distinction was that seven(!) of his students, including his son, also became Nobel Prize winners.

[8]The corresponding angular momentum of the field has to be quantized to $\hbar n$: This is another way to derive the same Dirac quantization condition we discussed above.

[9]More generally, it is a motion in relative coordinates, with the so-called reduced mass.

Lorentz force

$$m\ddot{\vec{x}} = g[\vec{E} \times \dot{\vec{x}}] \tag{24.20}$$

proportional to the velocity of the monopole. From this, he observed that motion is happening on the surface of a cone, now known as the *Poincare cone*. Indeed, since the radius of a circle in a field gets smaller as the field gets stronger near the charge, $E \sim 1/r^2$, this focuses the path toward the cone's top $r \to 0$, and thus, a collision of both charges is inevitable. (This will have important consequences for QGP kinetics, as we explain later.) Recall that, for a typical two-body potential, one has classical motion in a collision *plane*: This is forced by the conservation of the angular momentum, normal to this plane. What is different in the Poincare problem at hand is precisely Thomson's effect: Not all the angular momentum is due to a moving particle; part of it is due to *the rotating field*!

By solving the equation of motion numerically in an electric field of a Coulomb charge, one finds trajectories such as those shown in Fig. 24.6. Proceeding further, one may wonder what would be monopole motion in the field of an electric dipole or other arrangements of few electric charges. We (Liao and Shuryak, 2007) proposed the so-called *magnetic scenario* for sQGP, in which it is considered to be a "dual plasma" made of electrically and magnetically charged particles. We started by investigating several

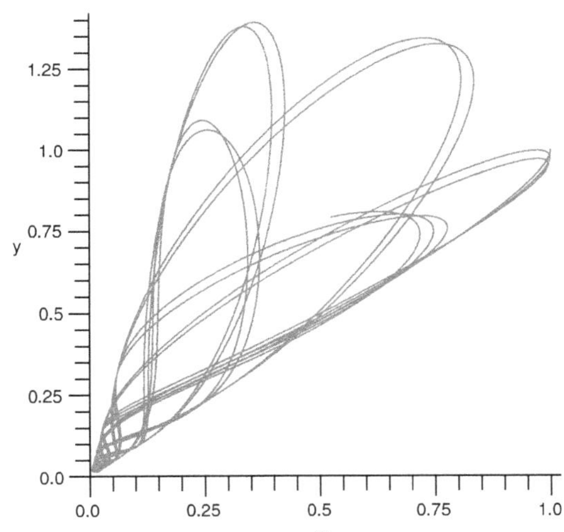

Fig. 24.6. Trajectory of a monopole in a static electric Coulomb field.

curious single-body motion in different settings. The Poincare problem, in a field of one static charge, was the first one. The second one was *two* static electric charges, e and $-e$, with a moving monopole. We found that the monopole can be trapped in between them, bouncing[10] from one to the other on a surface consisting of two connected Poincare cones, see Fig. 24.7, and colliding with the charges at the tops of the cones.

Our next configuration (not shown) was "a grain of salt," consisting of eight electric charges located at the corners of a 3D cube, with alternating plus and minus charges. It was found that a monopole can get out of such "cage" but only with great difficulty, suffering hundreds of collisions with the corners in the process. The mechanism for that is again due to the focusing mechanism just described.[11]

Finally, we took a few hundred charges, both electric and magnetic, with equal numbers of plus and minus charges of each type, to keep matter neutral. We then performed *molecular dynamics simulations*, solving the EOM for all of them numerically. Details can be found in Liao and Shuryak (2007). The outputs are calculations of the diffusion constant and viscosity from the corresponding "Kubo formulae." The qualitative conclusion is that

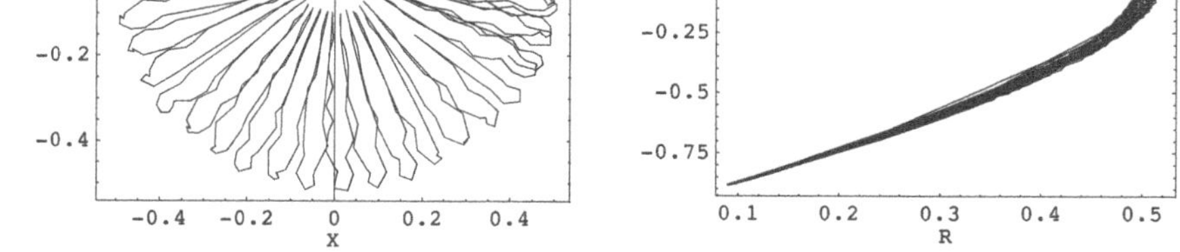

Fig. 24.7. Trajectory of monopole motion in a static electric dipole field (with charges at $\pm 1\hat{z}$) as projected on the (left panel) x-y and (right panel) R-z ($R^2 = x^2 + y^2$ planes).

[10] We called it "two charges playing ping-pong with a monopole, without even moving."
[11] We compared the monopole's behavior to that of a proverbial drunkard who cannot go home from the city square because there are lamp posts located at the corners, with which he keeps colliding.

each electric charge is trapped by a cell of magnetic ones around it and each magnetic charge by electric ones, as described above. The Lorentz nature of the forces between them does decrease particle (diffusion) and momentum (viscosity) transport significantly!

The history of the *quantum monopole-charge scattering problem* is also very interesting, although not widely known. In my opinion, it clearly belongs in any good QM textbook.

There are indications that Dirac, after his paper on monopoles in 1931, tried to solve it, but did not succeed. It took a long time until it was solved, a year after the non-Abelian monopoles were discovered, by two distinguished teams (Boulware *et al.*, 1976; Schwinger *et al.*, 1976).

First of all, let us think about the input parameters of the problem. The results of the quantum mechanical scattering problem are usually expressed in terms of the scattering amplitudes. They depend on the scattering phase shifts $\delta_l(k)$, as a function of angular momentum channel l and the collision momentum k. Note, however, that the Coulombic field of a charge is produced by the *dimensionless* $e^2/\hbar c$ coupling. A background point-like charge therefore cannot provide any length or momentum scale.[12]

Already, our conclusion that the (point-like) charge–monopole scattering has phase shifts *independent* of the scattering momentum k has an important implication. The general theory of non-ideal gases connects the contribution to the partition function with the number of states of a gas. Since scattering phases shift the phase of the wave function, the quantization of states in a box is modified, and extra states appear. Their contribution to the partition function is given by the so-called Beth–Uhlenbeck formula:

$$\Delta Z = \frac{1}{\pi}\sum_l (2l+1)\int_0^\infty dp \frac{d\delta_l(p)}{dp}\exp\left(-\frac{p^2}{mT}\right) \tag{24.21}$$

where an integration over all angular momentum values l and over all relative momenta is made. $\delta_l(p)$ is the so-called scattering phase in the channel l, a phase shift between the incoming and outgoing waves. The Boltzmann exponent does not have a factor of 2 in the denominator because, in the scattering of two particles of mass m, the relative motion has an effective mass of $m/2$.

[12]The usual charge–charge scattering does not have well-defined phase shifts at all due to the long-range nature of the Coulomb potential. However, in the charge–monopole scattering problem, the force is the dual Lorentz force, which is *not* a long-range force. Thus, the scattering phases are well defined.

Recall that $d\delta_l(p)/dp = 0$ for charge–monopole scattering. It thus follows that no new states are created in such a case; therefore, the thermodynamics of the "dual plasmas" is that of the ideal gas! (Yet, the kinetic coefficients are entirely different!)

The scattering phases do depend on the couplings, specifically on the product of electric and magnetic charges. Yet, according to the famous Dirac condition we discussed above, this product must be an integer n (for example, just 1). This integer n is *the only input* to the charge–monopole scattering problem, on which the scattering phase can depend!

A specific feature of it is that, after separation of the angular θ, ϕ and the radial r coordinates, it is the former part which happens to be most complex. To explain why, let me recall that, classically, the motion takes place on the Poincare cone rather than on a plane. The usual angular basis of spherical harmonics $Y_{l,m}(\theta, \phi)$ at large quantum numbers $l, m \gg 1$ is, in fact, planar. So, one has to select more complicated angular functions, located at the Poincare cone in the classical limit.

Omitting technical details, let me just give the answer for the scattering phases. Indeed, they depend on the integer total angular momentum l and do not depend on energy. The expression for a phase is

$$\delta(l') = \frac{\pi}{2} l', \quad l'(l'+1) = l(l+1) - n^2 \qquad (24.22)$$

where l' is to be found from the quadratic equation given above.[13] Note that, while the total angular momentum l and n (the integer from the Dirac condition) are *both integers*, their combination l' — apparently the angular momentum of a particle — is curiously *not* an integer! That is the only reason why a rather nontrivial angular distribution of scattering appears!

In QGP with monopoles, one needs to calculate the scattering amplitudes of quarks and gluons colliding with a monopole. The latter problem is especially tedious because orbital angular momentum, spin and color-spin of gluons are all mixed into a set of coupled channels. It has been solved by Ratti and myself (Ratti and Shuryak, 2009), where one can also find the resulting angular distribution and its contribution to the transport cross-section.

The results shown in Fig. 24.8 (together with those from other approaches and from phenomenology) do indeed indicate that

---

[13]It is not hard to solve it, but I think this expression is more instructive. It implies that the particle orbital momentum originates from the square of the total one minus that contained in the rotation of the field.

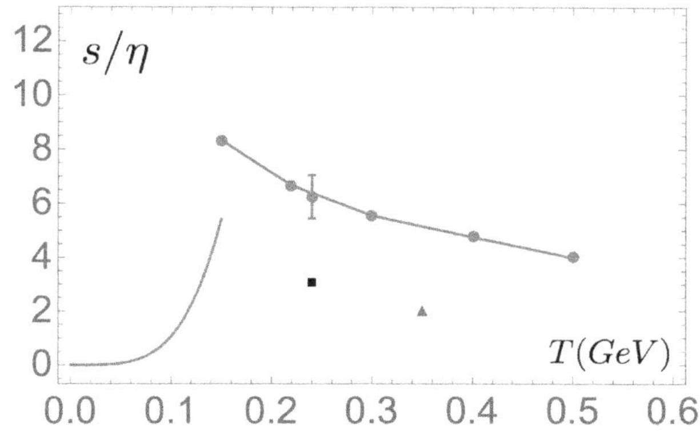

Fig. 24.8. The entropy density to shear viscosity ratio s/η versus the temperature T (GeV). The upper range of the plot, $s/\eta = 4\pi$ corresponds to the value in infinitely strongly coupled $N = 4$ plasma (Policastro *et al.*, 2001). The curve without points on the left corresponds to hadronic/pion rescattering according to chiral perturbation theory (Prakash *et al.*, 1993). The curve without points on the right corresponds to a weak coupling approximation (Arnold *et al.*, 2003). The single (red) triangle corresponds to a molecular dynamics study of classical strongly coupled colored plasma (Gelman *et al.*, 2006), and the single (black) square corresponds to numerical evaluation on the lattice (Nakamura and Sakai, 2005). The single point with the error bar corresponds to the phenomenological value extracted from the data, see the text for details. The series of points connected by a line on the right side correspond to gluon–monopole scattering (Ratti and Shuryak, 2009).

gluon–monopole scattering in sQGP dominates its kinetic properties near the QCD phase transition temperature T_c.

To explain why this is the case, one may recall that the kinetic theory (which is valid for weak coupling) would interpret this dimensionless ratio of the entropy density s to that of the shear viscosity η as the product of the particle density, transport cross-section and the inverse power of the mean velocity:

$$\frac{s}{\eta} \sim \frac{n\sigma_{\text{transport}}}{T\bar{v}} \tag{24.23}$$

or, in slightly simpler terms, as the ratio of interparticle separation to the mean free path. Note that the densities of quarks and gluons tend to zero at T_c. The density of (non-condensate) monopoles, on the other hand, does not vanish — the confinement is electric, not magnetic — and even has a peak there.

The peak in the shear viscosity correlates with similar peaks claimed for two more kinetic parameters: the heavy quark diffusion constant and the jet quenching parameter $\hat{q}$. Those are perhaps also due to gluon–monopole scatterings, see Liao and Shuryak (2009) and Xu *et al.* (2016).

24.6 Can QCD Flux Tubes Exist at $T > T_c$?

The T dependence of the string tension is rather nontrivial. In Fig. 24.9, one can see the lattice data for the tension (the force at a certain distance) for the free and internal energies. What is going on near T_c? Is there tension above T_c?

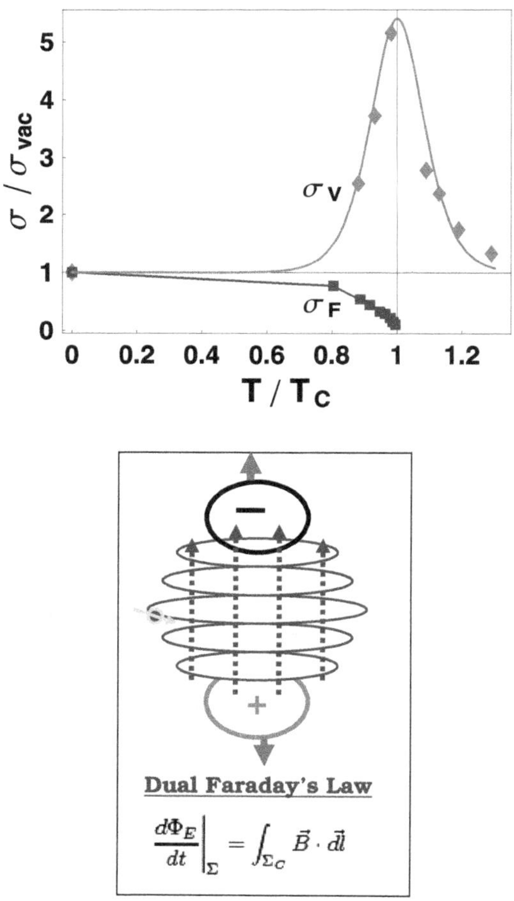

Fig. 24.9. (Left)

The application of the monopole model to the issue was done by Liao and Shuryak (2010). There are flux tubes in any plasma, such as on the Sun, visible through a modern telescope.[14]

The question put into the title of this section has also been addressed on the lattice. For the rest of the section, we follow a recent work by Cea *et al.* (2018). The principle observables are the mean components of the gauge fields correlated to the static $\bar{Q}Q$ pair explained in Fig. 24.10 (left).

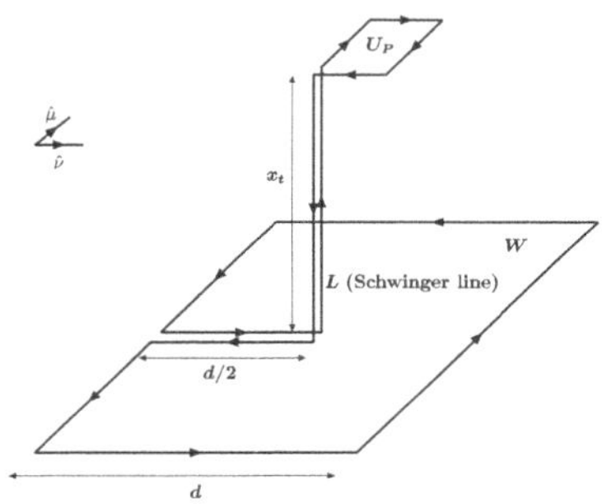

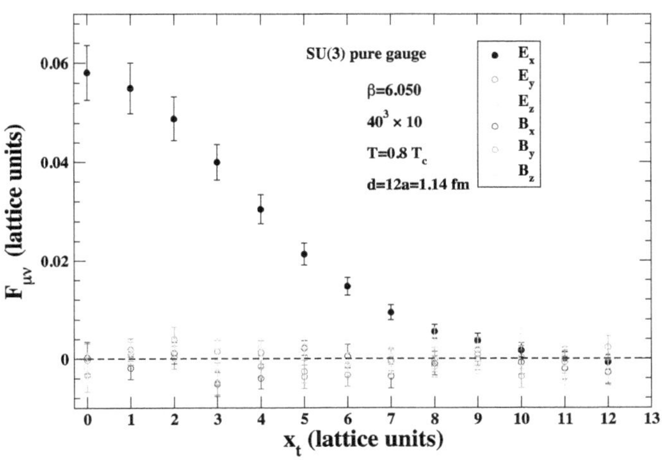

Fig. 24.10. (Left) The connected correlator between the plaquette U_P and the Wilson loop. (Right) The six components of the gauge fields as a function of transverse distance from the tube x_t.

---

[14]The black spots, studies of which gave Galileo so much trouble, are made of them. Those directed toward us look black, as they have less plasma inside, expelled by magnetic field pressure. Those which are visible outside the Sun's disk are, on the contrary, bright.

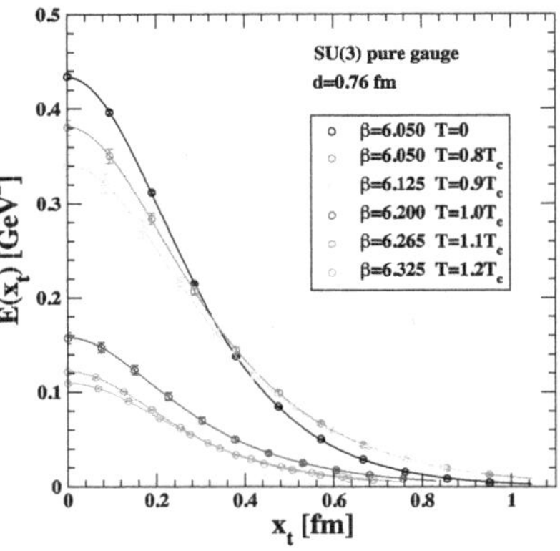

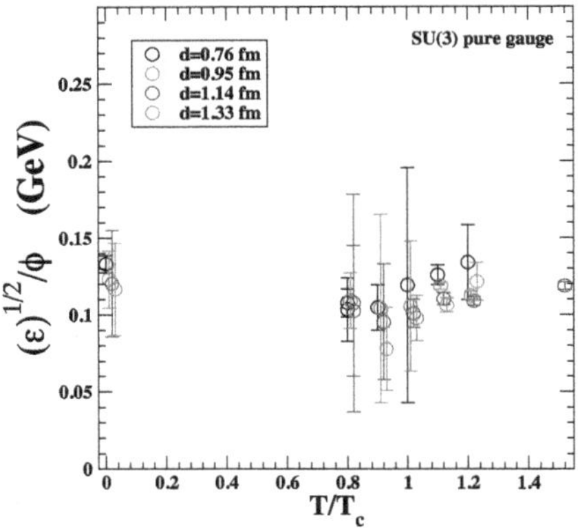

Fig. 24.11. (Left) Behavior of the longitudinal chromoelectric field El (in physical units) versus the distance x_t (in physical units) from the axis connecting the static sources, for several values of temperature across deconfinement. (Right). The energy per unit length in the flux tube as a function of temperature.

As it is demonstrated in the right plot, out of the six components of the gauge field, only the longitudinal electric field is present.

The main results of Cea *et al.* (2018) are shown in Fig. 24.11. On the left is the transverse profile of the longitudinal electric fields at different temperatures. The calculations are done for pure gauge theory, which — unlike physical QCD — has a first-order transition, in which many quantities experience a significant jump across the deconfinement, as does the transverse profile; however, the jump is about a factor of two, and by no means is it a jump to zero. Furthermore, as shown in the right plot, the energy per unit length in the flux tube shows very little temperature dependence, up to $T = 1.5T_c$.

These results confirmed the prediction made by Liao and Shuryak (2008a): Flux tubes do *not* need the supercurrent, and they in fact may exist in the "normal phase" of QCD, the QGP. Even the prediction that they would be there up to $1.5T_c$ seems to have been fulfilled.

Summary and connections: This chapter is devoted to a closer look at the QCD flux tube. We argued that its description by analogy to a "dual superconductor" implies the presence of a "magnetic Bose–Einstein condensate" of magnetic monopoles. We introduced their physics and argued that at $T > T_c$, in QGP, such monopoles are actually present as magnetically charged quasiparticles. A number of important effects are ascribed to them, especially kinetic properties, such as unusual temperature dependence of the viscosity, jet quenching parameter and heavy quark diffusion constant. We also argued that condensation is not required for flux tube formation, and in some temperature intervals above T_c, one can also see metastable flux tubes.

Chapter 25

HADRONIC SPECTROSCOPY

The first excited states of the nucleons were discovered in the 1950s, and in the 1960s, Gell-Mann and others developed the first successful theoretical framework for mass formulae based on the $SU(3)$ flavor symmetry, which appears if the masses of the u, d, s quarks be equal. The classification of mesons into octets and of baryons into octet and decuplet representations was completed by the experimental discovery of the triple-strange Ω baryon in 1964. Apart from masses, magnetic moments of baryons were successfully predicted by a nonrelativistic model with quarks having a "constituent mass" of $m_q \sim 1/3$ GeV and thus magnetic moments with $\sim$3 nuclear magnetons.

A revolution in hadronic spectroscopy happened in 1974 when a whole family of *charmonia* ($\bar{c}c$) was discovered, followed by the discovery of even heavier $\bar{b}b$ *bottomonia* in 1977. Another revolution has been happening since the past decade, with heavy-light tetraquark states discovered at the LHCb and b-factories.

In this chapter, rather than discuss the listings or properties of particular hadrons, we focus on what spectroscopic data tell us about central and spin-dependent quark *interactions*.

25.1 Heavy Quarkonia and Static Central Potential

Figure 25.1 (from the Review of Particle Properties) summarizes our current knowledge of $\bar{b}b$ spectroscopy. The first states discovered were those that could be directly produced via a virtual photon:

$$e^+e^- \to \gamma^* \to J/\psi, \Upsilon \ldots$$

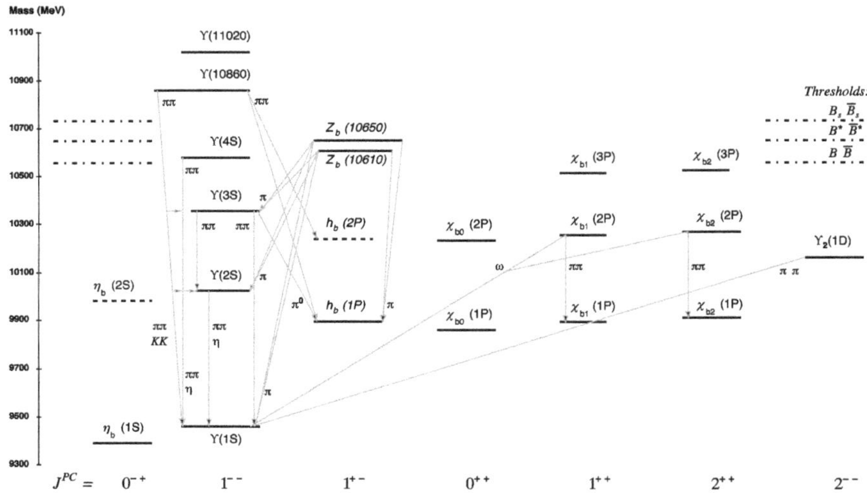

Fig. 25.1. RPP figure for bottomonium. Table masses are shown by solid lines; selected states not in the summary tables but with assigned quantum numbers, are shown with dotted lines. The arrows indicate the most dominant hadronic transitions.

The masses (in MeV) of the six lowest Upsilons are:

$$(1S)(9460), (2S)(10023), (3S)(10355), (4S)(10579), (5S)(10860), (6S)(11020)$$

some of whose features we will need in the following. Let us first note here the locations of the decay thresholds (shown on the right of Fig. 25.1) for "open bottom" $B\bar{B}$ channels: One can see that a good portion of the S states are below them. Therefore, they can only decay via heavy quark annihilation and are therefore very narrow.

Heavy quark mass is a large parameter, $m_Q \gg \Lambda_{\rm QCD}$. Ignoring first the spin-dependent effects $\sim 1/m_Q^2$, when one begins with a nonrelativistic Schrodinger equation and the celebrated Cornell potential,

$$V_{\rm Cornell}(r) = -\frac{4\alpha_s(r)}{3}\frac{1}{r} + \sigma_T r \tag{25.1}$$

one can get values for the quarkonia masses, to be compared with the data.[1] Using the Coulomb coefficient -0.713, the string tension $\sigma_T = (0.4\,{\rm GeV})^2$ and $m_b = 4.8$ GeV, one can easily get the results plotted in Fig. 25.2.

[1]Getting spectrum and wave functions in Mathematica is now simplified to one command, *NDEigensystem*, so it takes little effort to get all of Fig. 25.2. Do not forget to use the Laplacian $\psi'' + (d-1)\psi'/r$ for $d = 3$ and the "reduced" mass $m = m_b/2$.

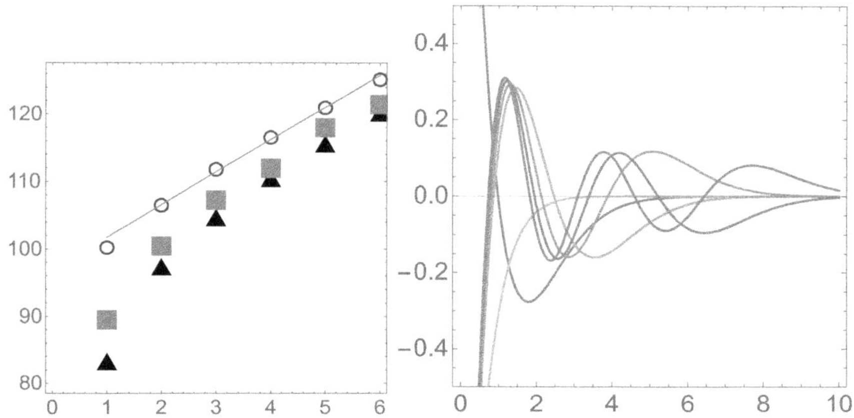

Fig. 25.2. (left) The red squares show squared masses M_n^2 (GeV^2) of six upsilons; the black triangles are masses calculated with the Cornell potentials and the blue circles are calculated with the linear potential without the Coulomb part. All are shown versus $n+1, n=0,\ldots,5$. (right) The Cornell potential wave functions versus $r\,(\mathrm{GeV})^{-1}$.

Note first from the right plot (wave functions) that, while the first one (no nodes) has a size of only about $R \sim 1\,\mathrm{GeV}^{-1} \approx 0.2$ fm, the sixth state extends to an order of magnitude larger size of $\sim\!10\,\mathrm{GeV}^{-1} \approx 2$ fm. Keep that in mind, as we will compare hadronic sizes to the sizes of vacuum field fluctuations.

Note then, from the left plot (masses squared), that the experiment results (red square) are above the theoretical prediction of the Cornell potential (triangles) by an amount which is decreasing with the quantum number n. That is because the (so far ignored) spin-dependent potential

$$\frac{1}{3m_1m_2}V_{SS}(r)(\vec{S}_1\vec{S}_2)$$

is *repulsive* for the Upsilons (states with $L = 0, S = 1, J = 1$).[2] The decrease with n is because the corresponding potential $V_{SS}(r)$ is *short-range*: Its contribution therefore scales as $1/\mathrm{volume}(n)$.

Finally, look at the blue circles, which are obtained when one switches off the Coulomb potential: In this case, one does find the celebrated

[2]But attractive for η_c states, with $L = 0, S = 0, J = 0$. Note also that standard squaring leads to $(\vec{S}_1\vec{S}_2) = \left[S(S+1) - 3/2\right]/2$. Unfortunately, only two of η_c are known, so I was not able to plot all six "spin-averaged."

linear Regge trajectory $M_n^2 \sim n$, which we will discuss extensively in the subsequent chapters.

Now, we turn to a more general approach to a theory of static potentials related to the *Wilson lines*:

$$W = P\exp\left[i\int d\tau g A_0^a(x)T^a\right] \tag{25.2}$$

Any fluctuations in vacuum gauge fields lead to some observable effects. Fluctuations of small amplitudes — described by plane waves — when summed produce what we call a "gluon propagator" $\sim 1/x^2$. If two static quarks are depicted by two Wilson lines along the time direction, their correlator $\langle W(0)W^+(\vec{x})\rangle$ gives the heavy quark potential. The lowest-order perturbative effect $\sim g^2$ is the one-gluon exchange, see Fig. 25.3. In the static case using Euclidean time is possible and much simpler. Integrating over the time difference of the propagator, one gets

$$g^2\int_0^{t_{\text{total}}}\frac{d\tau_1 d\tau_2}{(\tau_1-\tau_2)^2+\vec{x}^2} = g^2\frac{t_{\text{total}}}{|\vec{x}|} \tag{25.3}$$

which is nothing but the familiar Coulomb's law! For a weak coupling of g, one may retain only the lowest order. Further theoretical efforts through the development of effective nonrelativistic theories, such as nonrelativistic QCD (NRQCD) and perturbative nonrelativistic QCD (pNRQCD), have brought the spectroscopy of quarkonia close to that of atoms and nuclei in accuracy. A number of independent but universal (flavor-independent) central and spin-dependent potentials are defined in terms of certain

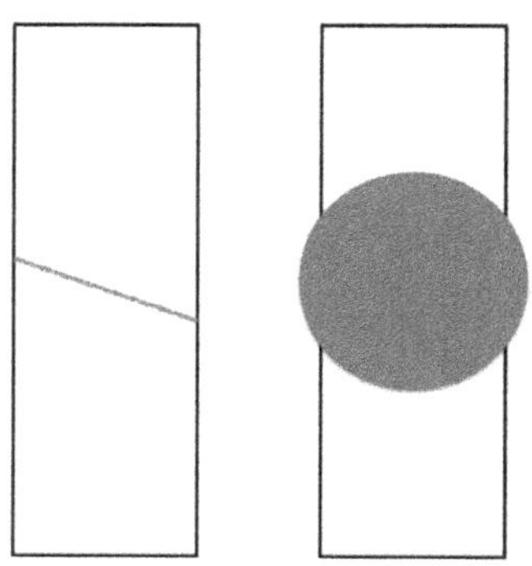

Fig. 25.3. Sketches of the one-gluon exchange (left) and instanton-induced contribution (right) to the static potential.

nonlocal correlators of vacuum gauge fields. For a short but concise summary; see, for example, Pineda and Vairo (2001).

Instantons are also quantum fluctuations of A_μ but of a large "non-perturbative" amplitude, $A_\mu \sim 1/g$, so there is no small parameter. The correlator (schematically shown in the right plot of Fig. 25.3) needs to be calculated without expansion of the field strength. Fortunately, for an instanton field, the Wilson line can be calculated analytically (Callan *et al.*, 1978):

$$W(\vec{r}) = \cos\left(\frac{\pi r}{\sqrt{r^2+\rho^2}}\right) + \frac{\vec{r}\vec{\tau}}{r}\sin\left(\frac{\pi r}{\sqrt{r^2+\rho^2}}\right) \tag{25.4}$$

Its contribution to heavy quark potential is then given by

$$\begin{aligned} V &= \int d\rho \frac{1}{\rho^2}\frac{dn(\rho)}{d\rho} W(x/\rho) \\ W(x/\rho) &= \frac{1}{3\rho^3}\int d^3r\, tr[1 - W(\vec{x}-\vec{r})W^+(-\vec{r})] \end{aligned} \tag{25.5}$$

It was also generalized to the velocity–velocity and spin–spin potentials.

Naively, the linear term of the Cornell potential represents the QCD flux tube (string) with a fixed tension. However, quantum mechanics tells us that the string must oscillate; therefore, the potential induced by the quantum string should rather be the so-called "Arvis potential":

$$V_{\text{Arvis}}(r) = \sigma_T r \sqrt{1 - \frac{\pi}{6}\frac{1}{\sigma_T r^2}} \tag{25.6}$$

which vanishes at $r \approx 1/3$ fm. You can see its effect in Fig. 25.4, which puts into doubt its interpretation at small distances.

The contribution of semiclassical vacuum fluctuations, *instantons*, gives a contribution (according to Shuryak and Zahed, 2023a) as shown in Fig. 25.4: It can complement the flux tube contribution up to distances $\sim 5\,\text{GeV}^{-1} \sim 1$ fm. To tell which type of vacuum field really works here, one should extend this calculation to spin-dependent forces. The V_{SS} mentioned above is proportional to the correlator of the Wilson lines complemented by two magnetic fields (because spins interact via magnetic moments):

$$V_{SS} \sim \int dt \langle\langle B^{ma}(\vec{0},0) B^{ma}(\vec{r},t)\rangle\rangle$$

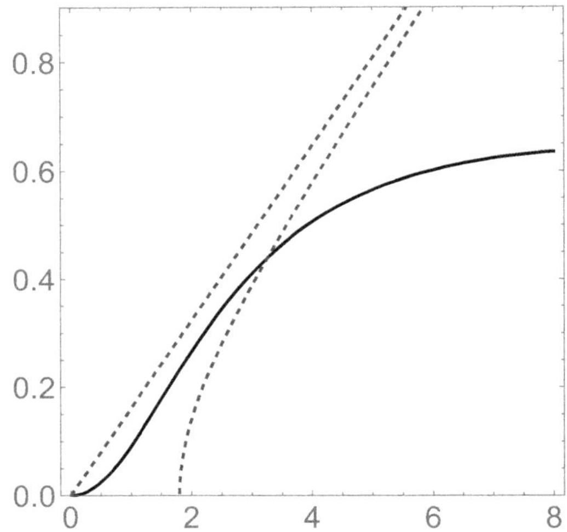

Fig. 25.4. Comparison of the central potentials $V_C(r)$ versus the inter-quark distance r in GeV^{-1} (so, $1\,\mathrm{fm} \approx 5\mathrm{GeV}^{-1}$ on this plot). The solid line is the one derived from the "dense" instanton ensemble. The potential produced by the confining flux tube is illustrated by two blue dashed lines: the upper linear line is the classical string, and the lower is the Arvis potential (25.6) accounting for quantum string vibrations.

where the double brackets indicate the inclusion of four appropriate Wilson lines and a vacuum average.

The QCD flux tube is electric, it does not have B, and it does not contribute to spin–spin forces.[3] The instantons are different: They are $selfdual$ $\vec{E} = \vec{B}$, so in a sense, their magnetic field is as large as possible.

According to Shuryak and Zahed (2023a), instantons do reproduce the nonperturbative part of the V_{SS} for charmonia well. Omitting the calculation itself, let me show in Fig. 25.5 how lattice, perturbative and instanton-induced effects in V_{SS} compare. Note that the area under perturbative and instanton-induced effects is comparable and that their sum roughly agrees with the lattice determination (Kawanai and Sasaki, 2015) of the spin–spin potential.[4]

---

[3]There is their contribution only to spin–orbit forces, caused by Thomas precession.

[4]The width of the instanton-induced one is larger than the lattice potential: we will return to this point in the following.

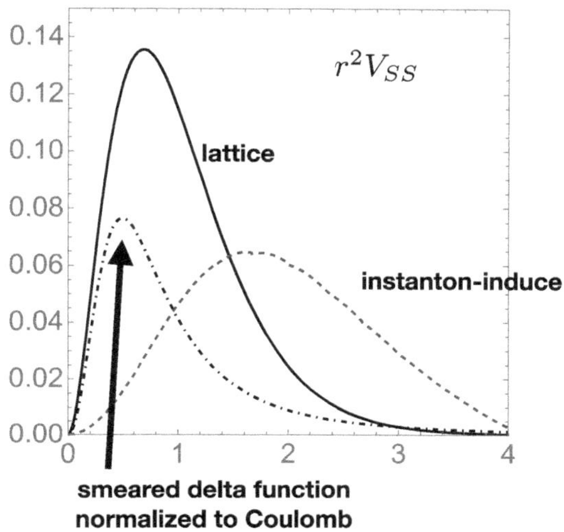

flavors	$M_\Upsilon - M_{\eta_b}$	$M_{J/\psi} - M_{\eta_c}$	$M(D^*) - M(D)$	$M(K^*) - M(K)$	$M(\rho) - M(\pi)$
Exp	61.	116.	137.	398.	636.
$\langle V_{SS}^{lat}/3M_1M_2\rangle$	46.	108.	98.	170.	
$\langle \vec{\nabla}^2 V_C/3M_1M_2\rangle$	28.	58.	48.	82.	
$\langle \vec{\nabla}^2 V_{inst}/3M_1M_2\rangle$	7.	30.	48.	90.	

Fig. 25.5. The spin–spin $r^2V_{SS}(r)$ (right) versus distance r, (GeV^{-1}). The lattice result is one of the parameterization given by Kawanai and Sasaki (2015), the perturbative is the Laplacian of the (regulated) Coulomb term.

25.2 Spin-Dependent Forces

The more general spin-dependent forces include five potentials (Eichten and Feinberg, 1981), each related to certain field-W correlators:

$$\begin{aligned} V_{SD} = &\left(\frac{S_Q \cdot L_Q}{2m_Q^2} - \frac{S_{\bar{Q}} \cdot L_{\bar{Q}}}{2m_{\bar{Q}}^2}\right)\left(\frac{1}{r}\frac{d}{dr}(V_C(r) + 2V_1(r))\right) \\ &+ \left(\frac{S_{\bar{Q}} \cdot L_Q}{m_Q m_{\bar{Q}}} - \frac{S_Q \cdot L_{\bar{Q}}}{m_{\bar{Q}} m_{\bar{Q}}}\right)\left(\frac{1}{r}\frac{d}{dr}V_2(r)\right) \\ &+ \frac{(3S_Q \cdot \hat{r} S_{\bar{Q}} \cdot \hat{r} - S_Q \cdot S_{\bar{Q}})}{3m_Q m_{\bar{Q}}} V_3(r) + \frac{1}{3}\frac{S_Q \cdot S_{\bar{Q}}}{m_Q m_{\bar{Q}}} V_4(r) \end{aligned} \tag{25.7}$$

$\vec{S}_{Q,\bar{Q}}$ and $\vec{L}_{Q,\bar{Q}}$ are the spin and orbital angular momenta of the $\bar{Q}Q$ pair. $V(r)$ is the central static potential, $V_1(r)$ and $V_2(r)$ are obtained by inserting a chromo-electric or chromo-magnetic field on the temporal Wilson loop. The spin–spin and tensor contributions $V_{3,4}(r)$ follow from the insertion of two chromo-magnetic fields on the Wilson loop. All of them can be expressed via certain Wilson lines with gauge field strengths, potentially calculable on the lattice.

Let me explain how one can get these terms separated using quarkonium phenomenology by considering masses of the next four states in the 1P shell, with the inclusion of the spin–orbit and tensor potentials. The empirical information to be used on several mesonic families is listed in Table 25.1. The theoretical estimates follow from the 1P wave functions derived with the standard Cornell potential. For both components of the central potential, perturbative and instanton-induced, we calculated the matrix elements of the three spin-dependent potentials using the formulae given above and then calculated their matrix elements in the 1P shell. The results are listed in Table 25.2. With $L = 1$, all three spin-dependent terms come into play, and with four masses, one has three differences, which allow us to solve for each individual contribution. We define the differences from the χ to h states as

$$a\langle(\vec{S}\cdot\vec{L})\rangle + b\langle S_{12}\rangle + c\langle(\vec{S}_1\cdot\vec{S}_2)\rangle$$

We can solve for a, b, c by using the corresponding quantum numbers for these states, namely

$$\begin{aligned} M(\chi_{b2}) - M(h_b) &= a - (2/5)b + c, \\ M(\chi_{b1}) - M(h_b) &= -a + 2b + c, \\ M(\chi_{b0}) - M(h_b) &= -2a - 4b + c \end{aligned} \tag{25.8}$$

Table 25.1. The first column is the flavor composition. The next four columns refer to the four states of the $n = 1, P$-shell, with the quantum numbers explained in the text. The last three columns are matrix elements of the spin–orbit, tensor and spin–spin terms, in MeV.

	$S=1, J=2$	$S=1, J=1$	$S=1, J=0$	$S=0, J=1$	a	b	c
$\bar{b}b$	$\chi_{b2}(9912)$	$\chi_{b1}(9893)$	$\chi_{b0}(9859)$	$h_b(9899)$	13.7	3.3	0.5
$\bar{c}c$	$\chi_{c2}(3556)$	$\chi_{c1}(3511)$	$\chi_{c0}(3415)$	$h_c(3525)$	35.0	10.0	0.1
$\bar{c}q$	$D_2^*(2461)$	$D_1(2430)$	$D_0^*(2300)$	$D_1(2421)$	34.0	16.0	12.0
$\bar{s}q$	$K_2(1430)$	$K_1(1403)$	$K_0(1430)$	$h_s(1270)$	6.7	−5.6	151.0
$\bar{q}q$	$a_2(1317)$	$a_1(1230)$	$a_0(1450)$	$b_1(1235)$	0.5	−36.0	68.0

Note that in this table, we included not only heavy quarkonia but their analogues with other flavors, including the light ones. Yes, the five potentials are universal; however, the masses in front are not, and also, the wave functions (sizes) of all these hadrons do depend on quark masses.

Completing this section, we need to say that spin-dependent forces are still far from being well understood. One particular issue which remains unexplained is the apparent absence of spin–orbit forces in baryons, as noted by Isgur and Karl as early as the 1980s.

25.3 Diquarks

Baryonic phenomenology has given multiple hints on nucleon substructure, in the quark–diquark form. Let me just mention one of them: Mesonic (e.g. ρ, ω) and nucleon Regge trajectories (in the M^2-J plane) look nearly identical and are explained using a picture of two colored objects connected by a single QCD string.

Dynamical calculations of baryonic correlators (Schäfer *et al.*, 1994) in the framework of interacting instanton liquids (IILs) have revealed further surprising facts. In the instanton vacuum, the nucleon was shown to be made of a "constituent quark" plus a *deeply bound diquark*, with a mass that is only slightly higher than that of constituent quarks. On the other hand, decuplet baryons (such as Δ^{++}) had shown no "good" scalar diquarks, remaining a weakly bound set of three constituent quarks. To my knowledge, it was the first dynamical explanation of deeply bound scalar diquarks. While being a direct consequence of the 't Hooft Lagrangian, this phenomenon has been missed for a long time. It also led to the realization that diquarks can become Cooper pairs in dense quark matter, see Schäfer and Shuryak (2001) for a review on "color superconductivity."

Another general theoretical reason for the lightness of the scalar–isoscalar diquark state follows from the special Pauli–Gursey symmetry of two-color QCD. In this theory (sometimes referred to as the "small Nc limit" of QCD), the scalar diquarks are actually massless Goldstone bosons. For general N_c, the instanton (gluon-exchange) in the qq channel is $1/(N_c - 1)$ down relative to the $\bar{q}q$ channel. So (Schäfer and Shuryak, 2001), the real world with $N_c = 3$ is half-way between $N_c = 2$, with a relative weight of 1, and $N_c = \infty$, with a relative weight of 0. In rough terms, the

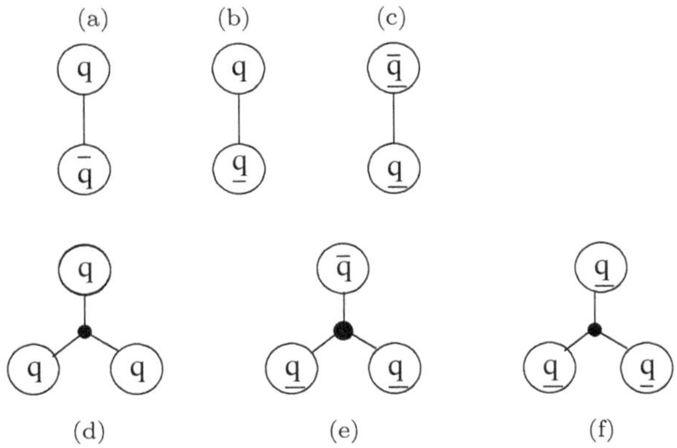

Fig. 25.6. Schematic structure of (a) ordinary mesons, (b) quark–diquark or octet baryons, (c) diquark–antidiquark states or tetraquarks, (d) decuplet baryons, (e) pentaquarks and (f) dibaryons. Underlined q indicate diquarks.

scalar–isoscalar diquarks are *half Goldstone bosons* with a binding energy about equal to that of constituent quark mass.

The idea of approximate quark–diquark symmetry[5] was developed by Shuryak and Zahed (2004), see Fig. 25.6. Note that the difference in masses between ρ, ω and nucleons is not that large and can be considered a "symmetry breaking term." This quark–diquark symmetry can be seen as the basis for the meson–baryon "quasi-supersymmetry" introduced by Brodsky *et al.* (2016) based on a holographic model.

25.4 Multiquark States

The puzzling question of why tetraquarks $qq\bar{q}\bar{q}$, pentaquarks $qqqq\bar{q}$ and dibaryons $qqqqqq$ (obviously not counting loosely bound nuclear states) were not seen[6] has been asked repeatedly. With the inclusion of heavy quarks, some of those were found during the past decade.

First, to make sense of this new phenomenology, one needs to understand the names and quantum numbers. The name X is applied to charge-zero states, such as $\bar{Q}Q\bar{q}q$, except those in the $J^{PC} = 1^-$ channel, which can

[5]Historically, it was based on the experimental announcement of the pentaquark $\Theta^+(1540)$, which was later not confirmed. Yet, the idea of a diquark is very much alive and is still used by spectroscopists in relation to multiquark hadrons.

[6]We will not go into the discussion of pentaquarks, whose discovery was not confirmed in later experiments.

couple to a photon and are called Y. If the light quark pair is electrically charged (e.g. $\bar{u}d$), they are called Z_c or Z_b, according to their heavy quark flavor.

The name P is used for pentaquarks, such as $\bar{Q}Quud$ (resonances in the $J/\psi + p$ channel). One type of such state is *hadro-charmonium*, the lowest $\bar{c}c$ state J/ψ bound to some light meson. Another possibility is a *molecular heavy-light* structure, e.g. a bound state of two ($c\bar{u}$ and $\bar{c}u$) D mesons. One more component can be the *diquark–antidiquark* ones, made of cu and $\bar{c}\bar{u}$. A mixture of those can be the case, for example, in one of the first states, $X(3872)$, discovered by BELL in 2003, in the $J/\psi\pi^+\pi^-$ channel.

The $X(3872)$ mass is close to the $D^0 + \bar{D}^{*0}$, $(1864.83 \pm 0.05) + (2006.85 \pm 0.05) = (3871.68 \pm 0.07)$ MeV threshold with very high accuracy. This may be interpreted as evidence in favor of its molecular structure. However, CMS has seen the production of $X(3872)$ in pp collisions at very large transverse momenta (up to 23 GeV) completely incompatible with the production of deuterons, suggesting that hard production of $\bar{c}c$ pairs "dressed up" in fragmentation. This is an argument for the hadro-charmonium component.

One more puzzle is that a b-quark analogue in the decay mode $\Upsilon\pi^+\pi^-$ at corresponding threshold is not observed(?).

The pentaquarks $P_c(4450)^+$ and $P_c(4380)^+$ were first seen in the $J/\psi p$ decay channel by LHCb, with a very high statistical significance.

The discoveries go on: Recently, Aaij *et al.* (2022) LHCb[7] discovered a state T_{cc} of (explicitly exotic) composition $cc\bar{u}\bar{d}$ (note: double charmed!) and mass 3875 MeV in the $D^0D^0\pi^+$ final state. Its width is quite unusual, only 0.4 MeV! The studies of such tetraquarks (e.g. Wagner *et al.*, 2022) have concluded that the meson–meson and diquark–diquark components of their wave functions are comparable.

Questions are asked about states beyond those observed, in particular with three, four and even six charm (or beauty) quarks. Their possible masses have been debated for a long time. Lately, lattice practitioners have taken over this debate. As an example, let me mention the lattice study of scattering of two $\Omega_c = ccc$ with variable momenta, from which the effective potential was deduced, followed by the existence of the bound state and its binding. One might think that, because a charm quark has two spin and three color states, 6 of them is a kind of magic number and that there would be significant binding, such as that with four nucleons. This is not so, as it

[7]This work is dedicated to the memory of our dear friend Simon Eidelman, who made significant contributions to this discovery but passed away in June 2021.

was found by Lyu *et al.* (2021) that the binding found is only about 5 MeV, and even this is negated if repulsion due to the electromagnetic charge of dibaryon ($Q = 4$) is included. Quite a similar situation is for a dibaryon with six strange quarks. The $6c$ or b quark states have not yet been studied, to my knowledge.

The lattice studies of light diquark properties (e.g. Francis *et al.*, 2022) did confirm that a "good" scalar diquark is different from others, although its extra binding ~170 MeV found is smaller than in earlier instanton-based calculations.

25.5 Static Potentials in Baryons

We have so far discussed quark–antiquark static potentials related to a correlator of two (oppositely directed) Wilson lines. The simplest conclusion about its nonperturbative part was that it can be approximated by a simple linear potential $\sigma_T r$.

Now, we are turning to interquark forces in baryons, related to correlators of three Wilson lines:

$$\langle \mathbf{W}(\vec{r}_1)\mathbf{W}(\vec{r}_2)\mathbf{W}(\vec{r}_3)\rangle$$

We start by enumerating the stringy models discussed in the literature:

1. Y-model: The simplest baryon configuration, following the quarkonium Cornell potential, consists of a perturbative Coulomb term plus a linear string potential. This is the Y (or "star") model, with three strings going from quarks to a certain point where the "string junction" is located.
2. V-model: In the V-model, one of the quarks sits directly on the color junction; therefore, there are only two strings in a baryon.
3. A-model: This is a somewhat mysterious model which is nevertheless widely used by spectroscopists. We call it the A-model as it corresponds to the following *ansatz*:

$$V_A = \frac{1}{2}\big(V_{q\bar{q}}(r_{12}) + V_{q\bar{q}}(r_{13}) + V_{q\bar{q}}(r_{23})\big) \tag{25.9}$$

in which $V_{q\bar{q}}(r_{12})$ is the usual quarkonium potential, summed over all three pairs. Its mysterious element is the factor $(1/2)$ in front, separating it from the so-called Δ model. It is justified at small distances, as the ratio of the perturbative color Casimir for $q\bar{q}$ and qq pairs in baryons. At large distances, the potential is nonperturbative, and (25.9) subsumes Casimir scaling.

Theoretical arguments using the *vortex piercing* picture of confinement have been given in support for it (Cornwall, 1996), which extend also to large N_c

$$\frac{1}{2} \to \frac{1}{N_c - 1},$$

much like the one-gluon exchange. Indeed, we know on general grounds that for $N_c = 2$, mesons and diquarks should have the same potential. Also, for large N_c, the qq force is down by $1/N_c$ in comparison to the $q\bar{q}$ force. So, perhaps this factor is correct for *all* nonperturbative effects.

These models provide certain predictions for 3-q potentials, which vary depending on the geometry of the three-quark locations. We will discuss those as we proceed and compare them with both numerically obtained potentials from lattice simulations and with our calculations using a "dense instanton liquid" vacuum model.

The three-quark static potentials have been assessed on the lattice since the 1980s. For definiteness, we use the data compiled in the appendix to the work of Koma and Koma (2017) for different geometries of the quark locations. As these authors have shown, none of the models listed above provide a fully satisfactory fit to the data. Nevertheless, they contain important lessons, a few of which we would like to formulate in this section.

Let us start with the most symmetric setting, the *equilateral* triangle. Following Koma and Koma (2017), three quarks are assigned to the points $(x, 0, 0), (0, x, 0), (0, 0, x)$ and will be characterized by the distance x. All three sides are $\sqrt{2}x$, but all plots below them will be given as a function of x.

The constants in front of the linear terms depend on the geometry of the quark locations and are different for different models. Here, we start with the well-known Y-model, with three strings going from the quarks to the center of mass of the triangle, where the "string junction" is located. In this model, the force should be three times the string tension times the ratio of the distance to the center in units of x. For an equilateral triangle, it is

$$\frac{dV_Y}{dx} = 3\sigma\sqrt{\frac{2}{3}} = \sqrt{6}\sigma \tag{25.10}$$

In the V-model, one of the quarks sits directly on the color junction. In this case, there are *two* strings rather than three. For the particular triangle under consideration, we have

$$\frac{dV_V}{dx} = 2\sigma\sqrt{2} = \sqrt{8}\sigma \tag{25.11}$$

Finally, for the A-model, we have the smallest force

$$\frac{dV_A}{dx} = \frac{1}{2}\,3\sigma\sqrt{2} \tag{25.12}$$

with $\sigma_A/\sigma_Y \approx 0.866 < 1$, regardless of the triangular configuration.

In Fig. 25.7, the filled points correspond to the string tension σ for a quark–diquark (identical to a quark–antiquark) using the same lattice configurations. For the Y-mode, the ratio is

$$\frac{\sigma}{\sigma_Y} = \frac{\sigma}{\sqrt{6}\sigma} \approx 0.854 \tag{25.13}$$

instead of 1. For the V-model, we have $\sqrt{8/6} \approx 1.15$, which is even larger. In the A-model, it is comparable to σ:

$$\frac{\sigma}{\sigma_A} \approx \frac{0.854}{0.866} \approx 1 \tag{25.14}$$

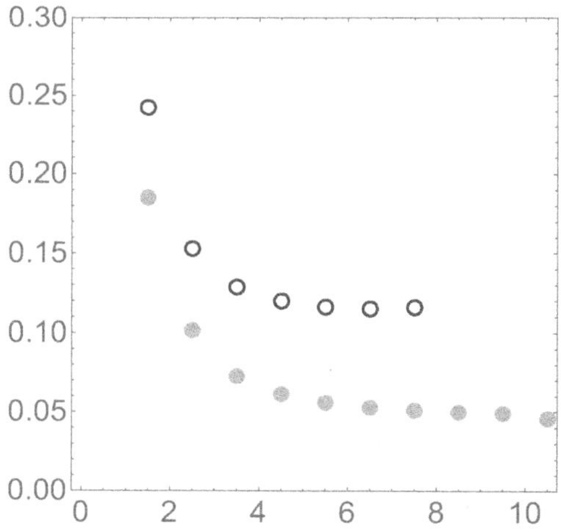

Fig. 25.7. Forces dV/dx versus x in lattice units, calculated from the lattice potentials, as done by Koma and Koma (2017). The open points are for an equilateral triangle, and the closed points are for a linear quark–diquark setting (the same as the standard quark–antiquark one).

The lesson about the three-quark lattice potentials is that the string-based Y and V models overpredict the forces at large distances, while the *ansatz* A-model reproduces the force (for an equilateral triangle). Of course, one should consider other geometries and probe intermediate distances as well: It can be found in the original paper. Still, the rule is that the most successful model for the three-quark potential is the *ansatz* A-model.

Before proceeding to the static three-quark potentials from the three models presented earlier, using instantons, we first make a few remarks that suggest why the A-model would be favored. In baryons made of three quarks, we first observe that the color wave function ϵ^{abc} involves all three colors, while the instanton fields are $SU(2)$ valued in the space of three colors. This suggests that the main potential is best approximated by binary potentials.

Furthermore, since the instanton fields do not generate confinement, the pertinent potentials are asymptote constants for the binary potential $V(r_{ij} \to \infty) = 2\Delta M$, and for the triple potential $3\Delta M$, etc. Since there are three binary potentials, these large distance limits can only be reconciled if there is a factor of $(1/2)$ in front, as in *ansatz* A-model.

The calculation was reported by Shuryak and Zahed (2023c). Here, we cannot present it in detail but just mention some main steps. One of them is the introduction of a "random orientation matrix" describing the instanton $SU(2)$ fields inside the $SU(3)$ colors, which is then averaged over an appropriate measure. The results of the calculation of the instanton-induced potentials for the three triangular shapes are shown as filled circles in Fig. 25.8. The instanton size was taken to be the usual $\rho = \frac{1}{3}$ fm and the density parameter to be $\kappa = \pi^2 n_{I+\bar{I}}\rho^4 = 1$ "dense instanton liquid"). So, the calculation has only one additive free parameter, which we have set by requiring all potentials to vanish for small triangles. In Fig. 25.8, the potential is plotted versus x defined via quark positions as follows (same as on the lattice):

$$r_i = (x, 0, 0)(0, x, 0), (0, 0, x) \quad \text{“equilateral”} \tag{25.15}$$

$$r_i = (x, 0, 0)(0, x, 0), (0, 0, 0) \quad \text{“direct”} \tag{25.16}$$

$$r_i = (x, 0, 0)(0, a, 0), (0, 0, a) \quad \text{“long”} \tag{25.17}$$

In the plots, the dashed and solid straight lines refer to the linear predictions of the models V and A, discussed at the beginning of this

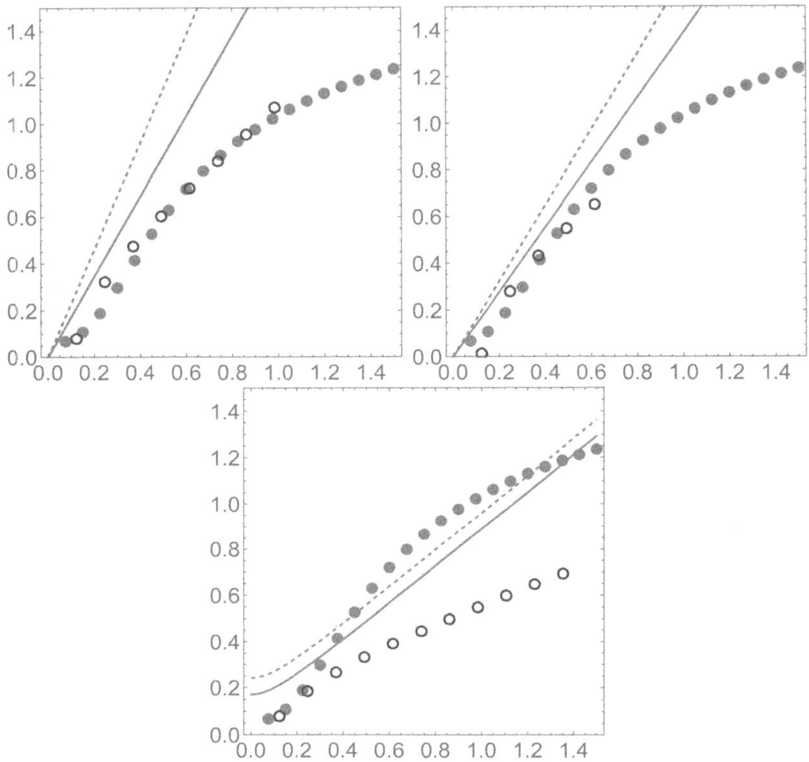

Fig. 25.8. The three-quark potentials V (GeV) versus x (fm) defined in (25.15) for "equilateral," "direct" and "long" triangles from top to bottom. The open points correspond to lattice simulation, and the closed points to our calculation in the "dense instanton liquid" model. Linear predictions of the models V and A are shown by the dashed and solid lines, respectively.

section, respectively. They correspond to

$$\begin{aligned} &V_V = \sqrt{8}\sigma x, \quad V_A = \frac{3}{\sqrt{2}}\sigma x \quad \text{"equilateral"} \\ &V_V = 2\sigma x, \quad V_A = \sigma x(1 + 1/\sqrt{2}) \quad \text{"direct"} \\ &V_V = \sigma\left(\sqrt{x^2 + a^2} + a\sqrt{2}\right), \\ &V_A = \sigma\left(\sqrt{x^2 + a^2} + a/\sqrt{2}\right) \quad \text{"long"} \end{aligned} \tag{25.18}$$

The open circles in these plots are from the lattice simulation tables reported by Koma and Koma (2017), and the lattice site size is $a = 0.123$ fm.

In conclusion, good agreement is shown between the instanton-induced and lattice potentials for large-enough triangles, but not for "long" ones. Of all the three models discussed, Y, V, and A discussed, the model A seems to be closer both to the lattice results and the instanton-induced potentials.

25.6 Dipole–Dipole Potentials and Scattering Amplitudes

For completeness, let us also mention the potential between two static *dipoles* given by Shuryak and Zahed (2000). Before describing it, let me recall that in QED (one photon exchange), the potential is the famous Casimir–Polder potential

$$V(r) = -\frac{\alpha_1 \alpha_2}{r^7} \tag{25.19}$$

where α_i are the so-called polarizabilities.[8] Similar results hold for QCD; note, however, that colorless dipoles have to emit not one but *two* gluons. Let me show how this result follows in the Euclidean setting we use. For simplicity, let us only consider the case when dipoles are small, $d \ll R, \rho$. In this case, the dipole approximation is justified, both gluons (photons) are emitted at close times, and the correlator to consider is

$$\langle \vec{E}^2(\tau_1)\vec{E}^2(\tau_2)\rangle \sim \frac{1}{[R^2 + (\tau_1 - \tau_2)^2]^4} \tag{25.20}$$

Its integral over relative time leads to the Casimir–Polder answer just mentioned.

It is also straightforward to calculate the instanton-induced contribution to the dipole–dipole potential (Shuryak and Zahed, 2000). We would not give here lengthy expressions but just note a general trend: The higher the order of the effect in perturbation theory, the more important the instanton-induced effects.

Note also that one can evaluate the correlators of the Wilson lines, which are not all going in the same (time) direction. If they make an angle θ

[8] Note that it is not a square of the dipole field $\sim 1/r^6$, and the difference comes from the additional integration over the time delay.

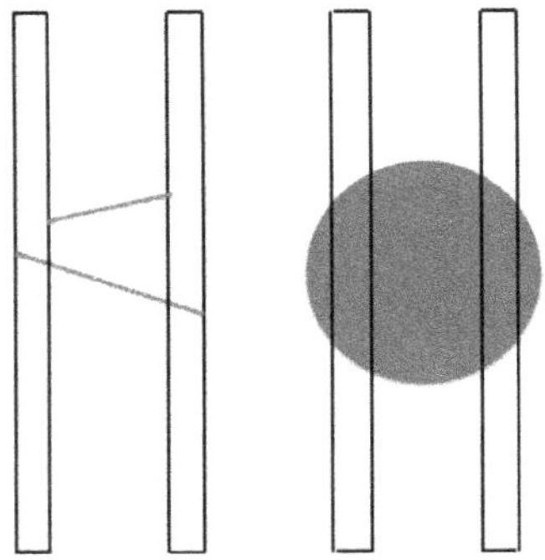

Fig. 25.9. The setting of the dipole–dipole potential calculation: two-gluon exchange and instanton-induced.

in Euclidean space-time, one can calculate it and then perform analytic continuation toward the Minkowskian description via

$$\theta \to iy \tag{25.21}$$

where y is the relative *rapidity* of two objects, say quarks or dipoles; see details in Shuryak and Zahed (2000).

Summary and discussion: The bottom line of this section is that the static confining potential is approximately a linear function of $q\bar{q}$ separation r, up to nearly one fm. The main question related to this, which we discussed, was whether it is due to: (i) electric flux tube, created by monopoles rotating around an electric field, or (ii) instantons.

Yet, following a philosophy stemming from "Poisson duality," the answer perhaps should be "both." It states that while monopoles and instantons are very different beasts, existing in different settings, they do lead to the *same* partition function! This implies that each and every nonperturbative phenomenon should have two *dual explanations*, one in terms of monopoles and the other in terms of instantons.

Duality does not imply that two explanations should be equally simple or complex. For example, we argued above that instantons provide a simple explanation for correlators containing magnetic fields since, for instantons, those are large, simple (they are "self-dual") and lead to straightforward evaluations of the spin-related forces. The results do match what we learned from lattice and phenomenology. Yes, it is nice; however, perhaps tomorrow, somebody can come up with some monopole-based theory to the same end (?)

What is actually needed is as many observables as possible. As a "discussion" item, let me point out one issue not yet addressed, at least not in a satisfactory way. It is a question about the dependence of all the potentials on the topological (CP-odd) angle θ, from θ-vacua.

Taking it naively, all amplitudes induced by instantons should be given a factor $\exp(i\theta)$ and antiinstantons $\exp(-i\theta)$. If so, one may speculate that the vacuum energy and, for instance, the string tension should be proportional to $\cos(\theta)$ and therefore vanish at some critical deconfinement value θ_c. The pioneering study of the issue (Nakamura and Schierholz, 2021) indeed finds that string tension decreases with θ in a distinctly remarkable manner, suggesting even stronger dependence of the static potentials on θ.

Here, however, is an alternative idea suggested as an explanation of the same effect but using monopoles. The authors (Nakamura and Schierholz, 2021) related it to the "Witten effect," a mixing between electric and magnetic charges and fields at a nonzero θ. They argue that if at small θ monopoles have small electric charges, any nonzero density of them would *screen* static charges rather than produce a growing confining potential.

Concluding this discussion, let me repeat a cliche statement: We do need more data, such as on the θ dependence, on which all proposed ideas and models can be tested.

Chapter 26

HADRONS AND THEIR HIGH-ENERGY COLLISIONS

26.1 High-Energy Hadronic Collisions and Reggeons

High-energy hadronic collisions have been the focus of high-energy physics since the 1960s in the hope that they reveal something important about strong interaction dynamics. And they did: We learned about the existence of what we now call *the QCD strings* from the Regge poles — families of hadrons — and the structure of the scattering amplitude (which described the data until now). Attention to cross-sections led us to Pomerons, with their perturbative and/or stringy evolutions.

Regge trajectories originated from rather formal mathematical ideas about potential scattering in quantum mechanics. Let me briefly mention only one textbook example related to the Coulomb potential $V = -e^2/r$ to explain its meaning. A hydrogen atom has bound states given by

$$E_{n,l} = -\frac{2m'\pi^2 e^4}{(n_r + l + 1)^2}, \quad m' = \frac{mM}{M + m} \tag{26.1}$$

where the radial number n_r and the orbital momentum l are both integers starting from zero. Expressing in terms of these,

$$l(E) = -n_r + g(E), \quad g(E) = -1 + i\pi e^2\sqrt{\frac{2m'}{E}} \tag{26.2}$$

one finds a function defining the corresponding Regge trajectory. When $E < 0$, i in front of the root is canceled out, and if l is an integer, it returns the spectrum. When $E > 0$, by analytic continuation, it describes the scattering S-matrix, which can be written as

$$S = \frac{\Gamma(l - g(E))}{\Gamma(l + g(E))} e^{i\pi l} \tag{26.3}$$

with characteristic poles.

Of course, we are not going to go into atomic physics here: Our basic problem would be two (or more) quarks connected by QCD strings. Let us focus first on the *radial* excitations of the light mesons and consider the relation between the (square of) masses M_n^2 and the radial quantum number n. For that, consider the reduced or *half* of the system in question, with one quark of mass m_Q and *half* the string. Classically, energy is a sum of kinetic and potential energies, which we write in relativistic form as

$$E = V(r) + \sqrt{\vec{p}^2 + m_Q^2} \tag{26.4}$$

or

$$(E - V(r))^2 = \vec{p}^2 + m_Q^2 \tag{26.5}$$

To quantize it, we use the standard substitution of the square of momentum,

$$\vec{p}^2 \to -\frac{\partial^2}{\partial r^2} - \frac{2}{r}\frac{\partial}{\partial r} + \frac{\hat{J}^2}{r^2}$$

with the radial Laplacian (the first two terms) and the angular Laplacian substituted by the angular momentum $\hat{J}^2$.

A proper quantization of such a relativistic Klein–Gordon equation is not as simple as for the usual nonrelativistic Schrodinger equation. In particular, if one just takes the linear form of the potential $V(r) = \sigma_H r$, the problem is identical to that of a particle in a constant electric field. In such a case, particle pair production takes place through the Schwinger mechanism (the Klein paradox); therefore, in precise terms, states with fixed particle numbers do not exist. One can, however, avoid all these difficulties by using semiclassical quantization.

Note the presence of a *turning point* at which momentum vanishes, $p(r_*) = 0$. Obviously, particle motion is different for $r < r_*$ and $r > r_*$ (classically forbidden region). To simplify formulae, consider the massless case, $m = 0$. Then, one simply has $E_n - \sigma_T r_* = 0$, and the turning point is easy to obtain.

As usual, for $r < r_*$, the momentum is real, and the standard WKB approach can be applied. We do so in its (original) Bohr's form:

$$n\hbar = 4\int_0^{r_*} p(r, E_n)dr \tag{26.6}$$

where the number 4 appears because the integral covers a *quarter* of the period.

For $J = 0$ and the linear potential $V = \sigma_T r$, this can be analytically calculated for any mass, with the result being

$$n = a_M + \alpha' E_n^2 \left[\sqrt{1-b^2} + b^2 \log\left(\frac{1-\sqrt{1-b^2}}{b} \right) \right] \tag{26.7}$$

where the "Reggeon slope" is $\alpha' = 1/2\pi\sigma_T$. For two particles, E_n is the total energy of both, and $b = 2m_Q/E_n$. In the massless case (nothing is attached to the string ends!), which is easy to invert, and one finds a linear Regge trajectory as $E_n^2 = (n - a_M)/\alpha'$.

For an arbitrary mass, the right-hand side of (26.7) is a complicated function of energy E_n, which cannot be inverted analytically. Instead of using some approximate formulae, one can invert it numerically. In Fig. 26.1, the closed points correspond to the square of the masses of the radial excitations (in GeV^2), with $n+1 = 1-10$, for a realistic constituent quark mass of $m_Q = 0.35$ GeV and a string tension of $\sigma_T = (0.42\,\mathrm{GeV})^2$. The results are not far from those obtained from the asymptotic massless formula, as indicated by the thick straight line. One can see that these mass corrections are actually rather small.

Here, we added a parameter a_M, known as a "quantum shift," which is not well determined. In practical applications (such as those reported by Sonnenschein and Weissman, 2014), it is used as a free parameter, different for various quantum numbers and Regge trajectories. We do, however, use the standard WKB recipe, taking $a_M = 1/2$: This value is correct for all light vector mesons ρ, ω, K^*, ϕ, but take different values for other channels. Of course, confinement is not the only term of the Hamiltonian, and certain shifts in the trajectories can be expected. One may also see that all vector mesons seem to have minimal "residual interactions" and thus negligible shifts.

In Fig. 26.1, we compare theoretical expectations (Bohr–Sommerfeld quantization) for two quarks with a string, shown by the solid line and blue disks, with the experimental masses for $\omega, L = 0$ and $\omega_3, L = 2$ mesons. The lesson from it is that the solution to this problem takes the form of "linear Regge trajectories" $M^2(n, L)$, both in the principal quantum number n and the orbital momentum L.[1]

In a QFT setting, the notion of Reggeons first originated from diagrams with an exchange of some particles with spin J in the t-channel, resulting in amplitudes of the form $A \sim s^J/(M_J^2 - t)$. It was then generalized to

[1]Recall that ω mesons are states with $J = 1, L = 0$ and ω_3 has $J = 3, L = 2$.

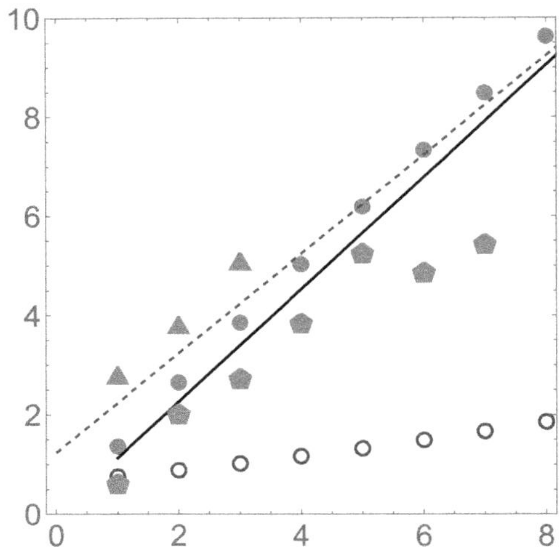

Fig. 26.1. The squared masses of the $(n + 1)$th states $E_{n+1}^2\,\mathrm{GeV}^2$ versus the radial quantum number $n+1$. The closed points are semiclassical results, for constituent quarks with mass of 0.35 GeV. The solid line corresponds to the massless limit with $E_n^2 = (n + 1/2)/\alpha'$. The five-angle polygons in red are the experimental data for the ω mesons, and the red triangles denote the ω_3 mesons listed in the particle data tables. The open points show the corresponding values from the Jia–Vary Hamiltonian, with their recommended value being $\kappa = 0.227$ GeV. The dashed line corresponds to the same expression with $\kappa = 0.5\,\mathrm{GeV}$.

noninteger J and to the so-called "Reggeon exchange," which summed up the contribution of all states belonging to a given trajectory. Delving deeper along such lines led to the famous Veneziano amplitudes (which I would not go into), which expanded these expressions further, with a s–t–u "duality." They included not only all t-channel exchanges of resonances but also those of s, u channel resonances.

The main lesson from Regge trajectories is that, while scattering happens for negative $t < 0$, an analytic continuation of the trajectories to positive t and integer J reveals all physical states with mass M and angular momentum J. Relativistic amplitudes are related to those in quantum mechanics by

$$T(s,t) = 8\pi\sqrt{s}f(s,\cos(\theta)) \tag{26.8}$$

and a contribution of a Regge trajectory to it has the form

$$T(s,t) = f_{1,3}(t)f_{2,4}(t)s^{\alpha(t)}\eta(\alpha(t)) \tag{26.9}$$

where

$$\eta(\alpha(t)) = -\frac{1 \pm \exp(-i\pi\alpha(t))}{\sin(\pi\alpha(t))}$$

is a signature factor (related to separate analytic continuations for even and odd J). The main feature here is the power energy dependence $s^{\alpha}(t)$. The total cross-section, via the optical theorem, is related to the imaginary part of the *forward* ($t = 0$) amplitude; therefore, $\sigma_{\text{tot}} \sim (1/s)\text{Im}A_{hh}(s, 0)$.

In 1960, Geoffrey Chew and Steven Frautschi conjectured, from very limited data at the time, that the strongly interacting particles had a very simple dependence of the square of mass on the angular momentum: The particles fall into families where the Regge trajectory functions are straight lines, see Fig. 26.2. Since a Regge description implied that the particles were bound states, Chew and Frautschi concluded that none of the strongly interacting particles were elementary.[2]

What they found were approximately straight trajectories, possessing various "intercepts" and universal "slopes":

$$\alpha(t) = \alpha_P(0) + \alpha'(0)t + \cdots \tag{26.10}$$

These straight-line Regge trajectories were later understood as arising from rotating relativistic strings with massless endpoints. The squares of their masses are given by the following simple expression:

$$M^2 = (2\pi\sigma_T)(J + n_r + \text{const}) \tag{26.11}$$

where J is angular momentum and the integer n_r is the radial excitation number. (The first term is classical, the others are quantum and $n_r > 0$ for an excited string.)

As one can see from Fig. 26.2, adapted from Kaidalov (2001), the highest meson trajectory has the intercept $\alpha(0) = 1/2$, thus the cross-sections decreasing with energy as $\sigma \sim 1/s^{1/2}$.

The Pomeron, introduced by Gribov and named after Pomeranchuck, is the Reggeon, which has the highest "intercept" $\alpha_P(0)$ of *all* trajectories. Since it is larger than one, $\alpha_P(0) > 1$, all total hadronic cross-sections *grow* with energy, at high s. A compilation of cross-sectional data is given in Fig. 26.3.[3]

---

[2]Remember that it was more than a decade before the QCD discovery in 1973.

[3]Note that the units are such that 100 mb = 10 fm^2. When I was a student, the *pp* cross-section was decreasing and was numerically about 50 mb, but by now (both for me and the nucleon)m it has nearly doubled!

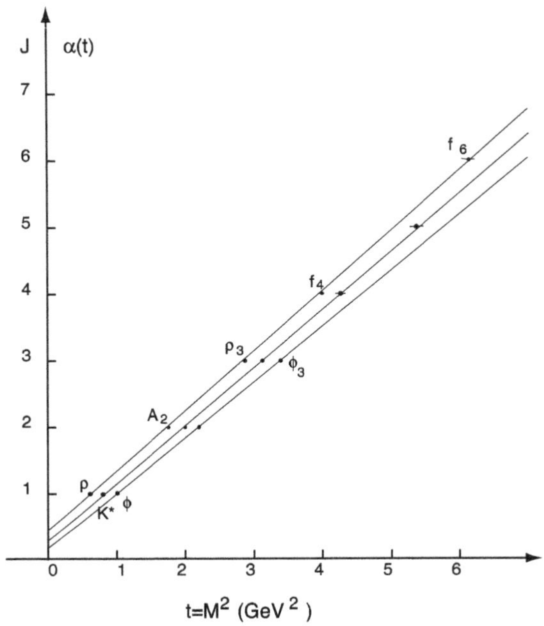

Fig. 26.2. Regge trajectories for vector and axial vector mesons made of light quarks.

Elastic amplitude can be related to the so-called "profile" (or eikonal) function by Fourier–Bessel[4] transform (here, we use the momentum transfer $q^2 = -t$):

$$T(s, q) = s \int_0^\infty dbbJ_0(bq)F(s, b) \tag{26.12}$$

Note that $T(s, q) \sim e^{Cq^2}$ is also a Gaussian function of b. Let me show the shape of the profiles (some parameterization of the data) in Fig. 26.4.

Note that the relatively small energy of ISR profile is approximately Gaussian, while the LHC one has a "black disk." It originates from the fact that the Pomeron amplitude becomes larger than one, while the probability should remain less than one: This means a resummation of multiple Pomerons is needed. The general rules for Reggeon diagrams are given by Gribov Reggeon diagrams; however, in b-representation, it is more

[4]Two dimensional Fourier is transformed into such an expression after the integral over the angle.

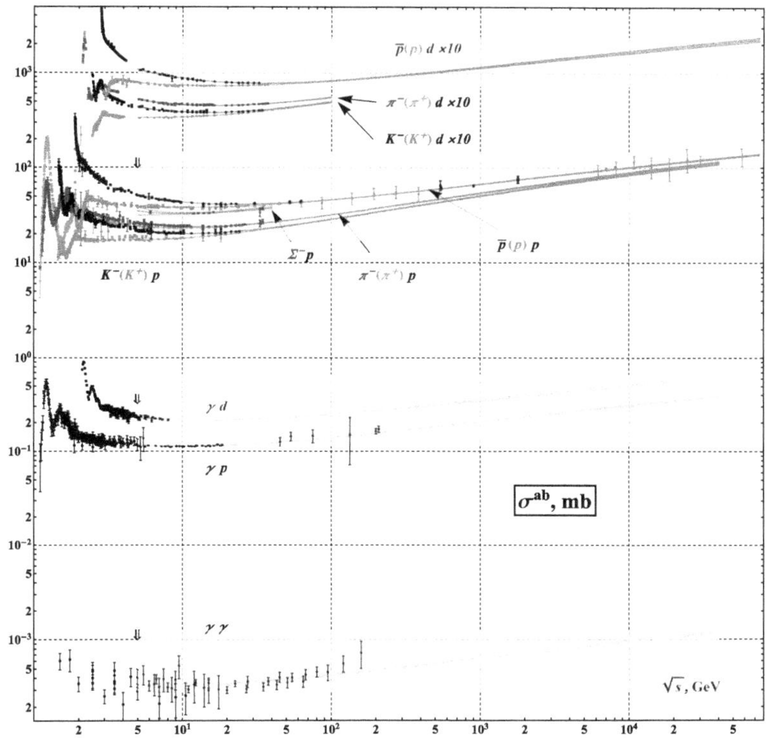

Fig. 26.3. Compilation of hadronic cross-sections as a function of energy $\sqrt{s}$, from PDG 2016.

instructive to use the Glauber resummation of the scattering profile $T(b)$:

$$(1 - e^{-T}) \approx T - T^2/2 + \cdots < 1 \tag{26.13}$$

When $T(b)$ becomes large, the multiple terms cancel each other out, and only one of them remains. This type of scattering is called "the black disk regime": Its appearance at LHC energies is illustrated by the picture.

At the edge of the regime R_{bd}, one obtains

$$\Delta \log(s) - \frac{R^2}{\log(s)} \approx 0 \tag{26.14}$$

It follows that the black disk cross-section grows as log squared:

$$\sigma_{\text{bd}} \sim R_{\text{bd}}^2 \sim (\log(s))^2 \tag{26.15}$$

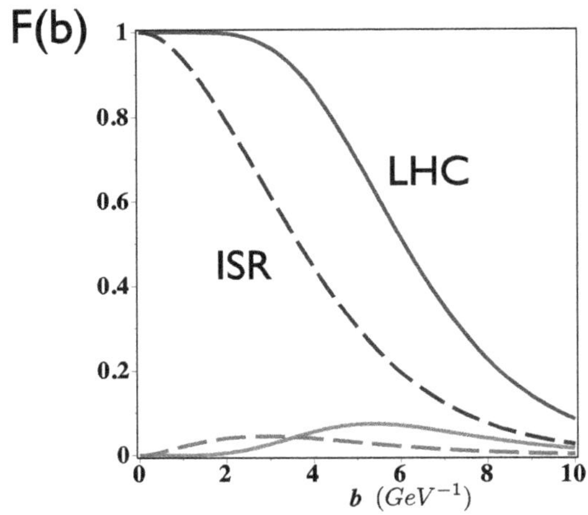

Fig. 26.4. The imaginary (upper) and real (down) parts of the profile function $F(s, b)$ versus $b(\mathrm{GeV}^{-1})$ for LHC $\sqrt{s} = 7\,\mathrm{TeV}$ (solid) and ISR $\sqrt{s} = 63\,\mathrm{GeV}$ (dashed) energies.

This section about the Reggeon phenomenology would be incomplete without mentioning the "odd brother" of the Pomeron, the *Odderon*. It is defined as the highest C-odd trajectory. It defines the asymptotic behavior of, among other things, the *difference* between the cross-sections of the pp and the $\bar{p}p$ scatterings. Let me first mention that since it is small and since there has never been accurate enough data on pp and $\bar{p}p$ cross-sections, especially not at the same s, the Odderon phenomenology remains rather murky.

The theoretical situation of Odderon is quite curious and is worth a (personal) comment. Available lattice data on C-odd glueballs, for example, from Meyers, which we used already, are shown in Fig. 26.5. Six (and perhaps even seven) out of eight data points can be united into groups, displaying the same slope. (I find it curious enough.) Furthermore, the slope $\alpha'_O \approx 0.32\,\mathrm{GeV}^2$ turns out to be exactly one-third of the mesonic slope. Thus, the tension seems to be three times stronger: Is it therefore produced by some C-odd three-string states?

Now, can one obtain the intercepts of these Odderons? If the lines are straight, then they should be negative, which means we will never see the effect at present (LHC) energies. However, if the leading trajectory is curved — as we know the Pomeron one is — then it can perhaps reach $\alpha_O(0) \approx 1$ or so, as the dashed line shows.

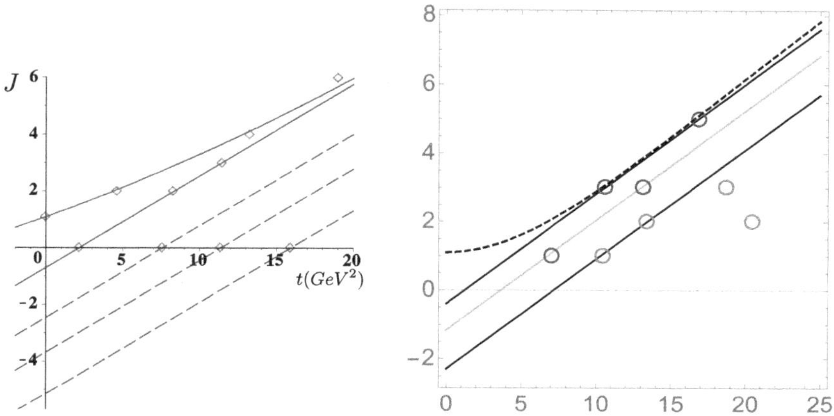

Fig. 26.5. Glueball Regge trajectories with positive and negative C parity, from lattice results for pure gauge theory.

26.2 Theories of the Pomeron and the Odderon

In this section, we briefly review the pQCD prediction for the Pomeron (BFKL) at small b and then compare it with the (more recent) stringy version at large b.

Reggeon physics in the context of weakly coupled QCD was successfully obtained by resumming rapidity-ordered Feynman graphs. In Fig. 26.6(a) two hadrons, schematically represented by a color dipole, interact via two gluon exchange. Two gluons are needed for elastic scattering, in which both the initial and final states are colorless. The gluons create two quark–antiquark pairs, which is called "splitting of the one dipole into two." The horizontal dotted line indicates the "unitarity cut," which splits the imaginary part of the elastic amplitude into the *square of the modulus* of the inelastic amplitude. Each of those includes only one gluon exchange.

In fact, studies of high-energy collisions in QFT settings started in the 1960s with "multiperipheral models" using ladder diagrams with some meson exchanges. The idea was that large logarithms, stemming from rapidity gaps between subsequent exchanges ("steps" of the ladders) could compensate for even weak coupling. When the logarithms are such that

$$\alpha_s \log\left(\frac{p_{i+1}}{p_i}\right) \sim 1 \qquad (26.16)$$

these diagrams need to be resummed. So, in reality, one needs to have ladders with multiple steps, thus the name of the model.

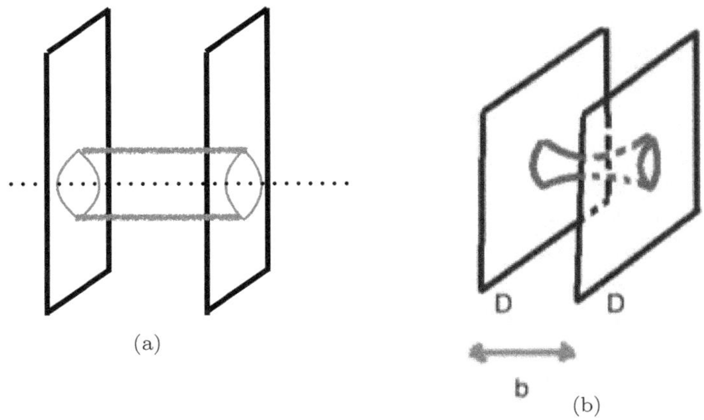

Fig. 26.6. Splitting the dipole into two in elastic scattering: (a) shows the perturbative process of two-gluon exchange. The horizontal dotted line indicates the unitarity cut. (b) The same process in terms of string world volume. It can be treated as a closed string (glueball) exchange.

With the advent of QCD, these ladders naturally switched to gluonic exchanges. Consistent resummation of them led to the so-called *hard Pomeron* by Kuraev *et al.* (1977), also known as BFKL. In the leading order, it gives the intercept proportional to the coupling[5]:

$$\Delta_{\text{BFKL}} = \alpha_P(0) - 1 = \alpha_s \frac{12}{\pi} \ln(2) + O(\alpha_s^2) \tag{26.17}$$

Phenomenologically, at "soft" collisions with min-biased momenta, the Pomeron intercept is about $\Delta_{\text{hard}} \approx 0.08$ or so; however, with increasing t (where one expects to see "hard" Pomeron), the intercept grows and reaches about $\Delta_{\text{hard}} \approx 0.15$ or so. If BFKL expression is applicable, it can give us an idea about the coupling involved.

As for the Pomeron slope α', the BFKL theory, and pQCD in general, cannot predict the slope since they lack dimensional quantities. (Remember that the slope is related to the tension of the flux tube, a very nonperturbative concept.)

The idea behind "soft" or "stringy" Pomeron is illustrated in Fig. 26.6(b). Note that it is also based on the idea of one dipole split into two, with a developing hole in between. The hole is made by charges

[5]Higher-order corrections calculated later seem to be negative and relatively large, so one does not really see a convergence.

making a circle under constant acceleration, as they are pulled by a string. For brevity, we refer to this construction as a "tube" in the following. It is known as the BKYZ Pomeron, for Basar *et al.* (2012) derived a semiclassical expression for the amplitude. We will discuss a derivation of its version later in a holographic setting, and for now, we only point out its basic features. Its profile is Gaussian in terms of the impact parameter b:

$$F(s,b) \sim e^{-\frac{\pi \sigma_T b^2}{\log(s)} + \frac{D_\perp}{12}\log(s)} \tag{26.18}$$

and it is supposed to be valid at large b, when the action in the exponent is large enough.

The second term in the exponent contains the number of transverse (to the beam and time) dimensions: It is two in a 4D flat space, but three in a holographic setting in AdS_5. The meaning of the first term is the classical action of a stringy tube, while the second comes from quantum corrections related to the string Luscher term. So, its main predictions for Pomeron parameters are the following values for the intercept and the slope:

$$\Delta_{\rm BKYZ} = \frac{D_\perp}{12}, \quad \alpha'_P = \frac{1}{2}\alpha'_M \tag{26.19}$$

Phenomenologically, the scattering data give a twice smaller $\alpha'_P \approx 0.25\,\mathrm{GeV}^2$ which we relate to the fact that the Pomeron trajectory, unlike others, is not linear but curved.

It is remarkable that both approaches (perturbative gluons and non-perturbative strings) lead to the same form of the amplitude. Perturbative growth of the hadronic sizes with energy is explained by *Gribov diffusion* in the transverse plane as a function of effective "time" (i.e. rapidity) $\log(s)$. The width of the Gaussian is such that

$$R^2 \sim \log(s) \tag{26.20}$$

Yet, quantum strings also have diffusive motion.[6] This dependence of the hadron effective size on energy is confirmed by the data (corrected for multi-Pomerons and the black disk).

Perturbative studies of the Odderon by Bartels *et al.* (2000) concluded that its intercept should be

$$\alpha_O(0) = 1 \tag{26.21}$$

Studies of the stringy mechanism (Zahed and myself, unpublished), on the other hand, seem to conclude that both Pomeron and Odderon may actually

[6]Do not confuse this expression with that for the edge of the black disk above.

have the same intercept of $\alpha_O(0) = \alpha_P(0)$. If so, the smallness of the Odderon contribution must be simply numerical. Yet, this question is by no means settled, neither empirically nor theoretically.

Given the two different pictures of the scattering process at small (gluon ladders) and large (stringy tube) impact parameter $\vec{b}$, one may wonder if we can also identify where and how a transition between them takes place. Remarkably, that seems to be related to the thermal properties of the glue, and to explain this connection, one needs to take a small digression.

Taken literally (unlike heavy ion collisions or pp at high multiplicity we discussed above), the min-bias hadronic collisions *do not* seem to possess any macroscopic properties, such as *entropy* and/or *temperature*. Indeed, they do not produce a fireball, which can be considered a macroscopic body in local equilibrium.

And yet, some interesting set of ideas has shown that even those can possess effective entropy and even temperature, defined in a very curious way.

Even for a BFKL Pomeron — a cascade of gluons described by a certain kinetic equation — according to Kharzeev and Levin (2017), one can define an entropy, growing in a Boltzmannian sense to its highest possible value:

$$S = \Delta_{\text{BFKL}} \log(s) \tag{26.22}$$

In Shuryak and Zahed (2018), we showed that stringy Pomeron also has the entropy, given by the same expression (with $\Delta_{\text{BFKL}} \to \Delta_{\text{BKYZ}}$, of course). In our previous paper (Shuryak and Zahed, 2014), we also defined an effective temperature, $T_{\text{eff}}(s, b) = \log(s)/2\pi b$. When it reaches the critical temperature of the glue ($T_v \sim 270$ MeV according to lattice data for pure gauge theories), string fluctuations become so high that they undergo a phase transition into a gluon plasma. We suggested that, therefore, when the impact parameter reaches a particular value, that would also be the location of the transition from stringy to perturbative Pomeron amplitude.

We will not discuss the properties of inelastic pp collisions. In brief, since we cannot select impact parameters, those can be described by a combination of perturbative gluon cascades and subsequent string breaking. Event generators based on the so-called Lund model can describe those processes. The main extra parameter in those is the "string breaking rate," which defines all masses of "primary clusters" produced, which sequentially decay into lighter ones and finally into a number of pions. The mass of primary clusters, obtained from rapidity correlation measurements, is of the order of 2 GeV or so. There are too many mesons around this mass to know in greater detail which ones are more or less involved in this process.

Summary: The elastic (as well as diffractive) collisions at large energies s are described by Pomeron exchanges. Microscopic theory is developed in terms of a gluon ladder at small b and/or QCD stringy tube at large b. The transition between the two regimes was related to the thermal phase transition between the stringy confining phase and the perturbative gluonic plasma.

26.3 Dynamics of Single QCD Flux Tube

The origins of confinement and the mechanisms of "deconfinement" transition from hadronic gas to QGP were already discussed in Chapter 5 for a finite T and confinement 24.

Here, we recall the basic idea that at $T < T_c$, in a confining phase of QCD, all electric fluxes associated with quark color charges are expelled into *flux tubes*, going from one quark to another. If their separation r is large, tubes are long and create a linear potential of

$$V = \sigma_T r \tag{26.23}$$

between color charges. In pure gauge theory, these flux tubes cannot decay, while in QCD with light quarks, this linear potential is only valid up to some distance, at which it is higher than the sum of the masses of two heavy-light mesons, e.g. $\bar{D}D$.

The studies of lattice gauge theories began with this potential by Creutz, in 1979, and have been going on for nearly 40 years now. The increased accuracy, as we will see, has paid off, and we continue to learn some interesting details about the static potential $V(r)$.

At *small distances*, the quark–antiquark[7] pQCD potential has the form of Coulomb attraction:

$$V(r \to 0) = -\frac{4}{3}\frac{\alpha_s(r)}{r} \tag{26.24}$$

proportional to the running coupling constant at the relevant scale $\alpha_s(r) \sim 1/\log(1/\Lambda r)$.

[7]For a quark–antiquark pair of the same colors, as is the case in mesons. In other cases, there is also $\cos(\theta_{\text{color}})$ of the angle between the color vectors. For example, three quarks making a baryon have the attractive Coulomb potential with a coefficient twice smaller because the color factor contains $\cos(2\pi/3) = -1/2$. Quarks with identical colors have the repulsive potential with the same coefficient of 4/3 as quarks–antiquarks.

At *large distances*, the string generates a linear potential. String quantum motion leads to terms containing inverse powers of r. Since the QCD string is an extended composite object and not a fundamental point-like strings of string theory, its description in general is not simple. It can be done in terms of "effective string theory" (EST), organized in increasing powers of derivatives. These contributions are generically split into the bulk M and the boundary ∂M terms. The external coordinates of strings in physical space are denoted by $X^i, i = 1, \ldots, D$, while their two-dimensional internal space τ, σ is indicated explicitly with Greek indices. Derivatives over τ are denoted by a dot and over σ by a "prime." The main kinetic term and the next one have the forms

$$S_{\text{bulk}} = -\sigma_T \int_M d^2x(1 + \partial_\alpha X^i \partial^\alpha X_i) \tag{26.25}$$

$$+\frac{1}{\kappa}\int d^2x \left(\ddot{X}^\mu \ddot{X}_\mu + 2\dot{X}'^\mu \dot{X}'_\mu + X''^\mu X''_\mu\right) + \cdots \tag{26.26}$$

$$S_{\text{boundary}} = \int_{\partial M} d^2x \left(\mu + b_2(\partial_0\partial_1 X^i)^2\right) \tag{26.27}$$

All contributions (26.25–26.27), in principle, contribute to the static potential. The linear potential stems from the string space-time volume times the tension σ_T. Other terms describe string vibrations. Keeping only the first part of the action, one expects expansion in the inverse powers of $\sigma_T r^2$

$$\sigma_T r\left(1 + \frac{V_0}{\sigma_T r^2}\right) + \cdots \tag{26.28}$$

It is the famous Luscher universal term with $V_0 = -\pi/12$ in four dimensions. Using certain string dualities, Luscher and Weisz (2004) have shown that the next contribution is universal (i.e. unaffected by other terms) as well:

$$\sigma_T r\left(1 + \frac{V_0}{\sigma_T r^2} - \frac{1}{8}\left(\frac{V_0}{\sigma_T r^2}\right)^2\right) \tag{26.29}$$

Using the first (Nambu–Goto) term in action, higher-order contributions were resummed by Arvis (1983):

$$V_{\text{Arvis}}(r) = \sigma_T r\sqrt{1 - \frac{\pi}{6}\frac{1}{\sigma_T r^2}} \tag{26.30}$$

However, from $1/r^{-5}$, these terms are affected by other corrections, such as boundary terms of the action.

To order $1/r^4$, we will need the contributions to the potential

$$V(r) \approx \sigma_T r - \mu - \frac{\pi}{12r} - \left(\frac{\pi}{12}\right)^2 \frac{1}{2\sigma_T r^3} - \frac{\pi^3}{30}\frac{b_2}{r^4} + \cdots \tag{26.31}$$

The coefficient of the last term is also called

$$\tilde{b}_2 = -\frac{\pi^3}{30} b_2.$$

One of the most interesting/natural questions is whether one can tell, from the potential itself, *where the transition from pQCD to stringy description takes place?* The answer seems to have been given relatively recently.

There are two logical options: either (i) the perturbative running stops and the $1/r$ coefficients smoothly tend to the Luscher term from above, or (ii) it overshoots the value, and then turns around and reaches it from above. The small distance potential studied on the lattice has shown that the latter is the case.

Let me jump over many papers and years of work to lattice simulations by Brandt (2017), who has identified several terms at large r. One can clearly see in his work that both universal terms $1/r, 1/r^3$ are indeed there. It then also convincingly finds the next correction, order $1/r^4$, with the opposite sign. In order to see these structures, one needs a "magnifying glass" since they are rather small in magnitude, $\sim 10^{-3}$. Therefore, the combination used and shown in the figures is the potential with the linear and Luscher terms *subtracted*:

$$V^{\text{RS}} = \left(\frac{V(r) - V_0}{\sqrt{\sigma}} - r\sqrt{\sigma}\right) \frac{r\sqrt{\sigma}}{\pi} + \frac{1}{24} \tag{26.32}$$

It is shown in Fig. 26.7, adapted from Brandt (2017).

So, the next non-universal term is $1/r^4$. Its magnitude fixes the b_2 coefficient:

$$\begin{aligned} b_2^{SU(2)} \sigma_T^{3/2} &= -\,0.0257(3)(38)(17)(3) \\ b_2^{SU(3)} \sigma_T^{3/2} &= -\,0.0187(2)(13)(4)(2) \end{aligned} \tag{26.33}$$

(for the details and explanation regarding the procedure and meaning of the errors, we refer to Brandt, 2017). Note that the overall contribution of this term to the potential is positive, which is the opposite of other corrections.

Note that at $r \to 0$, the limit is $V^{\text{RS}} \to 1/24$, the value being much larger than the scale of the plot.

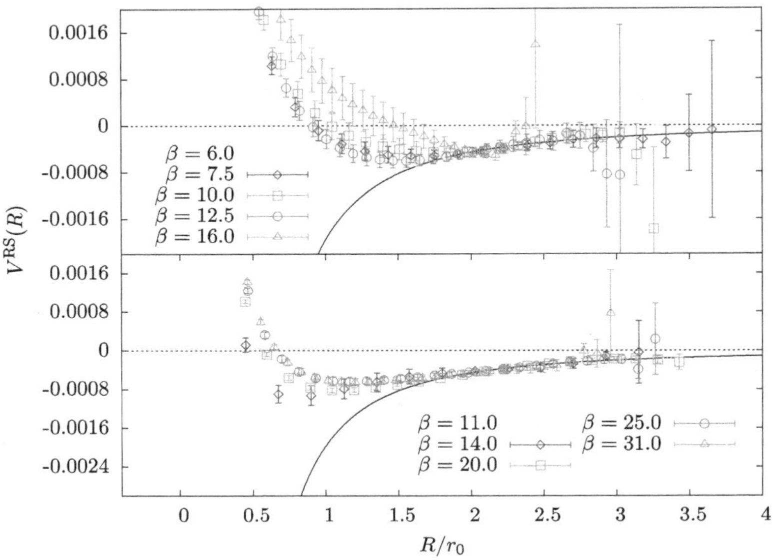

Fig. 26.7. Results for the static potential (in its rescaled form) for different lattice spacings versus the distance R/r_0, in units of the Sommer parameter $r_0 \approx 0.5$ fm defined via $r_0^2 F(r_0) = 1.65$. Plots are for SU(2) (top) and SU(3) (bottom) gauge theories. The black continuous line is the Arvis potential using the continuum extrapolated string tension.

Summary on static $\bar{Q}Q$ potential: It does contain pQCD running at small r and three universal terms, $\sim r, 1/r$, and $1/r^3$, as expected. It also seems to have the $1/r^4$ term, adding an interesting "whiggle" structure at $r \sim r_0 \approx 0.5$ fm. Perhaps, we see here the location of the hypothetical transition to the large-r stringy regime. If so, note that it is comparable to the size of various quarkonia states and also comparable to the transition from an instanton-induced one, as speculated in Chapter 24.

Chapter 27

SEMICLASSICAL THEORY OF DECONFINEMENT AND CHIRAL PHASE TRANSITIONS

We have described *monopoles* in Chapter 24 and *instantons* in Chapter 23 and shown how they can help us understand such important nonpertubative properties as *confinement* and *chiral symmetry breaking*, respectively. Yet, neither of them were able to describe *both* in a combined way.

Another important distinction between them is that, while instantons are classical solutions of the Yang–Mills equation, providing semiclassical description of tunneling through a topological landscape, the monopoles still require an *adjoint scalar field* for their construction. While such fields are present in some of QCD's cousins (e.g. the Seiberg–Witten $\mathcal{N} = 2$ supersymmetric theories), they are not available in QCD. Therefore, analytic studies of monopole-related effects were done only in such theories, while in QCD-like gauge theories, monopoles were defined only via certain *lattice definitions*. They are based on gauge fixing and certain topological features but lack classical solutions.

Needless to say, a formulation of nonperturbative phenomena in terms of a systematic semiclassical theory, with order-by-order parametric improvements being well defined[1] would be a very desirable goal.

That was achieved only relatively recently using a kind of hybrid of these two topological animals called the *instanton–dyons* (also known as instanton–monopoles) (Kraan and van Baal, 1998; Lee and Lu, 1998). Technically, if they are far from each other, they can be described as monopoles, which use the A_0 component of the gauge field instead of the

[1] Let me in passing mention recent progress in two- and three-loop semiclassics, worked out in quantum mechanical examples by Escobar-Ruiz *et al.* (2015, 2016). Similar calculations for gauge theory instantons and instanton–dyons are yet to be done.

adjoint scalar of the Georgi–Glashow model, involved in the 't Hooft–Polyakov monopole construction. When they overlap, they can still be followed analytically. When the centers of $N_c = 3$ of them happen to be located at the same spatial point, their superposition turns out to be nothing else but the BPST instanton.

The details of the solution and some early applications for supersymmetric theories can be found in my book (Shuryak, 2021). Significant work has been done using mean field methods for instanton–dyon ensembles, followed by extensive numerical simulations of those. We will only show few results, summarizing what is known at the time of this writing.

This chapter begins with Section 27.1, discussing the Polyakov line concept, its roots in mathematics and physics, and its particular use for studies of the deconfinement transition. Then, we provide some details about instanton–dyons themselves in Section 27.2 and their extraction from the lattice via "fermionic filters" in Section 27.3. Ensembles of the instanton–dyons in various settings are described in Section 27.4. We conclude the explanation of evidence for the "Poisson duality" between monopoles and instanton–dyon descriptions in Section 27.5.

27.1 The *Holonomy* and the Polyakov Line

Let us begin by emphasizing the special role of time-directed gauge potential A_0 in the finite-temperature theory, defined on a circle $\tau \in C^1$ with the Matsubara circumference $\hbar/T$. For manifolds such as this, there exists a phenomenon known in mathematics as *holonomy*.[2] Imagine that this circle is embedded in some higher-dimensional space in which magnetic fields are present, and our circular loop gets some nonzero magnetic flux through it. Since gauge transformations inside our manifold cannot change external (higher-dimensional) fields, such fluxes should be considered *external parameters* of the theory. Those are related to the integral of the gauge potential over the closed loop and are therefore gauge invariant. Their values need to be specified as part of the field quantization setting.

The so-called *Polyakov line* is

$$\hat{P}(\vec{x}) = P\exp\left[i\int_C d\tau A_0^a \hat{T}^a\right] \tag{27.1}$$

[2] In physics, its existence goes back to the famous paper by Aharonov and Bohm, describing charged particles on a circle.

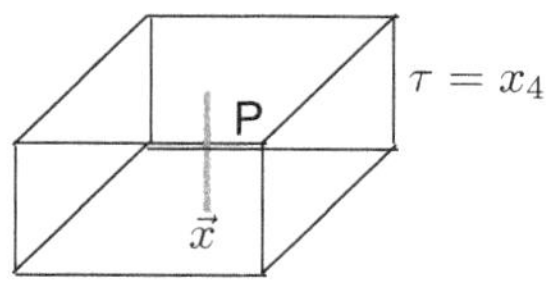

Fig. 27.1. The Polyakov line $P(\vec{x})$ on the lattice. The vertical direction is that of Euclidean time $x_4 = \tau$. Its periodicity means that the fields on the upper and lower planes are identified.

($\hat{T}^a$ is a color generator). Again, since the loop is closed by the periodicity of the thermal formulation, it is a gauge-invariant operator. The key point here is that the contour C in it is a *non-contractable contour.* For the standard $T \neq 0$ lattice setting, it is the Euclidean time $\tau = x_4$, as shown in Fig. 27.1. The lower and upper planes ($\tau = 0$ and $\tau = 1/T$, respectively) are identified, so the gauge field A_μ must be periodic.[3] Normally, $P\exp$ integrals between two points gauge transform via gauge factors at their two ends; however, for this contour, periodicity requires gauge factors to cancel out. Therefore, if P has some nonzero value, it cannot be changed by gauge transformations.[4]

$\hat{P}(\vec{x})$ is thus a unitary matrix defined at any spatial point $\vec{x}$. Its average, $\sim \int d^3x\hat{P}$, defines the expectation value $\langle\hat{P}\rangle(T)$ for gauge configuration. Its values and distributions were extensively studied on the lattice. Its color trace $P = (1/N_c)\mathrm{Tr}(\hat{P})$ has a direct physical meaning, being the exponent of the effective free energy of a static quark in the heat bath

$$\langle P\rangle = \exp\left(-\frac{F_Q}{T}\right) \tag{27.2}$$

Since it is a unitary $SU(N_c)$ matrix, one can define its eigenvalues, denoted by $\exp(i\mu_i), i = 1, \ldots, N_c$. These phases μ_i are called *holonomies* $\mu_i(T)$: They prescribe the magnitude of the homogeneous fields $A_0^a(T)$ which would generate those values.

[3]In nearly 50 years since Polyakov's paper, there have been several papers claiming that periodicity is not required and can be abolished. Among the consequences of that would be that gluons stopped being Bosons and the thermal features of QGP changed. Without getting into polemics, let me just state that those confusing papers are simply wrong.

[4]In particular, this consideration shows that at finite T, one cannot apply the $A_0 = 0$ gauge.

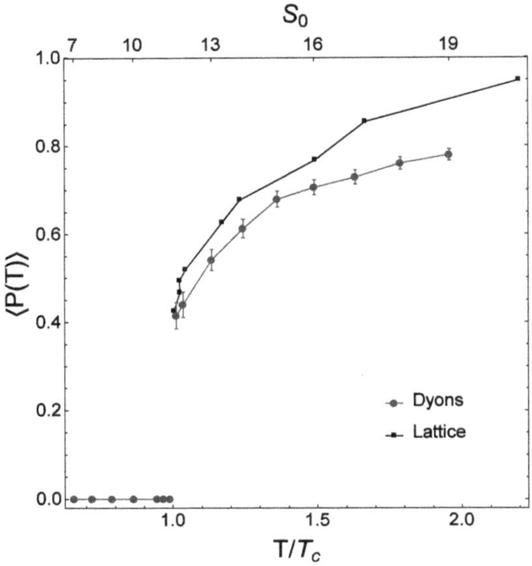

Fig. 27.2. Lattice data on the average value of the renormalized Polyakov line $\langle P\rangle(T)$, as a function of the temperature T, in pure gauge $SU(3)$ theory. Its finite jump implies first-order transitions. The curve marked "dyons" on it corresponds to semiclassical theory, to be discussed later in the chapter.

In a confining phase, the latter is infinite and $\langle P\rangle(T) = 0$, while in a quark-gluon plasma phase, it is finite and $\langle P\rangle(T) \neq 0$. So, the Polyakov line value is the *order parameter* of deconfinement.

Let us give two examples of the $\langle P\rangle(T)$ dependence on T. The first, from DeMartini and Shuryak (2021b), shown in Fig. 27.2, is the deconfinement first-order transition in pure gauge $SU(3)$ theory, in which it jumps as it is typical for the first-order transition. The second is for physical QCD with light quarks, shown in Fig. 27.3; here, there is no jump. There is also no jump in the derivative. So, in QCD, this transition is actually a smooth crossover. Yet, a small $\langle P\rangle$ at low T means that the effective quark-free energy is large and still provides strong (but not absolute) suppression.

27.2 Instanton–Dyons

In the previous section, we introduced A_0 potentials which have nonzero constant value all over the system (parameterized by N_c phases μ_i), which cannot be gauged away and are thus physical. In a semiclassical setting, one

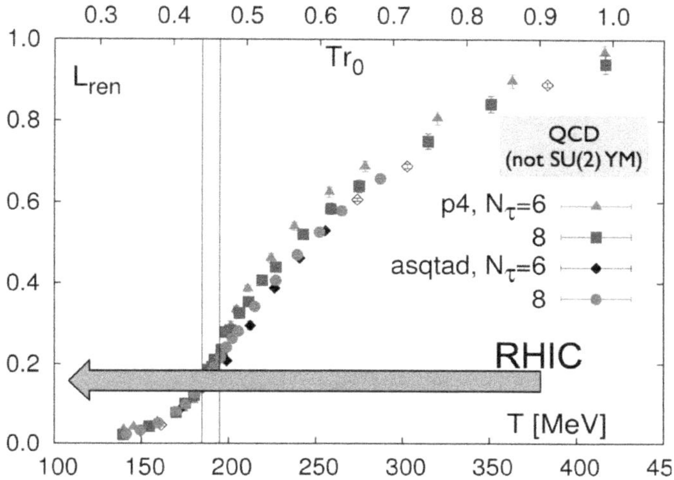

Fig. 27.3. Lattice data on $\langle P\rangle(T)$ in full QCD. Different points correspond to different lattice actions. Two vertical lines indicate the location of the critical point, following studies of the thermodynamical observables. The yellow arrow indicates the range of temperatures studied at the Relativistic Heavy Ion Collider (RHIC).

has to look for appropriate solitons: solutions of classical YM equations at finite temperatures, which, at large distances $\vec{r} \to \infty$, tend to corresponding fixed values:

$$A_4 = (2\pi T)\mathrm{diag}(\mu_1, \mu_2, \ldots, \mu_{N_c}) \tag{27.3}$$

Assuming holonomy phases are ordered in magnitude, $\mu_1 < \mu_2 < \cdots$ and introducing one more phase $\mu_{N_c+1} = 1 + \mu_1$, one can define their subsequent differences as

$$\nu_m = \mu_{m+1} - \mu_m \tag{27.4}$$

If μ_i are points on a unit circle, then ν_i are the *lengths of the segments* between subsequent points.

A solution to the Yang–Mills equation of this type for instantons was found (Kraan and van Baal, 1998), and (only after its actions were plotted), it was recognized that it describes continuous deformation, from one 4D spherical instanton into N_c-independent objects.[5]

[5]As it often happens, it was then realized that, two decades earlier, Gross *et al.* (1981) classified monopoles using A_0 for adjoint scalars. Yet, to my knowledge, nobody used those in any way.

Those objects are monopoles and are therefore still connected by a kind of *umbilical cord*, the "Dirac strings." Although those are invisible for gauge-independent observables, they still play a crucial role by allowing us to avoid topological classification of fields and allowing the instanton–dyons to possess *arbitrary* values of the topological charges Q_i (and thus actions proportional to them). Only the sum of such charges $\sum_i Q_i$ must be an integer.

Let us enumerate some of their properties. Like instantons, they are (anti-)selfdual $\vec{E} = \pm\vec{B}$ and live in Euclidean space-time. So, they are not really particles since they do not exist in the Minkowski world. Similar to instantons, they have nonzero topological charges $Q \sim \int d^4x(\vec{E}\vec{B})$. Unlike instantons, however, those charges are *not quantized to integers*: $Q_i, i = 1, 2, 3$ can take any values, except that their sum is still $|\sum_1^3 Q_i| = 1$. These Q_i (equal to their actions S_i) are proportional to the differences $\nu_i = \mu_{i+1} - \mu_i$ of the eigenvalues of the Polyakov line (the holonomies). So, $S_i = \nu_i S_{\text{inst}}$, where the coefficient is the "instanton action":

$$S_{\text{inst}} = \frac{8\pi^2}{g(T)^2} = \left(\frac{11}{3}N_c - \frac{2}{3}N_f\right)\log\left(\frac{T}{\Lambda_{\text{QCD}}}\right) \tag{27.5}$$

Going ahead of ourselves, let me comment on *where the semiclassical approximation can be used.* Clearly, at large T, the action S_{inst} is large. At $T \sim T_c$ (to be mainly discussed in this chapter), numerical value is still large $S_{\text{inst}}(T_c)/\hbar \sim 13$. Even per dyon, $S \sim S_{\text{inst}}/N_c \sim 4\hbar$, so the exponents $\exp(-S) \sim e^{-4} \ll 1$. The perturbative corrections scale as powers of $1/S$ and can be assumed to be under qualitative control. Although (as for instantons), even two-loop corrections have not yet been calculated.

For the simplest $SU(2)$ gauge group, there are two phases μ_1, μ_2 on a unit circle. Let us assume $P = \big(\exp(i\mu_1) + \exp(i\mu_2)\big)/2$ to be real, so eigenvalues are complex conjugate to each other, or $\mu_2 = -\mu_1$. At very high T, the Polyakov loop VEV gets trivial $P = 1$ and $\mu_2 = \mu_1 = 0$. Thus, $\nu_1 = 1, \nu_2 = 0$. One dyon (called L) has an action equal to that of the instanton, another (called M) has zero action. As T goes down in QGP, the first segment of the circle decreases, while the other increases, up to $T = T_c$ $P \to 0$, corresponding to $\mu_1 = \pi/2, \mu_2$. The same values remain for all $T < T_c$. Here, L and M dyons have equal actions, half of the instanton.

The QCD case, of course, has the gauge group $SU(3)$, so there are three holonomy phases μ_i. Let us, for example, put them three as follows: one phase at zero $\mu_1 = 0$ and two others being $\mu_3 = -\mu_2 = \mu$, so there remains

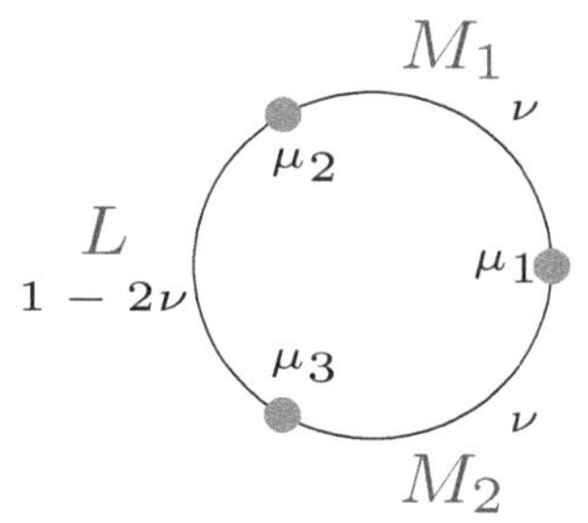

Fig. 27.4. Holonomies for $SU(3)$ gauge groups. L, M_1, M_2 are dyon types: Their actions are corresponding fractions of the instanton action S_{inst}.

only one free parameter ν. For a definition of this setting, see Fig. 27.2. If so,

$$P = \frac{1}{3}\left(1 + e^{i\mu} + e^{-i\mu}\right) \tag{27.6}$$

These three points split a unit circle into three sectors: two equal (corresponding to dyons called M_1, M_2) and another corresponding to a dyon called L. Again, at $T \to \infty$, only L has a nonzero action, while at $T \to T_c$, all three have equal actions, $S_{\text{inst}}/3$. Another set of solutions is *anti-instanton–dyons* $\bar{M}_1, \bar{M}_2, \bar{L}$ with opposite signs of their magnetic charges.

Although explicit expressions for instanton–dyons are quite complex, let me present some of them. The $M, \bar{M}$ dyons are the "ordinary" BPS dyons we already discussed, see in particular (24.15). In the spherical "hedghog," gauge is

$$A_4^a = \mp n_a v \Phi(vr) A_i^a = \epsilon_{aij} n_j \frac{1 - R(vr)}{r}$$

where $n_a = r_a/r$ is the radial unit vector and the "$\mp$" denotes self (antiself) dual solutions, and the two functions are

$$\begin{aligned} \Phi(vr) &= \left[coth(vr) - \frac{1}{vr}\right] \to \mp n_a v \left[1 - \frac{1}{vr}\right] + O(\exp(-vr)) \\ R(vr) &= vr/\sinh(vr) \end{aligned} \tag{27.7}$$

The corresponding magnetic field has the transverse and longitudinal structures, which contain derivatives of these functions but can be rewritten without them as follows:

$$B_i^a = (\delta_{ai} - n_a n_i)[-v\Phi(vr)R(vr)/r] + n_a n_i[R^2(vr) - 1]/r^2 \qquad (27.8)$$

While the former is exponentially decreasing with the distance, the longitudinal one exhibits a $1/r^2$ behavior, indicating a nonzero magnetic charge. It of course matches the electric charge.

Construction of the $L, \bar{L}$ dyons begins with the same expressions, in which the substitution

$$v \to \bar{v} = 2\pi T - v \qquad (27.9)$$

is made. But then, it has the "wrong" asymptotics of the Higgs field A_4; this, however, can be remedied by the so-called "twist gauge transformation." It is done with a *time-dependent* matrix

$$U = \exp\left[i2\pi T x^4(\tau^3/2)\right] \qquad (27.10)$$

so that its time derivative subtracts the unwanted $2\pi T$ phase from the holonomy. Note that the color matrix τ_3 commutes with the diagonal (Abelian) part of the dyon field, leaving it as before. However, the "core" of the L dyon is made of charged fields with colors 1 and 2. Thus, the L-dyon core is actually time-dependent. So, the $L, \bar{L}$ dyons are not static, and thus, these objects can only be defined in the $3+1$ finite-T theory with Matsubara time!

The explicit solution obtained for the single caloron in the Kraan–van Baal paper (Kraan and van Baal, 1998) led to this conclusion after plotting the action distribution. The solution itself can be described using the so-called *prepotential* scalar function[6]:

$$\psi(r) = (1/2)\mathrm{tr}(A_N \cdots A_1) - \cos(2\pi r_4) \qquad (27.11)$$

where here and below the temperature and circumference of the Matsubara circle are temporarily set at $T = 1/\beta = 1$. The matrices need to be multiplied in the order written, right to left, in the order corresponding to the magnitude of μ_i are ordered. The following 2×2 matrices A_m (not to

[6]In order to explain how the gauge potential A_μ was constructed, one needs to understand Nahm construction, which I cannot describe better than done in Kraan-van Baal papers. Without that, the formulae for A_μ and $G_{\mu\nu}$ look like pure magic. I decided not to go into it, keeping only the definition of ψ and the action distribution defined in terms of this function alone.

be confused with the gauge potential A_μ) are defined by the product of two matrices

$$A_m = \begin{pmatrix} 1 & (\vec{y}_m - \vec{y}_{m+1})/r_m \\ 0 & r_{m+1}/r_m \end{pmatrix} \begin{pmatrix} c_m & s_m \\ s_m & c_m \end{pmatrix} \tag{27.12}$$

with $c_m = \cosh(2\pi\nu_m r_m), s_m = \sinh(2\pi\nu_m r_m)$. Here, $r_m = |\vec{r} - \vec{y}_m|$ is the distance from the observation point to the mth dyon center $\vec{y}_m$, and $\vec{y}_m - \vec{y}_{m+1}$ are vector distances between dyons. Note further that the first matrix is "pure geometry" independent of holonomies, and the second is some hyperbolic rotation matrix by the angle including the mth holonomy and the distance to the mth dyon.

The action density can be expressed in terms of a prepotential in a surprisingly simple way:

$$\mathrm{Tr}G_{\mu\nu}^2 = \partial^2\partial^2 \log(\psi) \tag{27.13}$$

with two four-dimensional Laplacians. Note that there are four derivatives because the matrices define the "prepotential" in terms of which field the

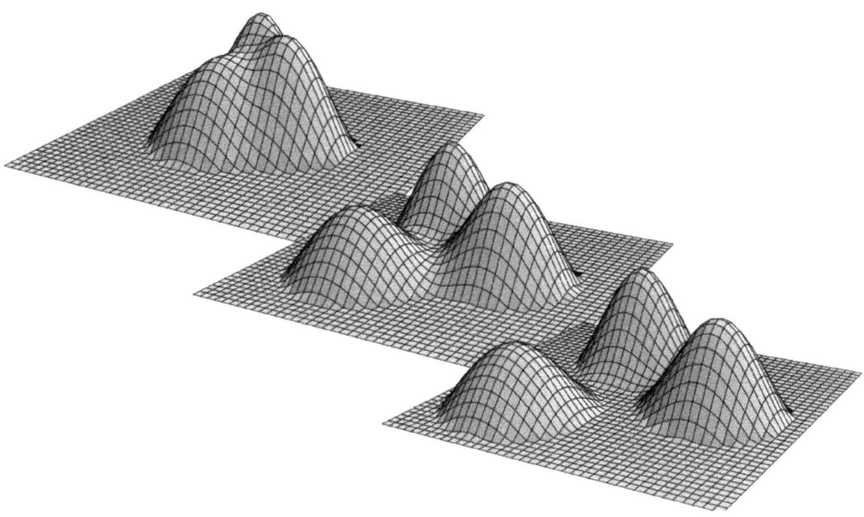

Fig. 27.5. Action densities for the SU(3) caloron at $x0 = 0$ in the plane defined by the centers of the three constituents for $1/T = 1.5$, 3 and 4 (increasing temperature from top to bottom). We choose mass parameters $(\nu_1, \nu_2, \nu_3) = (0.4,\ 0.35,\ 0.25)$, implemented by $(\mu_1, \mu_2, \mu_3) = (-17/60, -2/60, 19/60)$. The constituents are located at $\vec{y}_1 = (-1, 1, 0)$, $\vec{y}_2 = (0, 1, 0)$ and $\vec{y}_3 = (1, -1, 0)$, in units of T. The profiles are given on equal logarithmic scales, cut-off at an action density below $1/e$.

potentials already have a derivative. Note also that the matrices A_m do not include the time $\tau = r_4$, which is solely located in the cosine function in the last term of ψ (thus, the periodicity in the Matsubara box is explicit). The construction itself is based on ADHM multi-instanton construction and the so-called Nahm version of it, generalizing it to the monopoles. It is too technical to be presented here, see the original paper by Kraan and van Baal (1998).

Before we can proceed, we need to clarify one more puzzle related to fermionic zero modes of instanton–dyons. An instanton with the topological charge 1 has *one* fermionic zero mode. If it gets split into three separate instanton–dyons, one may naturally ask how this zero mode can be shared between them. The answer (also due to van Baal and collaborators) is that the fermionic zero mode is centered near *one of the three*: which one depends on the interrelation between holonomy phases μ_i and quark periodicity phases called z_f, where the index f denotes the flavor, $u, d, s, \ldots$. It is not possible to cover further details in this short review.

27.3 Instanton–Dyons on the Lattice

The information given in the previous section should be sufficient to understand how one can "hunt" for these objects on the lattice. One method is the "cooling" of vacuum fields. Better still is "constrained minimization" (Langfeld and Ilgenfritz, 2011), which preserves the value of $\langle P \rangle$: This revealed selfdual clusters of topological charges. (Unlike instantons, those integrate to their *non-integer* values.)

However, the best is the "fermionic filter," developed by Gattringer, Ilgenfritz and collaborators. It is based on hunting for zero (or quasi-zero) modes of the quark Dirac operator. In Fig. 27.6, we show an example from Larsen *et al.* (2019, 2020) in which it was used. QCD simulations with realistic masses, performed by large collaborations on supercomputers, provided a set of configurations to these authors. These calculations are especially expensive since they use the so-called *domain wall fermions* that provide a very accurate chiral symmetry of lattice fermions. Yet, Larsen *et al.* used even better ones, called the *overlap fermions*, for which chiral symmetry is exact even at finite lattices (without continuum limit $a \to 0$ taken). Those possess *exact* zero modes $\lambda = 0$ and configurations have exactly integer topological charges. What is shown in the top part of Fig. 27.6 is a typical landscape of the zero-mode densities $|\psi_0(x)|^2$

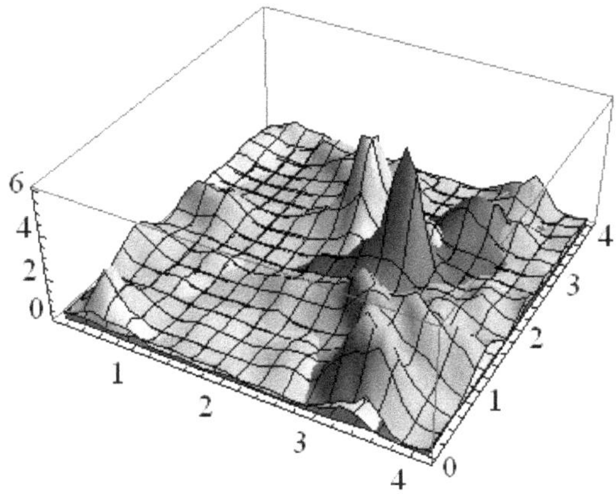

Fig. 27.6. Space slice of density of an exact zero mode from QCD simulations at $T = T_c$. The three colors refer to dyons of three different types.

in two spatial dimensions. Red, blue and green colors show those for three different fermionic periodicity phases, identifying three instanton–dyon types (for $N_c = 3$) they were able to locate. The peaks of the density seen correspond to the locations and sizes of the individual zero modes in these field configurations. One can get convinced that those are instanton–dyons because their shapes are well described by analytic formulae, as derived by van Baal and collaborators, with a few percent accuracy. Furthermore, it is true not only for well-separated ones but also for overlapping ones! The gauge field configurations are for T a little bit above deconfinement T_c, in quark-gluon plasma possessing zillions of thermal quarks and gluons; yet, the instanton–dyons are apparently undisturbed by them! (For clarity, we do not mean here the *values of the topological charge Q or the number of zero modes*, protected by certain mathematical theorems. The observed space-time shapes of the Dirac eigenmodes are not protected by any theorems known to us.)

This (and many similar plots) extracted from simulations of the QCD vacuum should convince the reader that instanton–dyons are well-identified objects in terms of which one can attempt to describe the underlying gauge field configurations. If so, perhaps a dream that has been alive for half a century since the 1970s — to develop a consistent semiclassical theory of deconfinement and chiral transitions — can still be realized.

27.4 Instanton–Dyon Ensembles

Let us begin by reminding ourselves of the main task we want these instanton–dyons to accomplish. If they are absent, there is a pQCD-based effective action derived by Gross *et al.* (1981) and known as GPY potential. For the $SU(3)$ case at hand, it has the form

$$V_{\mathrm{GPY}} = \frac{4\pi^2 T V_3}{3}\left[2\nu^2(1-\nu)^2 + 4\nu^2(1-2\nu)^2\right] \tag{27.14}$$

$V_{\mathrm{GPY}}(\mu_i)$, with a minimum at $\nu = 0$. It is because of this potential that, at high T, there is no confinement.[7] As T decreases and dyon density grows, they eventually overwhelm the GPY potential and put the minimum at the confining value of $\nu = 1/3$. Studies of dyon ensembles do show that it indeed takes place.

We will not discuss in any detail the partition function for dyons and their interactions: please look at the original papers. We will also not go into the description of the ensemble following the *mean field methods* without (Liu *et al.*, 2015a) and with light quarks (Liu *et al.*, 2015b, 2016). In it, the minimum of the free energy leads to certain "gap equations" for parameters, which are solved by various methods.

Another approach is straightforward statistical Monte Carlo simulations of the corresponding partition function. Those were performed first for the SU(2) gauge theory (Larsen and Shuryak, 2015) and then for the SU(3) (DeMartini and Shuryak, 2021b), first without dynamical fermions and then with them (Larsen and Shuryak, 2016; DeMartini and Shuryak, 2021a). We cannot present here the results of these works but just state that they compare well with the locations and properties of QCD phase transitions, which we now know from lattice simulations. Note that those works were done on laptops or ordinary PCs, not supercomputers, and yet the number of topological objects in them is counted by hundreds, while very expensive lattice simulations have only a few of them (as one can see from the example above).

Numerical simulations provide the free energy of the dyons. When the GPY potential is added, one finds a dependence of the total free energy on

[7]It was shown (Unsal, 2008) that two adjoint fermions with a bosonic periodicity phase can flip the GPY potential upside down, and thus create "perturbative confinement" in the weak coupling regime, all the way to $T \to \infty$. He and collaborators conjectured that there could be a continuity between this "perturbative confinement" and that at $T = 0$. There are, however, lattice evidence that it is not correct and that there are phase transitions separating these two limiting cases.

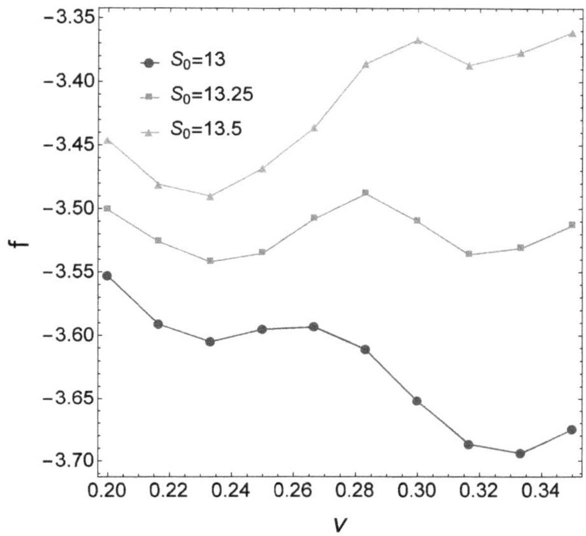

Fig. 27.7. The free energy of the dyon ensemble near the critical temperature versus holonomy ν. Three values of the parameter $S_0 = S_{\text{inst}}$ are given on the plot.

parameters, such as holonomy. In Fig. 27.7 adapted from DeMartini and Shuryak (2021b), one sees that in pure gauge $SU(3)$ theory, it is a typical first-order transition, in which the system jumps from one minimum to another. The corresponding value of the Polyakov line jumps, from 0 to about 0.4: we have already shown it in Fig. 27.2.

Adding light quarks is done via a "hopping" matrix, which is the method developed for instantons in the 1980s. The partition function in such a case is proportional to the determinant for each light flavor. The key observable is the quark condensate: If quarks are massless, then chiral symmetry is exact, and for $T > T_c$, its order parameter should vanish, $\langle \bar{q}q(T > T_c) \rangle = 0$. Another order parameter is the *gap* in the Dirac eigenvalue spectrum,[8] which is nonzero at $T > T_c$: Both are used to define the location of the chiral phase transition.

Here, we follow the simulations for the $SU(3)$ theory with two massless flavors (DeMartini and Shuryak, 2021a), from which we show Fig. 27.8.

[8]A well-known technical complication is the fact that in any finite volume, there is a nonzero gap $\Delta \sim 1/V_4$. Thus, careful study of several volumes is mandatory to get the infinite volume extrapolation. In the dyon simulations, the volume is orders of magnitude larger than on the lattice, which helps a lot.

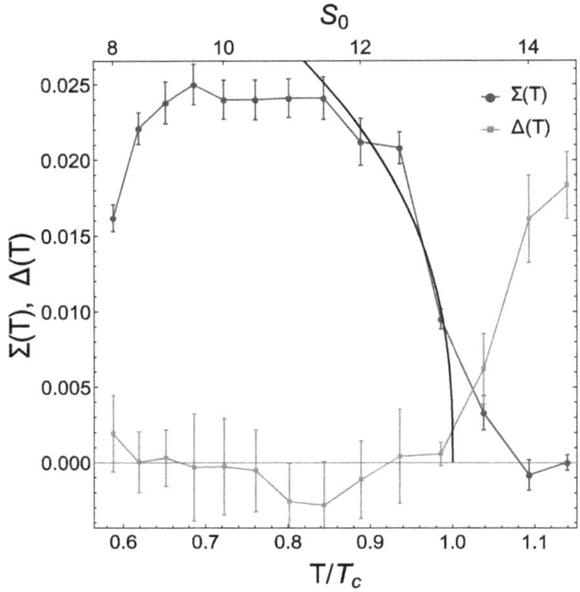

Fig. 27.8. The chiral quark condensate $\Sigma(T)$ and the spectral gap $\Delta(T)$ versus the temperature. The solid curve is the form of the condensate near the critical point, as expected for the O(4) universality class. The shown temperature range is roughly 80–150 MeV.

As one can see, the critical value of the action $S_0 \approx 13$. However, the plot for the Polyakov loop $P(T)$ (not shown) is, in this case, smooth and displays no (statistically significant) jumps: This again corresponds well to lattice data for two dynamical massless quarks.

The old question of whether the chiral and deconfinement transitions coincide or not can be further clarified with certain types of "QCD deformation" applied both on the lattice and in instanton–dyon ensembles. One of them adds extra operators with the powers of the Polyakov line to the gauge action. By changing their strengths, one can affect the location and strength of the deconfinement phase transition. Another type of QCD deformation makes quarks obey the modified periodicity condition on the Matsubara circle, making quark statistics to be *anyons*, interpolating between fermions and bosons. This deformation was shown to affect mostly the location and strength of the chiral phase transition. Studies of these deformations tell us that these two phase transitions in general should not be coincident, as they are in QCD. Again, one can apply such a deformation on the lattice or in the instanton–dyon semiclassical models: So far, the results match, which is encouraging.

Concluding this section, let me make the following comment. The semiclassical theory gives an approximate and limited description of the gauge fields, while the lattice theory is based on the first principles and includes everything. On the other hand, the former uses simulations with several hundreds of instanton–dyons, while in lattice configurations, there are only a few of them. It has a big effect on large-volume extrapolation and critical properties. At the same time, the former are done on laptops or ordinary PCs, while the latter are done on large supercomputers.

27.5 "Poisson Duality" between Theory of Monopoles and Instanton–Dyons

We had given two descriptions of confinement: one in terms of monopole BEC, and the other in terms of the instanton–dyon ensemble driving the Polyakov line. While being quite different, they must somehow be related. This relation is the duality to be discussed in this section.

Indeed, these objects themselves are very different: the monopoles are *particles*, their magnetic charges are not tied to the Euclidean finite-T setting, unlike those of the instanton–dyons.[9] Yet, thermal ensembles of them are in fact related!

A gradual understanding of this statement began some time ago, but remained rather unnoticed by the larger community. One reason for that was the setting in which it was shown, which was based on extended supersymmetry. Only in these cases was one able to reliably derive *both* partition functions — in terms of monopoles and instanton–dyons — and show them to be equal (Dorey and Parnachev, 2001; Poppitz and Unsal, 2011; Poppitz *et al.*, 2012). Furthermore, they were not summed up to an analytic answer but shown instead to be related by the so-called "Poisson duality."

Instead of reproducing here all those formulae, let me follow Ramamurti *et al.* (2018) in showing that such a duality exists in much wider settings and is related neither to supersymmetry nor to even gauge theory. The setting is a *rotator*, a charged particle on the circle, introduced by Aharonov and Bohm (1959).

[9]So to speak, one can put a monopole into a pocket, but it cannot be done with an instanton–dyon!

Its location is defined by the angle $\alpha \in [0, 2\pi]$, and its (initial) action contains only the kinetic term

$$S = \oint dt \, \frac{\Lambda}{2} \, \dot{\alpha}^2 \tag{27.15}$$

with $\Lambda = mR^2$, the corresponding moment of inertia for rotation. All possible paths are naturally split into topological homotopy classes, defined by their *winding numbers*. The paths belonging to different classes cannot be continuously deformed to each other. Therefore, a fundamental question arises: *How should one normalize those disjoint path integrals over classes of paths?* Clearly, there is no natural way to define their relative normalization, or rather, their relative *phase*. A certain device (existing in extra dimensions invisible to the rotator) creates a nonzero magnetic field flux $\Phi \neq 0$ through the circle. Stokes theorem relates it to the circulation of the gauge field

$$\oint d\alpha A_\alpha = \int \vec{B} d\vec{S}$$

While $A_\mu(x)$ is gauge-dependent, its circulation (called holonomy) is gauge-invariant since it is related to the field flux.[10]

The extra phase is thus physical. Furthermore, it propagates into the energy spectra and the partition function. One can write it in a Hamiltonian way, as the sum over states with the angular momentum m at temperature T

$$Z_1 = \sum_{m=-\infty}^{\infty} \exp\left(-\frac{m^2}{2\Lambda T} + im\omega \right) \tag{27.16}$$

where ω is the holonomy phase, which is so far arbitrary.

Although physical, the effect is invisible at the classical level. This can be seen from the inclusion of the additional term in the action $\sim (\omega/2\pi) \int d\tau \dot{\alpha}$, which would "explain" the holonomy phase. This term in the Lagrangian is, however, a full derivative, $\dot{\alpha}$, so the action depends on the endpoints of the paths only and is insensitive to its smooth deformations. It therefore generates no contribution to the classical equations of motion, thus failing to "exert any force" on the particle in the classical sense. In summary, the

---

[10]For non-Abelian cases, there is no Stokes theorem, but the gauge invariance of all closed paths is still true: It follows from the direct calculation of the gauge transformation of path-ordered exponents.

appearance of the holonomy phase is our first nontrivial quantum effect not coming from the classical action.

Now, one can also use the Lagrangian approach, looking for paths periodic in Euclidean time on the Matsubara circle. The classes of paths which make a different number n of rotations around the original circle can be defined as "straight" classical periodic paths,

$$\alpha_n(\tau) = 2\pi n \frac{\tau}{\beta} \tag{27.17}$$

plus small fluctuations around them. Carrying out a Gaussian integral over them leads to the following partition function:

$$Z_2 = \sum_{n=-\infty}^{\infty} \sqrt{2\pi\Lambda T} \exp\left(-\frac{T\Lambda}{2}(2\pi n - \omega)^2\right). \tag{27.18}$$

The key point here is that these quantum numbers, m used for Z_1 and n for Z_2, are very different in nature. The dependence on the temperature is different. Also, for Z_1, each term of the sum is periodic in ω, while for Z_2, this property is also true but recovered only after summation over n.

Despite such differences, both expressions are in fact the same! In this toy model, it is possible to do the sums numerically and plot the results. Furthermore, one can also derive the analytic expressions, expressible in terms of the elliptic theta function of the third kind:

$$Z_1 = Z_2 = \theta_3\left(-\frac{\omega}{2}, \exp\left(-\frac{1}{2\Lambda T}\right)\right), \tag{27.19}$$

which is plotted in Fig. 27.9 for a few values of the temperature T.

(In order to prove that, one may use a quite nontrivial Jacobi identity:

$$\theta_3(z,t) = (-it)^{-1/2} e^{z^2/i\pi t} \theta_3(z/t, -1/t))$$

The main point (Ramamurti *et al.*, 2018) is that two statistical sums are *equal*, related by the Poisson summation formula

$$\sum_{n=-\infty}^{\infty} f(\omega + nP) = \sum_{l=-\infty}^{\infty} \frac{1}{P}\tilde{f}\left(\frac{l}{P}\right) e^{i2\pi l\omega/P} \tag{27.20}$$

where $f(x)$ is some function, $\tilde{f}$ is its Fourier transform, and P is the period of both sums as a function of the "phase" ω. In this particular example, the function is Gaussian, with the Fourier transform being a periodic Gaussian; however, we will later encounter examples of the Poisson duality with other functions as well.

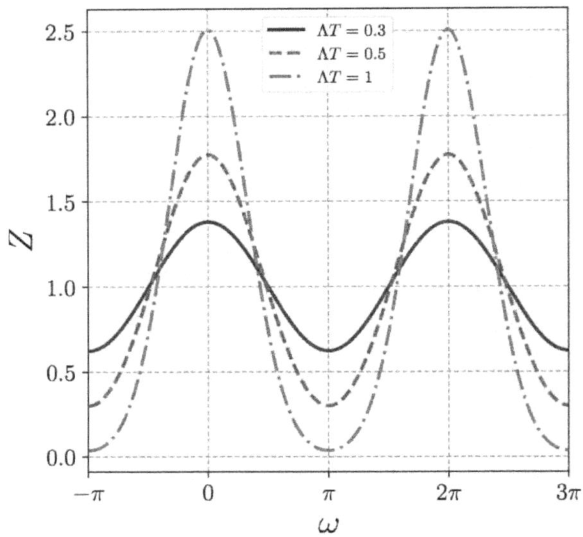

Fig. 27.9. The partition function Z of the rotator as a function of the external Aharonov–Bohm phase ω (two periods are shown to emphasize its periodicity). The (blue) solid, (red) dashed and (green) dash-dotted curves are for $\Lambda T = 0.3, 0.5, 1.0$, respectively.

Ramamurti *et al.* (2018) went further, performing the Poisson duality transformation over the semiclassical sum over twisted instanton–dyons. The resulting expression for the partition function is

$$Z_{\text{inst}} = \sum_n e^{-\left(\frac{4\pi}{g_0^2}\right)|2\pi n - \omega|} \tag{27.21}$$

It is periodic in the holonomy, as it should be. Note that, unlike in Eq. (27.21), it has a modulus rather than a square of the corresponding expression in the exponent. This is due to the fact that the sizes of L_n and their masses are all defined by the same combination $|2\pi n - \omega|T$; therefore, the moment of inertia $\Lambda \sim 1/|2\pi n\beta - v|$.

Using the Poisson relation (27.20), the Fourier transform of the corresponding function appears in the sum in Eq. (27.21), and it reads

$$\begin{aligned} F\left(e^{-A|x|}\right) &\equiv \int_{-\infty}^{\infty} dx\, e^{i2\pi\nu x - A|x|} \\ &= \frac{2A}{A^2 + (2\pi\nu)^2} \end{aligned} \tag{27.22}$$

Therefore, the monopole partition function is

$$Z_{\text{mono}} \sim \sum_{q=-\infty}^{\infty} e^{iq\omega - S(q)} \tag{27.23}$$

where

$$\begin{aligned} S(q) &= \log\left(\left(\frac{4\pi}{g_0^2}\right)^2 + q^2\right) \\ &\approx 2\log\left(\frac{4\pi}{g_0^2}\right) + q^2\left(\frac{g_0^2}{4\pi}\right)^2 + \cdots \end{aligned} \tag{27.24}$$

where the last equality is for $q \ll 4\pi/g_0^2$.

The resulting partition function can be interpreted as being generated by *moving and rotating monopoles.* The results are a bit surprising. First, the action of a monopole, although still formally large in weak coupling, is only a logarithm of the semiclassical parameter; these monopoles are therefore quite light. The second is the issue of monopole rotation. The very presence of an object that admits rotational states implies that the monopole core is not spherically symmetric. The Poisson-rewritten partition function has demonstrated that the rotating monopoles are *not* rigid rotators because their action, Eq. (27.24), depends on the angular momentum q and is quadratic only for small values of q. The slow (logarithmic) increase in the action with q implies that the dyons are in fact shrinking with increased rotation. In the moment of inertia, this shrinkage is more important than the growth in the mass, as the size appears quadratically. As strange as it sounds, it reflects the corresponding behavior of the instanton–dyons L_n with increasing n.

Such rotations are well recognized as Julia–Zee dyons.[11] They were studied mostly in theories with extended supersymmetries. To our knowledge, the existence of multiple rotational states of monopoles has not yet been explored in monopole-based phenomenology. In particular, one may wonder how the existence of multiple rotational states affects their Bose condensation at $T < T_c$, the basic mechanism behind the deconfinement transition. The electric charges of the rotating monopoles should, therefore, also contribute to the jet quenching parameter $\hat{q}$ and the viscosity, which were not yet included in literature.

[11]With *real* electric charge, unlike that of the instanton–dyons, which only exist in the Euclidean world.

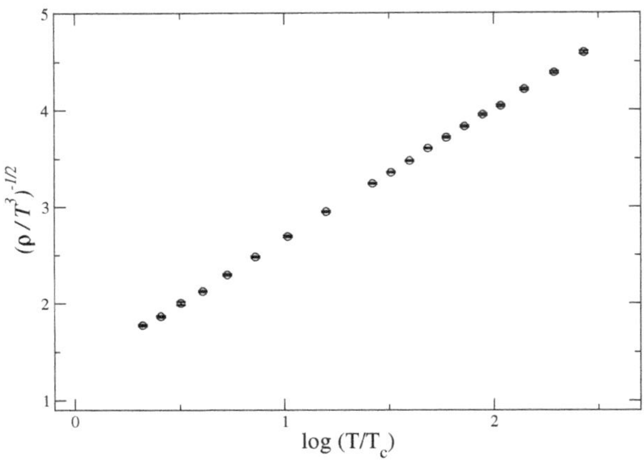

Fig. 27.10. The normalized monopole density in SU(2) gauge theory in the power of $-1/2$, $(\rho/T^3)^{-1/2}$ versus $\log(T/T_c)$ shows an apparent linear dependence.

Summary and connections: The Euclidean formulation of gauge theories at finite temperatures possesses selfdual semiclassical solitons, instanton–dyons or instanton–monopoles. We have shown that those can be well identified on the lattice via their fermionic zero modes. It has been found that, multiple thermal gluons notwithstanding, their shape is well described by the Dirac zero modes of the unperturbed dyons. The statistical mechanics of the ensemble does reproduce lattice data on deconfinement and chiral transitions in QCD and its deformations. Particle–monopoles and instanton–dyons provide "Poisson-dual" descriptions with the same statistical sum. This idea describes, in particular, why magnetic monopoles in QCD-like theories without scalars do not have classical solutions.

Chapter 28

FROM HADRONIC SPECTROSCOPY ON THE LIGHT FRONT TO PARTONIC OBSERVABLES

The physics of hadrons consists of two rather disconnected parts. Hadronic spectroscopy describes mesons and baryons as bound states of two or three massive *constituent* quarks.[1] The other part describes hadrons in terms of *partons*, fundamental (near-massless) quarks and gluons. Their number is in principle indefinite, and the description aims at hard inclusive processes, such as deep inelastic (lepton–hadron) scattering (DIS), or hard exclusive reactions, such as elastic scattering with large momentum transfer.

Spectroscopy operates in terms of certain Hamiltonians, providing masses and wave functions of hadronic states as their eigenstates. The partonic models operate in terms of nonlocal matrix elements of certain operators, taken over the QCD vacuum state along some light-front (LF) lines.

In this chapter, we review efforts to "bridge the gap" between them. This task is not easy for many reasons. It can hardly be done by just "boosting" the wave functions of the hadrons from their center-of-mass (CM) frame to the LF since the boost operator is not simpler than the Hamiltonian. Following a program developed in a sequence of papers (Shuryak and Zahed, 2023b, 2023c, 2023d, 2023e, 2023f), it can be done in three steps:

(i) putting hadronic spectroscopy on the LF;
(ii) adding nonminimal sectors (such as 4-q components of the mesons and 5-q components of the baryons) via "chiral evolution" toward the $Q^2 \sim 1\,\mathrm{GeV}^2$ scale;

[1] Their effective mass is due to chiral symmetry breaking in the QCD vacuum, see Chapter 25.

(iii) further evolving observables *upward* to higher scales by (modified) pQCD evolution.

As the reader will see, at the time of this writing, only the first step can be considered nearly fully accomplished, while the two others are only treated in Shuryak and Zahed (2023d) at the level of estimates. Indeed, proper variables and quantum-mechanical methods to treat appropriate few-body LF Hamiltonians have been developed in detail for mesons and baryons. Development is quite systematic in the sense that it starts with the simpler cases (e.g. quarkonia and flavor symmetric baryons) and proceeds toward light quark hadrons and nucleons, with diquark correlations. Not only the ground states are obtained but a certain number of excited states as well: Those are, for example, properly orthogonal to each other, as they should be, and their masses form appropriate Regge trajectories.

The channels are studied in the same order as they are in spectroscopy, namely *from the simplest to the most complex*. In particular, studies of mesons start with heavy quarkonia, first without and then with spin forces, then lighter and lighter ones, ending with *pions* — a very special case in which instanton-induced forces dominate. Similarly, baryons start with flavor-symmetric ones, bbb, ccc, sss, uuu, then those with a single light ud diquark, and only then the nucleon, in which diquark pairing occurs in two channels at once.[2]

Another "bridge" — between partonic observable and correlation functions in Euclidean space-time — has received a lot of attention lately, after a proposal (Ji and Jung, 2001) to define *quasi-parton distribution functions* (quasi-PDFs), in which operators on the LF are approximated by operators with large momenta. It is crucial that this approach starts in Euclidean space-time, in which a nonperturbative QCD vacuum is present with all its properties (confinement and chiral symmetry breaking).

28.1 Kinematics and Dynamics on the Light Front

In the CM frame, hadrons made of (*heavy*) quarks are treated in the same familiar nonrelativistic way as atoms or nuclei. One starts with a Hamiltonian, calculating its spectrum and the corresponding wave functions. Those are "pure" states with zero entropy, not density matrices like PDFs and other partonic observables. The number of variables used

[2]Note that in the partonic community, the emphasis is placed in the opposite direction, on the main hadrons available for experiments, the nucleons and the pions.

to describe hadrons made of n quarks is $N_{\text{var}} = 3(n-1)$. For example, a baryon ($n = 3$) needs six variables, all quark momenta minus that of the CM motion properly eliminated. Moving to another frame, such as LF, does not change this: The only difference is that in CM $\sum \vec{p}^{\,i} = 0$, while on the LF

$$\sum_i \vec{p}_\perp^{\,i} = 0, \quad \sum_i p_l^i = \left(\sum_i x_i\right) P = P \tag{28.1}$$

The LF is reached when the total momentum $P \to \infty$. Assuming quarks are (on-shell) particles with mass m_q, the expression for their energy is

$$\sqrt{(\vec{p}_\perp)^2 + m_q^2 + x_i^2 P^2} \to x^i P + \frac{(\vec{p}_\perp)^2 + m_q^2}{2x^i P} \tag{28.2}$$

Here comes a crucial point: In CM, in order to use the nonrelativistic Schrödinger equation, one needs masses to be larger than momenta $\vec{p}^{\,2} \ll m_q^2$. On the LF, transverse momenta and masses squared appear both as a sum. It *does not matter* whether one of them is larger than the other, so the same setting may be used for heavy and light quarks. In practice, it is a very significant advantage, allowing us to treat, say, mesons from heavy quarkonia to pions.

The LF Hamiltonian comes from the product of generators P^+ and P^-, so its eigenvalues are *squared* hadronic masses, M_n^2. As we already discussed in the chapter on hadronic scattering, we know that hadronic squared masses are (approximately) described by *linear* Regge trajectories: $M_{n,l}^2 \sim (C_1 n + C_2 l)$. This feature alone hints that H_{LF} should be close to a *harmonic oscillator*, a famed Hamiltonian giving a linear spectrum.

As we will elaborate in the following, it can indeed be arranged for transverse motion with a relatively simple trick. The complications come from the longitudinal fraction x_i in the denominator of (28.2). The usual usage of Schrödinger equation in *coordinate representation* is quite difficult: Substituting momenta by derivatives $\vec{p} \to i\partial/\partial\vec{r}$, one needs to define those, for example, for inverse momenta $1/x_i$, not a well-defined operator. We use instead the *momentum representation*, treating this kinetic energy (28.2) as a certain multidimensional *potential*. Note that, at the kinematical boundaries, where $x_i \to 0$, this potential becomes large, cutting the wave functions off.

The interquark interactions due to linear confining potentials can, through certain tricks, be approximated by forms *quadratic* in coordinates. As coordinates morph into *derivatives over momenta* $\vec{r} \to i\partial/\partial\vec{p}$, one finds

Laplacian-like operators, leading to conventional Hamiltonians, yet with the roles of momenta and coordinates interchanged. Its transverse part is indeed close to the desired 2d oscillator, which provides a natural functional basis for use. In particular, for a ground state, this immediately suggests wave function dependence on $\vec{p}_\perp$ to be Gaussian, in good correspondence with the empirical data on hadrons.

The longitudinal derivatives are diagonalized by standing waves on appropriate simple manifolds.[3] In the case of a meson, with a single x variable ($x_2 = 1 - x$), they are standing waves on an interval $x \in [0, 1]$. In the case of a baryon, $n = 3$, they are standing waves on an equilateral triangle defined by $x_1 + x_2 + x_3 = 1$ (see the following), for $n = 4$ on a tetrahedron, for $n = 5$ on five-link, etc.

In summary, while $H_{\rm LF}$ takes time to get used to, its application is logically not so different from, for instance, the derivation of few-body wave functions for light nuclei. One can take various approximations, use the Hamiltonian as a matrix in a specified functional subspace, or use variational wave functions, etc. For example, we consider addition to a baryon as $n = 3$ quark state, another $n = 5$ sector by adding a quark–antiquark pair. Similar logic is used in atomic and nuclear physics when "simple" states of some mean field basis are complemented by particle–hole pairs.

What is important is that when the LFWF $\psi_{n,l,\ldots}(p^i_\perp, x^i)$ is defined (inside certain boundaries), it gives a complete description of the states.

The very existence of LFWF leads to important relations between partonic observables. The first of those, from 1969,[4] is the Drell–Yan relation (Drell and Yan, 1970) between the hard limits of the form factor $Q^2 \to \infty$ and the PDF at $x \to 1$. It states that those are related as follows:

$$PDF(x) \sim (1-x)^p \leftrightarrow F_1(Q^2) \sim \left(\frac{1}{Q^2}\right)^{(p+1)/2} \tag{28.3}$$

[3]Here, we deviate from the path taken by Vary and collaborators, who use a different strategy for deriving light front Hamiltonians and wave functions (LFWFs). First, they do not eliminate the CM motion, keeping $3n$ variables instead of $3(n-1)$. Second, they quantize longitudinal *coordinates* in a box, proceeding to a basis state with rational values of (rescaled) momenta. These states are not mutually orthogonal, unlike the ones we will use.

[4]It is not a random year: This was when the SLAC DIS experiment confirmed Bjorken scaling and inspired Feynman's parton model.

Here, p was an unspecified power; however, if the power on the r.h.s. is $1/Q^4$, then[5] $p = 3$, as indeed was approximately derived from pioneering SLAC data more than 50 years ago.[6]

At very high Q, one can factorize form factors into a convolution of its *hard block*, and we will discuss the so-called distribution amplitudes (DAs) in Section 28.6. There are two of them: for incoming and outgoing hadrons. Basically, those are the LFWS compressed into *zero size*[7] $\vec{r}_\perp \to 0$ of a hadron.

PDFs, describing DIS, to be discussed in Section 28.5, are *probabilities* of finding a "struck" quark with its longitudinal fraction $x^{i=i\text{struck}}$. They are *squared* LWFS integrated over all variables except this one.

Of course, by devising more complex experiments, one can define other observables. For example, one can study *deeply virual Compton scattering* (DVCS) and derive generalized parton distributions (GPDs), which provide information on correlations of $x^{i\text{struck}}$ with its transverse momentum (and thus transverse location).

Let me emphasize here the key distinction between LFWFs and all these other observables. The former provide a full description of a quantum state and, by integration, can provide all other ones. This integration cannot be turned around. Given experimental PDFs, one can, of course, get certain restrictions on LFWFs but it cannot be determined by them. We already mentioned that PDFs are density matrices and have a *nonzero entropy*. This entropy does not characterize the nucleon but the DIS experiments. It is an *entanglement* entropy (Kharzeev and Levin, 2017), describing the interaction of one part of the system, e.g. the stuck quark, with the rest of the system. The entropy we gain is the *information lost*. It can only be recovered from observations *more complex than DIS*.

Now, suppose LFWS are all found, for example, we would have (accurate in some practical sense) the one for the proton as made of three quarks, or a function of six variables. In order to confront its prediction with actual data, one would still need to do what is known as "evolution."

---

[5]The power p need not be an integer, and in fact, its value changes with DGLAP evolution (depending on the normalization scale).

[6]In the 1970s, QCD was discovered, and the leading power was indeed predicted to be $1/Q^4$, but as we will discuss in the section on form factors, we are still not in the "hard domain" in which their pQCD description is valid.

[7]Represented in the momentum representation we use by just an integral over transverse moments.

The reason for that is as follows: The notion of "quanta" depends on kinematics and resolution. Looking, say, at a hydrogen atom in the CM frame, we may think of it as described by a certain wave function of the electron, with three variables. A static Coulomb field of the nucleus has no variables. Yet, on the LF, some harmonics of this field become resolved as "photons." Including QED diagrams of higher order, one finds that they can even interact with "vacuum." Small QED coupling notwithstanding, the calculation of radiative corrections in hydrogen is a nontrivial task.

The situation is similar in QCD. At least in their perturbative version, radiative corrections can be reduced to emission and absorption of soft gluons. When a nucleon is put under a microscope (hard process) with higher and higher resolutions, one discovers that it consists of a larger and larger number of gluons. As Q^2 grows, such $n_{\text{partons}} \to \infty$. We do not even attempt to describe partonic processes by wave functions and use PDFs instead.

So, how are the two pictures of the nucleon connected? Well, going downward in resolution to a certain *low scale*, one could undo the radiative correction and at some point reach the description *without* gluons as partons, in which hadrons are describable by the minimal number of quarks needed to explain their quantum numbers.

28.2 Light Front Hamiltonians and Wave Functions

Quantization on the light front was suggested by Dirac long ago, and since then, there have been multiple attempts at its practical application for certain field theories. Not being an expert in that, let me jump over this history to the pioneering paper by Jia and Vary (2019), in which mesons were treated on the LF by a Hamiltonian containing three basic elements:

(i) kinetic energy containing "constituent quark masses" m_q
(ii) confining term, with a coupling κ
(iii) NJL four-quark "residual interaction," with a coupling G_{NJL}.

Since that time, Vary and collaborators have worked out many other applications of their approach. We will not, however, discuss those, following instead our papers (Shuryak and Zahed, 2023a, 2023b, 2023c, 2023d, 2023f).

There are too many differences between the approaches by Jia and Vary (2019) and ours, some technical, some not. To mention one, transverse coordinates (and thus momenta) are combined with longitudinal momentum fractions into some mysterious combinations, which also appear in the (confining part of the) Hamiltonian:

$$H_{\rm conf}^{{\rm Vary}\,et al.} = \kappa^4 x(1-x)\vec{r}_\perp^{\,2} - \frac{\kappa^4}{2m_q}\partial_x x(1-x)\partial_x \tag{28.4}$$

which is just postulated, not derived. The Hamiltonians are from certain field theory settings, not just guessed in some form. Technically, we do eliminate CM motion, using (modified) Jacobi coordinates. We use basis sets defined by the eigenstates of some Laplacians on manifolds instead of non-orthogonal basis sets. Last but not least, we start with the most basic problems and gradually add new elements, rather than jump directly on fits to experimental partonic data for the nucleons.

We first derived our $H_{\rm LF}$ in Shuryak and Zahed (2023b) using the relativistic string model, of which we only mention the (relatively obvious) result. In 1+1-dimensional QCD, the LF Hamiltonian takes the form

$$H \equiv P^- = \frac{m_Q^2}{2p_q^+} + \frac{m_Q^2}{2p_{\bar{q}}^+} + \sigma_T |r_{\bar{q}}^- - r_q^-| \tag{28.5}$$

with a linear confining term.[8] In 1+3 dimensions, the string-induced Hamiltonian takes the form

$$M^2 = 2P^+P^- = \frac{m_Q^2 + k_\perp^2}{x\bar{x}} + 2\sigma_T\left(\left|i\frac{\partial}{\partial x}\right|^2 + M^2\vec{r}_\perp^{\,2}\right)^{\frac{1}{2}} \tag{28.6}$$

where, in the confining term, the derivatives over the longitudinal momentum fraction are actually the longitudinal coordinates, added in quadrature to the transverse coordinates. Of course, one can also substitute those

$$\vec{r}_\perp \to i\frac{\partial}{\partial \vec{p}_\perp}$$

if usage of the momentum representation for $\psi_{\rm LFWF}(x, \vec{p}_\perp)$ is intended.

Here, there are two difficulties: (i) the differential operator under a square root; (ii) the fact that the total mass M^2 here appears both as the answer on the l.h.s. and also on the r.h.s. These difficulties are solved by the so-called *einbein trick*,[9] which represents the Hamiltonian as

$$M^2(a,b) = \frac{m_Q^2 + p_\perp^2}{x\bar{x}} + \sigma_T\left(\frac{|i\partial/\partial x|^2 + bx_\perp^2}{a} + a\right) \tag{28.7}$$

with the auxillary parameters a and b. Note that minimization over a would return us to the original Hamiltonian with the square root; however, one

[8]This problem with perturbative linear confinement was actually solved by 't Hooft, which we will not discuss here.

[9]I do not know its exact origin. It is a kind of theoretical folklore, used for example in eliminating the square root in relativistic kinetic energy and in many other settings.

can postpone this minimization to be done *afterword* once the problem is solved and Hamiltonian is diagonalized. The parameter b, in general, would be iteratively selected to reach self-consistency when $b \to M_{nl}^2$: For the heavy quarkonia example, we start with $b \to (2m_Q)^2$.

The next step is to represent this Hamiltonian as a sum of two terms, $H_0 + V$:

$$H_0 = \frac{\sigma_T}{a}\left(-\frac{\partial^2}{\partial x^2} - b\frac{\partial^2}{\partial p_\perp^2}\right) + \sigma_T a + 4(m_q^2 + \vec{p}_\perp^2) \tag{28.8}$$

Note that the last term is artificially added: It is then subtracted from the "potential" defined by

$$V(x, \vec{p}_\perp) \equiv (m_q^2 + p_\perp^2)\left(\frac{1}{x\bar{x}} - 4\right) \tag{28.9}$$

The H_0 part is just an oscillator in the transverse direction *momentum*, which selects a convenient functional basis set. The set for longitudinal x we use is simply harmonic $\sin\big((2n+1)\pi x\big)$. The "potential" V we call the "cup potential," as it forces the wave function to vanish at kinematical boundaries $x \to 0$ and $\bar{x} = 1 - x \to 0$.

So, we have constructed $H_{\rm LF}$, which generates the Schrödinger equation in the momentum representation.[10] In the case of mesons, there are three variables $x, \vec{p}_\perp$, and one can directly solve the Schrödinger equation numerically. Another well-known method to solve the problem is to calculate numerically the matrix elements of H_{LF} in a certain (truncated) set of basis functions, with subsequent diagonalization.[11]

Let us start with $\bar{b}b$ mesons (upsilons), which are very well studied experimentally and theoretically. In Fig. 28.1 (left), the red squares show the dependence of the *experimental* squared masses of Upsilons ($\bar{b}b$ vector mesons). The triangles are *theoretical* values for a Cornell potential: They deviated from the experiment because spin-related forces were excluded. Finally, blue circles denote masses if only the linear potential is included and the Coulomb term is excluded: The straight line connecting them is here to remind us about linear Regge trajectories[12] for our basic problem

[10]Its slightly unusual feature is that derivatives come from the confining forces and potential from kinetic energy.

[11]Both can rather effectively be done in Mathematica: We compared the wave functions obtained by them and found agreement inside the width of the line, even for rather modest functional sets.

[12]Note that it is not a linear function of the hadron spin J, but of the principle quantum number n.

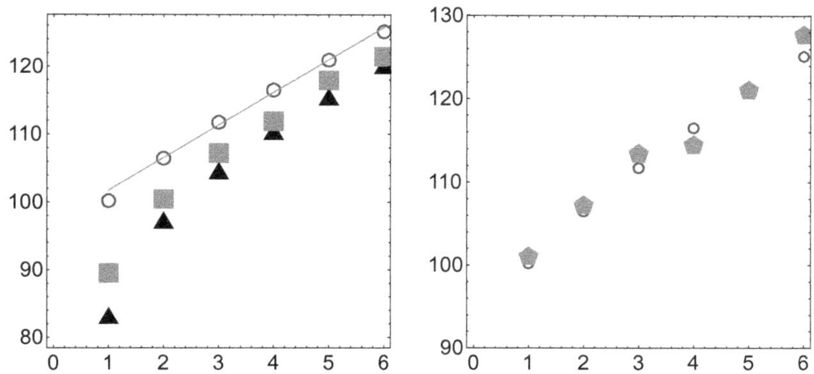

Fig. 28.1. (left) The squared masses of the nth state $E_n^2(\mathrm{GeV}^2)$ versus the radial quantum number $n = 1, \ldots, 6$ for Upsilon mesons. Red squares are experimental masses, from the PDG table. Open blue circles are from the Schrödinger equation with the linear potential only; the blue triangles are also from the Schrödinger equation, but with the Cornell potential including the Coulomb interaction. (right) A comparison between the solution of the Schrödinger equation for the $\bar{b}$ pair connected by a QCD string (blue circles) with masses we obtain from the LF Hamiltonian (red five-point polygons).

of two massive quarks connected by a string (which we discussed in Chapter 26.

These blue circles are the same on the right plot of Fig. 28.1, and they are compared to masses one obtains using the LF Hamiltonian. In this calculation, the set of functions in the transverse and longitudinal directions included just 12 functions: that is why it displays some irregularity between the $n = 3$ and $n = 4$ states. For a larger basis set, it moves to a larger n.

Let us emphasize that quarkonium masses can, of course, be calculated through a standard nonrelativistic approach in the CM frame. The main objective is that not only do we get them from H_{LF} as well, but we also get LFWFs as a nontrivial function of the three variables $\vec{p}_\perp, x$.

The larger point is that proceeding to light mesons is now straightforward, as are generalization to baryons and many-quark states, the inclusion of spin-dependent forces, etc.

28.3 Baryons on the Light Front

With quark–antiquark systems, the coordinates used were naturally their relative coordinates $\vec{r}_q - \vec{r}_{\bar{q}}$. For three-quark systems, one would like to eliminate the *center-of-mass* motion, requiring the momentum to be constrained:

$$\vec{p}_\perp^{\,\rm tot} = \vec{p}_{1\perp} + \vec{p}_{2\perp} + \vec{p}_{3\perp} = 0$$
$$x_1 + x_2 + x_3 = 1 \tag{28.10}$$

so the system is in fact six-dimensional. Different ways of solving the issue were used in the literature, but the simplest one, due to Shuryak and Zahed (2023c), is based on (modified) Jacobi coordinates:

$$\vec{p}_{\rho\perp} = \frac{1}{\sqrt{2}}(\vec{p}_{1\perp} - \vec{p}_{2\perp}), \quad \vec{p}_{\lambda\perp} = \frac{1}{\sqrt{6}}(\vec{p}_{1\perp} + \vec{p}_{2\perp} - 2\vec{p}_{3\perp}) \tag{28.11}$$

in terms of which

$$\begin{aligned}\vec{p}_{1\perp} &= (\sqrt{6}\vec{p}_{\lambda\perp} + 3\sqrt{2}\vec{p}_{\rho\perp})/6,\\ \vec{p}_{2\perp} &= (\sqrt{6}\vec{p}_{\lambda\perp} - 3\sqrt{2}\vec{p}_{\rho\perp})/6,\\ \vec{p}_{3\perp} &= -\sqrt{6}\vec{p}_{\lambda\perp}/3\end{aligned} \tag{28.12}$$

Now, the total transverse momentum $\vec{p}_{\rm tot} = \vec{p}_1 + \vec{p}_2 + \vec{p}_3$ vanishes automatically.

The three-quark static potentials we discussed in Section 25.5: Let me remind the reader that models with different arrangements of the flux tubes were tested there versus lattice and instanton-based numerical calculations. The conclusion was that the best of these models is the so-called A-*ansatz* (25.9). The longitudinal momentum fractions are defined similarly:

$$\begin{aligned}x_1 &= (\sqrt{6}\lambda + 3\sqrt{2}\rho + 2X)/6\\ x_2 &= (\sqrt{6}\lambda - 3\sqrt{2}\rho + 2X)/6\\ x_3 &= (-\sqrt{6}\lambda + X)/3\end{aligned} \tag{28.13}$$

Note that $X = x_1 + x_2 + x_3$: Unlike in the transverse direction, for which $X = 0$, here X should be set to 1. Therefore, the physical domain is the three-dimensional cube $x_i \in [0, 1]$ *cut by a plane* $X = 1$, leaving as a physical domain a triangle in λ, ρ coordinates, inside which the parton fractions are all positive $x_i > 0$. The three corners of the triangles correspond to parton configurations, with one quark having a fraction of 1 and the two others having zero. The longitudinal part of LFWF should therefore be defined *on this triangle.*

For the Y (or star) model, this amounts to

$$\sigma_T(|\vec{r}_1| + |\vec{r}_2| + |\vec{r}_3|) \to \frac{\sigma_T}{2}\left(\frac{3}{a} + a(\vec{r}_1^{\,2} + \vec{r}_2^{\,2} + \vec{r}_3^{\,2})\right)$$
$$\to \frac{\sigma_T}{2}\left(\frac{3}{a} + a(\vec{r}_\lambda^{\,2} + \vec{r}_\rho^{\,2})\right)$$

with all einbein parameters set equal to a by the saddle point. For *ansatz* A (25.9), this amounts to

$$\frac{\sigma_T}{2}(r_{12} + r_{23} + r_{13}) \to \frac{\sigma_T}{4}\left(\frac{3}{a} + a(r_{12}^2 + r_{23}^2 + r_{13}^2)\right)$$
$$\to \frac{\sigma_T}{4}\left(\frac{3}{a} + 3a(\vec{r}_\lambda^{\,2} + \vec{r}_\rho^{\,2})\right)$$

Note that when a is to be eliminated by minimization, the minimum is at $a^* = \sqrt{3/(r_\lambda^2 + r_\rho^2)}$ in the former case, but $a^* = \sqrt{1/(r_\lambda^2 + r_\rho^2)}$ in the latter one. Substituting those values back to the original expressions, one finds that confining potentials in Y in the Jacobi coordinates are the same as A-*ansatz*, after simple rescaling of the string tension:

$$\sigma_T \to \frac{\sqrt{3}}{2}\sigma_T. \tag{28.14}$$

Note that the rescaling factor $\frac{\sqrt{3}}{2} \approx 0.866 < 1$, so the tension (and thus all baryonic masses) corresponding to model A are lower than in model Y, as expected.

The quantum replacement

$$\vec{r}_\lambda \to i\frac{\partial}{\partial \vec{p}_\lambda}, \quad \vec{r}_\rho \to i\frac{\partial}{\partial \vec{p}_\rho}$$

yields a 6D Laplacian in $\vec{p}_\lambda, \vec{p}_\rho$, for both the Y and A models. The kinetic term contains the sum of the corresponding squares of momenta: So, in this approximation, there appears to be 6D spherical symmetry[13] in the Jacobi coordinates.

After we defined the coordinates and the confining part, let us focus on the kinetic energy. Following what was done for mesons, we use

$$\frac{p_{1\perp}^2+m_Q^2}{x_1}+\frac{p_{2\perp}^2+m_Q^2}{x_2}+\frac{p_{3\perp}^2+m_Q^2}{x_3}$$
$$=3\sum_i(p_{i\perp}^2+m_i^2)+\sum_i(p_{i\perp}^2+m_i^2)\left(\frac{1}{x_i}-3\right)$$

Let us focus for now on longitudinal momenta. In variable λ,ρ,X (28.13), the line element defining the metric tensor in the new coordinates is diagonal and simple:

$$dl^2=d\lambda^2+d\rho^2+dX^2/3 \tag{28.15}$$

The Laplacian (which we encounter in the confining term of the Hamiltonian) in the original coordinates also takes a simple form:

$$\nabla^2=\sum_i\frac{\partial^2}{\partial x_i^2}\rightarrow\frac{\partial^2}{\partial\lambda^2}+\frac{\partial^2}{\partial\rho^2}+3\frac{\partial^2}{\partial X^2} \tag{28.16}$$

Since we work on the constant $X=1$, we need only the first two terms.

Therefore, the first problem we encounter is defining the eigenfunctions of the Laplacian on the triangular physical domain in the $\lambda-\rho$ variable. In Shuryak and Zahed (2023c), this problem has been solved analytically by a certain combination of interfering standing waves. Laplacian eigenvalues are numerated by two quantum numbers m_L,n_L and are proportional to $\left(m_L-\frac{n_L}{2}\right)^2+\frac{3}{4}n_L^2$. Let us give as an example the simplest nondegenerate case $m_L=2n_L$, with normalized eigenstates

$$\varphi^D_{2n_L,n_L}(\lambda,\rho)=\frac{2^{\frac{3}{2}}}{L\,3^{\frac{3}{4}}}\left[2\cos\left(\frac{2\pi n_L\rho}{L}\right)\sin\left(\frac{2\pi n_L\tilde{\lambda}}{\sqrt{3}L}\right)-\sin\left(\frac{4\pi n_L\tilde{\lambda}}{\sqrt{3}L}\right)\right] \tag{28.17}$$

with $L=\sqrt{2}$, $\tilde{\lambda}=\lambda+\sqrt{2/3}$. Two of those are shown in Fig. 28.2.

[13]Use of this symmetry for the ground state, assuming it has no dependence on angles, can be found in the original paper (Shuryak and Zahed, 2023c). For few-body systems, the idea itself has been well known since the 1960s.

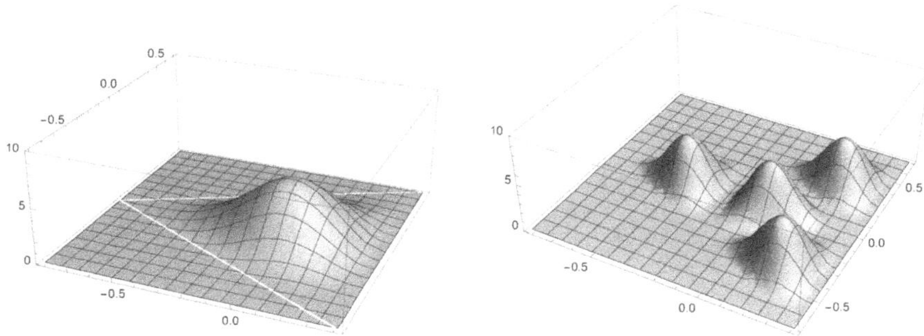

Fig. 28.2. Probability distribution $|\phi^D_{2n_L,n_L}|^2$ for $n_L = 1$ (upper) and $n_L = 2$ (lower) in the λ, ρ plane, with manifest mirror symmetry in ρ.

The non-factorizable potential $\tilde{V}$ still tends to infinity at all sides of the triangle, and one can include it on the triangle and solve the Schrödinger equation on the triangle numerically. Finally, let us recall that all the calculations were done for the traditional "star" or Y model of confinement. Transition to (perhaps more accurate) *ansatz* A leads to the same Hamiltonian with a string tension rescaling (28.14), downward by about 13%, as one has observed it in static potentials.

Our results for squared masses of "flavor-symmetric" baryons (bbb, ccc, sss, uuu) are indicated by black symbols in Fig. 28.3. For comparison, we show the experimental masses as red hexagons. The blue hexagons are available model predictions for ccc and bbb baryons. Since a constant in the LF Hamiltonian remains undefined, we fixed one constant for all masses, so that the mass of the uuu baryon Δ^{++} is set to the experimental one. For definiteness, the plots corresponding to the effective masses of u, s, c, b quarks to be $0.28, 0.45, 1.5, 4.8$ GeV, respectively, not specially fitted to this plot but inherited from meson studies. Finally, let us remember that all the calculations were done for the traditional "star" or Y model of confinement. Transition to (perhaps more accurate) *ansatz* A leads to the same Hamiltonian with a string tension rescaling (28.14), downward by about 13%, as one has observed it in static potentials.

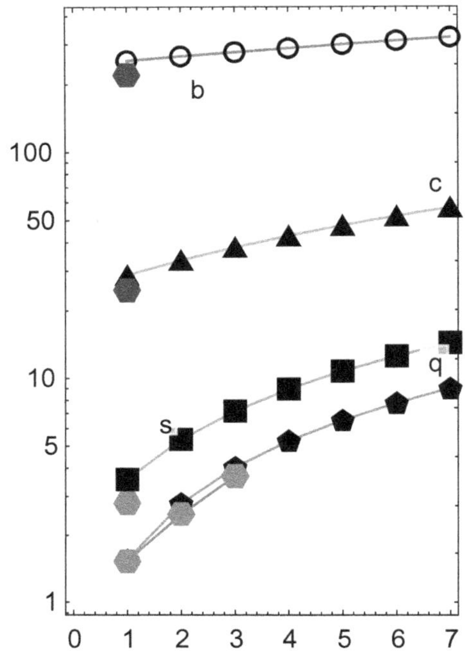

Fig. 28.3. Squared masses of baryons $M^2_{n+1}(Q, \frac{3}{2})$ in GeV^2 versus the principal quantum number $n + 1 = 1, \ldots, 7$. The black circles, triangles, squares and pentagons are the results of our calculations for the flavors b, c, s, q, respectively. The red hexagons are the experimental values of three Δ^{++} and one Ω^- masses from PDG. The two blue hexagons are model predictions for the masses of *ccc* and *bbb* baryons from Table I.

28.4 Diquarks in the Nucleons, Form Factors and Generalized Parton Distributions

We already had a general discussion of *diquarks* in the chapter on spectroscopy. Now, we return to it with the purpose of including the attraction in the "good diquark" channel in H_{LF} and seeing its effect on the wave functions, following Shuryak and Zahed (2023d). The approximation used is via representing instanton-induced 't Hooft Lagrangian as a point-like four-fermion vertex.

The simplest cases are heavy-light $\Lambda_Q = Q(ud)_{0^+}$ baryons, in which the ud diquark is "good" (well bound), as compared to the $\Sigma_Q = Q(uu)_{1^+}$ case, in which, due to symmetry, the 't Hooft vertex is absent. Not going into details, we show in Fig. 28.4 the difference between LFWFs, which, for heavy quark being the charm quark c.

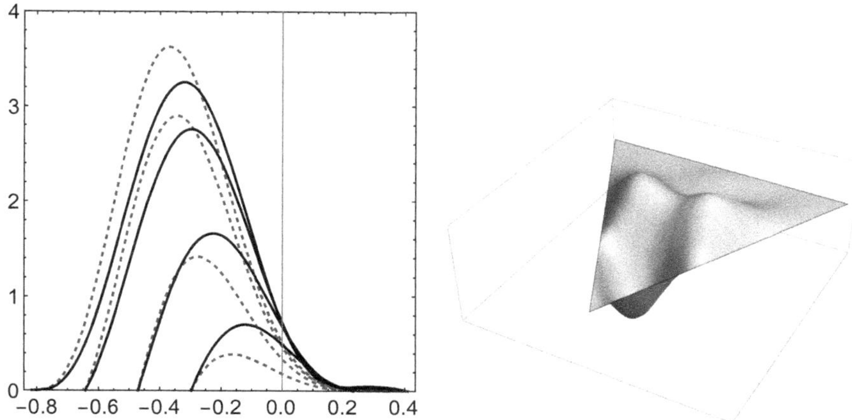

Fig. 28.4. (left) the longitudinal wave functions $\Psi_\Sigma(\rho,\lambda)$ of Σ_c (blue dashed) and $\Psi_\Lambda(\rho,\lambda)$ of Λ_c (black solid), as functions of λ. The four pairs of curves are for $\rho = 0.0, 0.1, 0.2, 0.3$, top to bottom. (right) 3D plot shows their *difference*, $\Psi_\Lambda(\rho,\lambda)-\Psi_\Sigma(\rho,\lambda)$.

The next comparison is done between the proton (uud) and $\Delta = uuu$: In the former case, quasi-local forces act for both 1–3 and 2–3 ud pairs. In order to calculate its matrix elements, one has to change variables, rotating our equilateral triangle by $\pm 60°$. We will not go into excited states (which were calculated up to $n = 12$) and just show a comparison of the wave functions on the ρ, λ triangle, see Fig. 28.5. The main message here is that the diquark pairing is not small, as it leads to visibly wider LFWFs.

With these wave functions, one can calculate and compare the form factors. In particular, the helicity preserving the Dirac form factor is (Drell and Yan, 1970; West, 1970; Lepage and Brodsky, 1980)

$$F_1(q^2) = \langle \vec{P} + \vec{q}, \uparrow | J^+/2P^+ | \vec{P}, \uparrow \rangle \tag{28.18}$$

To evaluate (28.18), we select the momentum transfer to be q in the transverse x-direction and the struck quark to be number 3 (d quark). More specifically, the transverse momenta in the struck baryon (with prime) are related to those in the non-struck one (without prime) by

$$\begin{aligned} k_x^{1'} &= k_x^1 - x_1 q, \quad k_x^{2'} = k_x^2 - x_2 q, \\ k_x^{3'} &= k_x^3 + (1 - x_3) q \end{aligned} \tag{28.19}$$

The three transverse momenta $\vec{k}^i$ and longitudinal fractions x_i need to be reexpressed in terms of two Jacobi momenta, in our notations $\vec{p}_\lambda, \vec{p}_\rho$ and

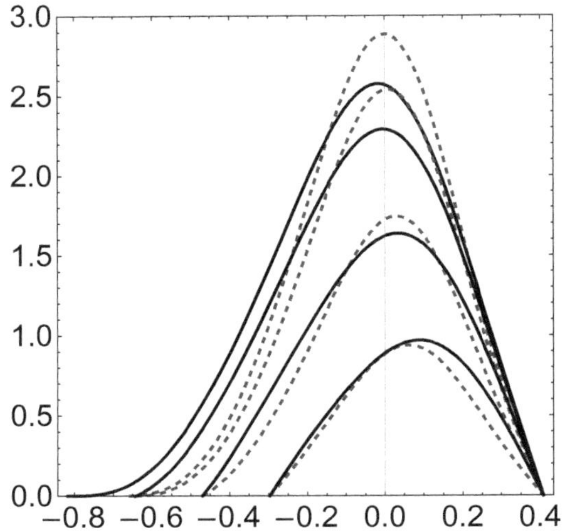

Fig. 28.5. LFWFs for the lowest Delta (dashed lines) and N (solid lines). The plots are shown versus the Jacobi coordinate λ, for fixed $\rho = 0, 0.1, 0.2, 0.3$, top to bottom.

ρ, λ, on which LFWFs depend. We now show the form factors calculated with the longitudinal wave functions for the Delta and Proton, following from our analysis in the preceding section. In order to better see the most interesting region of large q^2, we plot $Q^4 F_1^d(Q^2)$ in Fig. 28.6.

Experiments are, of course, performed with protons and neutrons, but using them, one can extract separate form factors for u or d quarks. This was done, for example, by Qattan and Arrington (2012), and the red circles in Fig. 28.6 are from Fig. 8 of this work. (For clarity, we do not show the data points in the range $Q^2 < 1\,\text{GeV}^2$, as the error bars for these points are ± 0.02 on average.) From the plot, we see that this Q^4 times the form factor does not appear to reach a constant limit at large Q^2, with the measured points slowly decreasing toward the right-hand side. Old dipole parametrization $Q^4 F = Q^4/(1 + Q^2/m_\rho^2)^2$ asymptotes a constant at large Q^2 from below.

Remarkably, our longitudinal proton wave functions reproduce such a trend, and (with the parameter $A = 4\,\text{GeV}^{-2}$) they follow the shape indicated by the data rather well. The calculated form factor for the case when the struck quark is u has a similar shape. Unfortunately, according to Qattan and Arrington (2012), the experimental trend is different, the

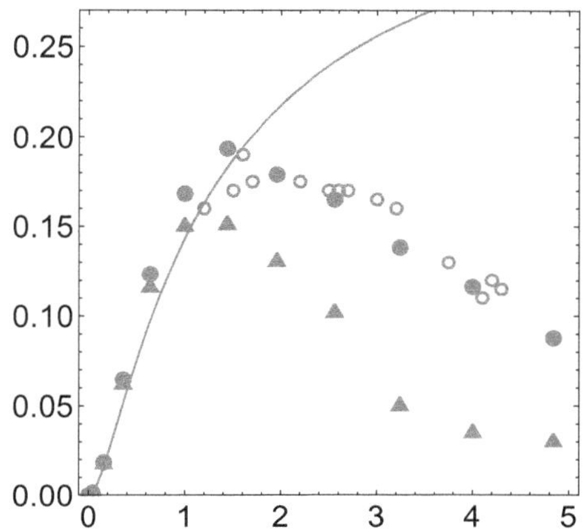

Fig. 28.6. $Q^4 F_1^d(Q^2)$, (GeV^4) versus the momentum transfer Q^2 (GeV^2). The triangles and closed points correspond to the Delta and Proton LFWFs, respectively. The red circles are extracted from the experimental data on the p and n form factors mentioned in the text. The solid line shown for comparison corresponds to the dipole ansatz $Q^4/(1+Q^2/m_\rho^2)^2$.

constant at $Q^2 \to \infty$ is approached from below. According to the Drell–Yan relation, this flavor difference is also seen in the PDFs of u and d at $x \to 1$. Flavor asymmetry must be related to the asymmetry of the spin–orbit part of the wave function, which in our approximations has so far been ignored.

Note that the corresponding form factor for Delta (triangles) is significantly softer, as one would expect from the size of the wave function. Recall that the large difference between the Delta and Proton form factors (so well seen in this plot) is completely due to the 't Hooft quasi-local pairing ud interaction. While we do not have Delta targets for experiments, perhaps its form factor can be calculated on the lattice or in other models.

GPDs are defined by *off-forward* matrix elements of the leading quark and gluon operators on the LF, and they provide a more detailed "tomography" of hadrons, combining the x-dependence of PDFs with the Q-dependence of certain form factors. GPDs are accessible by semi-inclusive processes through deeply virtual Compton scattering (DVCS) and deeply virtual meson production. DVCS has been pursued by the CLAS collaboration at JLab and the COMPASS collaboration at CERN, with more experiments planned at future electron ion colliders such as the Electron-Ion Collider (EIC) now under construction at BNL.

We cannot address the physics of GPDs in any detail here, but just show how they can be calculated from LFWFs, following some results from Shuryak and Zahed (2023e). At nonzero "skewness" ξ, the GPD functions are related to the following matrix elements:

$$\frac{\sqrt{1-\xi}}{1-\frac{1}{2}\xi}H(x,\xi,t)-\frac{1}{4}\frac{\xi^2}{(1-\frac{1}{2}\xi)\sqrt{1-\xi}}E(x,\xi,t) \tag{28.20}$$

$$=\int_P \delta(x-x_1)\,\psi^{+*}([x_i',k_{i\perp}',\lambda_i])\psi^{+}([x_i,k_{i\perp},\lambda_i]) \tag{28.21}$$

and

$$-\frac{q_L}{2m_N}\frac{1}{\sqrt{1-\xi}}E(x,\xi,t)=\int_P \delta(x-x_1)\,\psi^{+*}([x_i',k_{i\perp}',\lambda_i])\psi^{-}([x_i,k_{i\perp},\lambda_i])$$

The kinematical arrangement is as follows: (a) *active quark* $i=1$: $x_1'=(x_1-\xi)/(1-\xi)$, $k_{1\perp}'=k_{1\perp}+((1-x_1)/(1-\xi))\Delta_\perp$ and $\lambda_1=\pm 1$; (b) *passive quarks* $i=2,3$: $x_i'=x_i$ and $k_{i\perp}'=k_{1\perp}-(x_i/(1-\xi))\Delta_\perp$.

At zero "skewness" $\xi=0$, the formulae simplify and the function E gets dropped out. The integral over x of GPD is the electromagnetic form factor:

$$F_1(t)=\int dx H(x,0,t) \tag{28.22}$$

so $H(x,0,t)$ can be seen as a "form factor at particular x." Another integral weighted with x

$$A(t)=\int dx\, xH(x,0,t) \tag{28.23}$$

gives the "gravitational form factor" related to the correlator of stress tensors (instead of an electromagnetic current).

For the proton with quark assignment uud, we assume that the struck quark is d, with a longitudinal momentum fraction x_3. In our (modified) Jacobi coordinates, this momentum fraction is directly related to the longitudinal variable λ. For the unpolarized d-quark GPD in (28.20), this amounts to integrating the off-forward LFWFs over five variables, the transverse momenta $\vec{p}_\rho, \vec{p}_\lambda$ and ρ for the nucleon. To proceed further, we approximate the dependence on the transverse momenta $\vec{p}_\rho, \vec{p}_\lambda$ using Gaussians, which is quite accurate, and carry out the integrals analytically. The remaining integration over ρ is performed numerically. Nucleon zero-skewness GPD calculated without LFWF is shown in Fig. 28.7.

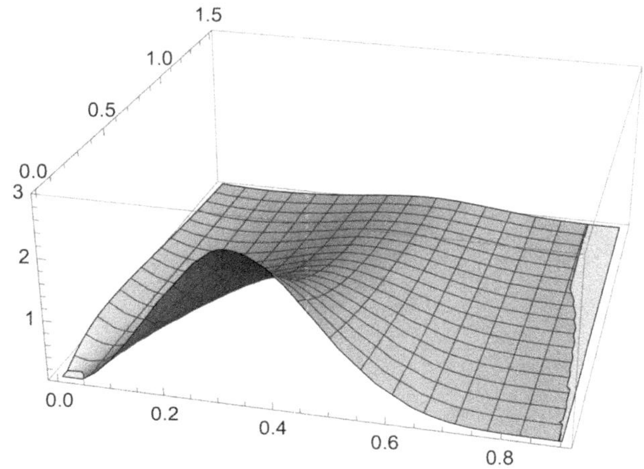

Fig. 28.7. The nucleon GPD function $H_d^N(x, \xi = 0, Q^2)$ for a struck d-quark.

The GPDs we found can be approximated generically by a "Gaussian *ansatz*" in the momentum transfer Q^2, with a width and a pre-exponent that are x-dependent:

$$\text{GPDs}(x, Q^2) \sim f_1(x) \exp\left[\frac{-Q^2 f_2(x)}{\sqrt{1-\xi^2}}\right] \tag{28.24}$$

In Fig. 28.8, we show the effective Q-slope of the unpolarized d-quark GPD, for the nucleon (filled-points) and Δ-isobar (open-points). The r.m.s. size $R_{\text{r.m.s.}}(x)$ is maximal at $x \sim \frac{1}{3}$, and is numerically about 0.6 fm. It decreases sharply for $x \sim 1$, and moderately for $x \sim 0$. This can be explained by the fact that for $x \sim \frac{1}{3}$, all three quarks carry about the same longitudinal momentum on the LH, which corresponds to three quarks at rest in the CM or rest frame. Semiclassically, this means a configuration in which a struck quark is near a "turning point" of the wave function, with their QCD strings maximally stretched. In contrast, x away from $\frac{1}{3}$ corresponds to a struck quark rapidly moving in the CM frame, which must happen near the hadron center, the corresponding size is therefore small. Note that for $x \to 1$, the distribution of the struck d-quark becomes nearly point-like, with slopes close to zero. Our LFWFs show that the magnitude of this effect is different for the nucleon and the Δ-isobar, sensitive to the diquark substructure of the former.

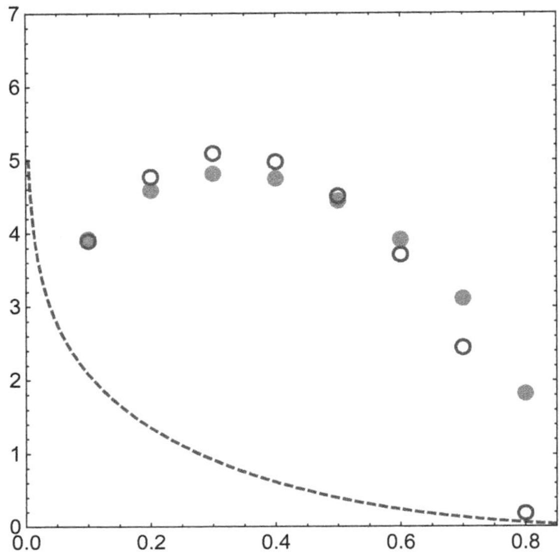

Fig. 28.8. Closed and open points correspond to slopes $-d\log[\text{GPD}]/dQ^2$ (at $Q^2 \approx 1\,\text{GeV}^2$, in GeV^{-2}) calculated from our LFWF for the nucleon and Delta baryons, respectively. The dashed line shown for comparison is the *ansatz* (28.24).

The dashed line in Fig. 28.8 corresponds to the *ansatz* by de Teramond *et al.* (2018):

$$f(x) = \frac{1}{4\lambda}\left((1-x)\ln\left(\frac{1}{x}\right) + a(1-x)^2\right),$$

with the parameters $a = 0.53$ and $\lambda = (0.548\,\text{GeV})^2$ shown for comparison. The sharp rise at small x is due to the Reggeon contribution, due to a multi-parton cloud around a baryon. It is absent in our approach, which is limited to the lowest Fock state. The decrease at large x is in qualitative agreement with our results. However, at intermediate x, there is a significant disagreement with our baryon r.m.s. size $R^2_{\text{r.m.s.}}(x)$, which is significantly larger and reaching $\sim 5\,\text{GeV}^{-2}$.

28.5 PDFs and Their "Evolution"

PDFs are distributions of certain *partons* (quarks, antiquarks or gluons) which possess a certain fraction of hadron longitudinal momentum x. Bjorken scaling (confirmed by the SLAC experiment of 1969) and the

Feynman parton model led, around 1970, to the successful phenomenological description of the DIS, Drell–Yan process (large mass dilepton from parton annihilation), culminating in jet and heavy gauge boson production. For large momentum transfer (say, $Q^2 > 10\,\text{GeV}^2$), one can rely on *QCD factorization theorems*, which state that corresponding cross-sections can be split into a convolution of (i) a "hard block" calculable via perturbative QCD, plus two nonperturbative objects; (ii) the *structure functions* of the incoming hadrons; and (iii) the *fragmentation functions* of outgoing ones. in the, while PDFs are extracted from a compilation of various reactions using *factorization theorems.* PDFs have the form $f(x, \mu^2)^\lambda$, where μ is the so-called normalization scale at which it is defined, and λ is the helicity of a parton. Phenomenological PDFs for unpolarized and polarized proton, at $\mu^2 = 100\,\text{GeV}^2$ (Ethier and Nocera, 2020) are shown in Fig. 28.9.

Note that valence distributions peak at $x \sim 0.2$ but have a tail toward small x. The strange and charm quarks belong to the "sea," with antiquarks, and they strongly grow toward small x due to quark pair production. This growth strongly depends on the normalization scale μ, which is at $\mu^2 = 100\,\text{GeV}^2$ in this figure; however, at $\mu^2 \sim 1\,\text{GeV}^2$, gluons and the sea disappear. Perturbatively, it is described by a *gluon* $\to \bar{q}q$ process, whereas nonperturbatively, there are other effects to which we will return later.

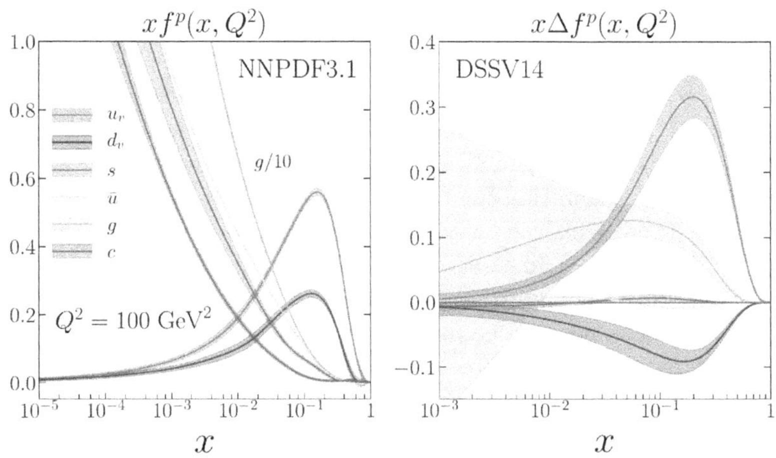

Fig. 28.9. Parton distribution functions versus momentum fraction x; u_v, d_v are "valence" distribution on top of "sea" quark pairs at $\mu^2 = 100\,\text{GeV}^2$.

Moments of PDFs are certain integrals: For example, the mean momentum of u quark in a proton would be defined as

$$\int_0^1 dx x u(x,\mu^2) = \langle x \rangle_{u,\mu^2} \tag{28.25}$$

The polarized PDFs use differences of two helicities $\Delta f = f^+ - f^-$; for example, the famous "quark spin" part of the proton spin is

$$\Delta\Sigma = \sum_{q=u,d,s} \int dx[\Delta q + \Delta\bar{q}] \tag{28.26}$$

Lattices involve operator product expansion (OPE) to evaluate moments of PDFs (for a good review, see, for example, Lin *et al.*, 2018).

A theoretical definition of PDFs is given using the LF matrix element

$$q(x) = \frac{1}{4\pi}\int dy^- \exp(-ixy^- p^+)\langle p|\bar{\psi}(0,y^-,0_\perp)\gamma^+[]\psi(0,0,0_\perp)|p\rangle \tag{28.27}$$

where the empty square brackets stand for P-exponent with the gauge field along the LF line connecting two points. It is obviously needed to make the bi-local operator gauge invariant. The plus and minus components of the vectors are defined as 0 ± 3 components, and the 1 and 2 components are called transverse $\perp$. In the polarized Δq definition, one should change $\gamma^+ \to \gamma^+\gamma_5$.

Let us note at this point that PDFs are the *density matrices* "traced" over many variables (such as the momenta of other quarks and gluons). Such quantities are known to have "entanglement entropy." Dependence on normalization scale μ can be calculated using the Dokshitzer–Gribov–Lipatov–Altarelli–Parisi (DGLAP) equation, including radiation (or absorption) of gluons as the scale changes. As a kinetic equation for probabilities, it has its own entropy production and equilibrium — the maximum of entropy.

A nonlocal LF definition can be written in terms of local operators using the OPE method, which is basically a Taylor expansion of the definition. The so-called *twist* is defined as the dimension minus spin of the operator. At large Q^2, the dominant are the "leading" twist two operators containing one gamma matrix and some number of covariant derivatives. The matrix elements of such operators are kinematically related to the powers of momentum times the constant of interest:

$$\langle N(p)|\bar{q}\gamma_{\mu_1} iD_{\mu_2}\dots iD_{\mu_n} q|N(p)\rangle = p_{\mu_1}\dots p_{\mu_n}\langle O_n\rangle \tag{28.28}$$

For example, the mean momentum $\langle x\rangle$ is related to the quark stress tensor operator (dimension 4, spin 2) $\bar{q}\gamma_\mu iD_\nu q$, (where D is assumed to act to the right and the left with (1/2)). A fixed spin of the operator implies that all traces (lower spins) are subtracted.

The zeroth moment of the polarized distribution is related to the dimension 3, spin 1 operator $\bar{q}\gamma_\mu\gamma_5 q$ (axial current), which is related to coupling to weak interactions and thus related to a certain "sum rule" restricting the integral of Δq.

Next twist corrections to stress tensor and axial current operators were calculated by Shuryak and Vainshtein (1982a, 1982b), who included gauge field together with $\bar{q}q$ or even four-fermion operators $(\bar{q}q)(\bar{q}q)$ and, by dimension, one power of $1/Q^2$. In principle, those would give important information about nucleon structure; however, four decades later, their matrix elements are still known with large error bars.

Lattice studies are done in Euclidean space-time, but matrix elements of local operators can be studied using standard three-point (source–operator–sink) construction. Many moments of PDFs found this way can be found in Lin *et al.* (2018); let me just mention that gluon momentum fraction $\langle x \rangle_g \approx 0.4$ at $\mu = 2$ GeV and that matrix element of axial current — nucleon axial charge — is in excellent agreement with experiments: $g_{\text{exp}} = 1.2723(23)$.

28.6 Distribution Amplitudes

Another group of phenomenologically important matrix elements of various operators between the vacuum and the hadron are known as distribution amplitudes (DAs). They appear in hard exclusive processes, e.g. the form factors of elastic scattering, which we will discuss in the following.

In the case of the pion, the DA is $\pi^+(p) \to q_{if\alpha}(k)\bar{q}_{jg\beta}(k-p)$, which corresponds formally to the connected amplitude

$$\int_{-\infty}^{+\infty} \frac{p^+ dz^-}{2\pi} e^{ixp\cdot z} \left\langle 0 \left| \bar{d}_\beta(0)[0,z] u_\alpha(z) \right| \pi^+(p) \right\rangle$$
$$= \left(+ \frac{if_\pi}{4}\gamma^5 \left(\not{p}\,\varphi_{\pi^+}(x) - \chi_\pi \varphi_\pi^P(x) + i\chi_\pi \sigma_{\alpha\beta} \frac{p^\alpha p'^\beta}{p\cdot p'} \frac{\varphi_\pi'^T(x)}{6} \right) \right)_{\alpha\beta} \tag{28.29}$$

where there are three functions of x, called axial, pseudoscalar and tensor ones, named after the gamma matrices in the definition. Note that the definition is bi-local in the LF direction, but local in the transverse plane:

the wave functions at the origin are thus represented by two constants f_π, χ_π.

In the form factor, there are two pions — in-going and out-going — plus the so-called *hard block* sandwiched in between. The second pion is described by conjugate DAs, which we give explicitly by

$$\int_{-\infty}^{+\infty} \frac{p'^- dz^+}{2\pi} e^{-ixp'\cdot z} \left\langle \pi^+(p') \left| \overline{u}_\beta(z)[z,0] d_\alpha(0) \right| 0 \right\rangle$$
$$= \left(-\frac{if_\pi}{4}\gamma^5 \left(\not{p}' \varphi_{\pi^+}(x) + \chi_\pi \varphi_\pi^P(x) - i\chi_\pi \sigma_{\alpha\beta} \frac{p^\alpha p'^\beta}{p\cdot p'} \frac{\varphi_\pi'^T(x)}{6} \right) \right)_{\alpha\beta} \tag{28.30}$$

up to twist-3. The bracketed $[x, y]$ again refers to the P-exp gauge link, and $\sigma_{\alpha\beta} = \frac{i}{2}[\gamma_\alpha, \gamma_\beta]$.

Note that the four-vector p'_μ appears in the DA of a pion with four-vector p_μ, in reference to the conjugate light-cone direction, with generally no relation to the second pion. In the DA of a pion with momentum p'_μ, the exchange $p \leftrightarrow p'$ needs to be enforced, effectively flipping the sign of the last term. Also, note that we dropped the contribution

$$\frac{f_\pi \chi_\pi}{4} \sigma_{\mu\nu} p^\mu \left[\left(\frac{\partial}{\partial k_{\perp\nu}} \frac{\varphi_\pi^T(x)}{6} \right) + \frac{\varphi_\pi^T(x)}{6} \frac{\partial}{\partial k_{\perp\nu}} \right] \tag{28.31}$$

as it involves the dependence on $k_\perp$, which we have ignored in both the soft and hard blocks.

Historically, the first DAs used were the so-called "asymptotic" ones, to which DGLAP-like evolution at $Q \to \infty$ leads to

$$\varphi_\pi \to \varphi_\pi^{\text{asymptotic}}(\xi) = \frac{3}{4}(1 - \xi^2) = 6x\bar{x} \tag{28.32}$$

Needless to say, this limit is very far from the realistic kinematic range of interest. Therefore, we neither use "asymptotic wave functions," nor restrict our analysis to the leading twist DAs. We would rather focus on the chiral structure of the DAs, making sure that all possible and large contributions are included.

The pion is a particular particle, a Nambu–Goldstone mode; therefore, its properties can be calculate under any theory in which chiral symmetry gets spontaneously broken. Historically, the NJL model and its nonlocal versions (some related to the instanton liquid model) have been used to calculate the pion LFWF (Broniowski and Ruiz Arriola, 2017; Petrov and Pobylitsa, 1997; Anikin *et al.*, 2001). Before we briefly discuss the results of

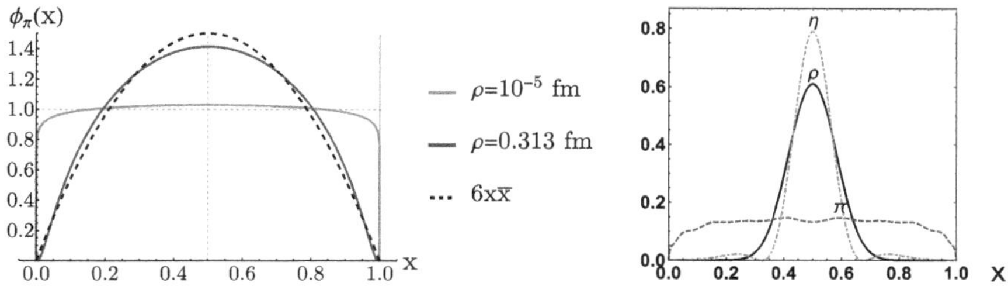

Fig. 28.10. Upper plot: The pion LF distribution function from Kock *et al.* (2020), ρ indicated is the instanton size. Lower plot: The light meson LF distribution functions from Shuryak (2019a), which also show DA for vector meson ρ and pseudoscalar (non-Goldstone) η'.

"realistic" models related to a larger set of hadronic wave functions, let us introduce some extreme cases. For example, in Ruiz Arriola and Broniowski (2002), a "flat" pion wave function was used as an "initial condition" for radiative evolution. Some typical shapes of the pion and other light meson wave functions stemming from some recent works are shown in Fig. 28.10.

28.7 Mesonic Form Factors and Instantons

The theory of *exclusive QCD reactions* began by following the same logic as that of DIS. The early works include Brodsky and Farrar (1973), Chernyak and Zhitnitsky (1977), and Radyushkin (1977). It is also based on two assumptions:

(i) the *separation of scales*, based on the assumption that the momentum transfer Q (the scale in the "hard block"), is large compared to the typical quark mass and transverse momenta inside hadrons;
(ii) the "hard block" can be calculated *perturbatively using gluon exchanges*.

However, the theory based on these two assumptions is insofar not successful, as there remains a wide gap between the semi-hard domain of Q in which experimental/lattice results are available and the "asymptotic" theory at $Q \to \infty$.

In order to show this, instead of real experimental data (which do not extend to a high enough Q^2), we first turn to lattice results. Before we do that, let us explain their terminology: The diagrams with "source" and "sink" operators are divided into "connected" and "disconnected" ones, see Fig. 28.11. The disconnected one is more noisy and more expensive

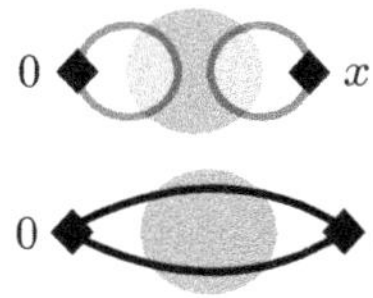

Fig. 28.11. Two diagrams for in-vacuum correlation functions. The rhombuses correspond to two operators $\langle \mathcal{O}(0)\mathcal{O}(x)\rangle$ inserted at Euclidean coordinates 0 and $x \sim 1/Q$. The blue circles refer to a nonperturbative background field.

to calculate, so the studies focus on mesons made of quarks of different flavors for which it does not exist. A set of such results from the HPQCD collaboration is presented in Fig. 28.12.

Kaons (e.g. made of u and $\bar{s}$) satisfy this requirement: Note that what is shown are two sets, corresponding to photons striking either the heavier s or the lighter u quark. Both show that $Q^2F(Q^2)$ is approximately constant. The artificial particle called η_s was invented. It is different from the physical η or η'. Its definition can be explained as follows: Imagine that there are *two more* species of strange quarks, s' and s'', which have the same mass as the standard m_s. Their additional properties are as follows: (i) They are not identical, so the pair $\bar{s}'s$" cannot annihilate, and the correlators consist only of the connected one-loop diagram. Again,

$$Q^2F_{\eta_s}(Q^2) \approx 0.6\,\text{GeV}^2 \tag{28.33}$$

is approximately constant, with no indication of reaching the pQCD asymptotic value, which is significantly smaller.

For the charmed η_c meson, the vector form factor was calculated to significantly higher momentum transfer $Q^2 \sim 25\,\text{GeV}^2$. Again, $Q^2F_{\eta_s}$ remains approximately constant, also about twice as large as the asymptotic value. Multi-gluon diagrams lead to a correction of the order $(1 + 1.18\alpha_s)$ (Melic *et al.*, 1999), which can partly help bridge the gap between the lattice results and the pQCD for the vector form factor.

However, this correction would not help explain the puzzling lattice data for the scalar form factor, which indicate that it (times Q^2) is *also about constant*, in contradiction with the chirality flip suppression rule predicted by the gluon exchange mechanism. Furthermore, the numerical value of this constant is *larger* than for the vector form factor, by another factor of two! The gluon exchange cannot create the necessary chirality flip, except through a penalty factor of m_f/Q: Some other mechanism is needed!

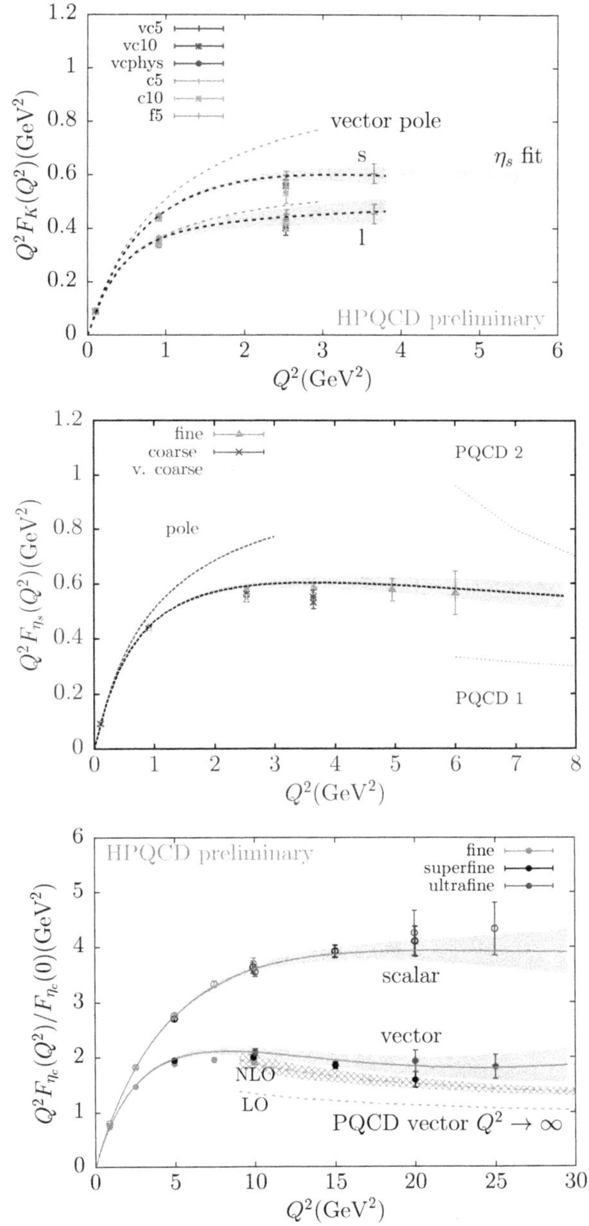

Fig. 28.12. Mesonic form factors calculated on the lattice by the HPQCD collaboration: K-meson (Davies *et al.*, 2018), η_s (Koponen *et al.*, 2017) and η_c (Davies *et al.*, 2018), top to bottom.

We have explored nonperturbative contributions to the "hard block" of mesonic form factors (Shuryak and Zahed, 2021). We calculated the photon, scalar, graviton and dilaton FFs for pion, rho and scalar a_0 (brother of η'). The field at the instanton center is rather strong, with $\rho = 1/3\,\text{fm}$ is large

$$G_{\text{rms}} \equiv \sqrt{\left(G^a_{\mu\nu}(0)\right)^2} = \sqrt{192}/\rho^2 \approx 5\,\text{GeV}^2,$$

comparable to Q^2 is the so-called semi-hard region studied.

One example, which was studied a lot experimentally, is the electromagnetic pion form factor, shown in Fig. 28.13. We accept the *factorization* of the hard block and DAs because the Q^2 scale is indeed large compared to the typical squared transverse momentum $\langle p_\perp^2 \rangle \sim 0.1\,\text{GeV}^2$ within a hadron, or the constituent quark mass $M_Q^2 \sim 0.1$–$0.15\,\text{GeV}^2$. In the Breit frame description of the form factors, conventional "collinear" kinematics should still hold.

Yet, we do include *nonperturbative contributions* to the hard block. At semi-hard momentum transfer Q, they are *not* at all negligible in comparison to gluon exchanges (diagram (a)).

It was found that they are mostly dominated by the diagram with three propagators made of nonzero Dirac modes (the left one on the

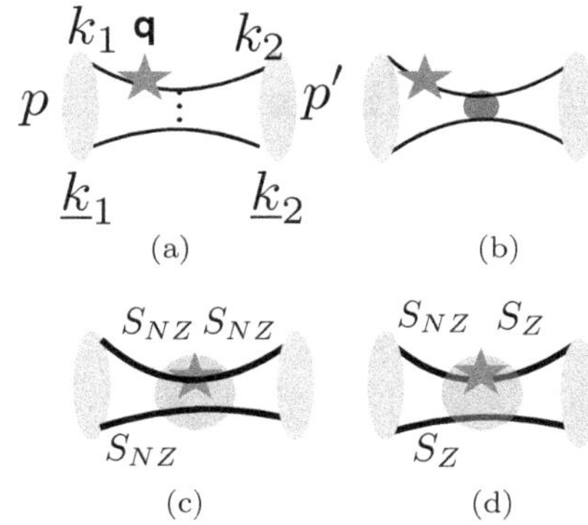

Fig. 28.13. The perturbative one-gluon exchange diagrams: (a) explains our notations for momenta of quarks and mesons. The thin solid lines are free quark propagators, the red star indicates the virtual photon (or scalar) vertex, bringing in large momentum q^μ, and the shaded ovals represent the mesonic DAs. The diagram (b) indicates a "Born-style contribution," in which the gluon propagator is substituted by the Fourier transform of the instanton field. The diagrams (c) and (d) contain three quark propagators in the instanton background (thick lines). In (c), all of them are S_{NZ}, made of nonzero Dirac modes, while in (d), two of them are S_Z made of quark zero modes. This last contribution will be referred to as the "'t Hooft-style term."

left), not the one with Dirac zero modes. Therefore, we included the contributions of the "dense instanton liquid," considering the diluteness parameter $\kappa = \pi^2 n_{I+\bar{I}}\rho^4$ as one, as shown in subsequent plots. As one can see, in this version, the sum of perturbative and instanton-induced form factor reproduces the data.

The expressions for perturbative, non-zero-mode and zero-mode parts are

$$\begin{aligned}
V_a^\pi(Q^2) &= \epsilon_\mu(q)(p^\mu + p'^\mu)\,(e_u + e_{\bar{d}}) \left[\left(\frac{2C_F\pi\alpha_s f_\pi^2}{N_c Q^2} \right) \right. \\
&\quad \times \int dx_1 dx_2 \left(\frac{1}{\bar{x}_1\bar{x}_2 + m_{\text{gluon}}^2/Q^2} \right) \left(\varphi_\pi(x_1)\varphi_\pi(x_2) \right. \\
&\quad + 2\frac{\chi_\pi^2}{Q^2} \left(\varphi_\pi^P(x_1)\varphi_\pi^P(x_2) \left(\frac{1}{\bar{x}_2 + E_\perp^2/Q^2} - 1 \right) + \frac{1}{6}\varphi_\pi^P(x_1)\varphi_\pi^{'T}(x_2) \right. \\
&\quad \left.\left.\left. \times \left(\frac{1}{\bar{x}_2 + E_\perp^2/Q^2} + 1 \right) \right) \right) \right]
\end{aligned} \tag{28.34}$$

$$\begin{aligned}
V_c^\pi &= \epsilon_\mu(q)(p^\mu + p'^\mu)\,(e_u + e_{\bar{d}}) \left[\frac{\kappa\pi^2 f_\pi^2 \chi_\pi^2}{N_c M_Q^2} \langle \rho^2 \mathbb{G}_V(Q\rho) \rangle \right. \\
&\quad \left. \times \int dx_1 dx_2 \bar{x}_1 \left(\varphi_\pi^P(x_1)\varphi_\pi^P(x_2) - \frac{1}{36}\varphi_\pi^{'T}(x_1)\varphi_\pi^{'T}(x_2) \right) \right]
\end{aligned} \tag{28.35}$$

$$\begin{aligned}
V_d^\pi &= -\epsilon_\mu(q)(p^\mu + p'^\mu)\,(e_u + e_{\bar{d}}) \left[\left(\frac{1}{N_c^2(N_c+1)} \right) \frac{4\kappa\pi^2 f_\pi^2 \chi_\pi^2}{3M_Q^2} \right. \\
&\quad \left. \times \left\langle \rho^2 \frac{K_1(Q\rho)}{Q\rho} \right\rangle \int dx_1 dx_2 \varphi_\pi^P(x_1)\varphi_\pi^{'T}(x_2) \right]
\end{aligned} \tag{28.36}$$

Next-twist distribution amplitudes $\varphi_\pi^P, \varphi_\pi^T$ are associated with pseudoscalar γ_5 and tensor $\sigma^{\mu\nu}$ ones, and the leading twist φ_π is the standard matrix element of the axial current $\gamma_\mu\gamma_5$. For the plot, all three are taken simply to be constants independent of x. (Therefore, all contributions of the tensor DA, appearing as a derivative over x, actually vanish.)

Let us start with the first — pertubative — expression, the first part of which has been known in the literature since the 1970s (Chernyak and Zhitnitsky, 1977; Lepage and Brodsky, 1979; Efremov and Radyushkin, 1980). It predicted that $Q^2F_\pi(Q^2)$ is a constant at large Q, which agrees with the data. Unfortunately, the predicted magnitudes do not agree by a large factor. Next-twist functions were defined and included by Geshkenbein and Terent'ev (Geshkenbein and Terentev, 1982), the twist-3 DAs of the pion are numerically enhanced, so that their contribution may in fact be larger than that of the leading twist in the semi-hard Q^2 range of interest. Chernyak and collaborators (Chernyak and Zhitnitsky, 1984), using the QCD sum rules, arrived at the pion wave function

$$\varphi_{\mathrm{CZ}}(x) = 30x(1-x)(2x-1)^2 \tag{28.37}$$

known as "the double-hump" one. However, since then, no support for this shape has materialized, and it also does not agree with lattice results on momentum fractions, so we will not discuss it.

The instanton contribution is plotted for a dense instanton ensemble with its "diluteness parameter" $\kappa = (\pi^2 n)I + \bar{I}\rho^4 = 1$, which includes the so-called instanton–antiitnstanton molecules. As seen from the plot, it reproduces an approximately constant $Q^2 F_\pi(Q^2)$ in the shown interval, although at high Q, the instanton contribution disappears as $\sim \exp(-Q\rho)$, for a fixed instanton size, or as a power if averaging over size distribution is made.

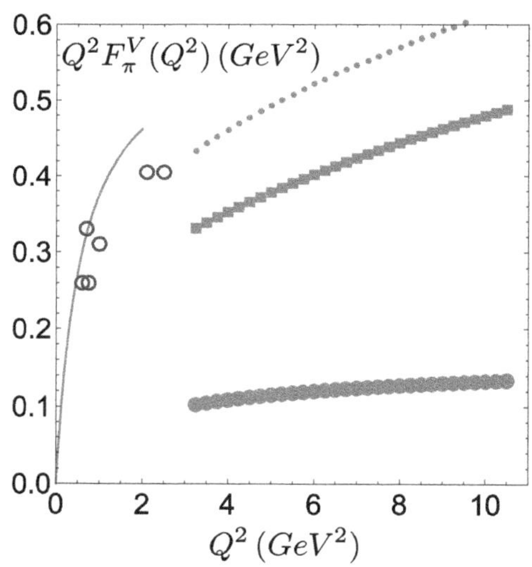

Fig. 28.14. The vector form factors of the pion times the square of momentum transfer, $Q^2F_\pi(Q^2)$ (GeV^2) versus Q^2(GeV^2). The closed disks show the total perturbative contribution. The squares correspond to the instanton contribution from the nonzero mode propagators S_{NZ}. The dotted line above is their sum. The solid line is the usual dipole fit with the rho meson mass, and the open points are from the experimental measurements. We do not show the multiple data points at smaller Q^2, where there is good agreement with the dipole formula.

28.8 Bridging the Gap between PDFs and LFWFs with a Focus on the Isospin Asymmetries of the Nucleon "Sea"

We already discussed that "hard" processes (electron–nucleon deep inelastic scattering, jet pair production, etc.) have led to rich "parton phenomenology" in the form of PDFs, transverse distribution functions (TDFs), etc. We also mentioned that phenomenology and lattice studies of PDF moments had explained rather well how partons share the nucleon momentum.

This is not so for nucleon spin. Naive expectations were that valence u, d quarks would share nearly all the spin of the nucleons, while gluons and quark–antiquark pairs of the "sea" would be mostly unpolarized. It was also expected that a admixture of states with nonzero angular momentum should be small. This turned out to be incorrect.

Theoretically, the problem is split into the polarizations of the sea and the gluons. **Puzzle one** is that (according to data and lattice) the antiquark polarization is *negative* (with respect to the proton spin).

Puzzle two is that the sea is also *polarized in isospin*, i.e. $\bar{u}$ and $\bar{d}$ distributions are significantly different, see Fig. 28.15.

A potential **puzzle three** is related to polarized gluon distribution $\Delta g(x)$: Its measurements at polarized pp collisions at the RHIC led to small values nearly consistent with zero. Multiple theoretical papers discussed the "chiral anomaly contribution" to the divergence of axial current (which gives the first moment of the quark polarized PDFs) related to the $G\tilde{G}$ operator and suggesting that instantons somehow may contribute to gluonic polarization. However, the data so far have error bars that are too large. A new project at Brookhaven — the Electron-Ion Collider — is expected to contribute to solving this puzzle.

In Shuryak (2019a), I have calculated LFWFs for several mesons and baryons, including the five-quark component of the latter related to the *isospin asymmetry puzzle*. Calculating admixture of 5-q states to the nucleon LFWF, I was able to quantify this effect and get the magnitude of the asymmetry and shape of first-generation antiquarks in qualitative agreement with the data. However, in this work, a complicated *nonlinear map* was used, with a rather complicated Laplacian and not-so-good convergence of the sums over basis functions. In a more recent paper (Shuryak and Zahed, 2023d), we suggested using (modified) Jacobi coordinates, similar to what is done for two and three quarks in this series of papers.

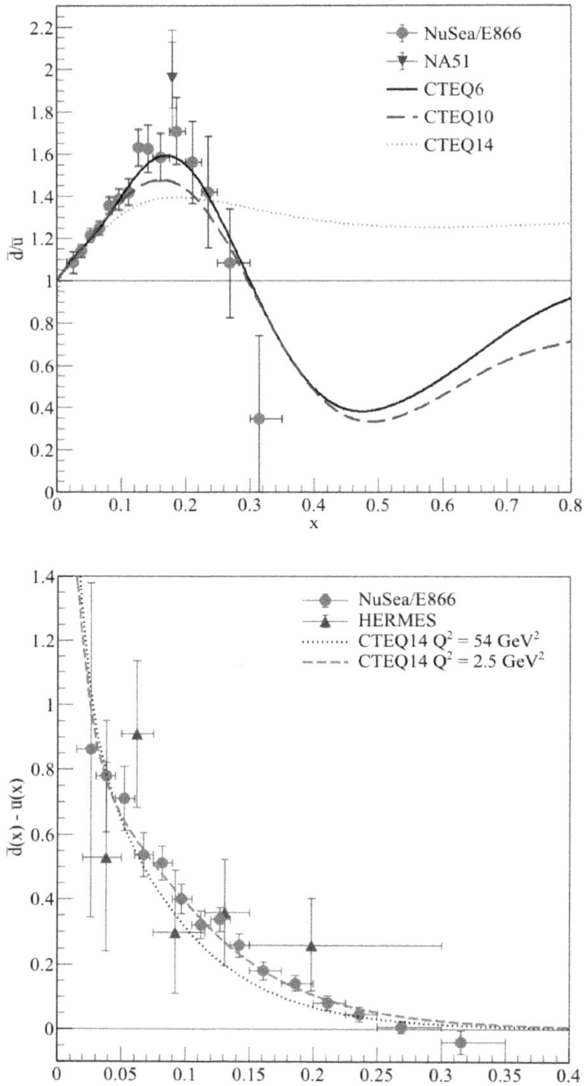

Fig. 28.15. Experimental data for the ratio $\bar{d}(x)/\bar{u}(x)$ (upper) and their difference (lower) versus x.

Let me skip the technicalities and go into the physical idea. The canonical process of quark pair production via gluons is "flavor blind," and yet, the antiquark sea of the nucleon is very asymmetric, as shown in Fig. 28.15 (right). As noted by Dorokhov and Kochelev (1993), the t'Hooft Lagrangian is strongly flavor-asymmetric; for example, the d quark can produce a $\bar{u}u$ pair (Fig. 28.15 (left)) but *not* $\bar{d}d$. If it were the only process, the $\bar{d}/\bar{u}$ ratio would be two, as there are two u quarks and only one d in the nucleon: not far from the data at certain x.

Calculating the 3-q LFWF for baryon, one finds that it is very different for octet (e.g. nucleons) and decuplet (e.g. $\Delta^{++} = uuu$), precisely because in the latter, the t'Hooft interaction does not contribute (it is flavor-asymmetric!). The nucleon wave function is not smooth and contains many harmonics with near-random coefficients.[14] By calculating 5-q wave functions, one can also determine the 5-q "tail" of the nucleons and thus *calculate* the antiquark sea. As shown in Fig. 28.16, it is similar in shape to the observed $\bar{d} - \bar{u}$ distribution shown in Fig. 28.15. Strong oscillations are perhaps due to an artificial restriction on the rather limited number of harmonics used.

Let me explain just the kinematical setting of both calculations, starting with mesons. For a two-body sector, one has $x_1 + x_2 = 1$; therefore, there

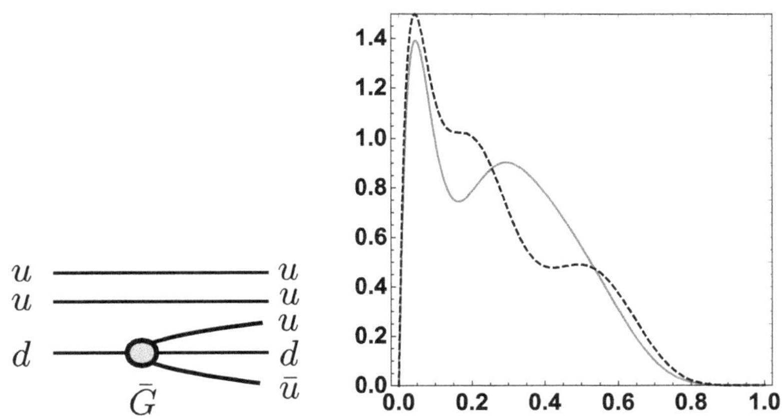

Fig. 28.16. Sea generation via the t'Hooft Lagrangian (left) and the $\bar{u}$ distributions in N (solid) and Δ (dashed) versus x.

[14]Which raises the question of whether three particles and two variables are enough for a "quantum billiard" of sorts.

is only one longitudinal variable s introduced, so that the total sum is 1:

$$x_1 = \frac{1+s}{2}, \quad x_2 = \frac{1-s}{2}$$

This map is still linear, common for both.

An idea of the *nonlinear* parameterization of x_i suggested by Shuryak (2019a) can be explained, for example, for five quarks:

$$x_1 = \frac{1}{2^4}(1+s)(1+t)(1+u)(1+w), \quad x_2 = \frac{1}{2^4}(1-s)(1+t)(1+u)(1+w)$$
$$x_3 = \frac{1}{2^3}(1-t)(1+u)(1+w), \quad x_4 = \frac{1}{2^2}(1-u)(1+w)$$
$$x_5 = 1 - x_1 - x_2 - x_3 - x_4 = \frac{1-w}{2}$$

Using these four variables, one can see that the integration measure factorizes into a product of factors depending on these four variables; therefore, one can use basis functions in the form of certain polynomials depending on them *separately*.

The Jacobi coordinates for this case would be defined as

$$\begin{aligned}
\vec{r}_1 &= \vec{x}_1 - \vec{x}_2, \quad \vec{r}_2 = \frac{1}{2}\vec{x}_1 + \frac{1}{2}\vec{x}_2 - \vec{x}_3, \\
\vec{r}_3 &= \frac{1}{3}\vec{x}_1 + \frac{1}{3}\vec{x}_2 + \frac{1}{3}\vec{x}_3 - \vec{x}_4, \\
\vec{r}_4 &= \frac{1}{4}\vec{x}_1 + \frac{1}{4}\vec{x}_2 + \frac{1}{4}\vec{x}_3 + \frac{1}{4}\vec{x}_4 - \vec{x}_5, \\
\vec{r}_5 &= \frac{1}{5}(\vec{x}_1 + \vec{x}_2 + \vec{x}_3 + \vec{x}_4 + \vec{x}_5)
\end{aligned} \tag{28.38}$$

with the determinant of this matrix — the Jacobian — being 1. The last coordinate $\vec{r}_5$ is the location of the center of mass if those are coordinates, or one-fifth of the total momentum if x_i are momentum fractions and is in any case redundant.

The physical domain of the five momentum fraction with a condition $\sum_1^5 x_i = 1$ is the 4d manifold called a *pentachoron* (or five-cell or four-simplex), which is one dimension higher than a tetrahedron. In Jacobian coordinates, the simplest "star-like" potential simplifies to a purely diagonal form:

$$\sum_1^5 \vec{x}_i^2 = \frac{1}{2}\vec{r}_1^2 + \frac{2}{3}\vec{r}_2^2 + \frac{3}{4}\vec{r}_3^2 + \frac{4}{5}\vec{r}_4^2 + 5\vec{r}_5^2 \tag{28.39}$$

Further *modification* of the Jacobi coordinates is done by a simple rescaling of $\vec{r}_i$

$$\vec{r}_1 \to \sqrt{2}\vec{\alpha},\ \vec{r}_2 \to \sqrt{3/2}\vec{\beta},\ \vec{r}_3 \to \sqrt{4/3}\vec{\gamma},$$
$$\vec{r}_4 \to \sqrt{5/4}\vec{\delta}, \vec{r}_5 \to \sqrt{1/5}\vec{\sigma} \tag{28.40}$$

which makes the "star-like" potential a sum of squares of Greek-letter coordinates. Another important potential, "*ansatz* A," is half the sum of 10 binary potentials. If it is assumed quadratic, it is proportional to the same sum of squares in terms of the Greek-letter coordinates:

$$V_A = \frac{1}{2}\sum_{i>j}(\vec{x}_i - \vec{x}_j)^2 \tag{28.41}$$

$$= \frac{5}{2}\left(\vec{\alpha}^2 + \vec{\beta}^2 + \vec{\gamma}^2 + \vec{\delta}^2\right) \tag{28.42}$$

except that the $\vec{\sigma}$ term is missing (which is unimportant, as it is a constant anyway). The four coordinates $\alpha, \beta, \gamma, \delta$ play the same role as ρ, λ for three quarks. We conjecture that the eigenstates of the Laplacian on the five-simplex can also be worked out analytically using some set of standing waves; see Shuryak and Zahed (2023f) for a discussion and for the ground state in this form.

Adding a quark–antiquark pair to 3-q baryons can be done in a number of ways. The straightforward one, used by Shuryak (2019a), is to add a 5-q sector to a 3-q one, calculate all matrix elements of the Hamiltonian and diagonalize it. However, with so many quantum numbers, this approach is, in practice, quite complex. In Shuryak and Zahed (2023d), two other approximations were pursued, close in spirit to the DGLAP evolution in the sense that extra $\bar{q}q$ are "radiated" with free phase space. The processes are illustrated in Fig. 28.17.

Let us begin by including the first order in the 't Hooft Lagrangian process, producing a quark pair. Let us denote by A_1 the amplitude of the $q\bar{q}$ pair production in the first order in the 't Hooft vertex, corresponding to the diagrams in Fig. 28.17 (c) or (d). Since it uses a process with the fermionic zero mode of the instanton, it should be proportional to a small instanton packing fraction, $\kappa = (\pi^2/2)\rho^4 n_{I+\bar{I}} \sim 1/10$, of the original instanton liquid model, which sets the order of magnitude of the effect.

All u- and d-quark emissions yield the final states of

$$u \to A_1^2\, u + A_1^2(d + \bar{d})$$
$$d \to A_1^2\, d + A_1^2(u + \bar{u}) \tag{28.43}$$

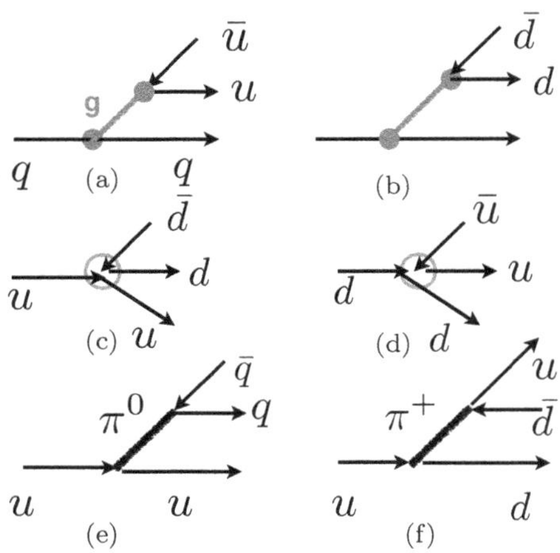

Fig. 28.17. Upper raw (a, b): Gluon-mediated quark pair production; middle raw (c, d): instanton-induced 't Hooft four-fermion interaction; lower raw (e, f): pion-mediated quark pair production, or iterated 't Hooft Lagrangian in s and t channels.

If we note that the probability for a quark to do nothing is $1 - A_1^2$, then the proton composition after one interaction is

$$(2 + A_1^2)u + (1 + 2A_1^2)d + A_1^2\bar{u} + 2A_1^2\bar{d} \tag{28.44}$$

which is valence quark preserving with $u_v = u - \bar{u} = 2$ and $d_v = d - \bar{d} = 1$. The parton distribution for the neutron is followed by isospin symmetry. In this schematic description of the proton and neutron sea contributions to the first order in A_1^2, the Gottfried sum rule reads

$$\frac{1}{3} + \frac{2}{3}(\bar{u} - \bar{d}) = \frac{1}{3} - \frac{2A_1^2}{3} \to 0.227 \tag{28.45}$$

Assuming the produced $\bar{d}$ in an unpolarized u, we calculate the *probability* of the process as a square of the amplitude defined by the old-fashioned perturbation theory:

$$dN_{\bar{d}/u} = \frac{G_{\text{tHooft}}^2}{2} \frac{\overline{|V_{u \to ud\bar{d}}|^2}}{(E_u - E_{\underline{u}} - E_d - E_{\bar{d}})^2} \times \frac{1}{2E_u} \frac{1}{2E_{\bar{d}}} \frac{1}{2E_{\underline{u}}} \frac{1}{2E_d} \frac{d^3K_{\underline{u}}}{(2\pi)^3} \frac{d^3K_d}{(2\pi)^3} \tag{28.46}$$

with the energy as the denominator and the bar on the matrix element referring to spin averaging over the the't Hooft vertex. In the vertex, there are terms proportional to the square of the constituent quark masse m_q^2 and to the square of momentad $\vec{p}^2$. The momenta are regulated by the instanton form factors; therefore, $\vec{p}^2 \sim 1/\rho^2$. For simplicity, we have ignored all terms with masses because $(\rho m_q)^2 \sim 1/4$ can be considered small. With this in mind, spin averaging gives

$$\overline{|V_{u\to ud\bar{d}}|^2} = \frac{1}{2}16(K_u \cdot K_{\underline{u}})\,(K_d \cdot K_{\bar{d}}) \tag{28.47}$$

with (28.46) taking the form

$$dN_{\bar{d}/u} = \frac{G^2_{\mathrm{tHooft}}}{4}\left[\frac{\frac{(\vec{k}_\perp+\vec{p}_\perp)^2}{2(1-x-z)}\left[\frac{\vec{k}^2_\perp}{2x}+\frac{\vec{p}^2_\perp}{2z}+\left(1-\frac{1}{x+z}\right)\frac{(\vec{k}_\perp+\vec{p}_\perp)^2}{2(1-x-z)}\right](x+z)}{\left(\frac{\vec{k}^2_\perp}{2x}+\frac{\vec{p}^2_\perp}{2z}+\frac{(\vec{k}_\perp+\vec{p}_\perp)^2}{2(1-x-z)}\right)^2}\right]$$
$$\times\,\frac{1}{z}\frac{dx}{x}\frac{dz}{(1-x-z)}\frac{d^2\vec{k}_\perp}{(2\pi)^3}\frac{d^2\vec{p}_\perp}{(2\pi)^3} \tag{28.48}$$

The sea distribution of $\bar{d}$ in an unpolarized constituent quark u is given by

$$N_{\bar{d}/u}(z,Q^2) = \int_{p_\perp}\int_{\vec{k}_\perp}\int_x \frac{dN_{\bar{d}/u}}{dz} \tag{28.49}$$

with the integrals carried sequentially in the ranges of $0 \le x \le 1$ and $0 \le p_\perp^2 \le Q^2$. The sea distribution $\bar{u}$ in an unpolarized constituent quark d is identical to $dN_{\bar{d}/u} = dN_{\bar{u}/d}$. Using the value of the coupling of the 't Hooft operator, one finds the probability of the process to be $A_1^2 \approx 0.11$, which sets the scale of the "primary sea" produced by $1 \to 3$ instanton-induced process. (The regulator on which the dependence is logarithmic is taken to be $\epsilon = 0.01$.) The next issue we address is the *shape* (x-dependence) of the PDFs of these sea (anti)quarks produced by the processes in Fig. 28.17 (c, d). Those can be obtained by convolution of this squared amplitude, treated as a "splitting function," with the original (valence) distributions $d^N(y), u^N(y)$ (e.g. those calculated from the LFWFs in the 3-q sector above). The unpolarized sea $\bar{u}, \bar{d}$ distributions in the nucleon are then

$$\bar{u}_N(x,Q^2) = \int_x^1 \frac{dy}{y}\,d^p(y,Q_0^2)N_{\bar{u}/d}\left(\frac{x}{y},Q^2\right)$$
$$\bar{d}_N(x,Q^2) = \int_x^1 \frac{dy}{y}\,u^p(y,Q_0^2)N_{\bar{d}/u}\left(\frac{x}{y},Q^2\right) \tag{28.50}$$

where $q^p(x)$ is the unpolarized flavor $f = u, d$ distribution in a proton p at the low-resolution point, which we argued above as $Q_0^2 \approx (2/\rho)^2$.

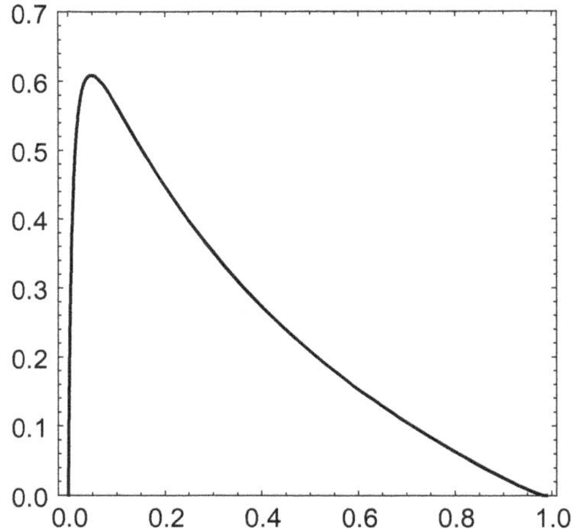

Fig. 28.18. The ratio of the "produced" sea PDF to the one which initiated it, $\bar{d}^N(x)/u^N(x)$, as a function of x following (28.50).

In Fig. 28.18. we plot the *ratio* of the produced sea PDF against the one which initiated it, $\bar{d}^N(x)/u^N(x)$. The sea is strongly shifted to small $x \sim 1/10$. At large x, if the initial PDF $u^N(x) \sim (1-x)^a$ has a certain power a, the produced one has power $a+1$. (While both of these features are clearly steps in the right direction, the reader is perhaps aware that the observed sea PDF has at small x negative power singularities and a much larger difference between powers of $(1-x)$ in $u^p(x)$ and $\bar{d}^p(x)$, being approximately three and seven, respectively. These features, however, are known to be generated by subsequent DGLAP evolution from Q_0 to the scale Q at which experiments are done.)

Another approximation sums up a sequence of the 't Hooft vertices, producing the pion. The "pion-induced" contributions to the PDFs, see Fig. 28.17 (e, f) (written in DGLAP-like form), were proposed by Eichten *et al.* (1992) three decades ago. Note that the π^0 diagrams (e) for the specific in and out quark pairs have a factor of $(1/2)^2$, while the π^+ diagram (f) leads to the flavor transition $u \to u\bar{d}$ with a larger factor 1. The former changes the flavor of the recoil quark, whereas the latter does not. Because these two diagrams have different probabilities, they also together lead to a flavor asymmetry of the sea. Note that these pion-induced diagrams can also be considered to be higher-order iterations of the 't Hooft vertex in different channels.

Here, we evaluate the contribution of the pion-induced antiquark production, following Eichten *et al.* (1992). Let the probability of the pion-generated pair production process be P_π, then all valence uud quarks together produce a sea with a probability of $P_\pi(7/4\bar{u} + 11/4\bar{d})$, or

$$\frac{\bar{d}}{\bar{u}} = \frac{11}{7} \approx 1.57 \tag{28.51}$$

which is in good agreement with the ratio reported experimentally. Using the absolute observed magnitude of $\bar{d} - \bar{u}$, one finds that in order to explain its integrated magnitude, one would need $P_\pi \approx 0.2$.

The expression for P_π from the chiral Lagrangian (Eichten *et al.*, 1992) (20, contains the following dimensionless combination of pion constants (in their notations)):

$$\frac{g_a^2 m_q^2}{8\pi^2 f_\pi^2} \approx 0.09 \tag{28.52}$$

times the certain integral, which is $O(1)$ (mildly depending on the upper cutoff Λ_{chiral}. As a result, P_π^{th}, given by the pion diagrams, gives about *half* of the empirical effect.

In summary, we conclude that the first order in the 't Hooft interaction and the iterated (pion) diagrams give comparable contributions to the integrated flavor asymmetry of the antiquark sea. Unfortunately, at this time, it is not possible to make a more quantitative evaluation that includes both.

The next question is how the antiquarks produced by the intermediate pions are distributed in x. For that, we adopt the expression (18) from Eichten *et al.* (1992), using the convolution of the quark PDFs in the nucleon $q^N(y)$ with the "splitting function" $P(z)$, followed by a convolution with the pion PDF $q^\pi(x/yz)$. Unlike Eichten *et al.*, however, we do not use here the PDFs fitted from experiments at some high Q^2 since our intention here is to build another — chiral — ark of the bridge. So, we take $q^N(y) = 12y(1-y)^2$ (approximately corresponding to the wave functions derived for three quarks with the quasi-local attraction above). We also take the symmetric PDF for the pion $q^\pi(x) = 6x(1-x)$, corresponding to a two-quark semicircular wave function. Convoluting those with the "splitting function" $P(z)$, we get the shape of the pion-induced antiquark PDF shown in Fig. 28.19 (lower). As one can see by comparing it to the upper experimental plot, it does reproduce the observed shape quite well.

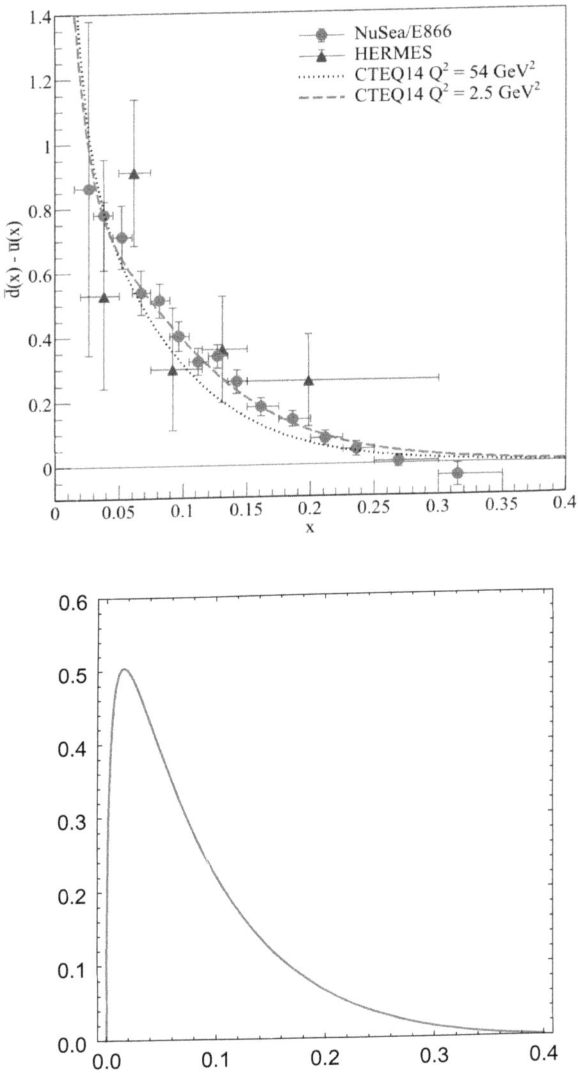

Fig. 28.19. Upper plot: The difference of the sea antiquark PDFs $\bar{d}(x) - \bar{u}(x)$ from experiments. Lower plot: The calculated shape of sea antiquark distribution (arbitrary units) described in the text.

28.9 Including Spin-Dependent Forces

So far, in the LF Hamiltonian, we have only included the kinetic energy, confinement and quasilocal "diquark pairing" via a 4-q scalar operator. Of course, there also exist the Coulomb and spin-dependent forces, and work is going on to determine their forms and calculate their effects on LFWFs.

Potentially, one needs to get the full spin–orbit structure of LFWFs. In order to explain what we mean, consider a meson with total *helicity* Λ (which is the longitudinal projection J_z). The total of two quark spins $\vec{S} = \vec{S}_1 + \vec{S}_2$ can be $S = 0$ or $S = 1$: In the former case, $\Lambda = L_z$, and in the latter, there are three cases: $L_z = \Lambda - 1, L_z = \Lambda, L_z = \Lambda + 1$. These four states (e.g. χ_b, h_b) are generally mixed by spin-dependent forces. Schematically, for the last three, it takes the form (the index of ψ is the longitudinal projection of orbital momentum, or M_L)

$$(\psi_{\Lambda-1}, \psi_\Lambda, \psi_{\Lambda+1}) \begin{pmatrix} V_{SS} & V_{SL} & V_T \\ V_{SL} & V_{SS} & V_{SL} \\ V_T & V_{SL} & V_{SS} \end{pmatrix} \begin{pmatrix} \psi_{\Lambda-1} \\ \psi_\Lambda \\ \psi_{\Lambda+1} \end{pmatrix}$$

because the spin–orbit V_{SL} changes L by ± 1 and the tensor term V_T by ± 2. So, in general, any meson has *three wave functions*, related by spin-dependent forces.

The case of baryons is even more complex: A nucleon (or other S = 1/2 baryon) may have up to *six* different functions, as defined by Ji *et al.* (2004). At the time of this writing, their calculation is not yet done.

Let us return to vector mesons: Interest in them is related to the fact that they constitute the simplest case in which a hadron should have a *quadrupole moment*. To explain why this structure is important, let me present an example. Let meson helicity $\Lambda = 1$ (so, it is, for example, the longitudinally polarized vector meson, $\Upsilon \ldots \rho$). The tensor force V_T term in the Hamiltonian relates ψ_0 and ψ_2, which therefore will get admixed together in eigenstates of fixed mass.[15] If the admixture is small, it is given by the first order of perturbation theory:

$$\epsilon_{02} = \sqrt{8} \frac{\int drr^2 \psi_0(r) V_T(r) \psi_2(r)}{M_2^2 - M_0^2} \tag{28.53}$$

[15] In the CM frame, such phenomena are well known in nuclear physics, for example, the $d = pn$ state has total $J = 1$. It is also a mixture of nonrelativistic $L = 0, S = 1$ and $L = 2, S = 1$ states, generating a nonzero quadrupole moment.

The first factor comes from the angular integral between two functions, as follows:

$$\int d\Omega (Y_0^0 \chi_1^1) S_{12} \left(\sqrt{\frac{3}{5}} Y_2^2 \chi_1^{-1} - \sqrt{\frac{3}{10}} Y_2^1 \chi_1^0 + \frac{1}{\sqrt{10}} Y_2^0 \chi_1^1 \right) = \sqrt{8} \quad (28.54)$$

The physical result of the mixing is that vector mesons obtain quadrupole moment:

$$\langle Q \rangle_\Upsilon = \frac{\epsilon_{02}}{5\sqrt{2}} \int dr r^4 \psi_0(r) \psi_2(r) \quad (28.55)$$

In the especially important case of $\Lambda = 0$, it simplifies to two functions: Following Ji *et al.* (2004), one has for pseudoscalar and vector mesons made of light u, d quarks the following expressions:

$$|\pi^+\rangle = \int d[1]d[2] \frac{\delta_{ij}}{\sqrt{3}} \left[\psi_0^\pi(x, k_\perp) \left(u_i^\uparrow(1) \bar{d}_j^\downarrow(2) - u_i^\downarrow(1) \bar{d}_j^\uparrow(2) \right) \right.$$
$$\left. + \psi_1^\pi(x, k_\perp) \left(k_\perp^- u_i^\uparrow(1) \bar{d}_j^\uparrow(2) + k_\perp^+ u_i^\downarrow(1) \bar{d}_j^\downarrow(2) \right) \right] \quad (28.56)$$

$$|\rho^+\rangle = \int d[1]d[2] \frac{\delta_{ij}}{\sqrt{3}} \left[\psi_0^\rho(x, k_\perp) \left(u_i^\uparrow(1) \bar{d}_j^\downarrow(2) + u_i^\downarrow(1) \bar{d}_j^\uparrow(2) \right) \right.$$
$$\left. + \psi_1^\rho(x, k_\perp) \left(k_\perp^- u_i^\uparrow(1) \bar{d}_j^\uparrow(2) - k_\perp^+ u_i^\downarrow(1) \bar{d}_j^\downarrow(2) \right) \right] \quad (28.57)$$

The subscripts of the functions, 0 and 1, indicate L_z, the z-projections of the orbital momentum. The $\Lambda = 0$ state of the vector mesons are called "transversely polarized." The notation used here, $k_\perp^\pm = k_\perp \exp(\pm i\phi)$, with azimuthal angle ϕ explicitly carries a longitudinal angular momentum of ± 1.

Summary and connections: In general, this chapter describes efforts to *bridge the gap* between hadronic spectroscopy and partonic observables. The *first ark of the bridge* (described in detail in these series of works) is to transfer such quark models from the CM frame to the LF. For some of the simplest cases, such as heavy quarkonia, it amounts to a transition from spherical to cylindrical coordinates, with subsequent transformation of longitudinal momenta into the Bjorken–Feynman variable x. However, in general, it is easier to start with LF Hamiltonians H_{LF} and perform their quantization. One of the benefits is that no nonrelativistic approximation is needed; therefore, heavy and light quarks are treated in the same way.

The *second ark of the bridge* is built via *chiral dynamics*, which seeds the quark sea by producing an extra quark–antiquark pair. We discussed how it can be done in the first order in the 't Hooft effective action as well as via intermediate pions. It brings observables to the upper chiral scale: $\mu^2_{\text{chiral}} \sim 1\,\text{GeV}^2$.

The *third ark of the bridge* was not really discussed: Let me remind the reader that its goal is to relate through (modified) DGLAP evolution the observed PDFs (and other partonic observables), down to the scale of μ_{chiral}, at which there are *no gluons*, and the $q\bar{q}$ sea is reduced to only the part generated by chiral dynamics (as discussed in the preceding section). The antiquark flavor asymmetry $\bar{d} - \bar{u}$ is the tool that allows us to identify gluon and chiral contributions, as it cannot be generated by "flavor blind" gluons.

Chapter 29

HOLOGRAPHIC MODELS OF QCD

In Chapter 21, we discussed the applications of holography to strongly coupled conformal plasmas based on the original AdS/CFT correspondence. Here, we return to holography with a new goal, following the studies which introduced *confinement* into the "bulk" world and model a wide range of QCD properties, from hadronic spectroscopy to the thermal equation of state of hadronic matter.

29.1 AdS/QCD Models

The original AdS/CFT correspondence relates the string theory (approximated by classical gravity) in the bulk to the $N = 4$ super-Yang–Mills theory on its flat 4D boundary. As we noted a few times, we were able to do so because we discussed only the high-temperature phase of this gauge theory, which is a strongly coupled, conformally invariant QGP. We argued that it should be rather similar to the sQGP of QCD, at least in some temperature range.

However, if one wants to apply holography to many other aspects of QCD, it is no longer possible to ignore the fact that $N = 4$ super-Yang–Mills and QCD are very different in many respects, in particular, they have entirely different vacuums. The former has a zero beta function and keeps its conformal symmetry at all scales, while QCD has a logarithmically running coupling. There is no confinement in $N = 4$ super-Yang–Mills, despite a large coupling constant, while in the QCD vacuum and at $T < T_c$, hadronic phase confinement is the most famous feature.

Recall the Maldacena dipole: When the distance between charges r becomes large, the string pending in the bulk also becomes large, sinking further into the z direction. The metric is such that the energy of this string is smaller at larger z, as the total string energy decreases as $V \sim 1/r$. In the confining theory, one would instead like to have a linear potential $V \sim r$. This may happen if the pending string finds some obstruction, preventing it from sinking into the z direction beyond a certain maximal value of $z_{\max}$.

The simplest way to do so, following Polchinski and Strassler (2002), is simply to cut off the holographic space, selecting the metric in such a way that its volume element vanishes at large z:

$$\sqrt{\det\left[g_{\mu\nu}(z \to z_{\max})\right]} \to 0$$

This will do the job: A string connecting two charges would stabilize if r got large, reaching what is called "the space bottom." If so, its energy would become linear, $\sim r$, at large r.

Of course, simply writing down some arbitrary metrics is not elegant and not consistent with GR. However, with extra fields, one can construct an action and then a GR solution to it, also with a similar "space bottom." One such solution, due to Karch *et al.* (2006a), includes a certain dilaton profile, not only reproducing confinement but also generating linear Regge trajectories.

Modern strong coupling models moved into what is collectively known as AdS/QCD; for a review, see, for example, Gursoy and Kiritsis (2008) and Gursoy *et al.* (2008). In contrast with the original AdS/CFT, the background metric is not conformally invariant and incorporates a specially tuned contribution of the *dilaton* (a scalar field associated with the trace of the metric). To keep the construction consistent, one can derive equations for the metric and dilaton from some model Lagrangian.

These models add several additional components to the bulk theory with the aim of modeling QCD:

- Scalar ("dilaton") field ϕ, which has some phenomenological potential. It is related directly to the 't Hooft coupling λ by

$$\lambda = N_c e^{\phi} \tag{29.1}$$

For convenience, the metric is rescaled to the "Einstein frame" metric

$$g^{E}_{\mu\nu} = \lambda^{-\frac{4}{3}} g_{\mu\nu} \tag{29.2}$$

The bulk action has the form

$$\int d^5x\sqrt{g}\left[R - \frac{4}{3}\frac{(\partial\lambda)^2}{\lambda^2} - \frac{\lambda^2}{2N_c^2} + (\partial a)^2 + V(\lambda)\right] \tag{29.3}$$

where a is another additional pseudoscalar (*axion*) field, which is mostly neglected later. The action is constructed in such a way that both *confinement* in the IR and the *asymptotic freedom* in the UV have the desired properties. Usually, one selects the metrics by

$$ds^2 = e^{2A(r)}[dr^2 + \eta^M_{\mu\nu} dx^\mu dx^\nu] \tag{29.4}$$

with only one function depending on the fifth coordinate r. Then, given a potential V, the action produces the equations of motion for $A(r), \lambda(r)$, which can be solved.

- If one would like to have quarks in *fundamental* color representation, one can be taken aback by the fact that those do not exist on the D_3 branes. Looking into the "brane dictionary," one would find that they are present in different branes, namely D_4.[1] By adding a stack of N_f (flavors) of those (to the original stack of N_c of the D_3 branes), it will model QCD.

Now, there is an option when considering the limits. One can either retain the original large-number-of-color limit

$$N_c \to \infty, N_f = \text{const}, \quad \text{(the 't Hooft limit)}$$

or set both the numbers of colors and flavors to be large, with some fixed ratio between them:

$$N_c \to \infty, N_f \to \infty, \frac{N_f}{N_c} \equiv x = \text{fixed}, \quad \text{(the Veneziano limit)}$$

The AdS/QCD models in the Veneziano limit are more complex since they include the backreaction of quarks on the vacuum background. They also include more fields, such as the quark–antiquark scalar field responsible for the quark condensate and chiral symmetry breaking. The vacuum corresponds to classical background fields, defined by the solution to Einstein and field theory equations. We will not discuss those in detail; for a good review, see, for example, Gursoy *et al.* (2011).

The next step involves writing down the equations for small perturbations. Keeping in the action only the second-order terms leads to linear equations of motion. The resulting sets of *graviton*-based excitations physically correspond to *radially* excited

$$J^{PC} = 2^{++}, 0^{++}, 0^{-+}$$

glueballs. Table 29.1 lists some results from Gursoy *et al.* (2011), which, when compared with lattice ($N_c = 3$ gluodynamics) (Chen *et al.*, 2006), demonstrate rather good agreement.

[1]The extra dimension of D_4 is selected to be five, which means the fermions do live in the bulk.

Table 29.1. Glueball mass ratios for AdS/CFT model and lattice. The first line is for the first excited scalar mass in units of the lowest scalar, etc.

	Holography	Lattice
$m^*_{0^{++}}/m_{0^{++}}$	1.61	1.56(11)
$m_{2^{++}}/m_{0^{++}}$	1.36	1.40(4)
$m_{0^{-+}}/m_{0^{++}}$	2.10	2.12(10)

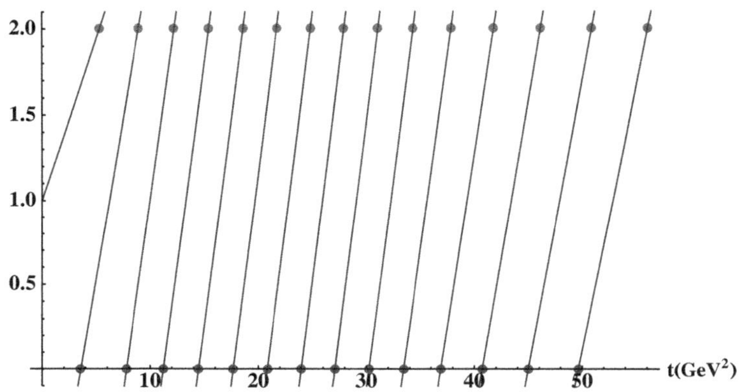

Fig. 29.1. Excited states of graviton-induced 2^{++} glueballs and dilaton-induced 0^{++} glueballs on a Regge plot of J versus $t = m^2$.

Many more excited states with $2^{++}, 0^{++}$ quantum numbers have been calculated in the same background model in our paper (Iatrakis *et al.*, 2016), see Fig. 29.1. As one can see from it, the slopes of the Regge trajectories (defined by two of them) are rather close. However, the slope of the first (Pomeron) trajectory may well be different from those of other Regge trajectories, as we lack the base point for it. (The value set at 1, as drawn, is only a guess.)

Using the setting of AdS/QCD in the Veneziano limit, let us focus on one particular phenomenon: a *mixing of the scalar mesons with scalar glueballs*, which would be needed for the following section.

In general, the mixing of scalar mesons is one of the most complex topics in light hadron spectroscopy. Without going into details, let me just say that the mixing is found to be small and the resulting masses of the (mixed glueball–meson) scalar states agree well with the experiment.

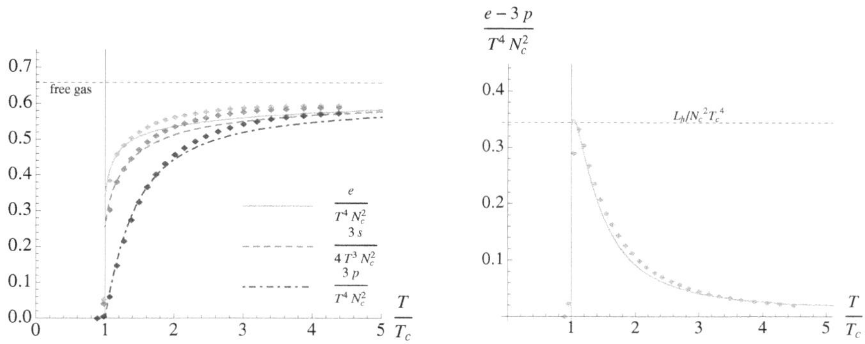

Fig. 29.2. (Left) Temperature dependence of the dimensionless thermodynamic variables in the QGP phase. The lines are AdS/QCD, and the points are lattice. (Right) The same for the "scale anomaly" combination.

In the setting under consideration, one can also describe (pure glue) thermodynamics in the QGP phase[2] ($T > T_c$) by generating numerically the corresponding black hole solutions with a variable location of the horizon, i.e. a variable T. Calculating the free energy of the solution, one finds $p(T)$ and thus all thermodynamical quantities. The comparison of the results for the simplest graviton-dilaton model with the lattice results for pure gauge theory is shown in Fig. 29.2.

29.2 Holography Describes QCD Strings and Their Interactions

AdS/CFT is a duality with string theory, so fundamental strings are naturally present in the bulk. Already, the first calculation of the "modified Coulomb law" in Maldacena's original paper was based on the evaluation of the shape and total energy of a "pending string," sourced by "quarks" on the boundary.

In AdS/QCD models with dilaton backgrounds, the bulk strings experience gravity force, acting toward large z, and dilaton force, acting in the opposite direction. As a result, they can "levitate" at some equilibrium position z^*, at which the downward gravity force is compensated by the uplifting dilaton gradient. The hologram of such a levitating string at the

[2]Note that both in AdS/QCD and gluodynamics, the phase transition is of the first order, and the hadronic phase is described in a completely different way.

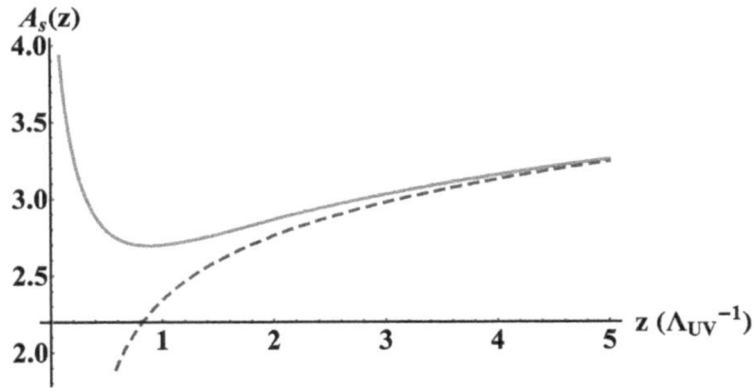

Fig. 29.3. The combination $A_s(z)$ as a function of the holographic coordinate z (solid) compared to its IR (large z) asymptotics (dashed). $A_s(z)$ has a minimum corresponding to the equilibrium scale of the QCD string.

boundary is perceived by an observer living in boundary gauge theory as a *QCD string*. Its tension, width and even distribution of all stress tensor components are all calculable as a hologram.

Since in the UV, these models possess a weak coupling regime, one can model a perturbative glasma by setting a certain density of color sources on the two planes, departing from each other. In such a setting, there would be a smooth transition between two alternative descriptions of the initial state that we discussed above — the perturbative glasma and a "spaghetti" made of the QCD strings. When the time τ is small (Fig. 21.11(a)), the strings are in the UV domain (at a small distance of $z \sim 1/Q_s$ from the boundary) and their hologram is Coulombic or glasma-like. When the strings fall further and reach the "levitation point" z^*, they begin oscillating around it (Fig. 21.11(b)) (Iatrakis *et al.*, 2015).

The AdS/QCD predictions for string–string interaction were derived by Iatrakis *et al.* (2015). In the QCD context, the long-range string–string attraction is dominated by the σ meson exchanges.[3]

In the AdS/QCD models, the bulk fields of this model — gravitons, dilaton and (quark-related scalar) "tachyon" — become quantized along the fifth coordinate, generating the Regge towers of four-dimensional hadronic states. In fact, from this approach, one can calculate not only the masses but also the *wave functions* in a scale space (i.e. in z). Because all of these fields interact with each other, the linearized equations describe the *mixing*

[3]Note that it is similar to that between nucleons in nuclear forces. However, there is no vector omega-based repulsive force for strings, as there are no corresponding vector charges.

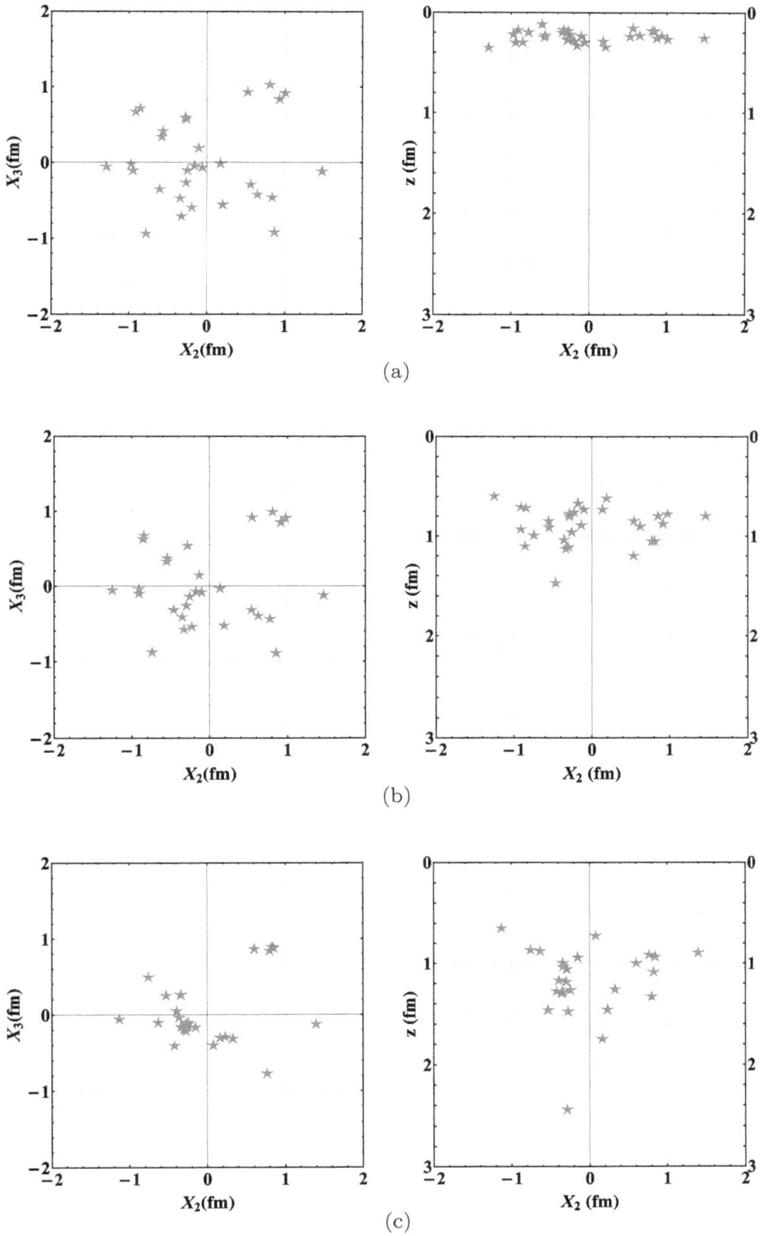

Fig. 29.4. Snapshots in the transverse (left) and holographic (right) planes of a falling 30 string configuration initially centered at $z = 0.44$ at $t =$ (a) 0, (b) 1.0 and (c) 1.32 fm/c.

between their excitations. The specific issue studied by Iatrakis *et al.* (2015) is that between mesonic flavorless scalars, which include the σ, with scalar glueballs. Without any changes in the setting of AdS/QCD, we found a very good description of the (quite involved) mixing pattern of the scalars, which puzzled spectroscopists for decades.

Since strings are gluonic objects and the σ interacts strongly with quarks, understanding such mixing is crucial for obtaining realistic string–string forces. Note that, so far, temperature or entropy have not appeared in the problem: The dynamics is given by the classical mechanics of strings moving in certain backgrounds. If the number (or density) of strings becomes high enough, as shown in Fig. 21.11(c), one should include the back reaction of their gravity and dilaton fields, or even include the mutual attraction of strings. Note that this AdS/QCD version of multi-string dynamics (Iatrakis *et al.*, 2015) is the holographic dual description of "spaghetti collapse," which was discussed without holography by Kalaydzhyan and Shuryak (2014a) in connection with pA central collisions.

29.3 Can Holography Be Used for Hard Processes and Partonic Observables?

Holography and pQCD have little in common, and their predictions do not agree in general. While the latter describes hard processes in the weak coupling regime, the former addresses QCD processes at strongly coupled and large N_c in the semi-hard and soft regimes.

In holography, a struck quark with large momentum does not create a jet but rather a very wide flow of energy (Lin and Shuryak, 2008a). In its evolution, all the partons migrate to small x (Polchinski and Strassler, 2002; Hatta *et al.*, 2008), where the physics is well captured by open and closed string excitations (Pomerons and Reggeons).

Inelastic DIS and *hard elastic* scattering are well defined in holography (Polchinski and Strassler, 2002), with structure functions and form factors. There is certain confusion in the literature about the relationship of these results to pQCD and partonic physics.

Holography and pQCD both have scattering rules for hard *elastic* processes, as originally noted by Polchinski and Strassler (2002). The hard elastic turn requires the involvement of the full hadron in the conformal limit features that are shared by both holography and QCD. In pQCD, this follows from the fact that all the partons in a given hadron have to undergo a "turn-around" under hard scattering, which implies a probability of $(1/Q^2)^{n-1}$, with n being the number of partonic constituents in the hadron (Brodsky and Farrar, 1973). In holography, a virtual photon has an extension of $1/Q$ in the holographic z-coordinate, and for a hard hit in bulk, the hadron also has to shrink to a size of $1/Q^{\tau-1}$ with a probability of $P \sim (1/Q^2)^{\tau-1}$. Here, τ is the "bulk" dimension of the field describing a hadron (of the corresponding double-trace 5D operators). For the elastic processes, weak and strong coupling scaling laws are both known, and they are similar to τ, identified as n (Polchinski and Strassler, 2002).

In *inelastic* DIS processes, the virtual photon of size $1/Q$ scatters on a quark parton with a form factor equal to one because any parton is point-like. In contrast, a scattering of a hadron can only happen if it shrinks to the same size, which happens with a probability given by the form factor. The corresponding wave function is $(1/Q^2)^{\tau-1}$. While τ is also sometimes called the "anomalous dimension," it depends on the hadron field behavior in 5D and has nothing to do with the "anomalous dimensions" of pQCD, which are the powers of α_s and the logarithms. Yet, neither $\alpha_s(Q^2)$ nor quarks or gluons are present in the bulk fields or actions. In DIS Bjorken, x (defined via $s = Q^2(1/x - 1)$) is constant. So, the $F_2(x)$ structure functions at large Q^2 and large $x \to 1$ are related. In fact, one can write the structure function in the generic form

$$F_2(x, Q^2) \sim Q^2 \left| \left(\frac{1}{Q^2}\right)^{\tau-1} \right|^2 \left(s = Q^2(1-x)\right)^{\alpha} \tag{29.5}$$

where α is *arbitrary*. To reproduce the hard scattering rule asymptotically, we *may* set it to some value, e.g. $\alpha = \tau - 2$,

$$F_2(x, Q^2) \to \left(\frac{1}{Q^2}\right)^{\tau-1} (1-x)^{\tau-2} \tag{29.6}$$

to reproduce the holographic DIS result on a nucleon with a "reasonable" $\tau = 3$, or for a meson with $\tau = 2$ (Polchinski and Strassler, 2002). Yet, there is still no Bjorken scaling at large Q^2!

Furthermore, the large x behavior in (29.6) is different from that expected from the Drell–Yan–West scaling rule (Drell and Yan, 1970; West, 1970). To reproduce it, one would require Q^2 independence of the structure function (29.5) as per Bjorken scaling, which fixes another power of $\alpha = 2\tau - 3$ in (29.6). Indeed, this rule follows from the existence of the LFWFs in terms of constituents, which are absent in holography.

A proposal to have a common setting in which weak and strong coupling regimes can be combined was made by Shuryak (2006a). The 5D "bulk" space is divided into two respective parts, divided by the *instanton wall* (corresponding to the instanton size) at which the coupling jumps. This setting allows for a smooth connection between wave functions of hadrons and virtual photons at the "wall," the border of the two parts of space, ensuring a smooth transition of the amplitudes of hard processes (form factors, etc.) from the soft to hard regime.

Summary and connections: In this chapter, we returned to the discussion of the holographic AdS/CFT correspondence that we began in Chapter 21. Its main idea is to introduce nontrivial background using the dilaton field with a certain profile, which will ensure the confining vacuum. We have shown examples in which quantized graviton fields lead to reasonable glueball spectra and Regge trajectories. The inclusion of fermions and holding to the Veneziano limit, in which the number of flavors and colors is equally large, produces a quark condensate and scalar and vector mesonic fields. We have focused on meson–glueball mixing and the collective interaction of multiple QCD strings in the "spaghetti" configuration. Many more applications of AdS/QCD can be found in the references given.

Chapter 30

SUMMARY AND DISCUSSION

30.1 Heavy Ion Collisions

30.1.1 *Equilibrium at the freeze-out*

Perhaps the most spectacular experimental fact from the heavy ion program is the very high accuracy of thermal equilibration reached at the chemical freeze-out stage. The yields of all secondaries, from pions ($m \approx 140$ MeV) to other mesons, baryons, hyperons and even light nuclei d, t, He^3, He^4 ($m \approx 2000$–4000 MeV), were found to be located on the same thermal exponent with $T_{\rm ch} \approx 150$ MeV. Chemical freeze-out seems to be universal for all collisions and centralities and is very close to the phase transition line.

We are a bit less certain about conditions at the final kinetic freeze-out, but we do know that T_f can go lower than 100 MeV for the central collisions of the heaviest ions. Evidence for that comes from the magnitude of the radial flow and also from femtoscopy. There is evidence for pion chemical potential building up between $T_{\rm ch}$ and T_f, which for pions seems to reach about half of the pion mass. Even the largest fireballs we can produce are apparently not large enough to cool a system until the Bose–Einstein condensation of the pions.

We know from lattice calculations the critical temperature $T_c \approx 155\,\text{MeV}$, which is remarkably close to $T_{\rm ch}$. It tells us that all inelastic reactions in the hadronic phase must disappear very quickly. Cascade "afterburners," traditionally used for freeze-out do not have a really good explanation for that. Perhaps, they are missing a huge number of multi-hadron processes and do not have perfect detailed balance, as they show much larger deviations from equilibrium than actually observed.

30.1.2 *Hydrodynamical explosion*

The major experimental discovery from the first years of RHIC was the observation of robust radial and elliptic flows in perfect agreement with hydrodynamical predictions.

Further quantitative studies of the elliptic flow had shown that elliptic anisotropy reaches very high values at its maximum, at $p_t \sim 3$ GeV. Theoretical studies of these data lead to the conclusion that the viscosity of QGP matter (or the mean free path, in kinetic language) must be remarkably small. This leads to (the often repeated) slogan that "*QGP is the most perfect fluid ever known.*"

The major experimental discovery from the first years of LHC operation was the observation of collective anisotropies in even "smaller systems": the central pA and even the high-multiplicity pp collisions. Their hydrodynamical description was further put to the test by dAu,[3] $HeAu$ collisions at the RHIC, which it passed excellently. There is, therefore, no doubt that such "small systems" do produce fireballs exploding hydrodynamically.

However, the opposite approach, looking for dynamically generated azimuthal asymmetries, must be able to predict their magnitude for the smaller multiplicity events. The models are the same as those used for minimally biased pp, and the issue is sometimes known as "*the shape of the Pomeron*" issue. Recall that the Pomeron amplitude originated from the pQCD ladder diagrams in weak coupling and was then derived by a string-based approach as well. Both need to be developed much further to predict v_n correlations. Experiments are looking for some sign of transitions as one goes from smaller to the highest multiplicity bins in the pA and pp collisions. The location of the micro-to-macro transition in pA and pp collisions, as a function of multiplicity, is debatable. Data on the mean $p_\perp$ and slopes shown in Fig. 20.7 indicate smooth growth of the radial flow; however, we now know that the radial flow can be "faked." At the same time, $v_2\{2\}$ as a function of multiplicity is rather flat. Its version $v_2\{n\}, n > 2$, from multi-particle correlations, shows changes, but their understanding is still missing. The calculations of v_n in the dynamical models of the Pomeron are in progress. Perhaps, they will later explain the low-multiplicity side.

A very positive theoretical development is that groups working on all scenarios are currently attempting to determine the limits of their approaches. Within hydrodynamics, for example, one study focuses on higher gradients and their effects. In the string-based picture, we discussed a string–string interaction which was ignored for a long time by event generators. The theoretical justification for a successful hydrodynamical description of small systems is getting under control. A number of examples show rather effective cancellations of all higher-gradient corrections. Navier–Stokes approximation seems to be rather accurate, even in situations in which one hardly expects it to work.

Meanwhile, phenomenologists describe the data. Hydrodynamical treatment of high-multiplicity pA, pp events seem to be rather successful. They require surprisingly small initial sizes with very high temperatures. However, we do not really understand how such systems can be produced. In particular, the case of central pA collisions is contested between the IP-glasma model and a string-based initial state picture. So far, one has very little theoretical control over the initial state of the high multiplicity pp: If it exists anywhere, glasma should be there. It is difficult to study this system for statistical reasons; having said that, since this is the highest density system we have, it should be pursued.

30.1.3 *Sounds*

The first triumph of hydrodynamics, at the onset of the RHIC program, was the description of the "Little Bang." The magnitude of the radial and elliptic flows was measured and calculated as a function of $p_\perp$, centrality, rapidity, particle type and collision energy. A spectacularly successful description of higher azimuthal harmonics of the flow, with $m = 3$–6, followed. As it has been repeatedly emphasized, these are sounds propagating on top of the exploding fireball. The damping of these modes agrees with acoustic-inspired formulae.

Another phenomenon, well known for the Big Bang perturbations, is due to the presence of "phase factors." Common freeze-out times for all harmonics imply m-dependent phases. As m grows, the phases rotate, so one should see maxima/minima in the power spectrum. These are perhaps observed in the Little Bang as well. As a maximum, it is a triangular flow $m = 3$ stronger than the elliptic one $m = 2$ for the ultra-central bin. As a minimum, it is observed at $m = 7-8$, with larger magnitudes for $m = 9, 10$.

Further studies of the highest harmonics can open up opportunities to observe the sources of sounds *not* originating from the initial state, in particular the long-expected fluctuations from the phase transition. Finally, even in equilibrium, there must be fluctuations emitting sounds. Those generate nontrivial "loop corrections" to hydrodynamics. The observation of "sound background" in hadronic matter is another challenge in the field.

30.1.4 *Electromagnetic processes*

After decades of building experimental methods and generations of detectors, we have clear-cut evidence for thermal dilepton production, with masses of 1–3 GeV from the QGP phase. It is seen best from the NA60 fixed-target muon pair detector at CERN and by both PHENIX and STAR in electron pairs at RHIC.

For masses of 0.4–1 GeV, these experiments do see a strong modification of the ρ spectral density, as theoretically predicted from meson rescattering. We also see that those dileptons are produced late, as indicated by the full-scale radial and strong elliptic flows.

The emission rates and our understanding of the space-time profile of the collision match the data. The photon yield following from standard rates is too low, by a factor of 2–3, as compared to the data. Also, the observed photon v_2 is about twice as large as the calculated value. Both indicate that perhaps many processes, especially at the late (hadronic) stage of the collision, are not yet correctly incorporated in the models.

30.1.5 *Jet quenching*

The theory of hard processes — jets, charm/bottom production — is based on the factorization theorems and the concept of structure functions. It is a solid but very restrictive foundation. When one asks questions about jets in high-multiplicity bins of pp collisions, one soon realizes that the "corresponding structure functions" do not exist: That concept has only been defined for the minimally biased (untouched) proton in a strictly inclusive setting. Universal structure functions, which are measured rather than calculated, have served us since the 1970s. However, they cannot be used anymore. If a certain fluctuation of a nucleon is selected, new models and much more measurements are needed.

Unfortunately, there are severe practical limits: Hard processes reduce probability by several orders of magnitude, on top of a 5% trigger for central pA. Yet, high LHC luminosity plus specialized triggers should be enough to get to some of those issues in the near future. Current jet quenching data, even for minimum bias pA, remains to be understood. Scaling arguments, such as the ones we used for hydro in smaller but hotter systems, can and should be developed and compared with the data.

We argued above that in AA collisions, the jet quenching parameter $\hat{q}$ seems to be strongly enhanced at the near-critical $T \approx T_c$ region.

Small systems are more explosive and rapidly pass the stage of near freeze-out. This should play a significant role in jet quenching.

30.1.6 *Heavy quarks and quarkonia*

The LHC data confirmed what has already been hinted at by the RHIC data: A significant fraction of the observed charmonia originates from the *recombination* of charm quarks at the chemical freeze-out. The charmonia which survived through the Little Bang continues to be reduced with increasing collision energies. Such a major change from "heavy-like" to "light-quark-like" behavior of charm is clearly an important discovery.

It remains true that c, b quarks are produced differently from the light ones, namely in the initial partonic processes. Yet, their interaction with ambient matter is strong. At large p_t, we observe quenching $R_{AA}^{c,b}$ comparable to that of gluons or light quarks. At small p_t, we observe an elliptic flow of open charm and changes in spectra.

Langevin/Fokker–Planck studies, however, suggest that c quarks are not moving with the flow. At an early time, c, b quarks are produced with large $p_\perp$ and start decelerating due to drag, while the matter is slowly accelerating due to pressure gradients: Their velocities move toward each other, yet they do not match, even by the end. As a result, charm radial/elliptic flows are *not* given by the Cooper–Frye expression. The recombinant charmonia may perhaps be an exception. Whether they actually co-move with the flow still needs to be established.

On the theoretical front, Langevin/Fokker–Planck studies introduced new conceptual developments. In particular, we discussed a new set of solutions of those for charmonia, the quasi-equilibrium attractors with constant particle flux. Those states, unlike the original bound states of $J/\psi, \psi'$, etc., provide a convenient basis for the evaluation of the speed of relaxation and out-of-equilibrium corrections to current charm hadronization models.

30.1.7 *The initial stage*

Have we also discovered glasma, another phase of QCD in which the gluonic field is highly excited? Can this be described by the classical chaotic solutions to Yang-Mills equations? Frankly, we don't know yet.

One thing we know for sure: Nuclei are made of nucleons, and the amplitudes of initial flow harmonics are correctly obtained by Glauber-like models, together with their distributions. We can even trace $O(1\%)$

deviations of nuclear shapes between two isobars, which are related to different distributions of protons and neutrons.

If the initial stage were structureless glasma, with many more partons than nucleons, these fluctuations would be reduced by a factor of $1/\sqrt{N}$. So, even models starting with partons have Glauber clustering included.

We also know that for *pp* and light nuclei collisions, the initial state is well represented by the "Lund model" with QCD strings. How exactly the transition from independent string breaking to QGP explosion happens needs to be studied further. According to the NA61 experiment, which involves the collision of various ions, this *string-to-fireball* transition is actually rather sharp, and it occurs at $A \sim 10$.

There are more puzzles associated with the early stage. What is the mechanism of quark production and its timing? Also, for how long do momenta distributions remain strongly anisotropic? When exactly does hydrodynamics (say, its version of the first order in gradients) become valid? To answer these questions, one needs more sophisticated experiments, for example, with angular distribution of dileptons originating in the early stages.

30.2 Nonperturbative QCD and Hadrons

30.2.1 *QCD phase transition: A competition between electric and magnetic objects*

Numerical simulation of the gauge fields on the lattice became indisputably the mainstream approach to nonperturbative QCD. It is based on the first principles, and over the years, it overcame severe technical limitations, now operating with dynamical quarks at masses close to their physical values. The hadronic spectra obtained are now in excellent agreement with experimental spectroscopy. The equation of state of hot matter calculated on the lattice is also confirmed by hydro calculations of the heavy ion collisions.

At the same time, we would like to not only reproduce the critical temperature of the deconfinement/chiral transition, or hadronic masses but also understand *why* confinement and chiral breaking happen.

In particular, we previously described an emerging view on confinement/deconfinement phase transition as being due to a "fight" between electric (quarks and gluons) and magnetic (monopoles) objects. Let me recall that at high T, the monopole density is suppressed as powers of $1/\log(T)$; however, if T is reduced toward T_c, it grows and becomes comparable to that of quark/gluon quasiparticles. As monopoles undergo Bose–Einstein condensation, they expel gluo-electric fields into "flux tubes." This is a "victory" for the (magnetically charged) monopole condensate.

One can follow this scenario theoretically in detail within the framework of "Seiberg–Witten" theories with extended supersymmetry. Unfortunately, in QCD, these magnetic monopoles are not semiclassical, and one has to rely on their lattice definitions and other data. Also, monopoles — 3D objects or quasiparticles — can only be modeled via their path integrals, which is technically difficult.

Fortunately, we learned that there exists a "Poisson dual" to the monopole theory, in terms of "instanton-dyons." Those are topological soliton solutions, and one can use semiclassical theory. The studies in terms of instanton-dyons do reproduce well the lattice data on deconfinement and chiral phase transitions.

To be sure that they are the correct degrees of freedom to use, it is useful to extend QCD studies to various versions of the *QCD deformations.* One of them, namely the introduction of some powers of the Polyakov line, shifts and modifies the order of deconfinement, and the other, namely the changing quark periodicity phases, affects the chiral transition. More details on those deformations are needed, both from the lattice and semiclassical models.

30.2.2 *Hadrons on the light front and partonic observables*

Since the 1970s, the parton description of "hard processes" has been successfully made by pQCD, in relation to their "hard blocks" and "evolution" (dependence on Q^2). Yet, two other ingredients — the PDFs and fragmentation functions — were basically determined from fits to the experimental data.

One may expect that at a certain low scale, at which gluons are not resolved and, say, a nucleon is mostly made of three constituent quarks, these PDFs, DAs, form factors, etc., should match those calculated via the models of hadronic spectroscopy. Although we discussed the ongoing work in this direction, it would perhaps be fair to conclude that, at this moment, "bridging the gap" between the two branches of hadronic physics is far from complete.

Let me conclude the summary of this (mostly theoretical) part of the book with remarks on future experimental developments. "Bridging the gap" between hadronic physics and partonic observables will be the key task of the *Electron Ion Collider* (EIC), currently under construction at the Brookhaven National Laboratory. This facility, together with the RHIC ion and proton beams and an electron beam, will provide high-quality data on partonic observables in the next decade.

30.2.3 *Holographic QCD*

Another analytic approach we discussed is the so-called "down-up" holographic models, which describe hadrons "holographically," by certain weakly coupled fields in (mostly) 5D *bulk* space-times. In particular, certain dilaton backgrounds ensure the confinement and quantization of these fields, resulting automatically in hadrons belonging to Regge trajectories.

The quantized dilaton and gravity fluctuations describe $J^P = O^+$ and 2^+ sets of glueballs, supplemented by extra scalar and vector fields for $O^-, 1^-$ mesons. By now, the corresponding effective Lagrangians have passed many tests, so one can say that basic hadronic interactions (traditionally delegated to chiral effective Lagrangians) are now extended beyond the physics of pions.

Another very attractive feature of these models is that they naturally have strings whose holograms are supposed to be the QCD flux tubes. We discussed what these models tell us about the collective interactions of multi-string settings. Also important is that high-energy hadron collisions, described by exchanges of *Pomerons*, can also be derived using holographic models with quite promising results.

BIBLIOGRAPHY

Aaboud, M. *et al.* (2017). Femtoscopy with identified charged pions in proton-lead collisions at $\sqrt{s_{\rm NN}} = 5.02\,\text{TeV}$ with ATLAS, *Phys. Rev. C* **96**(6), 064908. doi: 10.1103/PhysRevC.96.064908, arXiv:1704.01621 [hep-ex].

Aaij, R. *et al.* (2022). Observation of an exotic narrow doubly charmed tetraquark, *Nat. Phys.* **18**(7), 751–754. doi: 10.1038/s41567-022-01614-y, arXiv:2109.01038 [hep-ex].

Aamodt, K. *et al.* (2011a). Femtoscopy of pp collisions at $\sqrt{s} = 0.9$ and $7\,\text{TeV}$ at the LHC with two-pion Bose-Einstein correlations, *Phys. Rev. D* **84**, 112004. doi: 10.1103/PhysRevD.84.112004, arXiv:1101.3665 [hep-ex].

Aamodt, K. *et al.* (2011b). Higher harmonic anisotropic flow measurements of charged particles in Pb-Pb collisions at $\sqrt{s_{NN}} = 2.76\,\text{TeV}$, *Phys. Rev. Lett.* **107**, 032301. doi: 10.1103/PhysRevLett.107.032301, arXiv:1105.3865 [nucl-ex].

Abdallah, M. *et al.* (2022a). Observation of global spin alignment of ϕ and K^{*0} vector mesons in nuclear collisions. arXiv:2204.02302 [hep-ph].

Abdallah, M. *et al.* (2022b). Search for the chiral magnetic effect with isobar collisions at $\sqrt{s_{NN}} = 200\,\text{GeV}$ by the STAR collaboration at the BNL Relativistic Heavy Ion Collider, *Phys. Rev. C* **105**(1), 014901. doi: 10.1103/PhysRevC.105.014901, arXiv:2109.00131 [nucl-ex].

Abelev, B. *et al.* (2013a). Anisotropic flow of charged hadrons, pions and (anti-)protons measured at high transverse momentum in Pb-Pb collisions at $\sqrt{s_{NN}} = 2.76\,\text{TeV}$, *Phys. Lett. B* **719**, 18–28. doi: 10.1016/j.physletb.2012.12.066, arXiv:1205.5761 [nucl-ex].

Abelev, B. *et al.* (2013b). Centrality dependence of charged particle production at large transverse momentum in Pb–Pb collisions at $\sqrt{s_{\rm NN}} = 2.76\,\text{TeV}$, *Phys. Lett. B* **720**, 52–62. doi: 10.1016/j.physletb.2013.01.051, arXiv:1208.2711 [hep-ex].

Abelev, B. B. *et al.* (2014). Production of charged pions, kaons and protons at large transverse momenta in pp and Pb–Pb collisions at $\sqrt{s_{\rm NN}} = 2.76\,\text{TeV}$, *Phys. Lett. B* **736**, 196–207. doi: 10.1016/j.physletb.2014.07.011, arXiv:1401.1250 [nucl-ex].

Acharya, S. *et al.* (2020). Higher harmonic non-linear flow modes of charged hadrons in Pb-Pb collisions at $\sqrt{s_{\mathrm{NN}}}$ = 5.02 TeV, *JHEP* **05**, 085. 940 doi: 10.1007/JHEP05(2020)085, arXiv:2002.00633 [nucl-ex].

Adam, J. *et al.* (2021). Nonmonotonic energy dependence of net-proton number fluctuations, *Phys. Rev. Lett.* **126**(9), 092301. doi: 10.1103/PhysRevLett.126.092301, arXiv:2001.02852 [nucl-ex].

Adamczewski-Musch, J. *et al.* (2020). Proton-number fluctuations in $\sqrt{s_{NN}}$ = 2.4 GeV Au + Au collisions studied with the High-Acceptance DiElectron Spectrometer (HADES), *Phys. Rev. C* **102**(2), 024914. doi: 10.1103/PhysRevC.102.024914, arXiv:2002.08701 [nucl-ex].

Adamczyk, L. *et al.* (2017). Global Λ hyperon polarization in nuclear collisions: Evidence for the most vortical fluid, *Nature* **548**, 62–65. doi: 10.1038/nature23004, arXiv:1701.06657 [nucl-ex].

Adams, J. *et al.* (2004). Azimuthally sensitive HBT in Au + Au collisions at s(NN)**(1/2) = 200-GeV, *Phys. Rev. Lett.* **93**, 012301. doi: 10.1103/PhysRevLett.93.012301, arXiv:nucl-ex/0312009.

Adare, A. *et al.* (2008). Suppression pattern of neutral pions at high transverse momentum in Au+Au collisions at $\sqrt{s_{NN}}$ = 200 GeV and constraints on medium transport coefficients, *Phys. Rev. Lett.* **101**, 232301. doi: 10.1103/PhysRevLett.101.232301, arXiv:0801.4020 [nucl-ex].

Adare, A. *et al.* (2010). Azimuthal anisotropy of neutral pion production in Au+Au collisions at $\sqrt{(s_N N)}$ = 200 GeV: Path-length dependence of jet quenching and the role of initial geometry, *Phys. Rev. Lett.* **105**, 142301. doi: 10.1103/PhysRevLett.105.142301, arXiv:1006.3740 [nucl-ex].

Adare, A. *et al.* (2013). Neutral pion production with respect to centrality and reaction plane in Au+Au collisions at $\sqrt{s_{NN}}$ = 200 GeV, *Phys. Rev. C* **87**(3), 034911. doi: 10.1103/PhysRevC.87.034911, arXiv:1208.2254 [nucl-ex].

Adcox, K. *et al.* (2005). Formation of dense partonic matter in relativistic nucleus-nucleus collisions at RHIC: Experimental evaluation by the PHENIX collaboration, *Nucl. Phys. A* **757**, 184–283. doi: 10.1016/j.nuclphysa.2005.03.086, arXiv:nucl-ex/0410003.

Ade, P. A. R. *et al.* (2014a). Planck 2013 results. XV. CMB power spectra and likelihood, *Astron. Astrophys.* **571**, A15. doi: 10.1051/0004-6361/201321573, arXiv:1303.5075 [astro-ph.CO].

Ade, P. A. R. *et al.* (2014b). Planck 2013 results. XXIII. Isotropy and statistics of the CMB, *Astron. Astrophys.* **571**, A23. doi: 10.1051/0004-6361/201321534, arXiv:1303.5083 [astro-ph.CO].

Adler, S. S. *et al.* (2014). Transverse-energy distributions at midrapidity in p+p, d+Au, and Au+Au collisions at $\sqrt{s_{NN}}$ = 62.4–200 GeV and implications for particle-production models, *Phys. Rev. C* **89**(4), 044905. doi: 10.1103/PhysRevC.89.044905, arXiv:1312.6676 [nucl-ex].

Aharonov, Y. and Bohm, D. (1959). Significance of electromagnetic potentials in the quantum theory, *Phys. Rev.* **115**, 485–491. doi: 10.1103/PhysRev.115.485.

Alford, M. G., Rajagopal, K., and Wilczek, F. (1998). QCD at finite baryon density: Nucleon droplets and color superconductivity, *Phys. Lett. B* **422**, 247–256. doi: 10.1016/S0370-2693(98)00051-3, arXiv:hep-ph/9711395.

Alford, M. G., Rajagopal, K., and Wilczek, F. (1999). Color flavor locking and chiral symmetry breaking in high density QCD, *Nucl. Phys. B* **537**, 443–458. doi: 10.1016/S0550-3213(98)00668-3, arXiv:hep-ph/9804403.

Altinoluk, T., Armesto, N., Beuf, G., Kovner, A., and Lublinsky, M. (2015). Bose enhancement and the ridge, *Phys. Lett. B* **751**, 448–452. doi: 10.1016/j.physletb.2015.10.072, arXiv:1503.07126 [hep-ph].

Andersen, J. O., Braaten, E., and Strickland, M. (2000). Hard thermal loop resummation of the thermodynamics of a hot gluon plasma, *Phys. Rev. D* **61**, 014017. doi: 10.1103/PhysRevD.61.014017, arXiv:hep-ph/9905337.

Andrade, R., Grassi, F., Hama, Y., and Qian, W. L. (2010). A closer look at the influence of tubular initial conditions on two-particle correlations, *J. Phys. G* **37**, 094043. doi: 10.1088/0954-3899/37/9/094043, arXiv:0912.0703 [nucl-th].

Andronic, A., Braun-Munzinger, P., Redlich, K., and Stachel, J. (2018). Decoding the phase structure of QCD via particle production at high energy, *Nature* **561**(7723), 321–330. doi: 10.1038/s41586-018-0491-6, arXiv:1710.09425 [nucl-th].

Andronic, A., Braun-Munzinger, P., Friman, B., Lo, P. M., Redlich, K., and Stachel, J. (2019). The thermal proton yield anomaly in Pb-Pb collisions at the LHC and its resolution, *Phys. Lett. B* **792**, 304–309. doi: 10.1016/j.physletb.2019.03.052, arXiv:1808.03102 [hep-ph].

Andronic, A., Braun-Munzinger, P., Gündüz, D., Kirchhoff, Y., Köhler, M. K., Stachel, J., and Winn, M. (2021a). Influence of modified light-flavor hadron spectra on particle yields in the statistical hadronization model, *Nucl. Phys. A* **1010**, 122176. doi: 10.1016/j.nuclphysa.2021.122176, arXiv:2011.03826 [nucl-th].

Andronic, A., Braun-Munzinger, P., Köhler, M. K., Mazeliauskas, A., Redlich, K., Stachel, J., and Vislavicius, V. (2021b). The multiple-charm hierarchy in the statistical hadronization model, *JHEP* **07**, 035. doi: 10.1007/JHEP07(2021)035, arXiv:2104.12754 [hep-ph].

Anikin, I. V., Dorokhov, A. E., and Tomio, L. (2001). Pion distribution amplitude within the instanton model, *Phys. Atom. Nucl.* **64**, 1329–1336. doi: 10.1134/1.1389562.

Annala, E., Gorda, T., Kurkela, A., Nättilä, J., and Vuorinen, A. (2019). Constraining the properties of neutron-star matter with observations, *Mem. Soc. Ast. It.* **90**(1–2), 81–86. arXiv:1904.01354 [astro-ph.HE].

Anthony, P. L. *et al.* (1995). An accurate measurement of the Landau-Pomeranchuk-Migdal effect, *Phys. Rev. Lett.* **75**, 1949–1952. doi: 10.1103/PhysRevLett.75.1949.

Arnold, P. B. and Mattis, M. P. (1991). Peace in the valley: Concordant approaches to distorted instantons, *Phys. Rev. D* **44**, 3650–3660. doi: 10.1103/PhysRevD.44.3650.

Arnold, P. B. and Zhai, C.-X. (1995). The three loop free energy for high temperature QED and QCD with fermions, *Phys. Rev. D* **51**, 1906–1918. doi: 10.1103/PhysRevD.51.1906, arXiv:hep-ph/9410360.

Arnold, P. B., Moore, G. D., and Yaffe, L. G. (2000). Transport coefficients in high temperature gauge theories. 1. Leading log results, *JHEP* **11**, 001. doi: 10.1088/1126-6708/2000/11/001, arXiv:hep-ph/0010177.

Arnold, P. B., Moore, G. D., and Yaffe, L. G. (2003). Transport coefficients in high temperature gauge theories. 2. Beyond leading log, *JHEP* **05**, 051. doi: 10.1088/1126-6708/2003/05/051, arXiv:hep-ph/0302165.

Arvis, J. F. (1983). The exact $q\bar{q}$ potential in Nambu string theory, *Phys. Lett. B* **127**, 106–108. doi: 10.1016/0370-2693(83)91640-4.

Arzoumanian, Z. *et al.* (2021). Searching for gravitational waves from cosmological phase transitions with the NANOGrav 12.5-year dataset, *Phys. Rev. Lett.* **127**(25), 251302. doi: 10.1103/PhysRevLett.127.251302, arXiv:2104.13930 [astro-ph.CO].

Athenodorou, A., Boucaud, P., De Soto, F., Rodríguez-Quintero, J., and Zafeiropoulos, S. (2018). Instanton liquid properties from lattice QCD, *JHEP* **02**, 140. doi: 10.1007/JHEP02(2018)140, arXiv:1801.10155 [hep-lat].

Baier, R., Dokshitzer, Y. L., Mueller, A. H., and Schiff, D. (1998). Radiative energy loss of high-energy partons traversing an expanding QCD plasma, *Phys. Rev. C* **58**, 1706–1713. doi: 10.1103/PhysRevC.58.1706, arXiv:hep-ph/9803473.

Baier, R., Mueller, A. H., Schiff, D., and Son, D. T. (2001). 'Bottom up' thermalization in heavy ion collisions, *Phys. Lett. B* **502**, 51–58. doi: 10.1016/S0370-2693(01)00191-5, arXiv:hep-ph/0009237.

Bailin, D. and Love, A. (1984). Superfluidity and superconductivity in relativistic fermion systems, *Phys. Rep.* **107**, 325. doi: 10.1016/0370-1573(84)90145-5.

Bak, D., Karch, A., and Yaffe, L. G. (2007). Debye screening in strongly coupled N = 4 supersymmetric Yang-Mills plasma, *JHEP* **08**, 049. doi: 10.1088/1126-6708/2007/08/049, arXiv:0705.0994 [hep-th].

Balasubramanian, V., Bernamonti, A., de Boer, J., Copland, N., Craps, B., Keski-Vakkuri, E., Muller, B., Schafer, A., Shigemori, M., and Staessens, W. (2011). Holographic thermalization, *Phys. Rev. D* **84**, 026010. doi: 10.1103/PhysRevD.84.026010, arXiv:1103.2683 [hep-th].

Bali, G. S. (1998). The mechanism of quark confinement, in *3rd International Conference in Quark Confinement and Hadron Spectrum (Confinement III)*, pp. 17–36. arXiv:hep-ph/9809351.

Balitsky, I. I. and Yung, A. V. (1986). Collective — coordinate method for quasizero modes, *Phys. Lett. B* **168**, 113–119. doi: 10.1016/0370-2693(86)91471-1.

Barrois, B. C. (1977). Superconducting quark matter, *Nucl. Phys. B* **129**, 390–396. doi: 10.1016/0550-3213(77)90123-7.

Bartels, J., Lipatov, L. N., and Vacca, G. P. (2000). A new odderon solution in perturbative QCD, *Phys. Lett. B* **477**, 178–186. doi: 10.1016/S0370-2693(00)00221-5, arXiv:hep-ph/9912423.

Basar, G., Kharzeev, D. E., Yee, H.-U., and Zahed, I. (2012). Holographic Pomeron and the Schwinger mechanism, *Phys. Rev. D* **85**, 105005. doi: 10.1103/PhysRevD.85.105005, arXiv:1202.0831 [hep-th].

Basar, G., Kharzeev, D. E., and Shuryak, E. V. (2014). Magneto-sonoluminescence and its signatures in photon and dilepton production in relativistic heavy ion collisions, *Phys. Rev. C* **90**(1), 014905. doi: 10.1103/PhysRevC.90.014905, arXiv:1402.2286 [hep-ph].

Bauswein, A., Bastian, N.-U. F., Blaschke, D. B., Chatziioannou, K., Clark, J. A., Fischer, T., and Oertel, M. (2019). Identifying a first-order phase transition in neutron star mergers through gravitational waves, *Phys. Rev. Lett.* **122**(6), 061102. doi: 10.1103/PhysRevLett.122.061102, arXiv:1809.01116 [astro-ph.HE].

Baym, G., Friman, B. L., Blaizot, J. P., Soyeur, M., and Czyz, W. (1983). Hydrodynamics of ultrarelativistic heavy ion collisions, *Nucl. Phys. A* **407**, 541–570. doi: 10.1016/0375-9474(83)90666-8.

Bazavov, A. (2014). The QCD equation of state, *Nucl. Phys. A* **931**, 867–871. doi: 10.1016/j.nuclphysa.2014.08.073.

Bazavov, A. *et al.* (2019). Meson screening masses in (2+1)-flavor QCD, *Phys. Rev. D* **100**(9), 094510. doi: 10.1103/PhysRevD.100.094510, arXiv:1908.09552 [hep-lat].

Bearden, I. G. *et al.* (2001). One-dimensional and two-dimensional analysis of 3 pi correlations measured in Pb + Pb interactions, *Phys. Lett. B* **517**, 25–31. doi: 10.1016/S0370-2693(01)00978-9, arXiv:nucl-ex/0102013.

Bebie, H., Gerber, P., Goity, J. L., and Leutwyler, H. (1992). The role of the entropy in an expanding hadronic gas, *Nucl. Phys. B* **378**, 95–128. doi: 10.1016/0550-3213(92)90005-V.

Becattini, F. (2022). Spin and polarization: A new direction in relativistic heavy ion physics. arXiv:2204.01144 [nucl-th].

Becattini, F., Chandra, V., Del Zanna, L., and Grossi, E. (2013). Relativistic distribution function for particles with spin at local thermodynamical equilibrium, *Ann. Phys.* **338**, 32–49. doi: 10.1016/j.aop.2013.07.004, arXiv:1303.3431 [nucl-th].

Becattini, F. and Lisa, M. A. (2020). Polarization and vorticity in the quark–gluon plasma, *Ann. Rev. Nucl. Part. Sci.* **70**, 395–423. doi: 10.1146/annurev-nucl-021920-095245, arXiv:2003.03640 [nucl-ex].

Begun, V., Florkowski, W., and Rybczynski, M. (2014). Explanation of hadron transverse-momentum spectra in heavy-ion collisions at $\sqrt{s}_{NN} = 2.76$ TeV within chemical non-equilibrium statistical hadronization model, *Phys. Rev. C* **90**(1), 014906. doi: 10.1103/PhysRevC.90.014906, arXiv:1312.1487 [nucl-th].

Belavin, A. A., Polyakov, A. M., Schwartz, A. S., and Tyupkin, Y. S. (1975). Pseudoparticle solutions of the Yang-Mills equations, *Phys. Lett. B* **59**, 85–87. doi: 10.1016/0370-2693(75)90163-X.

Belenkij, S. Z. and Landau, L. D. (1955). Hydrodynamic theory of multiple production of particles, *Usp. Fiz. Nauk* **56**, 309. doi: 10.1007/BF02745507.

Bellwied, R., Borsányi, S., Fodor, Z., Guenther, J. N., Katz, S. D., Parotto, P., Pásztor, A., Pesznyák, D., Ratti, C., and Szabó, K. K. (2022). Quantifying corrections to the hadron resonance gas with lattice QCD, in *PoS LATTICE2021*, 186. doi: 10.22323/1.396.0186, arXiv:2112.02402 [hep-lat].

Bemfica, F. S., Bemfica, F. S., Disconzi, M. M., Disconzi, M. M., Noronha, J., and Noronha, J. (2019). Nonlinear causality of general first-order relativistic viscous hydrodynamics, *Phys. Rev. D* **100**(10), 104020. doi: 10.1103/PhysRevD.100.104020, arXiv:1907.12695 [gr-qc] [Erratum: (2022). *Phys. Rev. D* **105**, 069902].

Berges, J., Tetradis, N., and Wetterich, C. (2002). Nonperturbative renormalization flow in quantum field theory and statistical physics, *Phys. Rep.* **363**, 223–386. doi: 10.1016/S0370-1573(01)00098-9, arXiv:hep-ph/0005122.

Bhattacharyya, S., Hubeny, V. E., Loganayagam, R., Mandal, G., Minwalla, S., Morita, T., Rangamani, M., and Reall, H. S. (2008a). Local fluid dynamical entropy from gravity, *JHEP* **06**, 055. doi: 10.1088/1126-6708/2008/06/055, arXiv:0803.2526 [hep-th].

Bhattacharyya, S., Hubeny, V. E., Minwalla, S., and Rangamani, M. (2008b). Nonlinear fluid dynamics from gravity, *JHEP* **02**, 045. doi: 10.1088/1126-6708/2008/02/045, arXiv:0712.2456 [hep-th].

Biddle, J. C., Kamleh, W., and Leinweber, D. B. (2020a). Visualisations of centre vortices, *EPJ Web Conf.* **245**, 06010. doi: 10.1051/epjconf/202024506010, arXiv:2009.12047 [hep-lat].

Biddle, J. C., Kamleh, W., and Leinweber, D. B. (2020b). Visualization of center vortex structure, *Phys. Rev. D* **102**(3), 034504. doi: 10.1103/PhysRevD.102.034504, arXiv:1912.09531 [hep-lat].

Bilandzic, A., Snellings, R., and Voloshin, S. (2011). Flow analysis with cumulants: Direct calculations, *Phys. Rev. C* **83**, 044913. doi: 10.1103/PhysRevC.83.044913, arXiv:1010.0233 [nucl-ex].

Bjorken, J. D. (1982). Energy loss of energetic partons in quark — gluon plasma: Possible extinction of high p(t) jets in hadron — hadron collisions.

Bjorken, J. D. (1983). Highly relativistic nucleus-nucleus collisions: The central rapidity region, *Phys. Rev. D* **27**, 140–151. doi: 10.1103/PhysRevD.27.140.

Bjorken, J. D. (2000). Intersections 2000: What's new in hadron physics, *AIP Conf. Proc.* **549**(1), 211–229. doi: 10.1063/1.1345244, arXiv:hep-ph/0008048.

Bjorken, J. D., Brodsky, S. J., and Scharff Goldhaber, A. (2013). Possible multiparticle ridge-like correlations in very high multiplicity proton-proton collisions, *Phys. Lett. B* **726**, 344–346. doi: 10.1016/j.physletb.2013.08.066, arXiv:1308.1435 [hep-ph].

Blaizot, J.-P. and Iancu, E. (1993). Kinetic theory and quantum electrodynamics at high temperature, *Nucl. Phys. B* **390**, 589–620. doi: 10.1016/0550-3213(93)90490-G.

Blaizot, J. P., Iancu, E., and Rebhan, A. (1999). The entropy of the QCD plasma, *Phys. Rev. Lett.* **83**, 2906–2909. doi: 10.1103/PhysRevLett.83.2906, arXiv:hep-ph/9906340.

Borsányi, S., Fodor, Z., Hoelbling, C., Katz, S. D., Krieg, S., Ratti, C., and Szabo, K. K. (2015a). Recent results on the equation of state of QCD, *PoS* **LATTICE2014**, 224. doi: 10.22323/1.214.0224, arXiv:1410.7917 [hep-lat].

Borsányi, S., Fodor, Z., Katz, S. D., Pásztor, A., Szabó, K. K., and Török, C. (2015b). Static $\overline{Q}Q$ pair free energy and screening masses from correlators of Polyakov loops: Continuum extrapolated lattice results at the QCD physical point, *JHEP* **04**, 138. doi: 10.1007/JHEP04(2015)138, arXiv:1501.02173 [hep-lat].

Boucaud, P., De Soto, F., Le Yaouanc, A., Leroy, J. P., Micheli, J., Moutarde, H., Pene, O., and Rodriguez-Quintero, J. (2003). The strong coupling constant at small momentum as an instanton detector, *JHEP* **04**, 005. doi: 10.1088/1126-6708/2003/04/005, arXiv:hep-ph/0212192.

Boulware, D. G., Brown, L. S., Cahn, R. N., Ellis, S. D., and Lee, C.-k. (1976). Scattering on magnetic charge, *Phys. Rev. D* **14**, 2708. doi: 10.1103/PhysRevD.14.2708.

Boz, T., Giudice, P., Hands, S., and Skullerud, J.-I. (2020). Dense two-color QCD towards continuum and chiral limits, *Phys. Rev. D* **101**(7), 074506. doi: 10.1103/PhysRevD.101.074506, arXiv:1912.10975 [hep-lat].

Bozek, P. (2012). Collective flow in p-Pb and d-Pd collisions at TeV energies, *Phys. Rev.* **C85**, 014911. doi: 10.1103/PhysRevC.85.014911, arXiv:1112.0915 [hep-ph].

Bozek, P. and Broniowski, W. (2014). Collective flow in ultrarelativistic ^{3}He-Au collisions, *Phys. Lett.* **B739**, 308–312. doi: 10.1016/j.physletb.2014.11.006, arXiv:1409.2160 [nucl-th].

Bozek, P., Broniowski, W., and Moreira, J. (2011). Torqued fireballs in relativistic heavy-ion collisions, *Phys. Rev. C* **83**, 034911. doi: 10.1103/PhysRevC.83.034911, arXiv:1011.3354 [nucl-th].

Braaten, E. (1995). Solution to the perturbative infrared catastrophe of hot gauge theories, *Phys. Rev. Lett.* **74**, 2164–2167. doi: 10.1103/PhysRevLett.74.2164, arXiv:hep-ph/9409434.

Braaten, E. and Pisarski, R. D. (1990). Soft amplitudes in hot gauge theories: A general analysis, *Nucl. Phys. B* **337**, 569–634. doi: 10.1016/0550-3213(90)90508-B.

Brandt, B. B. (2017). Spectrum of the open QCD flux tube and its effective string description I: 3d static potential in SU(N = 2, 3), *JHEP* **07**, 008. doi: 10.1007/JHEP07(2017)008, arXiv:1705.03828 [hep-lat].

Brandt, A., Erhan, S., Kuzucu, A., Medinnis, M., Ozdes, N., Schlein, P. E., Zeyrek, M. T., Zweizig, J. G., Cheze, J. B., and Zsembery, J. (2002). A study of inclusive double pomeron exchange in $p\bar{p} \to pX\bar{p}$ at $\sqrt{s}$ = 630-GeV, *Eur. Phys. J. C* **25**, 361–377. doi: 10.1007/s10052-002-1031-x, arXiv:hep-ex/0205037.

Braun-Munzinger, P. and Stachel, J. (2000). (Non)thermal aspects of charmonium production and a new look at J / psi suppression, *Phys. Lett. B* **490**, 196–202. doi: 10.1016/S0370-2693(00)00991-6, arXiv:nucl-th/0007059.

Braun-Munzinger, P., Stachel, J., and Wetterich, C. (2004). Chemical freezeout and the QCD phase transition temperature, *Phys. Lett. B* **596**, 61–69. doi: 10.1016/j.physletb.2004.05.081, arXiv:nucl-th/0311005.

Braun-Munzinger, P. and Wambach, J. (2009). The phase diagram of strongly-interacting matter, *Rev. Mod. Phys.* **81**, 1031–1050. doi: 10.1103/RevModPhys.81.1031, arXiv:0801.4256 [hep-ph].

Brodsky, S. J. and Farrar, G. R. (1973). Scaling laws at large transverse momentum, *Phys. Rev. Lett.* **31**, 1153–1156. doi: 10.1103/PhysRevLett.31.1153.

Brodsky, S. J., de Téramond, G. F., Dosch, H. G., and Lorcé, C. (2016). Meson/baryon/tetraquark supersymmetry from superconformal algebra and light-front holography, *Int. J. Mod. Phys. A* **31**(19), 1630029. doi: 10.1142/S0217751X16300295, arXiv:1606.04638 [hep-ph].

Broniowski, W. and Ruiz Arriola, E. (2017). Nonperturbative partonic quasidistributions of the pion from chiral quark models, *Phys. Lett. B* **773**, 385–390. doi: 10.1016/j.physletb.2017.08.055, arXiv:1707.09588 [hep-ph].

Brown, G. E. and Rho, M. (1991). Scaling effective Lagrangians in a dense medium, *Phys. Rev. Lett.* **66**, 2720–2723. doi: 10.1103/PhysRevLett.66.2720.

Buchel, A. (2006). On jet quenching parameters in strongly coupled non-conformal gauge theories, *Phys. Rev. D* **74**, 046006. doi: 10.1103/PhysRevD.74.046006, arXiv:hep-th/0605178.

Callan Jr., C. G., Dashen, R. F., Gross, D. J., Wilczek, F., and Zee, A. (1978). The effect of instantons on the heavy quark potential, *Phys. Rev. D* **18**, 4684. doi: 10.1103/PhysRevD.18.4684.

Casalderrey-Solana, J., Heller, M. P., Mateos, D., and van der Schee, W. (2013). From full stopping to transparency in a holographic model of heavy ion collisions, *Phys. Rev. Lett.* **111**, 181601. doi: 10.1103/PhysRevLett.111.181601, arXiv:1305.4919 [hep-th].

Castorina, P., Plumari, S., and Satz, H. (2016). Universal strangeness production in hadronic and nuclear collisions, *Int. J. Mod. Phys. E* **25**(08), 1650058. doi: 10.1142/S0218301316500580, arXiv:1603.06529 [hep-ph].

Cea, P., Cosmai, L., Cuteri, F., and Papa, A. (2018). QCD flux tubes across the deconfinement phase transition, *EPJ Web Conf.* **175**, 12006. doi: 10.1051/epjconf/201817512006, arXiv:1710.01963 [hep-lat].

Chanfray, G., Rapp, R., and Wambach, J. (1996). Medium modifications of the rho meson at CERN SPS energies, *Phys. Rev. Lett.* **76**, 368–371. doi: 10.1103/PhysRevLett.76.368, arXiv:hep-ph/9508353.

Chatrchyan, S. *et al.* (2012a). Azimuthal anisotropy of charged particles at high transverse momenta in PbPb collisions at $\sqrt{s_{NN}} = 2.76\,\text{TeV}$, *Phys. Rev. Lett.* **109**, 022301. doi: 10.1103/PhysRevLett.109.022301, arXiv:1204.1850 [nucl-ex].

Chatrchyan, S. *et al.* (2012b). Study of high-pT charged particle suppression in PbPb compared to *pp* collisions at $\sqrt{s_{NN}} = 2.76\,\text{TeV}$, *Eur. Phys. J. C* **72**, 1945. doi: 10.1140/epjc/s10052-012-1945-x, arXiv:1202.2554 [nucl-ex].

Chatrchyan, S. *et al.* (2013). Observation of long-range near-side angular correlations in proton-lead collisions at the LHC, *Phys. Lett. B* **718**, 795–814. doi: 10.1016/j.physletb.2012.11.025, arXiv:1210.5482 [nucl-ex].

Chen, Y. *et al.* (2006). Glueball spectrum and matrix elements on anisotropic lattices, *Phys. Rev. D* **73**, 014516. doi: 10.1103/PhysRevD.73.014516, arXiv:hep-lat/0510074.

Chern, S.-S. and Simons, J. (1974). Characteristic forms and geometric invariants, *Ann. Math.* **99**, 48–69. doi: 10.2307/1971013.

Chernyak, V. L. and Zhitnitsky, A. R. (1977). Asymptotic behavior of hadron form-factors in quark model (In Russian), *JETP Lett.* **25**, 510.

Chernyak, V. L. and Zhitnitsky, A. R. (1984). Asymptotic behavior of exclusive processes in QCD, *Phys. Rep.* **112**, 173. doi: 10.1016/0370-1573(84)90126-1.

Chesler, P. M. (2016). How big are the smallest drops of quark-gluon plasma? *JHEP* **03**, 146. doi: 10.1007/JHEP03(2016)146, arXiv:1601.01583 [hep-th].

Chesler, P. M. and Yaffe, L. G. (2007). The wake of a quark moving through a strongly-coupled plasma, *Phys. Rev. Lett.* **99**, 152001. doi: 10.1103/PhysRevLett.99.152001, arXiv:0706.0368 [hep-th].

Chesler, P. M. and Yaffe, L. G. (2008). The stress-energy tensor of a quark moving through a strongly-coupled N=4 supersymmetric Yang-Mills plasma: Comparing hydrodynamics and AdS/CFT, *Phys. Rev. D* **78**, 045013. doi: 10.1103/PhysRevD.78.045013, arXiv:0712.0050 [hep-th].

Chesler, P. M. and Yaffe, L. G. (2014). Numerical solution of gravitational dynamics in asymptotically anti-de Sitter spacetimes, *JHEP* **07**, 086. doi: 10.1007/JHEP07(2014)086, arXiv:1309.1439 [hep-th].

Chesler, P. M., Kilbertus, N., and van der Schee, W. (2015). Universal hydrodynamic flow in holographic planar shock collisions, *JHEP* **11**, 135. doi: 10.1007/JHEP11(2015)135, arXiv:1507.02548 [hep-th].

Chirikov, B. V., Ford, J., and Vivaldi, F. (1980). Some numerical studies or Arnold diffusion in a simple model, *AIP Conf. Proc.* **57**, 323–340. doi: 10.1063/1.32115.

Chiu, C. B., Sudarshan, E. C. G., and Wang, K.-H. (1975). Hydrodynamical expansion with frame independence symmetry in high-energy multiparticle production, *Phys. Rev. D* **12**, 902. doi: 10.1103/PhysRevD.12.902.

Chiu, M., Hemmick, T. K., Khachatryan, V., Leonidov, A., Liao, J., and McLerran, L. (2013). Production of photons and dileptons in the glasma, *Nucl. Phys. A* **900**, 16–37. doi: 10.1016/j.nuclphysa.2013.01.014, arXiv:1202.3679 [nucl-th].

Chu, M. C., Grandy, J. M., Huang, S., and Negele, J. W. (1994). Evidence for the role of instantons in hadron structure from lattice QCD, *Phys. Rev. D* **49**, 6039–6050. doi: 10.1103/PhysRevD.49.6039, arXiv:hep-lat/9312071.

Cornwall, J. M. (1996). The baryon Wilson loop area law in QCD, *Phys. Rev. D* **54**, 6527–6536. doi: 10.1103/PhysRevD.54.6527, arXiv:hep-th/9605116.

Creutz, M. (1980). Monte Carlo study of quantized SU(2) gauge theory, *Phys. Rev. D* **21**, 2308–2315. doi: 10.1103/PhysRevD.21.2308.

D'Alessandro, A., D'Elia, M., and Shuryak, E. V. (2010). Thermal monopole condensation and confinement in finite temperature Yang-Mills theories, *Phys. Rev. D* **81**, 094501. doi: 10.1103/PhysRevD.81.094501, arXiv:1002.4161 [hep-lat].

Davies, C. T. H., Koponen, J., Lepage, P. G., Lytle, A. T., and Zimermmane-Santos, A. C. (2018). Meson electromagnetic form factors from lattice QCD, *PoS LATTICE2018*, 298. doi: 10.22323/1.334.0298, arXiv:1902.03808 [hep-lat].

de Cassagnac, R. (2014). Overview of CMS results at quark matter 2014, *Nucl. Phys. A* **931**, 22.

De Falco, A. (2015). Low mass dilepton production in pp, p–Pb and Pb–Pb collisions with ALICE, *EPJ Web Conf.* **97**, 00012. doi: 10.1051/epjconf/20159700012.

de Teramond, G. F., Liu, T., Sufian, R. S., Dosch, H. G., Brodsky, S. J., and Deur, A. (2018). Universality of generalized parton distributions in light-front holographic QCD, *Phys. Rev. Lett.* **120**(18), 182001. doi: 10.1103/PhysRevLett.120.182001, arXiv:1801.09154 [hep-ph].

DeMartini, D. and Shuryak, E. (2021a). Chiral symmetry breaking and confinement from an interacting ensemble of instanton dyons in two-flavor massless QCD, *Phys. Rev. D* **104**(9), 094031. doi: 10.1103/PhysRevD.104.094031, arXiv:2108.06353 [hep-ph].

DeMartini, D. and Shuryak, E. (2021b). Deconfinement phase transition in the SU(3) instanton-dyon ensemble, *Phys. Rev. D* **104**(5), 054010. doi: 10.1103/PhysRevD.104.054010, arXiv:2102.11321 [hep-ph].

DeMartini, D. and Shuryak, E. (2021c). Many-body forces and nucleon clustering near the QCD critical point, *Phys. Rev. C* **104**(2), 024908. doi: 10.1103/PhysRevC.104.024908, arXiv:2010.02785 [nucl-th].

DeMartini, D. and Shuryak, E. (2021d). Nucleon clustering at kinetic freezeout of heavy-ion collisions via path-integral Monte Carlo, *Nucl. Phys. A* **1016**, 122336. doi: 10.1016/j.nuclphysa.2021.122336, arXiv:2007.04863 [nucl-th].

Denicol, G. S., Heinz, U. W., Martinez, M., Noronha, J., and Strickland, M. (2014). Studying the validity of relativistic hydrodynamics with a new exact solution of the Boltzmann equation, *Phys. Rev. D* **90**(12), 125026. doi: 10.1103/PhysRevD.90.125026, arXiv:1408.7048 [hep-ph].

Detar, C. E. and Kogut, J. B. (1987). The hadronic spectrum of the quark plasma, *Phys. Rev. Lett.* **59**, 399. doi: 10.1103/PhysRevLett.59.399.

Dirac, P. A. M. (1931). Quantised singularities in the electromagnetic field, *Proc. R. Soc. Lond. A* **133**(821), 60–72. doi: 10.1098/rspa.1931.0130.

Dokshitzer, Y. L. and Kharzeev, D. E. (2001). Heavy quark colorimetry of QCD matter, *Phys. Lett. B* **519**, 199–206. doi: 10.1016/S0370-2693(01)01130-3, arXiv:hep-ph/0106202.

Dorey, N. and Parnachev, A. (2001). Instantons, compactification and S duality in N=4 SUSY Yang-Mills theory. 2. *JHEP* **08**, 059. doi: 10.1088/1126-6708/2001/08/059, arXiv:hep-th/0011202.

Dorokhov, A. E. and Kochelev, N. I. (1993). Instanton induced asymmetric quark configurations in the nucleon and parton sum rules, *Phys. Lett. B* **304**, 167–175. doi: 10.1016/0370-2693(93)91417-L.

Drell, S. D. and Yan, T.-M. (1970). Connection of elastic electromagnetic nucleon form-factors at large Q**2 and deep inelastic structure functions near threshold, *Phys. Rev. Lett.* **24**, 181–185. doi: 10.1103/PhysRevLett.24.181.

Dupuis, N., Canet, L., Eichhorn, A., Metzner, W., Pawlowski, J. M., Tissier, M., and Wschebor, N. (2021). The nonperturbative functional renormalization group and its applications, *Phys. Rep.* **910**, 1–114. doi: 10.1016/j.physrep.2021.01.001, arXiv:2006.04853 [cond-mat.stat-mech].

Efremov, A. V. and Radyushkin, A. V. (1980). Factorization and asymptotical behavior of pion form-factor in QCD, *Phys. Lett. B* **94**, 245–250. doi: 10.1016/0370-2693(80)90869-2.

Eichten, E. and Feinberg, F. (1981). Spin dependent forces in QCD, *Phys. Rev. D* **23**, 2724. doi: 10.1103/PhysRevD.23.2724.

Eichten, E. J., Hinchliffe, I., and Quigg, C. (1992). Flavor asymmetry in the light quark sea of the nucleon, *Phys. Rev. D* **45**, 2269–2275. doi: 10.1103/PhysRevD.45.2269.

Engels, J., Fromme, L., and Seniuch, M. (2003). Correlation lengths and scaling functions in the three-dimensional O(4) model, *Nucl. Phys. B* **675**, 533–554. doi: 10.1016/j.nuclphysb.2003.09.060, arXiv:hep-lat/0307032.

Escobar-Ruiz, M. A., Shuryak, E., and Turbiner, A. V. (2015). Three-loop correction to the instanton density. I. The quartic double well potential, *Phys. Rev. D* **92**(2), 025046. doi: 10.1103/PhysRevD.92.025046, arXiv:1501.03993 [hep-th] [Erratum: (2015). *Phys. Rev. D* **92**, 089902].

Escobar-Ruiz, M. A., Shuryak, E., and Turbiner, A. V. (2016). Quantum and thermal fluctuations in quantum mechanics and field theories from a new version of semiclassical theory, *Phys. Rev. D* **93**(10), 105039. doi: 10.1103/PhysRevD.93.105039, arXiv:1601.03964 [hep-th].

Ethier, J. J. and Nocera, E. R. (2020). Parton distributions in nucleons and nuclei, *Ann. Rev. Nucl. Part. Sci.* **70**, 43–76. doi: 10.1146/annurev-nucl-011720-042725, arXiv:2001.07722 [hep-ph].

Everett, D. *et al.* (2021). Multisystem Bayesian constraints on the transport coefficients of QCD matter, *Phys. Rev. C* **103**(5), 054904. doi: 10.1103/PhysRevC.103.054904, arXiv:2011.01430 [hep-ph].

Faccioli, P. and DeGrand, T. A. (2003). Evidence for instanton induced dynamics, from lattice QCD, *Phys. Rev. Lett.* **91**, 182001. doi: 10.1103/PhysRevLett.91.182001, arXiv:hep-ph/0304219.

Feinberg, E. L. (1976). Direct production of photons and dileptons in thermodynamical models of multiple hadron production, *Nuovo Cim. A* **34**, 391.

Feinberg, E. L. (1980). Hadron clusters and half dressed particles in quantum field theory, *Sov. Phys. Usp.* **23**, 629–650. doi: 10.1070/PU1980v023n10ABEH005036.

Fermi, E. (1951). Angular distribution of the pions produced in high energy nuclear collisions, *Phys. Rev.* **81**, 683–687. doi: 10.1103/PhysRev.81.683.

Feynman, R. (1972). *Statistical Mechanics: A Set of Lectures* (W.A. Benjamin Inc., Reading, M.A.).

Feynman, R. and Hibbs, H. (1965). *Quantum Mechanics and Path Integrals* (Mcgraw-Hill, New York).

Flambaum, V., Gribakina, A., Gribakin, F., and Kozlov, M. (1994). Structure of compound states in the chaotic spectrum of the Ce atom: Localization properties, matrix elements, and enhancement of weak perturbations, *Phys. Rev. A* **50**, 267.

Florkowski, W. and Ryblewski, R. (2011). Highly-anisotropic and strongly-dissipative hydrodynamics for early stages of relativistic heavy-ion collisions, *Phys. Rev.* **C83**, 034907. doi: 10.1103/PhysRevC.83.034907, arXiv:1007.0130 [nucl-th].

Francis, A., de Forcrand, P., Lewis, R., and Maltman, K. (2022). Diquark properties from full QCD lattice simulations, *JHEP* **05**, 062. doi: 10.1007/JHEP05(2022)062, arXiv:2106.09080 [hep-lat].

Fries, R. J., Muller, B., Nonaka, C., and Bass, S. A. (2003). Hadron production in heavy ion collisions: Fragmentation and recombination from a dense parton phase, *Phys. Rev. C* **68**, 044902. doi: 10.1103/PhysRevC.68.044902, arXiv:nucl-th/0306027.

Friess, J. J., Gubser, S. S., and Michalogiorgakis, G. (2006). Dissipation from a heavy quark moving through N=4 super-Yang-Mills plasma, *JHEP* **09**, 072. doi: 10.1088/1126-6708/2006/09/072, arXiv:hep-th/0605292.

Fukushima, K., Kharzeev, D. E., and Warringa, H. J. (2008). The chiral magnetic effect, *Phys. Rev. D* **78**, 074033. doi: 10.1103/PhysRevD.78.074033, arXiv:0808.3382 [hep-ph].

Gale, C., Jeon, S., Schenke, B., Tribedy, P., and Venugopalan, R. (2013). Event-by-event anisotropic flow in heavy-ion collisions from combined Yang-Mills and viscous fluid dynamics, *Phys. Rev. Lett.* **110**(1), 012302. doi: 10.1103/PhysRevLett.110.012302, arXiv:1209.6330 [nucl-th].

Gardim, F. G., Grassi, F., Luzum, M., and Ollitrault, J.-Y. (2012). Mapping the hydrodynamic response to the initial geometry in heavy-ion collisions, *Phys. Rev. C* **85**, 024908. doi: 10.1103/PhysRevC.85.024908, arXiv:1111.6538 [nucl-th].

Gasser, J. and Leutwyler, H. (1984). Chiral perturbation theory to one loop, *Ann. Phys.* **158**, 142. doi: 10.1016/0003-4916(84)90242-2.

Gavrielov, N., Leviatan, A., and Iachello, F. (2022). Zr isotopes as a region of intertwined quantum phase transitions, *Phys. Rev. C* **105**(1), 014305. doi: 10.1103/PhysRevC.105.014305, arXiv:2112.09454 [nucl-th].

Gazdzicki, M. and Mrowczynski, S. (1992). A method to study 'equilibration' in nucleus-nucleus collisions, *Z. Phys. C* **54**, 127–132. doi: 10.1007/BF01881715.

Gelman, B. A., Shuryak, E. V., and Zahed, I. (2006). Classical strongly coupled QGP. I. The model and molecular dynamics simulations, *Phys. Rev. C* **74**, 044908. doi: 10.1103/PhysRevC.74.044908, arXiv:nucl-th/0601029.

Geshkenbein, B. V. and Terentev, M. V. (1982). The enhanced power correction to the asymptotics of the pion form-factor, *Phys. Lett. B* **117**, 243–246. doi: 10.1016/0370-2693(82)90555-X.

Ghiglieri, J. and Moore, G. D. (2014). Low mass thermal dilepton production at NLO in a weakly coupled quark-gluon plasma, *JHEP* **12**, 029. doi: 10.1007/JHEP12(2014)029, arXiv:1410.4203 [hep-ph].

Glozman, L. Y. (2021). Three regimes of QCD, *Int. J. Mod. Phys. A* **36**(25), 2044031. doi: 10.1142/S0217751X20440315, arXiv:1907.01820 [hep-ph].

Glozman, L. Y., Lang, C. B., and Schrock, M. (2012). Symmetries of hadrons after unbreaking the chiral symmetry, *Phys. Rev. D* **86**, 014507. doi: 10.1103/PhysRevD.86.014507, arXiv:1205.4887 [hep-lat].

Goity, J. L. and Leutwyler, H. (1989). On the mean free path of pions in hot matter, *Phys. Lett. B* **228**, 517–522. doi: 10.1016/0370-2693(89)90985-4.

Goldhaber, G., Goldhaber, S., Lee, W.-Y., and Pais, A. (1960). Influence of Bose-Einstein statistics on the anti-proton proton annihilation process, *Phys. Rev.* **120**, 300–312. doi: 10.1103/PhysRev.120.300.

Gorda, T. and Romatschke, P. (2014). Precision studies of v_n fluctuations, *Phys. Rev. C* **90**(5), 054908. doi: 10.1103/PhysRevC.90.054908, arXiv:1406.6405 [nucl-th].

Goswami, J., Karsch, F., Mukherjee, S., Petreczky, P., and Schmidt, C. (2022). Conserved charge fluctuations at vanishing net-baryon density from Lattice QCD, *EPJ Web Conf.* **259**, 10010. doi: 10.1051/epjconf/202225910010, arXiv:2109.00268 [hep-lat].

Grandchamp, L. and Rapp, R. (2002). Charmonium suppression and regeneration from SPS to RHIC, *Nucl. Phys. A* **709**, 415–439. doi: 10.1016/S0375-9474(02)01027-8, arXiv:hep-ph/0205305.

Gross, D. J. and Wilczek, F. (1973). Ultraviolet behavior of nonabelian gauge theories, *Phys. Rev. Lett.* **30**, 1343–1346. doi: 10.1103/PhysRevLett.30.1343.

Gross, D. J., Pisarski, R. D., and Yaffe, L. G. (1981). QCD and instantons at finite temperature, *Rev. Mod. Phys.* **53**, 43. doi: 10.1103/RevModPhys.53.43.

Grosse-Oetringhaus, J. F. (2014). Overview of ALICE results at quark matter 2014, *Nucl. Phys. A* **931**, 22–31. doi: 10.1016/j.nuclphysa.2014.10.003, arXiv:1408.0414 [nucl-ex].

Grossi, E. and Wink, N. (2019). Resolving phase transitions with discontinuous Galerkin methods. arXiv:1903.09503 [hep-th].

Gubser, S. S. (2006). Drag force in AdS/CFT, *Phys. Rev. D* **74**, 126005. doi: 10.1103/PhysRevD.74.126005, arXiv:hep-th/0605182.

Gubser, S. S. (2010). Symmetry constraints on generalizations of Bjorken flow, *Phys. Rev. D* **82**, 085027. doi: 10.1103/PhysRevD.82.085027, arXiv:1006.0006 [hep-th].

Gubser, S. S. and Yarom, A. (2009). Linearized hydrodynamics from probe-sources in the gauge-string duality, *Nucl. Phys. B* **813**, 188–219. doi: 10.1016/j.nuclphysb.2008.12.032, arXiv:0803.0081 [hep-th].

Gubser, S. S. and Yarom, A. (2011). Conformal hydrodynamics in Minkowski and de Sitter spacetimes, *Nucl. Phys. B* **846**, 469–511. doi: 10.1016/j.nuclphysb.2011.01.012, arXiv:1012.1314 [hep-th].

Gubser, S. S., Klebanov, I. R., and Peet, A. W. (1996). Entropy and temperature of black 3-branes, *Phys. Rev. D* **54**, 3915–3919. doi: 10.1103/PhysRevD.54.3915, arXiv:hep-th/9602135.

Gubser, S. S., Klebanov, I. R., and Tseytlin, A. A. (1998). Coupling constant dependence in the thermodynamics of N=4 supersymmetric Yang-Mills theory, *Nucl. Phys. B* **534**, 202–222. doi: 10.1016/S0550-3213(98)00514-8, arXiv:hep-th/9805156.

Gubser, S. S., Gulotta, D. R., Pufu, S. S., and Rocha, F. D. (2008a). Gluon energy loss in the gauge-string duality, *JHEP* **10**, 052. doi: 10.1088/1126-6708/2008/10/052, arXiv:0803.1470 [hep-th].

Gubser, S. S., Pufu, S. S., and Yarom, A. (2008b). Entropy production in collisions of gravitational shock waves and of heavy ions, *Phys. Rev. D* **78**, 066014. doi: 10.1103/PhysRevD.78.066014, arXiv:0805.1551 [hep-th].

Gubser, S. S., Pufu, S. S., and Yarom, A. (2009). Off-center collisions in AdS(5) with applications to multiplicity estimates in heavy-ion collisions, *JHEP* **11**, 050. doi: 10.1088/1126-6708/2009/11/050, arXiv:0902.4062 [hep-th].

Gursoy, U. and Kiritsis, E. (2008). Exploring improved holographic theories for QCD: Part I, *JHEP* **02**, 032. doi: 10.1088/1126-6708/2008/02/032, arXiv:0707.1324 [hep-th].

Gursoy, U., Kiritsis, E., and Nitti, F. (2008). Exploring improved holographic theories for QCD: Part II, *JHEP* **02**, 019. doi: 10.1088/1126-6708/2008/02/019, arXiv:0707.1349 [hep-th].

Gursoy, U., Kiritsis, E., Mazzanti, L., Michalogiorgakis, G., and Nitti, F. (2011). Improved holographic QCD. Lecture Notes in Physics, Vol. 828, pp. 79–146. doi: 10.1007/978-3-642-04864-7_4, arXiv:1006.5461 [hep-th].

Gurzadyan, V. G. and Penrose, R. (2011). CCC-predicted low-variance circles in CMB sky and LCDM. arXiv:1104.5675 [astro-ph.CO].

Gyulassy, M., Kauffmann, S. K., and Wilson, L. W. (1979). Pion interferometry of nuclear collisions. 1. Theory, *Phys. Rev. C* **20**, 2267–2292. doi: 10.1103/PhysRevC.20.2267.

Hagedorn, R. (1965). Statistical thermodynamics of strong interactions at high-energies, *Nuovo Cim. Suppl.* **3**, 147–186.

Hagedorn, R. (1985). How we got to QCD matter from the hadron side: 1984. Lecture Notes in Physics, Vol. **221**, pp. 53–76. doi: 10.1007/978-3-319-17545-4_25.

Hanbury Brown, R. and Twiss, R. Q. (1954). A new type of interferometer for use in radio astronomy, *Phil. Mag. Ser. 7* **45**, 663–682. doi: 10.1080/14786440708520475.

Harada, M. and Sasaki, C. (2006). Dropping rho and A(1) meson masses at chiral phase transition in the generalized hidden local symmetry, *Phys. Rev. D* **73**, 036001. doi: 10.1103/PhysRevD.73.036001, arXiv:hep-ph/0511312.

Hart, A., Laine, M., and Philipsen, O. (2000). Static correlation lengths in QCD at high temperatures and finite densities, *Nucl. Phys. B* **586**, 443–474. doi: 10.1016/S0550-3213(00)00418-1, arXiv:hep-ph/0004060.

Hatta, Y., Iancu, E., and Mueller, A. H. (2008). Jet evolution in the N=4 SYM plasma at strong coupling, *JHEP* **05**, 037. doi: 10.1088/1126-6708/2008/05/037, arXiv:0803.2481 [hep-th].

Heinz, U. W. and Kuhlman, A. (2005). Anisotropic flow and jet quenching in ultrarelativistic U + U collisions, *Phys. Rev. Lett.* **94**, 132301. doi: 10.1103/PhysRevLett.94.132301, arXiv:nucl-th/0411054.

Heinz, U. and Snellings, R. (2013). Collective flow and viscosity in relativistic heavy-ion collisions, *Ann. Rev. Nucl. Part. Sci.* **63**, 123–151. doi: 10.1146/annurev-nucl-102212-170540, arXiv:1301.2826 [nucl-th].

Heller, M. P., Janik, R. A., and Witaszczyk, P. (2012). A numerical relativity approach to the initial value problem in asymptotically Anti-de Sitter spacetime for plasma thermalization — an ADM formulation, *Phys. Rev. D* **85**, 126002. doi: 10.1103/PhysRevD.85.126002, arXiv:1203.0755 [hep-th].

Herzog, C. P., Karch, A., Kovtun, P., Kozcaz, C., and Yaffe, L. G. (2006). Energy loss of a heavy quark moving through N=4 supersymmetric Yang-Mills plasma, *JHEP* **07**, 013. doi: 10.1088/1126-6708/2006/07/013, arXiv:hep-th/0605158.

Hietanen, A., Kajantie, K., Laine, M., Rummukainen, K., and Schroder, Y. (2009). Three-dimensional physics and the pressure of hot QCD, *Phys. Rev. D* **79**, 045018. doi: 10.1103/PhysRevD.79.045018, arXiv:0811.4664 [hep-lat].

Hirono, Y. and Shuryak, E. (2015). Femtoscopic signature of strong radial flow in high-multiplicity *pp* collisions, *Phys. Rev.* **C91**(5), 054915. doi: 10.1103/PhysRevC.91.054915, arXiv:1412.0063 [hep-ph].

Hirano, T., Heinz, U. W., Kharzeev, D., Lacey, R., and Nara, Y. (2006). Hadronic dissipative effects on elliptic flow in ultrarelativistic heavy-ion collisions, *Phys. Lett. B* **636**, 299–304. doi: 10.1016/j.physletb.2006.03.060, arXiv:nucl-th/0511046.

Hoang, T. F., Rhines, D., and Cooper, W. A. (1971). Inclusive anti-p p —> pi + anything at 2.32 gev/c, *Phys. Rev. Lett.* **27**, 1681–1685. doi: 10.1103/PhysRevLett.27.1681.

Huang, S. (2013). Measurements of identified particle anisotropic flow in Cu+Au and U+U collisions by PHENIX experiment, *Nucl. Phys. A* **904–905**, 417c–420c. doi: 10.1016/j.nuclphysa.2013.02.038, arXiv:1210.5570 [nucl-ex].

Hung, C. M. and Shuryak, E. V. (1995a). Hydrodynamics near the QCD phase transition: Looking for the longest lived fireball, *Phys. Rev. Lett.* **75**, 4003–4006. doi: 10.1103/PhysRevLett.75.4003, arXiv:hep-ph/9412360.

Hung, C. M. and Shuryak, E. V. (1995b). Hydrodynamics near the QCD phase transition: Looking for the longest lived fireball, *Phys. Rev. Lett.* **75**, 4003–4006. doi: 10.1103/PhysRevLett.75.4003, arXiv:hep-ph/9412360.

Hung, C. M. and Shuryak, E. V. (1998). Equation of state, radial flow and freezeout in high-energy heavy ion collisions, *Phys. Rev. C* **57**, 1891–1906. doi: 10.1103/PhysRevC.57.1891, arXiv:hep-ph/9709264.

Iancu, E., Leonidov, A., and McLerran, L. (2002). The color glass condensate: An introduction, in *Cargese Summer School on QCD Perspectives on Hot and Dense Matter*, pp. 73–145. arXiv:hep-ph/0202270.

Iatrakis, I., Ramamurti, A., and Shuryak, E. (2015). Collective string interactions in AdS/QCD and high-multiplicity pA collisions, *Phys. Rev. D* **92**(1), 014011. doi: 10.1103/PhysRevD.92.014011, arXiv:1503.04759 [hep-ph].

Iatrakis, I., Ramamurti, A., and Shuryak, E. (2016). Pomeron interactions from the Einstein-Hilbert action, *Phys. Rev. D* **94**(4), 045005. doi: 10.1103/PhysRevD.94.045005, arXiv:1602.05014 [hep-ph].

Ilgenfritz, E.-M. and Shuryak, E. V. (1989). Chiral symmetry restoration at finite temperature in the instanton liquid, *Nucl. Phys. B* **319**, 511–520. doi: 10.1016/0550-3213(89)90617-2.

Itoh, N. (1970). Hydrostatic equilibrium of hypothetical quark stars, *Prog. Theor. Phys.* **44**, 291. doi: 10.1143/PTP.44.291.

Janik, R. A. and Peschanski, R. B. (2006). Asymptotic perfect fluid dynamics as a consequence of Ads/CFT, *Phys. Rev. D* **73**, 045013. doi: 10.1103/PhysRevD.73.045013, arXiv:hep-th/0512162.

Ji, X.-D. and Jung, C.-W. (2001). Studying hadronic structure of the photon in lattice QCD, *Phys. Rev. Lett.* **86**, 208. doi: 10.1103/PhysRevLett.86.208, arXiv:hep-lat/0101014.

Ji, X.-D., Ma, J.-P., and Yuan, F. (2004). Classification and asymptotic scaling of hadrons' light cone wave function amplitudes, *Eur. Phys. J. C* **33**, 75–90. doi: 10.1140/epjc/s2003-01563-y, arXiv:hep-ph/0304107.

Jia, J. (2011). Measurement of elliptic and higher order flow from ATLAS experiment at the LHC, *J. Phys.* **G38**, 124012. doi: 10.1088/0954-3899/38/12/124012, arXiv:1107.1468 [nucl-ex].

Jia, S. and Vary, J. P. (2019). Basis light front quantization for the charged light mesons with color singlet Nambu–Jona-Lasinio interactions, *Phys. Rev. C* **99**(3), 035206. doi: 10.1103/PhysRevC.99.035206, arXiv:1811.08512 [nucl-th].

Kaidalov, A. B. (2001). Regge poles in QCD, pp. 603–636. doi: 10.1142/9789812810458_0018, arXiv:hep-ph/0103011.

Kajantie, K., Laine, M., Rummukainen, K., and Schroder, Y. (2001). How to resum long distance contributions to the QCD pressure? *Phys. Rev. Lett.* **86**, 10–13. doi: 10.1103/PhysRevLett.86.10, arXiv:hep-ph/0007109.

Kalaydzhyan, T. and Shuryak, E. (2014a). Collective interaction of QCD strings and early stages of high multiplicity pA collisions, *Phys. Rev. C* **90**(1), 014901. doi: 10.1103/PhysRevC.90.014901, arXiv:1404.1888 [hep-ph].

Kalaydzhyan, T. and Shuryak, E. (2014b). Self-interacting QCD strings and string balls, *Phys. Rev. D* **90**(2), 025031. doi: 10.1103/PhysRevD.90.025031, arXiv:1402.7363 [hep-ph].

Kalaydzhyan, T. and Shuryak, E. (2015). Gravity waves generated by sounds from big bang phase transitions, *Phys. Rev. D* **91**(8), 083502. doi: 10.1103/PhysRevD.91.083502, arXiv:1412.5147 [hep-ph].

Kapusta, J. I. (1979). Quantum chromodynamics at high temperature, *Nucl. Phys. B* **148**, 461–498. doi: 10.1016/0550-3213(79)90146-9.

Kapusta, J. I. and Shuryak, E. V. (1994). Weinberg type sum rules at zero and finite temperature, *Phys. Rev. D* **49**, 4694–4704. doi: 10.1103/PhysRevD.49.4694, arXiv:hep-ph/9312245.

Kapusta, J. I., Rrapaj, E., and Rudaz, S. (2020a). Hyperon polarization in relativistic heavy ion collisions and axial U(1) symmetry breaking at high temperature, *Phys. Rev. C* **101**(3), 031901. doi: 10.1103/PhysRevC.101.031901, arXiv:1910.12759 [nucl-th].

Kapusta, J. I., Rrapaj, E., and Rudaz, S. (2020b). Relaxation time for strange quark spin in rotating quark-gluon plasma, *Phys. Rev. C* **101**(2), 024907. doi: 10.1103/PhysRevC.101.024907, arXiv:1907.10750 [nucl-th].

Karch, A., Katz, E., Son, D. T., and Stephanov, M. A. (2006a). Linear confinement and AdS/QCD, *Phys. Rev. D* **74**, 015005. doi: 10.1103/PhysRevD.74.015005, arXiv:hep-ph/0602229.

Karch, A., Son, D. T., Katz, E., and Stephanov, M. A. (2006b). Linear confinement and AdS/QCD, in *7th Workshop on Continuous Advances in QCD*, pp. 96–102.

Karsch, F. and Redlich, K. (2011). Probing freeze-out conditions in heavy ion collisions with moments of charge fluctuations, *Phys. Lett. B* **695**, 136–142. doi: 10.1016/j.physletb.2010.10.046, arXiv:1007.2581 [hep-ph].

Karthein, J. M., Koch, V., Ratti, C., and Vovchenko, V. (2021). Constraining the hadronic spectrum and repulsive interactions in a hadron resonance gas via fluctuations of conserved charges, *Phys. Rev. D* **104**(9), 094009. doi: 10.1103/PhysRevD.104.094009, arXiv:2107.00588 [nucl-th].

Kawanai, T. and Sasaki, S. (2015). Potential description of charmonium and charmed-strange mesons from lattice QCD, *Phys. Rev. D* **92**(9), 094503. doi: 10.1103/PhysRevD.92.094503, arXiv:1508.02178 [hep-lat].

Khachatryan, V. *et al.* (2010). Observation of long-range near-side angular correlations in proton-proton collisions at the LHC, *JHEP* **09**, 091. doi: 10.1007/JHEP09(2010)091, arXiv:1009.4122 [hep-ex].

Kharzeev, D. E. and Levin, E. M. (2017). Deep inelastic scattering as a probe of entanglement, *Phys. Rev. D* **95**(11), 114008. doi: 10.1103/PhysRevD.95.114008, arXiv:1702.03489 [hep-ph].

Kharzeev, D. E., McLerran, L. D., and Warringa, H. J. (2008). The effects of topological charge change in heavy ion collisions: 'Event by event P and CP violation', *Nucl. Phys. A* **803**, 227–253. doi: 10.1016/j.nuclphysa.2008.02.298, arXiv:0711.0950 [hep-ph].

Khriplovich, I. B. (1969). Green's functions in theories with non-abelian gauge group. *Sov. J. Nucl. Phys.* **10**, 235–242.

Klinkhamer, F. R. and Manton, N. S. (1984). A saddle point solution in the Weinberg-Salam theory, *Phys. Rev. D* **30**, 2212. doi: 10.1103/PhysRevD.30.2212.

Koch, V., Shuryak, E. V., Brown, G. E., and Jackson, A. D. (1992). The propagation of quarks in the spatial direction in hot QCD, *Phys. Rev. D* **46**, 3169. doi: 10.1103/PhysRevD.46.3169, arXiv:hep-ph/9204236 [Erratum: (1993). *Phys. Rev. D* **47**, 2157].

Kock, A., Liu, Y., and Zahed, I. (2020). Pion and kaon parton distributions in the QCD instanton vacuum, *Phys. Rev. D* **102**(1), 014039. doi: 10.1103/PhysRevD.102.014039, arXiv:2004.01595 [hep-ph].

Koma, Y. and Koma, M. (2017). Precise determination of the three-quark potential in SU(3) lattice gauge theory, *Phys. Rev. D* **95**(9), 094513. doi: 10.1103/PhysRevD.95.094513, arXiv:1703.06247 [hep-lat].

Koponen, J., Zimermmane-Santos, A. C., Davies, C. T. H., Lepage, G. P., and Lytle, A. T. (2017). Pseudoscalar meson electromagnetic form factor at high Q^2 from full lattice QCD, *Phys. Rev. D* **96**(5), 054501. doi: 10.1103/PhysRevD.96.054501, arXiv:1701.04250 [hep-lat].

Kopylov, G. I. and Podgoretsky, M. I. (1973). Multiple production and interference of particles emitted by moving sources, *Yad. Fiz.* **18**, 656–666.

Kovtun, P. (2012). Lectures on hydrodynamic fluctuations in relativistic theories, *J. Phys. A* **45**, 473001. doi: 10.1088/1751-8113/45/47/473001, arXiv:1205.5040 [hep-th].

Kovtun, P. (2019). First-order relativistic hydrodynamics is stable, *JHEP* **10**, 034. doi: 10.1007/JHEP10(2019)034, arXiv:1907.08191 [hep-th].

Kovtun, P. K. and Starinets, A. O. (2005). Quasinormal modes and holography, *Phys. Rev. D* **72**, 086009. doi: 10.1103/PhysRevD.72.086009, arXiv:hep-th/0506184.

Kraan, T. C. and van Baal, P. (1998). Monopole constituents inside SU(n) calorons, *Phys. Lett. B* **435**, 389–395. doi: 10.1016/S0370-2693(98)00799-0, arXiv:hep-th/9806034.

Krzewicki, M. (2011). Elliptic and triangular flow of identified particles at ALICE, *J. Phys. G* **38**, 124047. doi: 10.1088/0954-3899/38/12/124047, arXiv:1107.0080 [nucl-ex].

Kuraev, E. A., Lipatov, L. N., and Fadin, V. S. (1977). The Pomeranchuk singularity in nonabelian gauge theories, *Sov. Phys. JETP* **45**, 199–204.

Lacey, R. A. (2015). Indications for a critical end point in the phase diagram for hot and dense nuclear matter, *Phys. Rev. Lett.* **114**(14), 142301. doi: 10.1103/PhysRevLett.114.142301, arXiv:1411.7931 [nucl-ex].

Lacey, R. A., Gu, Y., Gong, X., Reynolds, D., Ajitanand, N. N., Alexander, J. M., Mwai, A., and Taranenko, A. (2013). Is anisotropic flow really acoustic? arXiv:1301.0165 [nucl-ex].

Laine, M. and Vepsalainen, M. (2004). Mesonic correlation lengths in high temperature QCD, *JHEP* **02**, 004. doi: 10.1088/1126-6708/2004/02/004, arXiv:hep-ph/0311268.

Landau, L. D. (1953). On the multiparticle production in high-energy collisions, *Izv. Akad. Nauk Ser. Fiz.* **17**, 51–64.

Langfeld, K. and Ilgenfritz, E.-M. (2011). Confinement from semiclassical gluon fields in SU(2) gauge theory, *Nucl. Phys. B* **848**, 33–61. doi: 10.1016/j.nuclphysb.2011.02.009, arXiv:1012.1214 [hep-lat].

Larsen, R. and Shuryak, E. (2015). Interacting ensemble of the instanton-dyons and the deconfinement phase transition in the SU(2) gauge theory, *Phys. Rev. D* **92**(9), 094022. doi: 10.1103/PhysRevD.92.094022, arXiv:1504.03341 [hep-ph].

Larsen, R. and Shuryak, E. (2016). Instanton-dyon ensemble with two dynamical quarks: The chiral symmetry breaking, *Phys. Rev. D* **93**(5), 054029. doi: 10.1103/PhysRevD.93.054029, arXiv:1511.02237 [hep-ph].

Larsen, R. N., Sharma, S., and Shuryak, E. (2019). The topological objects near the chiral crossover transition in QCD, *Phys. Lett. B* **794**, 14–18. doi: 10.1016/j.physletb.2019.05.019, arXiv:1811.07914 [hep-lat].

Larsen, R. N., Sharma, S., and Shuryak, E. (2020). Towards a semiclassical description of QCD vacuum around T_c, *Phys. Rev. D* **102**(3), 034501. doi: 10.1103/PhysRevD.102.034501, arXiv:1912.09141 [hep-lat].

Laudicina, D., Dalla Brida, M., Giusti, L., Harris, T., and Pepe, M. (2022). QCD mesonic screening masses and restoration of chiral symmetry at high T, in *39th International Symposium on Lattice Field Theory*. arXiv:2212.02167 [hep-lat].

Lee, K.-M. and Lu, C.-H. (1998). SU(2) calorons and magnetic monopoles, *Phys. Rev. D* **58**, 025011. doi: 10.1103/PhysRevD.58.025011, arXiv:hep-th/9802108.

Leinweber, D. B. (1999). Visualizations of the QCD vacuum, in *Workshop on Light-Cone QCD and Nonperturbative Hadron Physics*, pp. 138–143. arXiv:hep-lat/0004025.

Lepage, G. P. and Brodsky, S. J. (1979). Exclusive processes in quantum chromodynamics: Evolution equations for hadronic wave functions and the form-factors of mesons, *Phys. Lett. B* **87**, 359–365. doi: 10.1016/0370-2693(79)90554-9.

Lepage, G. P. and Brodsky, S. J. (1980). Exclusive processes in perturbative quantum chromodynamics, *Phys. Rev. D* **22**, 2157. doi: 10.1103/PhysRevD.22.2157.

Liao, J. and Shuryak, E. (2007). Strongly coupled plasma with electric and magnetic charges, *Phys. Rev. C* **75**, 054907. doi: 10.1103/PhysRevC.75.054907, arXiv:hep-ph/0611131.

Liao, J. and Shuryak, E. (2008a). Electric flux tube in magnetic plasma, *Phys. Rev. C* **77**, 064905. doi: 10.1103/PhysRevC.77.064905, arXiv:0706.4465 [hep-ph].

Liao, J. and Shuryak, E. (2008b). Magnetic component of quark-gluon plasma is also a liquid! *Phys. Rev. Lett.* **101**, 162302. doi: 10.1103/PhysRevLett.101.162302, arXiv:0804.0255 [hep-ph].

Liao, J. and Shuryak, E. (2009). Angular dependence of jet quenching indicates its strong enhancement near the QCD phase transition, *Phys. Rev. Lett.* **102**, 202302. doi: 10.1103/PhysRevLett.102.202302, arXiv:0810.4116 [nucl-th].

Liao, J. and Shuryak, E. (2010). Static $\bar{Q}Q$ potentials and the magnetic component of QCD plasma near T_c, *Phys. Rev. D* **82**, 094007. doi: 10.1103/PhysRevD.82.094007, arXiv:0804.4890 [hep-ph].

Lin, S. and Shuryak, E. (2007). Stress tensor of static dipoles in strongly coupled N = 4 gauge theory, *Phys. Rev. D* **76**, 085014. doi: 10.1103/PhysRevD.76.085014, arXiv:0707.3135 [hep-th].

Lin, S. and Shuryak, E. (2008a). Toward the AdS/CFT gravity dual for high energy collisions. 2. The stress tensor on the boundary, *Phys. Rev. D* **77**, 085014. doi: 10.1103/PhysRevD.77.085014, arXiv:0711.0736 [hep-th].

Lin, S. and Shuryak, E. (2008b). Toward the AdS/CFT gravity dual for high energy collisions. 3. Gravitationally collapsing shell and quasiequilibrium, *Phys. Rev. D* **78**, 125018. doi: 10.1103/PhysRevD.78.125018, arXiv:0808.0910 [hep-th].

Lin, S. and Shuryak, E. (2008c). Toward the AdS/CFT gravity dual for high energy collisions: I. Falling into the AdS, *Phys. Rev. D* **77**, 085013. doi: 10.1103/PhysRevD.77.085013, arXiv:hep-ph/0610168.

Lin, S. and Shuryak, E. (2009). Grazing collisions of gravitational shock waves and entropy production in heavy ion collision, *Phys. Rev. D* **79**, 124015. doi: 10.1103/PhysRevD.79.124015, arXiv:0902.1508 [hep-th].

Lin, S. and Shuryak, E. (2011). On the critical condition in gravitational shock wave collision and heavy ion collisions, *Phys. Rev. D* **83**, 045025. doi: 10.1103/PhysRevD.83.045025, arXiv:1011.1918 [hep-th].

Lin, H.-W. *et al.* (2018). Parton distributions and lattice QCD calculations: A community white paper, *Prog. Part. Nucl. Phys.* **100**, 107–160. doi: 10.1016/j.ppnp.2018.01.007, arXiv:1711.07916 [hep-ph].

Linde, A. D. (1980). Infrared problem in thermodynamics of the Yang-Mills gas, *Phys. Lett. B* **96**, 289–292. doi: 10.1016/0370-2693(80)90769-8.

Lissauer, D. and Shuryak, E. V. (1991). K meson modification in hot hadronic matter may be detected via Phi meson decays, *Phys. Lett. B* **253**, 15–18. doi: 10.1016/0370-2693(91)91355-Y.

Liu, Y. and Zahed, I. (2021). Small size instanton contributions to the quark quasi-PDF and matching kernel. arXiv:2102.07248 [hep-ph].

Liu, Y., Shuryak, E., and Zahed, I. (2015a). Confining dyon-antidyon Coulomb liquid model. I. *Phys. Rev. D* **92**(8), 085006. doi: 10.1103/PhysRevD.92.085006, arXiv:1503.03058 [hep-ph].

Liu, Y., Shuryak, E., and Zahed, I. (2015b). Light quarks in the screened dyon-antidyon Coulomb liquid model. II. *Phys. Rev. D* **92**(8), 085007. doi: 10.1103/PhysRevD.92.085007, arXiv:1503.09148 [hep-ph].

Liu, Y., Shuryak, E., and Zahed, I. (2016). The instanton-dyon liquid model V: Twisted light quarks, *Phys. Rev. D* **94**(10), 105013. doi: 10.1103/PhysRevD.94.105013, arXiv:1606.02996 [hep-ph].

Lublinsky, M. and Shuryak, E. (2007). How much entropy is produced in strongly coupled quark-gluon plasma (sQGP) by dissipative effects? *Phys. Rev. C* **76**, 021901. doi: 10.1103/PhysRevC.76.021901, arXiv:0704.1647 [hep-ph].

Lublinsky, M. and Shuryak, E. (2009). Improved hydrodynamics from the AdS/CFT, *Phys. Rev. D* **80**, 065026. doi: 10.1103/PhysRevD.80.065026, arXiv:0905.4069 [hep-ph].

Lublinsky, M. and Shuryak, E. (2011). Universal hydrodynamics and charged hadron multiplicity at the LHC, *Phys. Rev. C* **84**, 061901. doi: 10.1103/PhysRevC.84.061901, arXiv:1108.3972 [hep-ph].

Luscher, M. and Weisz, P. (2004). String excitation energies in SU(N) gauge theories beyond the free-string approximation, *JHEP* **07**, 014. doi: 10.1088/1126-6708/2004/07/014, arXiv:hep-th/0406205.

Luscher, M. and Weisz, P. (2011). Perturbative analysis of the gradient flow in non-abelian gauge theories, *JHEP* **02**, 051. doi: 10.1007/JHEP02(2011)051, arXiv:1101.0963 [hep-th].

Lyu, Y., Tong, H., Sugiura, T., Aoki, S., Doi, T., Hatsuda, T., Meng, J., and Miyamoto, T. (2021). Dibaryon with highest charm number near unitarity from lattice QCD, *Phys. Rev. Lett.* **127**(7), 072003. doi: 10.1103/PhysRevLett.127.072003, arXiv:2102.00181 [hep-lat].

Mace, M., Schlichting, S., and Venugopalan, R. (2016). Off-equilibrium sphaleron transitions in the glasma, *Phys. Rev. D* **93**(7), 074036. doi: 10.1103/PhysRevD.93.074036, arXiv:1601.07342 [hep-ph].

Maezawa, Y., Aoki, S., Ejiri, S., Hatsuda, T., Ishii, N., Kanaya, K., Ukita, N., and Umeda, T. (2010). Electric and magnetic screening masses at finite temperature from generalized Polyakov-line correlations in two-flavor lattice QCD, *Phys. Rev. D* **81**, 091501. doi: 10.1103/PhysRevD.81.091501, arXiv:1003.1361 [hep-lat].

Makhlin, A. N. and Sinyukov, Y. M. (1987). The pion interferometry theory for the hydrodynamic stage of multiple processes, *Yad. Fiz.* **46**, 637–646.

Makhlin, A. N. and Sinyukov, Y. M. (1988). Hydrodynamics of hadron matter under pion interferometric microscope, *Z. Phys. C* **39**, 69. doi: 10.1007/BF01560393.

Maldacena, J. M. (1998a). The large N limit of superconformal field theories and supergravity, *Adv. Theor. Math. Phys.* **2**, 231–252. doi: 10.1023/A:1026654312961, arXiv:hep-th/9711200.

Maldacena, J. M. (1998b). Wilson loops in large N field theories, *Phys. Rev. Lett.* **80**, 4859–4862. doi: 10.1103/PhysRevLett.80.4859, arXiv:hep-th/9803002.

Mandelstam, S. (1976). Vortices and quark confinement in nonabelian gauge theories, *Phys. Rep.* **23**, 245–249. doi: 10.1016/0370-1573(76)90043-0.

Marrochio, H., Noronha, J., Denicol, G. S., Luzum, M., Jeon, S., and Gale, C. (2015). Solutions of conformal Israel-Stewart relativistic viscous fluid dynamics, *Phys. Rev. C* **91**(1), 014903. doi: 10.1103/PhysRevC.91.014903, arXiv:1307.6130 [nucl-th].

Martinez, M. and Strickland, M. (2010). Matching pre-equilibrium dynamics and viscous hydrodynamics, *Phys. Rev. C* **81**, 024906. doi: 10.1103/PhysRevC. 81.024906, arXiv:0909.0264 [hep-ph].

Matsui, T. and Satz, H. (1986). J/ψ suppression by quark-gluon plasma formation, *Phys. Lett. B* **178**, 416–422. doi: 10.1016/0370-2693(86)91404-8.

Mazeliauskas, A. and Teaney, D. (2015). Subleading harmonic flows in hydrodynamic simulations of heavy ion collisions, *Phys. Rev. C* **91**(4), 044902. doi: 10.1103/PhysRevC.91.044902, arXiv:1501.03138 [nucl-th].

McLerran, L. and Pisarski, R. D. (2007). Phases of cold, dense quarks at large N(c), *Nucl. Phys. A* **796**, 83–100. doi: 10.1016/j.nuclphysa.2007.08.013, arXiv:0706.2191 [hep-ph].

McLerran, L. D. and Venugopalan, R. (1994). Gluon distribution functions for very large nuclei at small transverse momentum, *Phys. Rev. D* **49**, 3352–3355. doi: 10.1103/PhysRevD.49.3352, arXiv:hep-ph/9311205.

Melic, B., Nizic, B., and Passek, K. (1999). Complete next-to-leading order perturbative QCD prediction for the pion form-factor, *Phys. Rev. D* **60**, 074004. doi: 10.1103/PhysRevD.60.074004, arXiv:hep-ph/9802204.

Mitchell, J. T., Perepelitsa, D. V., Tannenbaum, M. J., and Stankus, P. W. (2016). Tests of constituent-quark generation methods which maintain both the nucleon center of mass and the desired radial distribution in Monte Carlo Glauber models, *Phys. Rev. C* **93**(5), 054910. doi: 10.1103/PhysRevC.93. 054910, arXiv:1603.08836 [nucl-ex].

Molnar, D. and Gyulassy, M. (2002). Elliptic flow and freezeout from the parton cascade MPC, *Nucl. Phys. A* **698**, 379–382. doi: 10.1016/S0375-9474(01) 01384-7, arXiv:nucl-th/0104018.

Muto, R. *et al.* (2007). Evidence for in-medium modification of the phi meson at normal nuclear density, *Phys. Rev. Lett.* **98**, 042501. doi: 10.1103/ PhysRevLett.98.042501, arXiv:nucl-ex/0511019.

Nadeau, H., Kapusta, J. I., and Lichard, P. (1992). Parametrization of thermal photon emission rates from mesonic matter, *Phys. Rev. C* **45**, 3034–3036. doi: 10.1103/PhysRevC.45.3034.

Nagle, J., Adare, A., Beckman, S., Koblesky, T., Koop, J. O., McGlinchey, D., Romatschke, P., Carlson, J., Lynn, J., and McCumber, M. (2014). Exploiting intrinsic triangular geometry in relativistic He3+Au collisions to disentangle medium properties, *Phys. Rev. Lett.* **113**(11), 112301. doi: 10.1103/PhysRevLett.113.112301, arXiv:1312.4565 [nucl-th].

Nakamura, A. and Sakai, S. (2005). Transport coefficients of gluon plasma, *Phys. Rev. Lett.* **94**, 072305. doi: 10.1103/PhysRevLett.94.072305, arXiv:hep-lat/0406009.

Nakamura, Y. and Schierholz, G. (2021). The strong CP problem solved by itself due to long-distance vacuum effects. arXiv:2106.11369 [hep-ph].

Nambu, Y. (1974). Strings, monopoles and gauge fields, *Phys. Rev. D* **10**, 4262. doi: 10.1103/PhysRevD.10.4262.

Nambu, Y. and Jona-Lasinio, G. (1961a). Dynamical model of elementary particles based on an analogy with superconductivity. 1. *Phys. Rev.* **122**, 345–358. doi: 10.1103/PhysRev.122.345.

Nambu, Y. and Jona-Lasinio, G. (1961b). Dynamical model of elementary particles based on an analogy with superconductivity. II, *Phys. Rev.* **124**, 246–254. doi: 10.1103/PhysRev.124.246.

Natsuume, M. and Okamura, T. (2008). Causal hydrodynamics of gauge theory plasmas from AdS/CFT duality, *Phys. Rev. D* **77**, 066014. doi: 10.1103/PhysRevD.78.089902, arXiv:0712.2916 [hep-th] [Erratum: (2008). *Phys. Rev. D* **78**, 089902].

Neronov, A., Roper Pol, A., Caprini, C., and Semikoz, D. (2021). NANOGrav signal from magnetohydrodynamic turbulence at the QCD phase transition in the early Universe, *Phys. Rev. D* **103**(4), 041302. doi: 10.1103/PhysRevD.103.L041302, arXiv:2009.14174 [astro-ph.CO].

Nijs, G., van der Schee, W., Gürsoy, U., and Snellings, R. (2021). Bayesian analysis of heavy ion collisions with the heavy ion computational framework Trajectum, *Phys. Rev. C* **103**(5), 054909. doi: 10.1103/PhysRevC.103.054909, arXiv:2010.15134 [nucl-th].

Novikov, V. A., Shifman, M. A., Vainshtein, A. I., and Zakharov, V. I. (1981). Are all hadrons alike? *Nucl. Phys. B* **191**, 301–369. doi: 10.1016/0550-3213(81)90303-5.

Ostrovsky, D. M., Carter, G. W., and Shuryak, E. V. (2002). Forced tunneling and turning state explosion in pure Yang-Mills theory, *Phys. Rev. D* **66**, 036004. doi: 10.1103/PhysRevD.66.036004, arXiv:hep-ph/0204224.

Pang, L.-G., Hatta, Y., Wang, X.-N., and Xiao, B.-W. (2015). Analytical and numerical Gubser solutions of the second-order hydrodynamics, *Phys. Rev. D* **91**(7), 074027. doi: 10.1103/PhysRevD.91.074027, arXiv:1411.7767 [hep-ph].

Paquet, J.-F. (2017). Probing the space-time evolution of heavy ion collisions with photons and dileptons, *Nucl. Phys. A* **967**, 184–191. doi: 10.1016/j.nuclphysa.2017.06.003, arXiv:1704.07842 [nucl-th].

Parida, T. and Chatterjee, S. (2022). Directed flow of light flavor hadrons for Au+Au collisions at $\sqrt{S_{NN}} = 7.7 - 200\,\text{GeV}$. arXiv:2211.15659 [nucl-th].

Peshier, A., Kampfer, B., Pavlenko, O. P., and Soff, G. (1996). A massive quasiparticle model of the SU(3) gluon plasma, *Phys. Rev. D* **54**, 2399–2402. doi: 10.1103/PhysRevD.54.2399.

Petrov, V. Y. and Pobylitsa, P. V. (1997). Pion wave function from the instanton vacuum model, arXiv:hep-ph/9712203.

Pineda, A. and Vairo, A. (2001). The QCD potential at O $(1\,/\,m^2)$: Complete spin dependent and spin independent result, *Phys. Rev. D* **63**, 054007. doi: 10.1103/PhysRevD.64.039902, arXiv:hep-ph/0009145 [Erratum: (2001). *Phys. Rev. D* **64**, 039902].

Pinkenburg, C. *et al.* (1999). Elliptic flow: Transition from out-of-plane to in-plane emission in Au + Au collisions, *Phys. Rev. Lett.* **83**, 1295–1298. doi: 10.1103/PhysRevLett.83.1295, arXiv:nucl-ex/9903010.

Pisarski, R. D., Rennecke, F., Tsvelik, A., and Valgushev, S. (2021). The Lifshitz regime and its experimental signals, *Nucl. Phys. A* **1005**, 121910. doi: 10.1016/j.nuclphysa.2020.121910, arXiv:2005.00045 [nucl-th].

Polchinski, J. and Strassler, M. J. (2002). Hard scattering and gauge / string duality, *Phys. Rev. Lett.* **88**, 031601. doi: 10.1103/PhysRevLett.88.031601, arXiv:hep-th/0109174.

Policastro, G., Son, D. T., and Starinets, A. O. (2001). The shear viscosity of strongly coupled N=4 supersymmetric Yang-Mills plasma, *Phys. Rev. Lett.* **87**, 081601. doi: 10.1103/PhysRevLett.87.081601, arXiv:hep-th/0104066.

Policastro, G., Son, D. T., and Starinets, A. O. (2002). From AdS / CFT correspondence to hydrodynamics, *JHEP* **09**, 043. doi: 10.1088/1126-6708/2002/09/043, arXiv:hep-th/0205052.

Politzer, H. D. (1973). Reliable perturbative results for strong interactions? *Phys. Rev. Lett.* **30**, 1346–1349. doi: 10.1103/PhysRevLett.30.1346.

Polyakov, A. M. (1974). Particle spectrum in quantum field theory, *JETP Lett.* **20**, 194–195.

Polyakov, A. M. (1979). String representations and hidden symmetries for gauge fields, *Phys. Lett. B* **82**, 247–250. doi: 10.1016/0370-2693(79)90747-0.

Pomeranchuk, I. (1951a). On the theory of production of many particles.

Pomeranchuk, I. Y. (1951b). On the theory of multiple particle production in a single collision, *Dokl. Akad. Nauk SSSR* **78**(5), 889–891. doi: 10.1142/9789812815941_0009.

Poppitz, E. and Unsal, M. (2011). Seiberg-Witten and 'Polyakov-like' magnetic bion confinements are continuously connected, *JHEP* **07**, 082. doi: 10.1007/JHEP07(2011)082, arXiv:1105.3969 [hep-th].

Poppitz, E., Schäfer, T., and Unsal, M. (2012). Continuity, deconfinement, and (super) Yang-Mills theory, *JHEP* **10**, 115. doi: 10.1007/JHEP10(2012)115, arXiv:1205.0290 [hep-th].

Poskanzer, A. M. *et al.* (1984). Compression and expansion at the Bevalac.

Prakash, M., Prakash, M., Venugopalan, R., and Welke, G. (1993). Nonequilibrium properties of hadronic mixtures, *Phys. Rep.* **227**, 321–366. doi: 10.1016/0370-1573(93)90092-R.

Pratt, S. (2009). Resolving the HBT puzzle in relativistic heavy ion collision, *Phys. Rev. Lett.* **102**, 232301. doi: 10.1103/PhysRevLett.102.232301, arXiv:0811.3363 [nucl-th].

Qattan, I. A. and Arrington, J. (2012). Flavor decomposition of the nucleon electromagnetic form factors, *Phys. Rev. C* **86**, 065210. doi: 10.1103/PhysRevC.86.065210, arXiv:1209.0683 [nucl-ex].

Qiu, Z. and Heinz, U. W. (2011). Event-by-event shape and flow fluctuations of relativistic heavy-ion collision fireballs, *Phys. Rev. C* **84**, 024911. doi: 10.1103/PhysRevC.84.024911, arXiv:1104.0650 [nucl-th].

Radyushkin, A. V. (1977). Deep elastic processes of composite particles in field theory and asymptotic freedom. arXiv:hep-ph/0410276.

Rahman, R. M. A., Tawfik, A. N., El-Bakry, M. Y., Habashy, D. M., and Hanafy, M. (2022). Rapidity distribution within Landau hydrodynamical model and EPOS event-generator at AGS, SPS, and RHIC energies, *Phys. Scripta* **97**(6), 065305. doi: 10.1088/1402-4896/ac710d, arXiv:2109.07248 [hep-ph].

Ramamurti, A. and Shuryak, E. (2018). Extending the hydrodynamical description of heavy-ion collisions to the "outer edge" of the fireball. arXiv:1811.03655 [hep-ph].

Ramamurti, A., Shuryak, E., and Zahed, I. (2018). Are there monopoles in the quark-gluon plasma? *Phys. Rev. D* **97**(11), 114028. doi: 10.1103/PhysRevD.97.114028, arXiv:1802.10509 [hep-ph].

Rapp, R. (2013). Dilepton production in heavy-ion collisions, *PoS CPOD2013*, 008. doi: 10.22323/1.185.0008, arXiv:1306.6394 [nucl-th].

Rapp, R. and Shuryak, E. V. (2000). Thermal dilepton radiation at intermediate masses at the CERN — SPS, *Phys. Lett. B* **473**, 13–19. doi: 10.1016/S0370-2693(99)01367-2, arXiv:hep-ph/9909348.

Rapp, R. and Shuryak, E. V. (2001). Resolving the anti-baryon production puzzle in high-energy heavy ion collisions, *Phys. Rev. Lett.* **86**, 2980–2983. doi: 10.1103/PhysRevLett.86.2980, arXiv:hep-ph/0008326.

Rapp, R. and Shuryak, E. V. (2002). Anti-baryon puzzle in ultrarelativistic heavy ion collisions, in *30th International Workshop on Gross Properties of Nuclei and Nuclear Excitation: Hirschegg 2002: Ultrarelativistic Heavy Ion Collisions*, pp. 130–139. arXiv:nucl-th/0202059.

Rapp, R., Schäfer, T., Shuryak, E. V., and Velkovsky, M. (1998). Diquark Bose condensates in high density matter and instantons, *Phys. Rev. Lett.* **81**, 53–56. doi: 10.1103/PhysRevLett.81.53, arXiv:hep-ph/9711396.

Rapp, R., Schäfer, T., Shuryak, E. V., and Velkovsky, M. (2000). High density QCD and instantons, *Ann. Phys.* **280**, 35–99. doi: 10.1006/aphy.1999.5991, arXiv:hep-ph/9904353.

Rapp, R., Shuryak, E. V., and Zahed, I. (2001). A chiral crystal in cold QCD matter at intermediate densities? *Phys. Rev. D* **63**, 034008. doi: 10.1103/PhysRevD.63.034008, arXiv:hep-ph/0008207.

Ratti, C. (2017). Bulk properties of QCD matter from lattice simulations, *J. Phys. Conf. Ser.* **779**(1), 012016. doi: 10.1088/1742-6596/779/1/012016.

Ratti, C. and Shuryak, E. (2009). The role of monopoles in a gluon plasma, *Phys. Rev. D* **80**, 034004. doi: 10.1103/PhysRevD.80.034004, arXiv:0811.4174 [hep-ph].

Reichl, L. (1989). *The Transition to Chaos* (Springer-Verlag).

Rischke, D. H. (1996). Hydrodynamics and collective behavior in relativistic nuclear collisions, *Nucl. Phys. A* **610**, 88C–101C. doi: 10.1016/S0375-9474(96)00345-4, arXiv:nucl-th/9608024.

Romatschke, P. (2010). New developments in relativistic viscous hydrodynamics, *Int. J. Mod. Phys.* **E19**, 1. doi: 10.1142/S0218301310014613, arXiv:0902.3663 [hep-ph].

Romatschke, P. (2015). Collective flow without hydrodynamics: Simulation results for relativistic ion collisions, *Eur. Phys. J. C* **75**(9), 429. doi: 10.1140/epjc/s10052-015-3646-8, arXiv:1504.02529 [nucl-th].

Romatschke, P. (2017). Do nuclear collisions create a locally equilibrated quark–gluon plasma? *Eur. Phys. J. C* **77**(1), 21. doi: 10.1140/epjc/s10052-016-4567-x, arXiv:1609.02820 [nucl-th].

Rose, J.-B., Paquet, J.-F., Denicol, G. S., Luzum, M., Schenke, B., Jeon, S., and Gale, C. (2014). Extracting the bulk viscosity of the quark–gluon plasma, *Nucl. Phys. A* **931**, 926–930. doi: 10.1016/j.nuclphysa.2014.09.044, arXiv:1408.0024 [nucl-th].

Rothkopf, A. (2020). Heavy quarkonium in extreme conditions, *Phys. Rep.* **858**, 1–117. doi: 10.1016/j.physrep.2020.02.006, arXiv:1912.02253 [hep-ph].

Ruiz Arriola, E. and Broniowski, W. (2002). Pion light cone wave function and pion distribution amplitude in the Nambu-Jona-Lasinio model, *Phys. Rev. D* **66**, 094016. doi: 10.1103/PhysRevD.66.094016, arXiv:hep-ph/0207266.

Ryu, S., Jeon, S., Gale, C., Schenke, B., and Young, C. (2013). MUSIC with the UrQMD afterburner, *Nucl. Phys. A* **904–905**, 389c–392c. doi: 10.1016/j.nuclphysa.2013.02.031, arXiv:1210.4588 [hep-ph].

Sakai, T. and Sugimoto, S. (2005). Low energy hadron physics in holographic QCD, *Prog. Theor. Phys.* **113**, 843–882. doi: 10.1143/PTP.113.843, arXiv:hep-th/0412141.

Schäfer, T. and Shuryak, E. V. (1998). Instantons in QCD, *Rev. Mod. Phys.* **70**, 323–426. doi: 10.1103/RevModPhys.70.323, arXiv:hep-ph/9610451.

Schäfer, T. and Shuryak, E. V. (2001). Phases of QCD at high baryon density. Lecture Notes in Physics, Vol. **578**, pp. 203–217. arXiv:nucl-th/0010049.

Schäfer, T., Shuryak, E. V., and Verbaarschot, J. J. M. (1994). Baryonic correlators in the random instanton vacuum, *Nucl. Phys. B* **412**, 143–168. doi: 10.1016/0550-3213(94)90497-9, arXiv:hep-ph/9306220.

Schenke, B. and Venugopalan, R. (2014). Eccentric protons? Sensitivity of flow to system size and shape in p+p, p+Pb and Pb+Pb collisions, *Phys. Rev. Lett.* **113**, 102301. doi: 10.1103/PhysRevLett.113.102301, arXiv:1405.3605 [nucl-th].

Schwinger, J. S., Milton, K. A., Tsai, W.-Y., DeRaad Jr., L. L., and Clark, D. C. (1976). Nonrelativistic dyon-dyon scattering, *Ann. Phys.* **101**, 451. doi: 10.1016/0003-4916(76)90020-8.

Seck, F., Galatyuk, T., Mukherjee, A., Rapp, R., Steinheimer, J., Stroth, J., and Wiest, M. (2022). Dilepton signature of a first-order phase transition, *Phys. Rev. C* **106**, 014904. doi: 10.1103/PhysRevC.106.014904, arXiv:2010.04614 [nucl-th].

Seiberg, N. and Witten, E. (1994). Electric — magnetic duality, monopole condensation, and confinement in N=2 supersymmetric Yang-Mills theory, *Nucl. Phys. B* **426**, 19–52. doi: 10.1016/0550-3213(94)90124-4, arXiv:hep-th/9407087 [Erratum: (1994). *Nucl. Phys. B* **430**, 485–486].

Serot, B. D. and Walecka, J. D. (1986). The relativistic nuclear many body problem, *Adv. Nucl. Phys.* **16**, 1–327.

Shamir, L. (2020). Multipole alignment in the large-scale distribution of spin direction of spiral galaxies. arXiv:2004.02963 [astro-ph.GA].

Shifman, M. A., Vainshtein, A. I., and Zakharov, V. I. (1979). QCD and resonance physics. Theoretical foundations, *Nucl. Phys. B* **147**, 385–447. doi: 10.1016/0550-3213(79)90022-1.

Shnir, Y. M. (2005). *Magnetic Monopoles.* Text and Monographs in Physics (Springer, Berlin/Heidelberg). doi: 10.1007/3-540-29082-6. ISBN 978-3-540-25277-1, 978-3-540-29082-7.

Shuryak, E. V. (1972a). Multiparticle production in high energy particle collisions, *Yad. Fiz.* **16**, 395–405.

Shuryak, E. V. (1972b). The multibody production at mediate energy, *Phys. Lett. B* **42**, 357–360. doi: 10.1016/0370-2693(72)90508-4.

Shuryak, E. V. (1973a). Correlation of identical pions in multiple production reactions, *Yad. Fiz.* **18**, 1302–1308.

Shuryak, E. V. (1973b). The correlation of identical pions in multibody production, *Phys. Lett. B* **44**, 387–389. doi: 10.1016/0370-2693(73)90414-0.

Shuryak, E. V. (1974a). Final state interaction and composition of secondaries, *Yad. Fiz.* **20**, 549–553.

Shuryak, E. V. (1974b). High-energy multiple production of hadrons and Landau hydrodynamical theory.

Shuryak, E. V. (1976). Nonlinear resonance in quantum systems, *Zh. Eksp. Teor. Fiz.* **71**, 2039–2056.

Shuryak, E. V. (1978a). Quark-gluon plasma and hadronic production of leptons, photons and psions, *Phys. Lett. B* **78**, 150. doi: 10.1016/0370-2693(78)90370-2.

Shuryak, E. V. (1978b). Theory of hadronic plasma, *Sov. Phys. JETP* **47**, 212–219.

Shuryak, E. V. (1980). Quantum chromodynamics and the theory of superdense matter, *Phys. Rep.* **61**, 71–158. doi: 10.1016/0370-1573(80)90105-2.

Shuryak, E. V. (1982a). Hadrons containing a heavy quark and QCD sum rules, *Nucl. Phys. B* **198**, 83–101. doi: 10.1016/0550-3213(82)90546-6.

Shuryak, E. V. (1982b). The role of instantons in quantum chromodynamics. 3. Quark — gluon plasma, *Nucl. Phys. B* **203**, 140–156. doi: 10.1016/0550-3213(82)90480-1.

Shuryak, E. V. (1988a). *The QCD Vacuum, Hadrons and the Superdense Matter*, Vol. 8. doi: 10.1142/0161.

Shuryak, E. V. (1988b). Toward the quantitative theory of the 'instanton liquid' 4. Tunneling in the double well potential, *Nucl. Phys. B* **302**, 621–644. doi: 10.1016/0550-3213(88)90191-5.

Shuryak, E. V. (1991). Collective interaction of mesons in hot hadronic matter, *Nucl. Phys. A* **533**, 761–788. doi: 10.1016/0375-9474(91)90543-F.

Shuryak, E. V. (1992). Two stage equilibration in high-energy heavy ion collisions, *Phys. Rev. Lett.* **68**, 3270–3272. doi: 10.1103/PhysRevLett.68.3270.

Shuryak, E. V. (1993). Correlation functions in the QCD vacuum, *Rev. Mod. Phys.* **65**, 1–46. doi: 10.1103/RevModPhys.65.1.

Shuryak, E. V. (1998). Event per event analysis of heavy ion collisions and thermodynamical fluctuations, *Phys. Lett. B* **423**, 9–14. doi: 10.1016/S0370-2693(98)00127-0, arXiv:hep-ph/9704456.

Shuryak, E. V. (2000). High-energy collisions of strongly deformed nuclei: An old idea with a new twist, *Phys. Rev. C* **61**, 034905. doi: 10.1103/PhysRevC.61.034905, arXiv:nucl-th/9906062.

Shuryak, E. V. (2002). The azimuthal asymmetry at large p(t) seem to be too large for a 'jet quenching', *Phys. Rev. C* **66**, 027902. doi: 10.1103/PhysRevC.66.027902, arXiv:nucl-th/0112042.

Shuryak, E. V. (2004). *The QCD Vacuum, Hadrons and the Superdense Matter*, Vol. 71. doi: 10.1142/5367.

Shuryak, E. (2006a). Building a 'holographic dual' to QCD in the AdS(5): Instantons and confinement. arXiv:hep-th/0605219.

Shuryak, E. (2006b). Some thoughts on critical and tricritical points and deconfinement, *PoS CPOD2006*, 026. doi: 10.22323/1.029.0026, arXiv:nucl-th/0609011.

Shuryak, E. (2007). A 'domain wall' scenario for the AdS/QCD. arXiv:0711.0004 [hep-ph].

Shuryak, E. (2012a). Monitoring parton equilibration in heavy ion collisions via dilepton polarization. arXiv:1203.1012 [nucl-th].

Shuryak, E. (2012b). Shocks in quark-gluon plasmas, *Phys. Rev. C* **86**, 024907. doi: 10.1103/PhysRevC.86.024907, arXiv:1203.6614 [hep-ph].

Shuryak, E. (2016). Comment on measurement of the rotaion frequency and the magnetic field at the freezeout of heavy ion collisions. arXiv:1606.02915 [hep-ph].

Shuryak, E. (2017). Strongly coupled quark-gluon plasma in heavy ion collisions, *Rev. Mod. Phys.* **89**, 035001. doi: 10.1103/RevModPhys.89.035001, arXiv:1412.8393 [hep-ph].

Shuryak, E. (2019a). Light-front wave functions of mesons, baryons, and pentaquarks with topology-induced local four-quark interaction, *Phys. Rev. D* **100**(11), 114018. doi: 10.1103/PhysRevD.100.114018, arXiv:1908.10270 [hep-ph].

Shuryak, E. (2019b). *Quantum Many-Body Physics in a Nutshell* (Princeton University Press).

Shuryak, E. (2021). *Nonperturbative Topological Phenomena in QCD and Related Theories*. Lecture Notes in Physics, Vol. 977. doi: 10.1007/978-3-030-62990-8. ISBN 978-3-030-62989-2, 978-3-030-62990-8.

Shuryak, E. (2022a). Mesonic "screening masses" in high temperature QCD. arXiv:2212.11825 [hep-ph].

Shuryak, E. (2022b). The heavy ions are "preheated" prior to hight energy collisions. arXiv:2201.11064 [nucl-th].

Shuryak, E. and Staig, P. (2013). Sound edge of the quenching jets, *Phys. Rev. C* **88**(5), 054903. doi: 10.1103/PhysRevC.88.054903, arXiv:1307.2568 [nucl-th].

Shuryak, E. and Torres-Rincon, J. M. (2019). Baryon clustering at the critical line and near the hypothetical critical point in heavy-ion collisions, *Phys. Rev. C* **100**(2), 024903. doi: 10.1103/PhysRevC.100.024903, arXiv:1805.04444 [hep-ph].

Shuryak, E. and Torres-Rincon, J. M. (2020a). Baryon preclustering at the freeze-out of heavy-ion collisions and light-nuclei production, *Phys. Rev. C* **101**(3), 034914. doi: 10.1103/PhysRevC.101.034914, arXiv:1910.08119 [nucl-th].

Shuryak, E. and Torres-Rincon, J. M. (2020b). Light-nuclei production and search for the QCD critical point, *Eur. Phys. J. A* **56**(9), 241. doi: 10.1140/epja/s10050-020-00244-3, arXiv:2005.14216 [nucl-th].

Shuryak, E. V. and Vainshtein, A. I. (1982a). Theory of power corrections to deep inelastic scattering in quantum chromodynamics. 1. Q**2 effects, *Nucl. Phys. B* **199**, 451–481. doi: 10.1016/0550-3213(82)90355-8.

Shuryak, E. V. and Vainshtein, A. I. (1982b). Theory of power corrections to deep inelastic scattering in quantum chromodynamics. 2. Q**4 effects: Polarized target, *Nucl. Phys. B* **201**, 141. doi: 10.1016/0550-3213(82)90377-7.

Shuryak, E. V. and Xiong, L. (1993). Dilepton and photon production in the 'hot glue' scenario, *Phys. Rev. Lett.* **70**, 2241–2244. doi: 10.1103/PhysRevLett.70.2241, arXiv:hep-ph/9301218.

Shuryak, E. V. and Zahed, I. (2000). Instanton induced effects in QCD high-energy scattering, *Phys. Rev. D* **62**, 085014. doi: 10.1103/PhysRevD.62.085014, arXiv:hep-ph/0005152.

Shuryak, E. and Zahed, I. (2003a). Prompt quark production by exploding sphalerons, *Phys. Rev. D* **67**, 014006. doi: 10.1103/PhysRevD.67.014006, arXiv:hep-ph/0206022.

Shuryak, E. and Zahed, I. (2003b). Semiclassical double pomeron production of glueballs and eta-prime, *Phys. Rev. D* **68**, 034001. doi: 10.1103/PhysRevD.68.034001, arXiv:hep-ph/0302231.

Shuryak, E. V. and Zahed, I. (2003c). Jet quenching in high-energy heavy ion collisions by QCD synchrotron — like radiation, *Phys. Rev. D* **67**, 054025. doi: 10.1103/PhysRevD.67.054025, arXiv:hep-ph/0207163.

Shuryak, E. and Zahed, I. (2004). A schematic model for pentaquarks using diquarks, *Phys. Lett. B* **589**, 21–27. doi: 10.1016/j.physletb.2004.03.019, arXiv:hep-ph/0310270.

Shuryak, E. and Zahed, I. (2013). High-multiplicity pp and pA collisions: Hydrodynamics at its edge, *Phys. Rev. C* **88**(4), 044915. doi: 10.1103/PhysRevC.88.044915, arXiv:1301.4470 [hep-ph].

Shuryak, E. and Zahed, I. (2014). New regimes of the stringy (holographic) Pomeron and high-multiplicity *pp* and *pA* collisions, *Phys. Rev. D* **89**(9), 094001. doi: 10.1103/PhysRevD.89.094001, arXiv:1311.0836 [hep-ph].

Shuryak, E. and Zahed, I. (2018). Regimes of the Pomeron and its intrinsic entropy, *Ann. Phys.* **396**, 1–17. doi: 10.1016/j.aop.2018.06.008, arXiv:1707.01885 [hep-ph].

Shuryak, E. and Zahed, I. (2021). Nonperturbative quark-antiquark interactions in mesonic form factors, *Phys. Rev. D* **103**(5), 054028. doi: 10.1103/PhysRevD.103.054028, arXiv:2008.06169 [hep-ph].

Shuryak, E. and Zahed, I. (2023a). Hadronic structure on the light front. I. Instanton effects and quark-antiquark effective potentials, *Phys. Rev. D* **107**(3), 034023. doi: 10.1103/PhysRevD.107.034023, arXiv:2110.15927 [hep-ph].

Shuryak, E. and Zahed, I. (2023b). Hadronic structure on the light front. II. QCD strings, Wilson lines, and potentials, *Phys. Rev. D* **107**(3), 034024. doi: 10.1103/PhysRevD.107.034024, arXiv:2111.01775 [hep-ph].

Shuryak, E. and Zahed, I. (2023c). Hadronic structure on the light front. IV. Heavy and light baryons, *Phys. Rev. D* **107**(3), 034026. doi: 10.1103/PhysRevD.107.034026, arXiv:2202.00167 [hep-ph].

Shuryak, E. and Zahed, I. (2023d). Hadronic structure on the light front. V. Diquarks, nucleons, and multiquark Fock components, *Phys. Rev. D* **107**(3), 034027. doi: 10.1103/PhysRevD.107.034027, arXiv:2208.04428 [hep-ph].

Shuryak, E. and Zahed, I. (2023e). Hadronic structure on the light-front VI. Generalized parton distributions of unpolarized hadrons. arXiv:2301.12238 [hep-ph].

Shuryak, E. and Zahed, I. (2023f). Meson structure on the light-front III: The Hamiltonian, heavy quarkonia, spin and orbit mixing, *Phys. Rev. D* **107**(3), 034025. doi: 10.1103/PhysRevD.107.034025, arXiv:2112.15586 [hep-ph].

Shuryak, E. V. and Zhirov, O. V. (1979). Vacuum pressure effects in low p(t) hadronic spectra, *Phys. Lett. B* **89**, 253–255. doi: 10.1016/0370-2693(80)90023-4.

Shuryak, E., Sin, S.-J., and Zahed, I. (2007). A gravity dual of RHIC collisions, *J. Korean Phys. Soc.* **50**, 384–397. doi: 10.3938/jkps.50.384, arXiv:hep-th/0511199.

Siemens, P. J. and Rasmussen, J. O. (1979). Evidence for a blast wave from compress nuclear matter, *Phys. Rev. Lett.* **42**, 880–887. doi: 10.1103/PhysRevLett.42.880.

Sin, S.-J. and Zahed, I. (2005). Holography of radiation and jet quenching, *Phys. Lett. B* **608**, 265–273. doi: 10.1016/j.physletb.2005.01.020, arXiv:hep-th/0407215.

Sin, S.-J. and Zahed, I. (2007). Ampere's law and energy loss in AdS/CFT duality, *Phys. Lett. B* **648**, 318–322. doi: 10.1016/j.physletb.2007.01.074, arXiv:hep-ph/0606049.

Sirunyan, A. M. *et al.* (2021). Measurement of the azimuthal anisotropy of image 1 and image 2 mesons in PbPb collisions at sNN=5.02TeV, *Phys. Lett. B* **819**, 136385. doi: 10.1016/j.physletb.2021.136385, arXiv:2006.07707 [hep-ex].

Soltz, R. A., DeTar, C., Karsch, F., Mukherjee, S., and Vranas, P. (2015). Lattice QCD thermodynamics with physical quark masses, *Ann. Rev. Nucl. Part. Sci.* **65**, 379–402. doi: 10.1146/annurev-nucl-102014-022157, arXiv:1502.02296 [hep-lat].

Son, D. T. (1999). Superconductivity by long range color magnetic interaction in high density quark matter, *Phys. Rev. D* **59**, 094019. doi: 10.1103/PhysRevD.59.094019, arXiv:hep-ph/9812287.

Sonnenschein, J. and Weissman, D. (2014). Rotating strings confronting PDG mesons, *JHEP* **08**, 013. doi: 10.1007/JHEP08(2014)013, arXiv:1402.5603 [hep-ph].

Spaliński, M. (2016). Small systems and regulator dependence in relativistic hydrodynamics, *Phys. Rev. D* **94**(8), 085002. doi: 10.1103/PhysRevD.94.085002, arXiv:1607.06381 [nucl-th].

Specht, H. J. (2010). Thermal dileptons from hot and dense strongly interacting matter, *AIP Conf. Proc.* **1322**(1), 1–10. doi: 10.1063/1.3541982, arXiv:1011.0615 [nucl-ex].

Srivastava, D. K. and Kapusta, J. I. (1994). Nature of the photon correlation function for quark — gluon plasma, *Phys. Rev. C* **50**, 505–508. doi: 10.1103/PhysRevC.50.505.

Stachel, J., Andronic, A., Braun-Munzinger, P., and Redlich, K. (2014). Confronting LHC data with the statistical hadronization model, *J. Phys. Conf. Ser.* **509**, 012019. doi: 10.1088/1742-6596/509/1/012019, arXiv:1311.4662 [nucl-th].

Staig, P. and Shuryak, E. (2011a). The fate of the initial state fluctuations in heavy ion collisions. II. The fluctuations and sounds, *Phys. Rev. C* **84**, 034908. doi: 10.1103/PhysRevC.84.034908, arXiv:1008.3139 [nucl-th].

Staig, P. and Shuryak, E. (2011b). The fate of the initial state fluctuations in heavy ion collisions. III. The second act of hydrodynamics, *Phys. Rev. C* **84**, 044912. doi: 10.1103/PhysRevC.84.044912, arXiv:1105.0676 [nucl-th].

STAR Collaboration. (2022a). Beam energy dependence of the linear and mode-coupled flow harmonics in Au+Au collisions. arXiv:2211.11637 [nucl-ex].

STAR Collaboration. (2022b). Beam energy dependence of triton production and yield ratio ($N_t \times N_p/N_d^2$) in Au+Au collisions at RHIC. arXiv:2209.08058 [nucl-ex].

STAR Collaboration. (2022c). First observation of directed flow of hypernuclei ${}^3_\Lambda$H and ${}^4_\Lambda$H in $\sqrt{s_{\rm NN}} = 3$ GeV Au+Au collisions at RHIC. arXiv:2211.16981 [nucl-ex].

STAR Collaboration. (2022d). K^{*0} production in Au+Au collisions at $\sqrt{s_{\rm NN}} = 7.7, 11.5, 14.5, 19.6, 27$ and 39 GeV from RHIC beam energy scan. arXiv:2210.02909 [nucl-ex].

Stephanov, M. A. (2009). Non-Gaussian fluctuations near the QCD critical point, *Phys. Rev. Lett.* **102**, 032301. doi: 10.1103/PhysRevLett.102.032301, arXiv:0809.3450 [hep-ph].

Stephanov, M. A. (2011a). On the sign of kurtosis near the QCD critical point, *Phys. Rev. Lett.* **107**, 052301. doi: 10.1103/PhysRevLett.107.052301, arXiv:1104.1627 [hep-ph].

Stephanov, M. A. (2011b). On the sign of kurtosis near the QCD critical point, *Phys. Rev. Lett.* **107**, 052301. doi: 10.1103/PhysRevLett.107.052301, arXiv:1104.1627 [hep-ph].

Stephanov, M. A., Rajagopal, K., and Shuryak, E. V. (1998). Signatures of the tricritical point in QCD, *Phys. Rev. Lett.* **81**, 4816–4819. doi: 10.1103/PhysRevLett.81.4816, arXiv:hep-ph/9806219.

Stephanov, M. A., Rajagopal, K., and Shuryak, E. V. (1999). Event-by-event fluctuations in heavy ion collisions and the QCD critical point, *Phys. Rev. D* **60**, 114028. doi: 10.1103/PhysRevD.60.114028, arXiv:hep-ph/9903292.

Stodolsky, L. (1995). Temperature fluctuations in multiparticle production, *Phys. Rev. Lett.* **75**, 1044–1045. doi: 10.1103/PhysRevLett.75.1044.

Stoffers, A. and Zahed, I. (2013). Holographic Pomeron: Saturation and DIS, *Phys. Rev. D* **87**, 075023. doi: 10.1103/PhysRevD.87.075023, arXiv:1205.3223 [hep-ph].

't Hooft, G. (1974). Magnetic monopoles in unified gauge theories, *Nucl. Phys. B* **79**, 276–284. doi: 10.1016/0550-3213(74)90486-6.

't Hooft, G. (1976a). Computation of the quantum effects due to a four-dimensional pseudoparticle, *Phys. Rev. D* **14**, 3432–3450. doi: 10.1103/PhysRevD.14.3432 [Erratum: (1978). *Phys. Rev. D* **18**, 2199].

't Hooft, G. (1976b). Magnetic monopoles and non-abelian gauge groups. Lecture Notes in Physics, Vol. 50, pp. 3–10. doi: 10.1007/3-540-07789-8_1.

Teaney, D. (2002). Chemical freezeout in heavy ion collisions. arXiv:nucl-th/0204023.

Teaney, D. (2003). The effects of viscosity on spectra, elliptic flow, and HBT radii, *Phys. Rev. C* **68**, 034913. doi: 10.1103/PhysRevC.68.034913, arXiv:nucl-th/0301099.

Teaney, D. and Yan, L. (2012). Non linearities in the harmonic spectrum of heavy ion collisions with ideal and viscous hydrodynamics, *Phys. Rev. C* **86**, 044908. doi: 10.1103/PhysRevC.86.044908, arXiv:1206.1905 [nucl-th].

Teaney, D., Lauret, J., and Shuryak, E. V. (2001a). A hydrodynamic description of heavy ion collisions at the SPS and RHIC. arXiv:nucl-th/0110037.

Teaney, D., Lauret, J., and Shuryak, E. V. (2001b). Flow at the SPS and RHIC as a quark gluon plasma signature, *Phys. Rev. Lett.* **86**, 4783–4786. doi: 10.1103/PhysRevLett.86.4783, arXiv:nucl-th/0011058.

Tetradis, N. and Wetterich, C. (1994). Critical exponents from effective average action, *Nucl. Phys. B* **422**, 541–592. doi: 10.1016/0550-3213(94)90446-4, arXiv:hep-ph/9308214.

Thorne, K. S., Price, R. H., and Macdonald, D. A. (eds.) (1986). *Black Holes: The Membrane Paradigm.* ISBN 978-0-300-03770-8.

Tserruya, I. (2013). Photons and low-mass dileptons: Results from PHENIX, *Nucl. Phys. A* **904-905**, 225c–232c. doi: 10.1016/j.nuclphysa.2013.01.063, arXiv:1211.6002 [nucl-ex].

Tsypin, M. M. (1994). The universal effective potential for three-dimensional massive scalar field theory from the Monte Carlo study of the Ising model. arXiv:hep-lat/9401034.

Tuchin, K. (2013b). Time and space dependence of the electromagnetic field in relativistic heavy-ion collisions, *Phys. Rev. C* **88**(2), 024911. doi: 10.1103/PhysRevC.88.024911, arXiv:1305.5806 [hep-ph].

Unsal, M. (2008). Abelian duality, confinement, and chiral symmetry breaking in QCD(adj), *Phys. Rev. Lett.* **100**, 032005. doi: 10.1103/PhysRevLett.100.032005, arXiv:0708.1772 [hep-th].

Vanyashin, V. S. and Terentev, M. V. (1965). The vacuum polarization of a charged vector field, *Zh. Eksp. Teor. Fiz.* **48**(2), 565–573.

Velkovska, J. (2017). PHENIX results on elliptic and triangular flow in d+Au collisions from 19.6 to 200 GeV, *Nucl. Phys. A* **967**, 353–356. doi: 10.1016/j.nuclphysa.2017.05.022.

Verbaarschot, J. J. M. (1991). Streamlines and conformal invariance in Yang-Mills theories, *Nucl. Phys. B* **362**, 33–53. doi: 10.1016/0550-3213(91)90554-B [Erratum: (1992). *Nucl. Phys. B* **386**, 236–236].

Waeber, S., Rabenstein, A., Schäfer, A., and Yaffe, L. G. (2019). Asymmetric shockwave collisions in AdS_5, *JHEP* **08**, 005. doi: 10.1007/JHEP08(2019)005, arXiv:1906.05086 [hep-th].

Wagner, M., Bicudo, P., Peters, A., and Velten, S. (2022). Comparing meson-meson and diquark-antidiquark creation operators for a $\bar{b}\bar{b}ud$ tetraquark, *PoS LATTICE2021*, 380. doi: 10.22323/1.396.0380, arXiv:2108.11731 [hep-lat].

Weinberg, S. (1971). Physical processes in a convergent theory of the weak and electromagnetic interactions, *Phys. Rev. Lett.* **27**, 1688–1691. doi: 10.1103/PhysRevLett.27.1688.

West, G. B. (1970). Phenomenological model for the electromagnetic structure of the proton, *Phys. Rev. Lett.* **24**, 1206–1209. doi: 10.1103/PhysRevLett.24.1206.

Wicks, S., Horowitz, W., Djordjevic, M., and Gyulassy, M. (2007). Heavy quark jet quenching with collisional plus radiative energy loss and path length fluctuations, *Nucl. Phys. A* **783**, 493–496. doi: 10.1016/j.nuclphysa.2006.11.102, arXiv:nucl-th/0701063.

Wilson, K. G. and Fisher, M. E. (1972). Critical exponents in 3.99 dimensions, *Phys. Rev. Lett.* **28**, 240–243. doi: 10.1103/PhysRevLett.28.240.

Wilson, K. G. and Kogut, J. B. (1974). The renormalization group and the epsilon expansion, *Phys. Rep.* **12**, 75–199. doi: 10.1016/0370-1573(74)90023-4.

Xie, M., Ke, W., Zhang, H., and Wang, X.-N. (2022). Global constraint on the jet transport coefficient from single hadron, dihadron and γ-hadron spectra in high-energy heavy-ion collisions. arXiv:2208.14419 [hep-ph].

Xiong, L., Shuryak, E. V., and Brown, G. E. (1992). Photon production through A1 resonance in high-energy heavy ion collisions, *Phys. Rev. D* **46**, 3798–3801. doi: 10.1103/PhysRevD.46.3798, arXiv:hep-ph/9208206.

Xu, J., Liao, J., and Gyulassy, M. (2016). Bridging soft-hard transport properties of quark-gluon plasmas with CUJET3.0, *JHEP* **02**, 169. doi: 10.1007/JHEP02(2016)169, arXiv:1508.00552 [hep-ph].

Yang, C.-N. and Mills, R. L. (1954). Conservation of isotopic spin and isotopic gauge invariance, *Phys. Rev.* **96**, 191–195. doi: 10.1103/PhysRev.96.191.

Young, C. and Shuryak, E. (2009). Charmonium in strongly coupled quark-gluon plasma, *Phys. Rev. C* **79**, 034907. doi: 10.1103/PhysRevC.79.034907, arXiv:0803.2866 [nucl-th].

Yung, A. V. (1988). Instanton vacuum in supersymmetric QCD, *Nucl. Phys. B* **297**, 47. doi: 10.1016/0550-3213(88)90199-X.

Zahed, I. (2021). Mass sum rule of hadrons in the QCD instanton vacuum, *Phys. Rev. D* **104**(5), 054031. doi: 10.1103/PhysRevD.104.054031, arXiv:2102.08191 [hep-ph].

Zakharov, V., Lvov, V., and Falkovich, G. (1990). *Kolmogorov Spectra of Turbulence I. Wave Turbulence* (Springer Verlag). ISBN 3-540-54533-6.

Zelevinsky, V. and Volya, A. (2016). Chaotic features of nuclear structure and dynamics: Selected topics, *Phys. Scr.* **91**(3), 033006. doi: 10.1088/0031-8949/91/3/033006.

Zetocha, V. and Schäfer, T. (2003). Instanton contribution to scalar charmonium and glueball decays, *Phys. Rev. D* **67**, 114003. doi: 10.1103/PhysRevD.67.114003, arXiv:hep-ph/0212125.

Zhang, C. and Jia, J. (2022). Evidence of quadrupole and octupole deformations in Zr96+Zr96 and Ru96+Ru96 collisions at ultrarelativistic energies, *Phys. Rev. Lett.* **128**(2), 022301. doi: 10.1103/PhysRevLett.128.022301, arXiv:2109.01631 [nucl-th].

Zhao, W., Shen, C., Ko, C. M., Liu, Q., and Song, H. (2020). Beam-energy dependence of the production of light nuclei in Au + Au collisions, *Phys. Rev. C* **102**(4), 044912. doi: 10.1103/PhysRevC.102.044912, arXiv:2009.06959 [nucl-th].

Zhirov, O. V. and Shuryak, E. V. (1975). Multiple production of hadrons and predictions of the Landau theory, *Yad. Fiz.* **21**, 861–867.

INDEX

www.ingramcontent.com/pod-product-compliance
Lightning Source LLC
LaVergne TN
LVHW010338220926
848484LV00004BB/927
* 9 7 8 9 8 1 1 2 8 2 3 4 8 *